GIS and Geostatistical Techniques
for Groundwater Science

GIS and Geostatistical Techniques for Groundwater Science

Venkatramanan Senapathi

Research Professor, Department of Earth and Environmental Sciences,
Pukyong National University, Busan, South Korea
Research Collaborator, Department for Management of Science and Technology Development,
Faculty of Applied Sciences, Ton Duc Thang University, Ho Chi Minh City, Vietnam
Guest Lecturer, Department of Geology, Alagappa University, Karaikudi, Tamilnadu, India

Prasanna Mohan Viswanathan

Associate Professor, Department of Applied Geology, Faculty of Engineering and Science,
Curtin University Malaysia, Miri, Sarawak, Malaysia

Sang Yong Chung

Professor, Department of Earth and Environmental Sciences,
Pukyong National University, Busan, South Korea

ELSEVIER

Elsevier
Radarweg 29, PO Box 211, 1000 AE Amsterdam, Netherlands
The Boulevard, Langford Lane, Kidlington, Oxford OX5 1GB, United Kingdom
50 Hampshire Street, 5th Floor, Cambridge, MA 02139, United States

Notices
Knowledge and best practice in this field are constantly changing. As new research and experience broaden our understanding, changes in research methods, professional practices, or medical treatment may become necessary.

Practitioners and researchers must always rely on their own experience and knowledge in evaluating and using any information, methods, compounds, or experiments described herein. In using such information or methods they should be mindful of their own safety and the safety of others, including parties for whom they have a professional responsibility.

To the fullest extent of the law, neither the Publisher nor the authors, contributors, or editors, assume any liability for any injury and/or damage to persons or property as a matter of products liability, negligence or otherwise, or from any use or operation of any methods, products, instructions, or ideas contained in the material herein.

Library of Congress Cataloging-in-Publication Data
A catalog record for this book is available from the Library of Congress

British Library Cataloguing-in-Publication Data
A catalogue record for this book is available from the British Library

ISBN: **978-0-12-815413-7**

For information on all Elsevier publications
visit our website at https://www.elsevier.com/books-and-journals

Publisher: Candice Janco
Acquisition Editor: Marisa LaFleur
Editorial Project Manager: Tasha Frank
Production Project Manager: Prem Kumar Kaliamoorthi
Cover Designer: Greg Harris

Typeset by SPi Global, India

Working together
to grow libraries in
developing countries

www.elsevier.com • www.bookaid.org

Contents

5. A Comparative Study of Spatial Interpolation Technique (IDW and Kriging) for Determining Groundwater Quality

Prafull Singh and Pradipika Verma

6. Methods for Assessing the Groundwater Quality

R. Rajesh, L. Elango and K. Brindha

7. Spatial Prediction of Groundwater Level Using Models Based on Fuzzy Logic and Geostatistical Methods

Ata Allah Nadiri, Rahman Khatibi, Fatemeh Vahedi and Keyvan Naderi

8. Remote Sensing and Fuzzy Logic Approach for Artificial Recharge Studies in Hard Rock Terrain of South India

C.K. Muthumaniraja, S. Anbazhagan, A. Jothibasu and M. Chinnamuthu

9. Groundwater Quality Assessment Using Multivariate Statistical Methods for Chavara Aquifer System, Kerala, India

R. Anand Krishnan, Jamal Ansari,
M. Sundararajan, Cincy John and P.M. Saharuba

10. Overview, Current Status, and Future Prospect of Stochastic Time Series Modeling in Subsurface Hydrology

Priyanka Sharma, Deepesh Machiwal and
Madan Kumar Jha

11. Intelligent Prediction Modeling of Water Quality Using Artificial Neural Networking: Nambiyar River Basin, Tamil Nadu, India

C. Gajendran, K. Srinivasamoorthy and
P. Thamarai

Section C
Groundwater Quality Assessment Using GIS and Geostatistical Aspects

12. Simulation of Seasonal Rainfall and Temperature Variation—A Case Study of Climate Change Projection in Ponnaiyar River Basin, Southern India

A. Jothibasu and S. Anbazhagan

13. Impact of Urbanization on Groundwater Quality

K. Brindha and Michael Schneider

14. Groundwater and Surface Water Interaction

Haris Hasan Khan and Arina Khan

15. Evaluation of Vulnerability Zone of a Coastal Aquifer Through GALDIT GIS Index Techniques

G. Gnanachandrasamy, T. Ramkumar, J.Y. Chen, S. Venkatramanan, S. Vasudevan and S. Selvam

16. A Statistical Approach to Identify the Temporal and Spatial Variations in the Geochemical Process of a Coastal Aquifer, South East Coast of India

M.V. Prasanna, S. Chidambaram, R. Thilagavathi, C. Thivya, S. Venkatramanan and N. Murali Krishnan

21. Geostatistical Studies for Evaluation of Fluoride Contamination in Part of Dharmapuri District, South India

S. Anbazhagan, M. Rajendran and A. Jothibasu

22. Fluoride Contamination in Groundwater—A GIS and Geostatistics Reappraisal

Banajarani Panda, V. Dhanu Radha,
S. Chidambaram, M. Arindam,
R. Thilagavathi, S. Manikandan, C. Thivya,
A.L. Ramanathan and N. Ganesh

23. Arsenic Contamination

S. Selvam, K. Jesuraja and G. Gnanachandrasamy

24. Evaluation of Heavy-Metal Contamination in Groundwater using Hydrogeochemical and Multivariate statistical Analyses

Sang Yong Chung, R. Rajesh
and S. Venkatramanan

Contributors

Numbers in parentheses indicate the pages on which the authors' contributions begin.

R. Anand Krishnan (113), CSIR—National Institute for Interdisciplinary Science and Technology, Council Of Scientific and Industrial Research, Thiruvananthapuram, India

S. Anbazhagan (91, 167, 297), Centre for Geoinformatics and Planetary Studies, Periyar University, Salem, India

S. Anirudhan (349), Department of Geology, University of Kerala, Kariavattom Campus, Thiruvananthapuram, Kerala, India

Jamal Ansari (113), CSIR—National Institute for Interdisciplinary Science and Technology, Council Of Scientific and Industrial Research, Thiruvananthapuram, India

M. Arindam (309), Water Sciences Lab, University of Nebraska-Lincoln, Lincoln, NE, United States

K. Brindha (179, 57), Hydrogeology Group, Institute of Geological Sciences, Freie Universität Berlin, Berlin, Germany

J.Y. Chen (209), School of Geography and Planning; School of Earth Sciences and Engineering, Sun Yat-Sen University, Guangzhou, China

S. Chidambaram (223, 309), Water Research Center, Kuwait Institute for Scientific Research, Safat, Kuwait

M. Chinnamuthu (91), Centre for Geoinformatics and Planetary Studies, Periyar University, Salem, India

Sang Yong Chung (33, 331), Department of Earth and Environmental Sciences, Pukyong National University, Busan, South Korea

V. Dhanu Radha (309), Water Research Center, Kuwait Institute for Scientific Research, Safat, Kuwait

L. Elango (57, 283), Department of Geology, Anna University, Chennai, India

Hussam Eldin Elzain (33), Department of Earth and Environmental Sciences, Pukyong National University, Busan, South Korea

C. Gajendran (153), Department of Civil Engineering, Karunya Institute of Technology and Sciences, Coimbatore, India

N. Ganesh (309), Department of Earth Sciences, Annamalai University, Chidambaram, India

G. Gnanachandrasamy (209, 323), School of Geography and Planning; School of Earth Sciences and Engineering, Sun Yat-sen University, Guangzhou, China

S. Gopinath (237), Department of Earth Sciences, School of Physical, Chemical and Applied Sciences, Pondicherry University, Pondicherry, India

Cüneyt Güler (251), Department of Geological Engineering, Mersin University, Mersin, Turkey

K. Jesuraja (323), Department of Geology, V.O. Chidambaram College, Thoothukudi, India

Madan Kumar Jha (133), AgFE Department, Indian Institute of Technology Kharagpur, Kharagpur, India

Cincy John (113), CSIR—National Institute for Interdisciplinary Science and Technology, Council Of Scientific and Industrial Research, Thiruvananthapuram, India

A. Jothibasu (91, 167, 297), Centre for Geoinformatics and Planetary Studies, Periyar University, Salem, India

D. Karunanidhi (237), Department of Civil Engineering, Sri Shakthi Institute of Engineering and Technology, Coimbatore, India

Haris Hasan Khan (197), Department of Geology, Aligarh Muslim University, Aligarh, India

Arina Khan (197), Residential Coaching Academy, Aligarh Muslim University, Aligarh, India

Rahman Khatibi (79, 269), GTEV-ReX Limited, Swindon, United Kingdom

Deepesh Machiwal (133), ICAR-Central Arid Zone Research Institute, Regional Research Station, Bhuj, India

N.S. Magesh (283), National Centre for Antarctic and Ocean Research, Vasco da Gama, Goa, India

S. Manikandan (309), Department of Earth Sciences, Annamalai University, Chidambaram, India

A. Manisha (3), Department of Geology, V.O. Chidambaram College, Thoothukudi, India

N. Murali Krishnan (223), Media and Communication Department, Faculty of Humanities, Curtin University Malaysia, Miri, Malaysia

C.K. Muthumaniraja (91), Groundwater Division, Tamil Nadu Water Supply and Drainage Board (TWAD), Erode, India

Keyvan Naderi (79), Department of Earth Sciences, University of Tabriz, Tabriz, Iran

Ata Allah Nadiri (79, 269), Department of Earth Sciences, Faculty of Natural Sciences, University of Tabriz, Tabriz, Iran

Banajarani Panda (309), Department of Earth Sciences, Annamalai University, Chidambaram, India; Water Sciences Lab, University of Nebraska-Lincoln, Lincoln, NE, United States

C.R. Paramasivam (17, 23), Department of Geology, Alagappa University, Karaikudi, India

R. Prakash (237), Department of Earth Sciences, School of Physical, Chemical and Applied Sciences, Pondicherry University, Pondicherry, India

M.V. Prasanna (33, 223), Department of Applied Geology, Faculty of Engineering and Science, Curtin University Malaysia, Miri, Malaysia

M. Rajendran (297), Groundwater Division, Tamil Nadu Water Supply and Drainage Board (TWAD), Chennai, India

R. Rajesh (57, 331), Department of Civil Engineering, Indian Institute of Science, Bangalore, India

A.L. Ramanathan (309), School of Environmental Sciences, JNU, New Delhi, India

T. Ramkumar (209), Department of Earth Sciences, Annamalai University, Chidambaram, India

R.G. Rejith (349), Minerals Section, Materials Science and Technology Division, National Institute for Interdisciplinary Science and Technology (CSIR-NIIST), Council of Scientific & Industrial Research; Academy of Scientific and Innovative Research (AcSIR), CSIR-NIIST, Thiruvananthapuram, Kerala, India

P.M. Saharuba (113), CSIR—National Institute for Interdisciplinary Science and Technology, Council Of Scientific and Industrial Research, Thiruvananthapuram, India

K. Saravanan (237), Department of Earth Sciences, School of Physical, Chemical and Applied Sciences, Pondicherry University, Pondicherry, India

Michael Schneider (179), Hydrogeology Group, Institute of Geological Sciences, Freie Universität Berlin, Berlin, Germany

S. Selvam (3, 33, 209, 323), Department of Geology, V.O. Chidambaram College, Thoothukudi, India

S. Venkatramanan (3, 23, 33, 209, 223, 331), Department of Earth and Environmental Sciences, Pukyong National University, Busan, South Korea; Department for Management of Science and Technology Development; Faculty of Applied Sciences, Ton Duc Thang University, Ho Chi Minh City, Vietnam; Department of Geology, Alagappa University, Karaikudi, India

Priyanka Sharma (133), SWE Department, College of Technology and Engineering, Maharana Pratap University of Agriculture and Technology, Udaipur, India

Prafull Singh (43), Amity Institute of Geoinformatics and Remote Sensing, Amity University, Noida, India

K. Srinivasamoorthy (153, 237), Department of Earth Sciences, School of Physical, Chemical and Applied Sciences, Pondicherry University, Pondicherry, India

M. Sundararajan (113, 349), CSIR—National Institute for Interdisciplinary Science and Technology, Council Of Scientific and Industrial Research; Minerals Section, Materials Science and Technology Division, National Institute for Interdisciplinary Science and Technology (CSIR-NIIST), Council of Scientific & Industrial Research; Academy of Scientific and Innovative Research (AcSIR), CSIR-NIIST, Thiruvananthapuram, Kerala, India

P. Thamarai (153), Government College of Technology, Coimbatore, India

R. Thilagavathi (223, 309), Department of Earth Sciences, Annamalai University, Chidambaram, India

C. Thivya (223, 309), Department of Geology; School of Earth and Atmospheric Sciences, University of Madras, Chennai, India

Frank T.-C. Tsai (269), Department of Civil and Environmental Engineering, Louisiana State University, Baton Rouge, LA, United States

Fatemeh Vahedi (79), Department of Earth Sciences, University of Tabriz, Tabriz, Iran

S. Vasudevan (209), Department of Earth Sciences, Annamalai University, Chidambaram, India

Pradipika Verma (43), Amity Institute of Geoinformatics and Remote Sensing, Amity University, Noida, India

J. Vidhya (3), Department of Geology, V.O. Chidambaram College, Thoothukudi, India

Preface

This book discusses the power of GIS as a tool to develop solutions for groundwater resource problems and presents applications of GIS and geostatistical techniques in various field such as science, engineering, planning, and resource management. The modern applications of GIS and statistical approaches can be used to study issues in groundwater hydrogeology and determine water quality. The authors in this book explain the process of analyzing data, identification and parameter estimation tools, GIS applications incorporate with geostatistical techniques and the conjunctive use of groundwater techniques such as remediation, and monitoring techniques and analysis of groundwater quality management. Authors also deals with the groundwater management and then explore the impact of climate change on groundwater and aquifer scales for environmental impact studies.

The goal of this book is to bridge the gap in the application of GIS and geostatistics in groundwater studies. The chapters of this book have been written by students, researchers, and environmental engineers. The chapters present an introduction to GIS and the application of geostatistical techniques to hydrogeological sciences. The chapters in this book have been written in a style that can be easily understood by those who are completely new to this field. The chapters also provide a background of the concepts and terminology used as well as descriptions of the more advanced technologies in this field. The improved advanced methods are discussed and case studies are presented that are largely indicative of the more advanced GIS and geostatistical models available. The list of suggested further reading materials, which are mainly recent publications, textbooks, and review papers, given at the end of each chapter is a useful starting point for readers interested in obtaining more information on these topics. The chapters provide useful and up-to-date information on GIS and geostatistical techniques for the younger generation of students and researchers. We hope this book will help readers obtain a good understanding of the application of GIS and geostatistical techniques in groundwater science.

Acknowledgments

We would like to thank the following people for their encouragement in developing our thought processes and for their contributions to the GIS and geostatistical techniques of groundwater science. Some of those acknowledged may not recognize that they made a contribution, but we acknowledge them for sharing their approach to recent advanced GIS and geostatistical techniques of hydrogeological sciences. This book is dedicated to my beloved father, the late Mr. Senapathi, for his valuable inspiration and support throughout my life. I am ever grateful to my family members, my mother Paruvatham, wife Vaishali, daughter Avignaya, sister Uma Mangalam, uncle Subramanian, and niece Lakshmi, for their endless affection, kindness, cooperation, patience, encouragement, and gracious support during the writing of this book. I owe a deep sense of gratitude and am indebted to my research supervisor Prof. Ramkumar for his encouragement. I express my special thanks to Prof. Sang Yong Chung and Associate Prof. Prasanna Mohan Viswanathan for their valuable help in completing this book. It has been a great pleasure to work with many different people in many different countries, all of whom have motivated me in some way.

Venkatramanan Senapathi

Introduction and History of GIS and Geostatistical Techniques in Groundwater Science

Chapter 1

Fundamentals of GIS

S. Selvam*, A. Manisha*, J. Vidhya* and S. Venkatramanan[†,‡,§,¶]

*Department of Geology, V.O. Chidambaram College, Thoothukudi, India [†]Department of Earth and Environmental Sciences, Pukyong National University, Busan, South Korea [‡]Department for Management of Science and Technology Development, Ton Duc Thang University, Ho Chi Minh City, Vietnam [§]Faculty of Applied Sciences, Ton Duc Thang University, Ho Chi Minh City, Vietnam [¶]Department of Geology, Alagappa University, Karaikudi, India

Chapter Outline

1.1 INTRODUCTION

Geography is the scientific study of the Earth's surfaces and its various climates and natural resources, especially the surface features of an area and their variations in spatial location and time (Aronoff, 1991). Many questions arising regarding agricultural production are geographic in nature as the production depends on the environment and established socio-economic conditions; both of these conditions vary spatially and in time. The questions are related to natural resource management, exactitude agriculture, agro-ecological categorization for land use preparation, regional trends and patterns in technology adaptation, agricultural yield and profits, nonpoint source effluence from agricultural lands, etc. Answering these questions requires access to large volumes of multidimensional geographical (spatial) information regarding weather, soils, topography, water resources, socio economic status, etc. Furthermore, answering the simple question requires large volumes of data and these data are collected from various sources to be compiled in a consistent form. Geographical information systems (GIS) permit the representation and compilation of such spatial information.

The geographical information is represented in the form of two-dimension maps. Maps are graphic representations of the Earth's surface on plane paper, which show the shape in the way we see, assess, and analyze spatial information. A map consists of points, lines, and other area elements that are positioned with reference to a common coordinate system (usually latitude and longitude). They are drawn to specified scales and projections. In a map preparation the scales of the maps may differ from each other. The scale of a map depends upon the purpose for which the map is prepared. A map projection is one of many methods used to represent the three-dimensional surface of the Earth or other round body on a two-dimensional plane in cartography (mapmaking). This process is a mathematical procedure (some methods are graphically based). The map legend is useful to connect the nonspatial attributes (name, symbols, colors, thematic data) to the spatial data (Congalton, 1991). Maps are very useful to store the present data and provide information to the user. However, analogue maps (on paper) are cumbersome to construct and to use, because they contain a large amount of data, so it is very difficult to analyze. Computer-based GIS facilitates both map creation and use for various multifaceted analyses. It allows work associated with geographic data in a digital format and is helpful for decision making in resource management.

GIS is a common term that indicates the exploitation of computers to create and exhibit digital maps. The data that is illustrated in the different features presented in maps are related to the physical, chemical, biological, environmental, social, economic, and other properties associated with the Earth. GIS holds all databases in a single database and these large

GIS and Geostatistical Techniques for Groundwater Science. https://doi.org/10.1016/B978-0-12-815413-7.00001-8

quantities of different data are used in mapping, modeling, querying, analyzing, and displaying. These large quantities of data are integrated by its power and appeal stem and its capacity in terms of the environment and the wide range of tools it provides to explore the diverse data. The history and development of GIS equals runs parallel to the history of the development of computers (Tomlinson, 2007). GIS is used in two distinct ways: as a database management system and in cartography and the automation of map production. The development of GIS depends upon innovations made in several other disciplines, like geography, photogrammetry, remote sensing, civil engineering and statistics, etc.

GIS is used to produce and read maps. The major advantage of GIS is that it permits the identification of the spatial relationships between specific different map features. The maps can be produced using different scales, projections, and colors. But it is not just a map-making tool, it is an analytical tool that provides distinct methods of linking and analyzing data by projecting tabular data collected from different sources onto maps. Using this tool different sets of maps are created. These maps are prepared using different themes (soils, rainfall, temperature, relief, water sources, etc.).

At the advent of GIS, its technology of integration was based on two points of view: one its development and the other its uses. Any kind of geographic information is easily translated into the digital form in GIS, and is easy to copy, edit, analyze, manipulate, and transmit. This allows vital links to be made between apparently unrelated activities based on a common geographic location. This has led to decision making in resource management about the fundamental changes in a variety of situations, such as forest management; marketing management; utility management; transportation; and agricultural, environmental, and regional planning and management. Some potential agricultural applications where GIS can lead to better management decisions are: precision farming, land-use planning, watershed management, pest and disease management, irrigation management, resources inventory and mapping, crop area assessment and yield forecasting, biodiversity assessment, genetic resources management, etc.

1.2 DEFINITION OF GIS

A GIS is basically a computerized information system like any other database, but with an important difference: all information in GIS must be linked to a geographic (spatial) reference (latitude/longitude, or other spatial coordinates) (Fig. 1.1).

There are many different definitions of GIS, as different users stress different aspects of its use. For example:

> *ESRI defined as GIS has an large quantities of collections such include as computer hardware, software, geographic data and personnel designed to efficiently capture, store, update, manipulate, analyze and display geographically referenced information.*

ESRI (1990)

ESRI gives a simple definition of GIS as basically a computer system with the capacity for storing and using data that describes places on the Earth's surface.

Duecker defined GIS as a special type of information system where the database consists of observations on spatially distributed features, activities, or events that are definable in space as points, lines, or areas. A GIS manipulates data about these points, lines, or areas to recover data for ad hoc queries and analyses.

The United States Geological Survey (USGS) defines GIS as a computer hardware and software system that is designed to collect, manage, analyze, and display geographically (spatially) referenced data. This definition is fairly comprehensive and applicable to the agricultural applications of GIS (Foresman, 1998).

Note that GIS does not store a map or image. What it does store is a relational database from which maps can be created as and when they are needed. Relational database concepts are particularly crucial to the development of GIS. Each map (say a soil map) has the capacity to hold a layer of information. GIS works with several layers of such thematic data. This

FIG. 1.1 GIS flow chart.

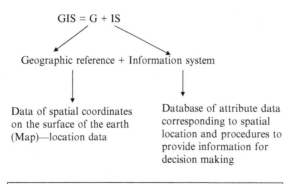

GIS = G + IS

Geographic reference + Information system

Data of spatial coordinates on the surface of the earth (Map)—location data

Database of attribute data corresponding to spatial location and procedures to provide information for decision making

GIS = IS with geographically referenced data

idea is useful when explaining that when comparing the different layers by overlaying them, the layers must be referenced to the same locations, i.e., location is the common key for all the thematic datasets. This ensures that every location (spatial reference point) is exactly matched to its location on other maps. Once the comparison process of different layers can be done this leads to the identification of spatial patterns and process. Thus GIS relates to other database applications, but with an important difference: all information in a GIS must be linked to a spatial reference. Other databases may contain locational information (addresses, pin codes, etc.), but the GIS uses georeferences as the primary means of storing and accessing information.

1.3 WHAT GIS CAN DO

There are five basic questions GIS can answer:

What exists at a particular location? Given a geographic reference (e.g., lat, long) for a location, the GIS must describe the features of that location (Goodchild et al., 1992).

Where can specific features be found? This is the contrary of the first question. For example, where are the districts with rainfall >500 and <750 mm?

What are the trends or what has changed over time? This involves answering the two previous questions. For example, at what locations are the crop yields showing declining trends?

What spatial patterns exist? If the occurrence of a pest is associated with a hypothesized set of conditions of temperature, precipitation, humidity, where do those conditions exist?

Modeling or What if …? This is a high-level application of GIS, answering questions like, what would be the nitrate distribution in groundwater over an area if fertilizer use was doubled?

1.4 GEOGRAPHIC REFERENCING CONCEPTS

A GIS can produce maps from different thematic layers (soils, land use, temperature, etc.). The maps are in two-dimensions whereas the Earth's surface is a three-dimensional ellipsoid. Every map has a specific projection and scale (Burrough, 1986).

To understand how maps are created by projecting the three-dimensional Earth's surface into a two-dimensional plane of an analogue map, we need to understand the georeferencing concepts. The concept of georeferencing involves two stages: locate the points on the Earth's surface by specifying the three-dimensional coordinate system known as the geographic coordinate system (GCS) and use the projected coordinate system to project into two dimensions and create an analogue map.

1.5 GEOGRAPHIC COORDINATE SYSTEM

The traditional way of representing locations on the surface of the Earth in the three-dimensional coordinate system is by its latitude and longitude (Maguire et al., 1991) (Fig. 1.2).

Note that the distance between two points on the three-dimensional Earth's surface varies with latitude. The three-dimensional system therefore does not provide a consistent measure of distances and areas at all latitudes (Heywood et al., 1998).

The true surface of the Earth is not a smooth ellipsoid shown in the figure, but is quite uneven and rugged. The GCS's version of the surface that is used for specifying the latitude and longitude of a point on the Earth's surface, is also an approximation and a three-dimensional model of the Earth (Konecny, 2002). GCS is defined by several standard models of the ellipsoid (WGS 84, Everest ellipsoid, etc.). These models vary depending upon their critical parameters (semi-major or equatorial axis and semi-minor or polar axis of the ellipsoid and the point of origin). The term datum is referred as calculate the latitude and longitude of the ellipsoid model. While we change the value of latitude and longitude the value of datum will be change.

Specifying the geographic coordinate system therefore requires specifying the datum. The datum is a fixed three-dimensional ellipsoid that is approximately the size and shape of the surface of the Earth, based on which the geographic coordinates (latitude and longitude) of a point on the Earth's surface are calculated. In fact, the description of a place by its latitude/longitude is not complete without specifying its datum. In India the Everest ellipsoid is used as the datum for the survey of India (SOI) maps.

The ideal spheroidal model of the Earth has both the correct equatorial and polar radii, and is centered at the actual center of the Earth. One would then have a spheroid that when used as a datum would accurately map the entire Earth.

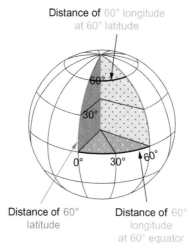

FIG. 1.2 Three dimensional view of coordinate system.

All latitude/longitude on all maps would agree. That spheroid, derived from satellite measurements of the Earth, is GRS80, and the WGS84. The datum matches this spheroid.

1.6 PROJECTED COORDINATE SYSTEM

The availability of maps on a paper (an analogue map) the development of GIS is started. This map therefore represents a projection of a three-dimensional GCS in two-dimensional form (Fig. 1.3).

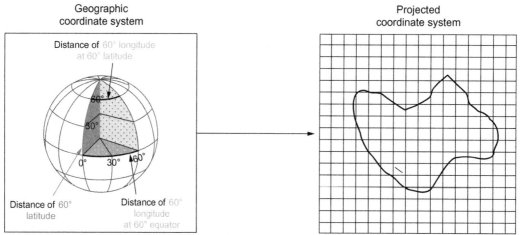

FIG. 1.3 A view of projected coordinated system.

1.7 ADOPTED FROM ESRI

Projection is a mathematical transformation used to project the real three-dimensional spherical surface of the Earth onto a plane sheet of paper in a two-dimensional form (Fig. 1.4). The projection causes distortions of one or more spatial properties (area, shape, distance, or direction) (Selvam et al., 2013).

There are many methods of map projections, since there are an infinite number of ways to project the three-dimensional Earth's surface onto a two-dimensional planar surface. The types of projection vary, such as azimuthal, conic, or cylindrical. These types of projection are based on three-dimensional to two-dimensional projections can be done onto a plane or onto the surface of a cone or cylinder (Fig. 1.5).

Map projections lead to distortions ... ..

Choice of projections depends on allowable distortions in:

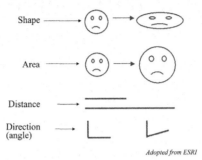

Adopted from ESRI

FIG. 1.4 The projection causes distortions of multispatial properties.

Types of projections

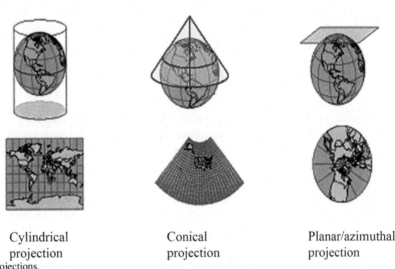

Cylindrical
projection

Conical
projection

Planar/azimuthal
projection

FIG. 1.5 A view of types of projections.

Depending on the scale and the agreeable tradeoffs with respect to distortions, a specific projection form is chosen. The different standard projections are adopted by different countries and these projections have different map scale. In India, the polyconic projection is commonly used by SOI. All SOI toposheets are in the form of polyconic projection.

1.8 MAP SCALE

Map scale is the ratio of distance on a map to the distance on the surface of the Earth (Fig. 1.6). It is specified in verbal, numeric, or graphical form on all standard maps (Campbell, 1996).

In GIS all the maps should contain a graphical scale as this ensures that any changes in scale in photocopying, etc., are accounted for (Table 1.1).

The following are the standard map scales:

SOI maps are available at all the above levels, except the micro level.

1.9 CREATING A GIS

As for any other information system, creating a GIS involves four stages:

(i) Data input

(ii) Data storage

(iii) Data analysis and modeling, and

(iv) Data output and presentation.

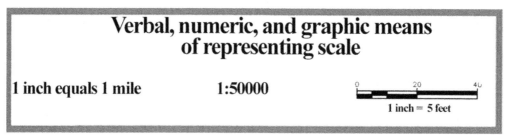

FIG. 1.6 A view of topographical map scale.

TABLE 1.1 Topographical Standard Map Scales

• 1:1,000,000	Country level or state level
• 1: 250,000	State or district level
• 1: 50,000	District level
• 1: 12,500	Micro level

There are, however, many differences from other information systems, one of the most important of which is the data-input method. There are two types of data input: spatial data (latitude/longitude for georeferencing the features on a map, e.g., soil units, administrative districts), and attribute data (descriptive data about the features, e.g., soil properties, population of districts, etc.) (Fig. 1.7)

GIS create analogue maps (soil map, land use map, administrative districts, map, agroecological zone map, etc.) or aerial photographs and satellite images with the use of spatial data. Data input is the process of programming analogue data in the form of maps, images, or photographs into a computer-readable digitized form and writing data into the GIS database.

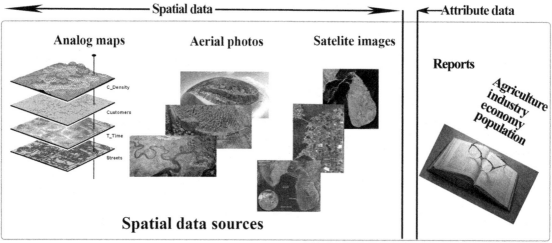

FIG. 1.7 Types of data input.

1.10   GIS DATA INPUT

Spatial-data capture (representing locations in a database) can be in two basic formats:

(i) Vector format
(ii) Raster format.

The vector format is represented as points, lines, and areas, and the raster format represented as grid of cells/pixels. The vector format is based on discrete object views of reality (analogue maps) and the raster format is based on continuous field views (photographs, imageries, etc.). In principle any real-world situation can be represented in digital form in both raster and vector formats (Fig. 1.8). The choice is up to the user. Each format has advantages and disadvantages (Arctur and Zeiler, 2004).

1.10.1   Vector Data Capture

This is generally used for capturing data for analogue maps. The map consist of three basic kinds of features based on observation (Huisman and De By, 2009). These are:

(i) point features
(ii) line features, and
(iii) polygon or area features.

Points do not have length, width, or area. They are described completely by their coordinates and are used to represent discrete locational information on the map to identify locations and features such as, cities, towns, well locations, rain-gauge stations, soil sampling points, etc.

A line consists of a set of ordered points. It has length, but no width or area. Therefore it is used to represent features such as roads, streams, or canals, which have too narrow a width to be displayed on the map at its specified scale.

A polygon or area is formed when a set of ordered lines form a closed figure the boundary of which is represented by the lines. Polygons are used to represent area features such as land parcels, lakes, districts, agro-ecological zones, etc. A polygon usually encloses an area that may be considered homogeneous with respect to some attribute. For example, in a soil map, each polygon will represent an area with a homogeneous soil type.

A vector-based system displays graphical data as points, lines, or curves, or areas with attributes. Cartesian coordinates (x, y) or geographical coordinates (latitude, longitude) define points in a vector system (Fig. 1.9).

Data are captured from a map in the form of known x-y coordinates or latitude-longitude by first discrediting the features on the map into a series of nodes (dots) and digitizing the points one by one directly after placing the map on a digitizer (Campbell, 1996). The digitizer can be considered as an electronic graph paper with a very fine grid. The map is placed on the digitizer and the lines and areas are discredited into a series of points. The digitizer's cursor is used to systematically trace over the points. The points on the map are captured directly as point coordinates. Line features are captured as a series of ordered points. Area features are also captured as an ordered list of points. By making the beginning and end points/nodes of the digitization the same for the area, the shape or area is closed and defined. The process of digitizing from a digitizer is

Vector and raster representations

FIG. 1.8   Vector and Raster data representations.

•Vector formats

–Discrete representations of reality

• Raster formats

– Square cells to model reality

**Reality
(a highway)**

Source: ESRI

Spatial data generation in vector format

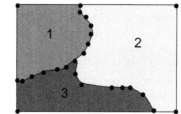

- Discretize lines into points (nodes) and digitize as straight-line segments called vectors or arcs.

- Data of *X,Y* coordinates of points and vectors and their connections (topology) are generated and stored in a database

- For areas, geometry (area, perimeter) data are generated

- Points, lines and areas have independent database tables

- Add attribute data to database

FIG. 1.9 Spatial data generation in vector format.

both time-consuming and painstaking. Alternately, the map can be scanned and the scanned image digitized on-screen with appropriate software tools. The latter process is relatively simpler, more accurate, and is often preferred.

Digitization is usually done feature by feature. For example, all point features presented in the map (say cities, towns, etc.) are digitized in one layer. Similarly, all line features (e.g., roads, rivers, drainage network, canal network, etc.) are digitized as a separate layer, and the polygon features (soils, districts, agro ecological zones, etc.) are also digitized as a separate layer (Fig. 1.10). For the point features the digitization process builds up a database of the points' identification numbers (IDs) and their coordinates. For the lines it builds up a database of their IDs, the starting and end nodes for the line

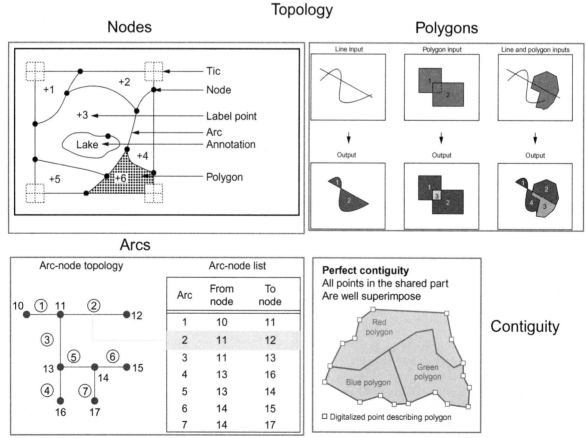

FIG. 1.10 Topology parameters in GIS.

and its length. In addition, the GIS also creates a database of the topology, that is, the spatial relationships between the lines. Also, for the polygons it develops a database of their ID, lines, or arcs, which comprise its topology and its area and perimeter.

The ID is considered to be key to the field of each database (points, lines, areas databases) as it can be used to relate the spatial data with the attribute data. The data resolution depends on the discretization of the digitized points on the initial map. Vector systems are capable of very high resolution (<0.001 in.) and graphical output is similar to hand-drawn maps. But it is less compatible with remote-sensing data, for which the raster system is preferred (FAO, n.d.).

1.10.1.1 Map Projections and Scale

Note that all standard maps that are to be digitized are drawn to specific *projection* and *scale*. However, the digitizer that facilitated the computerized map has its own scale and units, and the digitized maps are in these units and scale. Translating information from a digitized map into the real-world information of locations, lengths, and areas requires information about the mathematical equations used for the projection, as well as the scale in which the original analogue map was prepared. In case several map layers are to be digitized (topography, soils, districts, etc.), it is necessary to ensure that they are all assembled in the same projection and scale before any spatial analysis is done using them. Most standard GIS have the facility to convert from one map projection to another and to transform scales from the digitizer scale to map scale to ensure that all map layers have the same locational reference.

1.10.2 Raster Data Capture

A raster-based GIS locates and stores map data in the form of matrix of grid cells or pixels. Each cell or pixel is represented either at its corner or centroid by a unique reference coordinate (cell address). Each cell also has a discrete attribute data assigned to it.

The resolution of raster data is dependent on the pixel or grid cell size (Fig. 1.11). Data can be easily captured from remote sensing images, aerial photographs, and other images of the Earth's surface are available in a raster-data format. In this format, it helps to identify the various features by superposing the images over a fine rectangular grid of the Earth's surface. Raster-data captures the images, but it does not build topography, i.e., derive spatial relationships between the identified features. However, these raster data facilitate simple scalar operations on the spatial data, which a vector format does not permit. Before creating the topology and the spatial operations, the raster data must be converted into a vector format. The raster format requires more storage space on the computer compared with the vector format http://www.col orado.edu/geography/gcraft/notes.

Most standard GIS software has the facility to transform maps from raster to vector format and vice versa (Fig. 1.12).

1.10.3 Attribute Data

Attribute data are descriptive data of point, line, and area features. For points, the data are the name of the location, its elevation, etc. For lines, attribute data could be the name of a road or canal, and other descriptions associated with them. The categories of district, population, and area of specific crops in the district are categorized under polygons.

Spatial data generation in raster format

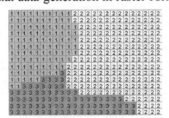

FIG. 1.11 Spatial data generation in raster format.

• Map is represented by rectangular or square cells

• Each cell is assigned a value based on what it represents

• Attribute data are assigned by user to cells

Adopted from FAO

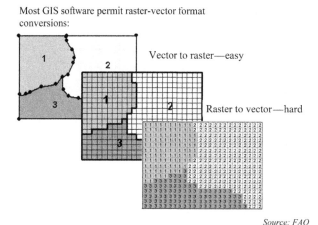

Most GIS software permit raster-vector format conversions:

Vector to raster—easy

Raster to vector—hard

Source: FAO

FIG. 1.12 Raster and Vector format conversions.

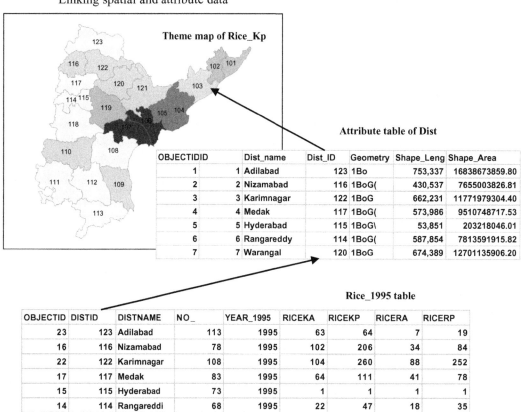

Linking spatial and attribute data

Theme map of Rice_Kp

Attribute table of Dist

OBJECTIDID		Dist_name	Dist_ID	Geometry	Shape_Leng	Shape_Area
1	1	Adilabad	123	1Bo	753,337	16838673859.80
2	2	Nizamabad	116	1BoG(	430,537	7655003826.81
3	3	Karimnagar	122	1BoG	662,231	11771979304.40
4	4	Medak	117	1BoG(	573,986	9510748717.53
5	5	Hyderabad	115	1BoG\	53,851	203218046.01
6	6	Rangareddy	114	1BoG(	587,854	7813591915.82
7	7	Warangal	120	1BoG	674,389	12701135906.20

Rice_1995 table

OBJECTID	DISTID	DISTNAME	NO_	YEAR_1995	RICEKA	RICEKP	RICERA	RICERP
23	123	Adilabad	113	1995	63	64	7	19
16	116	Nizamabad	78	1995	102	206	34	84
22	122	Karimnagar	108	1995	104	260	88	252
17	117	Medak	83	1995	64	111	41	78
15	115	Hyderabad	73	1995	1	1	1	1
14	114	Rangareddi	68	1995	22	47	18	35

FIG. 1.13 Linking spatial and attribute data.

Attribute data about point/line/area features can be entered into different database files. After digitizing, the files will be connected to the default spatial database for the purpose of creating an identification key for each database. In GIS this is a common process used to generate a spatial database after the process of digitization (Fig. 1.13).

Maps representing several layers of spatial and thematic or attribute information (soil map, rainfall map, agro-eco zone map, district map, state maps, etc.) can be digitized in this fashion independently (Longley et al., 1999).

1.11 DATA STORAGE AND RETRIEVAL

A GIS does not store maps. It stores data organized into a database. The locational data of different features (coordinates, topology) are generated during the digitization process. The attribute data of locations are created separately. The GIS must provide the link between locational and attribute data. The relational database model is most suitable to ensure such a link and the database query language can be used to retrieve data. Relational database concepts are therefore central to organizing and managing the data in GIS. The specific format of data storage varies with the GIS software. For example, spatial data were stored in a Geomedia GIS and attribute data were stored in a Microsoft Access databases. The feature-attribute database is created in a specific folder called the Warehouse during digitization. In the Geoworkspaces folder the file is created for the storage of map connections. In this folder the data retrieval is possible by employing the appropriate query language for the database model. Other attribute databases can be stored as MS Access files anywhere in the system and these can establish connections if they share a common ID with the feature-attribute table.

1.12   GEOGRAPHIC ANALYSIS

What makes the GIS different from the other databases or information systems are its spatial analysis functions. These functions use spatial and nonspatial data to answer questions about the real world. The answers could relate to a presentation of the current data (first level use), some patterns in the current data (second level use), and predictions of what the data could be in a different place or at a different time (third level use). Geographic analysis is carried out using the layers of map information created in vector or raster data formats and the associated attribute data to find solutions to specific problems. In each case the problem needs to be defined clearly before the relevant map layers and analysis procedures can be identified. For instance, if the problem is to find the most favorable locations for the siting of wells for conjunctive use in an irrigation project area, information about the geographical features influencing the groundwater recharge will be required to find the solution. These will include maps of existing well locations, rainfall, land use, soils, and the command area of the project, all of which influence recharge. Regions with recharge above a selected threshold value may be considered suitable for additional wells. Further, if the area is situated near the coast, a buffer zone may be required within which no wells can be sited to prevent sea-water intrusion. Similarly, buffer zones may be required on either side of canals to prevent draw of canal water by the wells. What could happen to the quality and levels of groundwater in the area if the present use is maintained or changed could be the subject of another study, where the GIS can help to provide more realistic answers.

Most standard GIS software comes with basic analytical tools that permit overlays of thematic maps, creation of buffers, etc., in addition to calculations of lengths and areas. Overlay operations permit overlaying one polygon over the other to generate a new map of their intersections that are new polygon combinations with the desired homogenous properties with respect to specified polygon attributes (Fig. 1.14).

Map layer overlay

FIG. 1.14   Map layer overlay analysis.

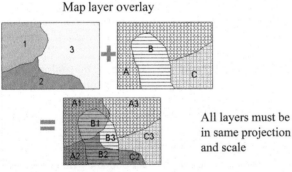

All layers must be in same projection and scale

Overlay generates homogenous units—e.g., agroecozones

Source: FAO

1.13   STEPS IN BUILDING A GIS

The way in which a GIS is built will depend on the way information will be used in the decision-making process (Clarke, 2001). Building a GIS proceeds through at least four stages:

(i) Defining the objectives
(ii) Building the spatial and attribute databases
(iii) Database management for geographic analysis
(iv) Presenting results in the form of maps, etc.

The definition of objectives or the problem to be solved using GIS is critical to the choice of spatial and attribute databases. Once the problem is defined and the relevant map layers and attribute data are identified, building databases involves:

(i) database design
(ii) entering spatial data
(iii) creating topology
(iv) entering attribute data.

Designing the database requires identifying:

(i) study area boundaries
(ii) coordinate system
(iii) data layers
(iv) features in each layer
(v) attributes for each feature type
(vi) coding and organizing attributes.

Depending on whether the map sources are two-dimensional maps of the area or remote sensing imageries, data are entered in vector or raster format. In the vector format, entering spatial data and creating topology are components of the overall digitization process. Raster data will need to be vectorized before topologies can be built. Attribute data are created in the form of database files with one field, the feature identification field, in common with the spatial database created during the spatial data-entry process (Selvam et al., 2014). Database management refers to translating the digitized map into real-world coordinates, identifying coverages for analysis and maintaining the database. Presenting the maps for decision making is facilitated by creating customized maps using the various facilities available in GIS software.

1.14 SUMMARY

A GIS is a computer-based tool used for analyzing the geographical information. It is not a simple digitized map, nor does it hold maps. It holds a database of spatial data and attributes or descriptive information about features on a map that can be used to create preferred maps. The crucial concept of GIS is the separation of spatial or geographic reference information and attributes or descriptive information of map features for data entry and database development and their linkage during analysis. Central to both spatial and attribute information is the database-management concept. The separation of the two types of information facilitates entering the spatial information (map) into computers in a digitized form and establishing connectivity (topology) between different stored map features (points, lines and polygons). The feature attributes data are entered independently taking care to introduce an identification variable for each feature that is shared with the spatial database. For geographic analysis, the spatial and attributes data are linked through this unique identifier variable common to the two types of databases.

Initially, spatial data capture is in spatial units and coordinates of the data capture tool. To translate the map information into real-world information of locations, distances, and areas these need to be translated to real-world units through appropriate transformations of scale and map projections.

The digitized maps and their associated feature attributes are the building blocks of the GIS. The maps can be created and stored in different layers, with each layer containing information about one feature. They can be overlaid onto each other to obtain new maps (coverages) with new polygons that are homogeneous with respect to specified feature attributes of maps that were used in the overlays. The overlay operations must be between maps with exact boundary fits. Exact fits are obtained between maps only if they are created in the same projection and scale. To make exact fits, in overlay operations map projection and scale transformation appropriate map projection and scale transformation operations will be needed before geographic analysis can be performed using overlay operations.

REFERENCES

Arctur, D., Zeiler, M., 2004. Designing Geodatabases: Case Studies in GIS Data Modeling. ESRI Press, Redlands, CA.

Aronoff, S., 1991. Geographic Information Systems: A Management Perspective. WDL Publications, Canada.

Burrough, P.A., 1986. Principles of Geographical Information Systems for Land Resources Assessment. Oxford University Press, Oxford. 193 pp.

Campbell, J.B., 1996. Introduction to Remote Sensing. Taylor and Francis, London, pp. 1–21.

Clarke, K., 2001. Getting Started With Geographic Information Systems, third ed. Prentice-Hall, NJ.

Congalton, R.G., 1991. Remote sensing and geographic information system data integration: error sources and research issues. Photogramm. Eng. Remote Sens. Colorado.edu.

ESRI, 1990. Understanding GIS.

FAO, n.d. http://www.fao.org/sd/eidirect/gis/EIgis000.htm

Foresman, T.W. (Ed.), 1998. The History of Geographic Information Systems: Prospectives from the Pioneers. Prince-Hall, Upper Saddle River, NJ.

Goodchild, M., et al., 1992. Integrating GIS and spatial data analysis: problems and possibilities. Int. J. Geogr. Inf. Syst. 6.

Heywood, I., Cornelius, S., Carver, S., 1998. An Introduction to Geographical Information Systems. Longman Pub. (279 pp).

Huisman, O., De By, R.A., 2009. Principles of Geographic Information Systems. ITC Educational Textbook Series. kartoweb.itc.nl.

Konecny, G., 2002. Geoinformation: Remote Sensing, Photogrammetry and Geographical Information Systems. Taylorfrancis.com.

Longley, P.A., Goodchild, M.F., Maguire, D.J., Rhind, D.W., 1999. Geographic Information Systems. vols. 1 & 2 Wiley Pub.

Maguire, D.J., Goodchild, M., Rhind, D. (Eds.), 1991. Geographic Information System. Longman Scientific, Harlow.

Selvam, S., Manimaran, G., Sivasubramanian, P., 2013. Hydrochemical characteristics and GIS-based assessment of groundwater quality in the coastal aquifers of Tuticorin corporation, Tamilnadu, India. Appl. Water Sci. 3, 145–159.

Selvam, S., Manimaran, G., Sivasubramanian, P., Balasubramanian, N., Seshunarayana, T., 2014. GIS-based evaluation of Water Quality Index of groundwater resources around Tuticorin coastal city, South India. Environ. Earth Sci. https://doi.org/10.1007/s12665-013-2662-y.

Tomlinson, R.F., 2007. Thinking About GIS: Geographic Information System Planning for Managers. books.google.com.

Chapter 2

Merits and Demerits of GIS and Geostatistical Techniques

C.R. Paramasivam

Department of Geology, Alagappa University, Karaikudi, India

Chapter Outline

2.1 INTRODUCTION

The geographical information system (GIS) is simply a collection of hardware, software, geographical information, and individual designs kept in an arranged order that is available for capture, storage, update, manipulation, analysis, and display in any required data form (Singh and Fiorentino, 1996; Alameen and Ramadan, 2015). According to Burrough (2001), GIS defines absolute location and has the ability to relate coordinates with particular domain information. This technology incorporates database operations such as query with statistical analysis derived from the maps. The functional operative system of GIS is able to carry out analysis through spatial and nonspatial data. Nonspatial data can be merged with spatial data using GIS as a platform with the help of relational database management system (RDBMS). Obviously, GIS provides the interactive graphic interface and also support with nonspatial database. The terminology geostatistical encompasses taking measurements from a map or sophisticated geocomputations-based analysis about geographic data.

GIS proves effective for original and uniformity calculations using a geostatic database. Geostatistics are widely used with GIS operations in many fields, such as spatial and nonspatial, adopted by interpolation techniques (Derya and Fatmagul, 2016). One interpolation technique, termed deterministic, obtains the surfaces that are created by measured points on the expansion of matches. This method of inverse distance weighted or the smoothing points is termed trend surface analysis. In addition, geostatistical interpolation based on prediction surface modeling, termed kriging, faces the potential for errors or uncertainty from the prediction. GIS and geostatistical methods are strong alternatives for interpolation and analysis of spatial data (Inna and Alksei, 2016).

GIS can be applied to any mode of life. It is significantly used in multidisciplinary subjects, such as nature, wildlife, culture, wells, springs, water bodies, fire hydrants, roads, streams, and so on. The support of attributes can evaluate also find the domain quantity and density for the selected area could be evaluated and displayed. Apart from these applications, GIS can be tailored according to the requirements and needs of end users.

Modern remote-sensing methods can be integrated with GIS technology to portray real-time environment. This involves various science disciplines adapting to different geographic approaches to satisfy the user community, as shown in Fig. 2.1.

The goal of this chapter is to understand GIS and geostatistical techniques and approaches for spatial and nonspatial analysis through which the advantages and disadvantages of GIS to store, select, manipulate, explore, analyze, and display georeferenced data can be known (Singh and Fiorentino, 1996). The geographical data models are represented as point, line, and polygons in an irregular sampling of databases with a set of spaced locations represented as x, y, and z coordinates. A number of continuity data points arranged in a closed manner is termed as polygon. It bounds the pair of x and y set of data (Longley et al., 2011). The regular spacing measurements that are carried out in the observations of a digital elevation model can be utilized or digitized as contours and isohyetal data.

FIG. 2.1 Interrelationship between GIS disciplines.

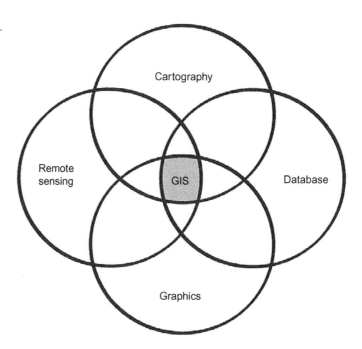

The remotely sensed data grouped as a irregular triangle system produces a TIN model of elevation (Singh and Fiorentino, 1996; Gupta, 2005; Hengl, 2007; Lillesand et al., 2009; Burrough, 2001; Longley et al., 2011). The geostatistical methods assume the spatial variation in a random process and also spatial auto correlation. The different location landscape of the continuous surface can be justified by interpolation (Anderson, 2014; Inna and Alksei, 2016; Clark, 1987; Danilov et al., 2018). The geostatistical analysis can be adapted to find the predicted landscape of group of and similar entities (Inna and Alksei, 2016).

2.2 BASIC PRINCIPLES OF GEOSTATISTICAL METHODS

Geostatistical analysis for interpolation can be achieved by similarity, i.e., inverse distance weighted values and radial functions as the degree of smoothing. Both methods can be used for a mathematical approach to create surface and uncertainty predictions for a better understanding of the surface based on the available information (Setianto and Triandini, 2013; Hengl, 2007; Burrough, 2001; Nas, 2009). Modern GIS modeling and analysis can be obtained through three-dimensional visualization and by understanding the quality of modeling (Jangwon et al., 2017; David, 1996).

The sample point's elevation distribution and surface characters will depend the predicted distance value from nearby points and if they form reasonable accuracy in interpolation (Fig. 2.2). This implies the concept of decrease and increase of weight of value with respect to location prediction. The interpolation method adapted by which accuracy prediction with the help of statistical relationship and auto correlation among the measured points (Clark, 1987).

Spatial measured points and predicted location can be utilized through auto correlation in geostatistics using the ordinary kriging method. Kriging analysis is similar to inverse distance weighting for which the surrounding measured values are derived in location prediction (Setianto and Triandini, 2013; Dobesch et al., 2007).

The semivariogram explains the relationship and difference between the measured and predicted values with the help of spatial auto correlation and distance measurement. The distance between the two locations can be calculated using distance (Hengl et al., 2007; Clark, 1987).

The formula to determine the semivariance at any given distance (Eq. 2.1):

$$\text{Semivariance} = \text{Slope}^* \text{Distance} \qquad (2.1)$$

where, Slope is the slope of the fitted model and Distance is the distance between the pairs of locations.

The empirical semivariogram defines a line that provides the best fit in the points. In addition, the line that is formed with the weighted square between each point represents the variations and distance which are derived from the quality of data (Fig. 2.3).

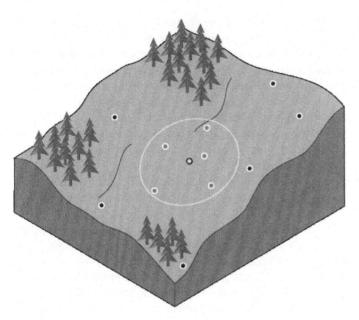

FIG. 2.2 Inverse distance weighting (IDW). *http://pro.arcgis.com.*

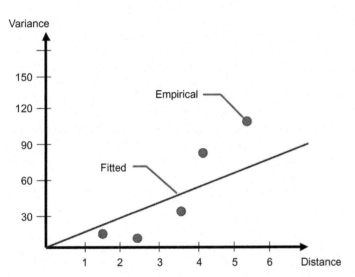

FIG. 2.3 Empirical semivariogram. *http://pro.arcgis.com.*

Splines are one of the mechanical interpolation techniques preferred for polynomial measurement, which includes an amount of smoothing (Hengl, 2007).

2.3 MERITS AND DEMERITS OF GIS

GIS is a computer-based application for mapping and analyzing geographically referenced data in the form of digitized three-dimensional operations. The use of GIS has become inevitable in almost every geospatial application. The end-user community is interested in exploring the economic and deliberate value of GIS due to the benefits of this rapidly growing technology. The fundamental advantages of GIS can be generalized as shown in the following section.

It is easy to visualize the spatial information represented by GIS output maps with clear legends and different groups of coloring and patterns. Thus the novice users of GIS could be comfortable with its application. Expert use of the GIS application is highly supportive in all kinds of environment.

GIS can build various themes and is supportive for database operations such as creation, updating, and manipulation. The map accuracy depends on the quality of input data.

GIS offers an influential decision-making tool in the education sector for its administration, policy making, and instruction. For administrators, GIS can offer an approach to visualize and manage systems in their entirety, including monitoring campus safety, mapping campus buildings, surveilling cable and other infrastructures, routing school buses, planning school closures and opening new ones, and outlining strategies for recruitment. For policymakers in education, GIS provides them with tools that can present patterns in educational achievement and guide the targeting of new programs.

Groundwater analysis uses GIS to interpret spatial correlation targeting potential groundwater resources as well as determining water quality. A logical approach can be adapted for the efficient management of water resources, such as surface and subsurface delineations. GIS can be applied to site selection, zoning, planning, and conservation measures.

Geophysical parameters on the subsurface conditions can be incorporated with spatial data and interpolation techniques can be applied to the exploration, assessment, and prediction of groundwater resources, as well as to the selection of an artificial recharge site. The subsurface flow and pollution model guides how to assess hazards and helps in planning the preventive measures with the help of GIS and geostatistical technologies.

GIS methods can predict, assess risk, and identify hazardous locations of natural resources. It further integrates the spatial and nonspatial data to enable better understanding of emergency conditions. It also supports analysis and the creation of preventive and mitigating solutions.

GIS can be used in criminology to identify facts. Some spatial analyses uses underlying social phenomena to identify rates of crime, which necessitates the use of boundary units such as census tracts and police beats. Although some researchers have also recognized that a simple spatial concentration of crime can also be valuable.

GIS creates employment opportunities in education, administration, research, government, and nonprofit organizations.

GIS technologies refine datasets, data models, and the relation between attributes. High standards are achieved with the scope for generation of newer objects on an on-demand basis. It also allows existing attributes to be linked with newly defined datasets (Singh and Fiorentino, 1996).

2.3.1 Disadvantages of Geographic Information Systems

- *Expensive*: GIS setup is complex, in addition to the cost of the equipment, there is the cost incurred in training. Frequent updating of datasets or data models may lead to errors in results.
- *Real-time parameters*: The handling of growing datasets is an overall challenge to the GIS system.
- *Geographical errors increase with larger-scale data*: The quality of the data collected directly affects the accuracy of the end system. Geographic errors will also affect net results since GIS handles large-scale data.
- *Relative loss of resolution*: Every technology has negotiable errors when deployed.
- *Positional accuracy and precision*: Accuracy and precision are the functions of the scale at which a map (paper or digital) is created. The nonspatial data linked to location may also be inaccurate or imprecise. Inaccuracies may result from mistakes of many sorts. Nonspatial data can also vary greatly in precision. Precise attribute information describes phenomena in great detail.
- *Violation of privacy*: The user community is not limited to authorized persons. So there is threat in the usage of data displayed from the GIS system.
- Error-prone interpretation could lead to failure of system implementation thus affecting the economic strategy of the implementer.
- There might be failures in initiating or additional effort required in order to fully implement the GIS but there might be large benefits to anticipate as well.
- There is a lack of trained teachers in the domain. Though GIS and remote sensing have been introduced in some universities across the country, still the subjects have not been taught to the fullest extent. Moreover, a link between secondary education and higher education must be established for a wide spread and its continuity in the system. Prior knowledge of GIS is a prerequisite to train the trainers.

2.3.2 Advantages of Geostatistics Techniques

- The use of geostatistics techniques enhances the distribution of spatial data.
- Kriging, one of the geostatistical techniques, offers convenient management of groundwater resources.
- Expert interpretation of water quality with the support of large-volume datasets as processed through these techniques.
- The model-based approach of these techniques optimizes the accuracy of the obtained results.
- The geostatistical techniques have the ability to reproduce the trend and provide continuity. This feature allows the user to be precise in their interpretation.

2.3.3   Disadvantages of Geostatistics techniques

- Spatial interpolation evaluates physical data in a continuous domain. The result depends on the correctness of the data input.
- The dataset used in the interpolation process might have errors.
- The choice of adapting the geostatistical method relies on several factors like budget, resource availability, user proficiency, etc.
- A systematic sampling pattern cannot be set up because of the changing cell size and missing/unsuitable data.
- Location of points may be a problem with random sampling distribution.
- Adjacent area coverage might not be supported.
- The interpolation method estimates the value for the center of each unmeasured grid cell with predicted assumption.

2.4   SUMMARY

Though GIS could benefit users from different disciplines, imparting the system is considered to be difficult. Hence it clarity is required for implementation in the health sector, for use in environmental and groundwater studies, and to extend into space science. GIS has improved the quality of research in the above areas through the provision of spatial attributes for the areas of interest.

Simple installation of a database connected to the GIS setup would serve the need. GIS and geostatistics can render solid knowledge and consistent accuracy in a geographic database. The operator failed in handling the geostatistical methods would reflect the visualization error in the data.

The geostatistical methods associated with interpolation techniques provide valuable maps with the effectiveness of kriging and develop semivariogram models. Besides the disadvantages of GIS technology, there is huge potential in various applications.

REFERENCES

Alameen, E.M., Ramadan, F.S., 2015. Geographical information systems applications using cloud computing technology. J. Human. Appl. Sci. (27), 1–13.

Anderson, F., 2014. Multivariate geostatistical model for groundwater constituents in Texas. Int. J. Geosci. (5), 1609–1617.

Burrough, P.A., 2001. GIS and geostatistics: essential patners for spatial analysis. In: Environmental and Ecological Statistics. vol. 8. Kluwer Academic Publishers, pp. 361–377.

Clark, I., 1987. Practical Geostatistics. Elsevier Applied Sciences, British Library Cataloging-in-Publication Data, p. 129.

Danilov, A., Pivovarova, I., Krotova, S., 2018. Geostatistical analysis methods for estimation of environmental data homogeneity. Hindwani Sci. World J. 1–7. https://doi.org/10.1155/2018/7424818.

David, J.U., 1996. GIS, spatial analysis and spatial statistics. Prog. Hum. Geogr. 20 (4), 540–551.

Derya, O., Fatmagul, K., 2016. Geostatistical approach for spatial interpolation of meteorological data. Ann. Braz. Acad. Sci. 88 (4), 2121–2136.

Dobesch, H., Dumolard, P., Dyras, I., 2007. Spatial Interpolation for Climate Data, The Use of GIS in Climatology and Meteorology. British Library of Congress Cataloging-in-Publication Data, . ISBN 978-1-905209-70-5p. 283.

Gupta, R., 2005. Remote Sensing Geology, second ed. Springer International Edition, p. 627.

Hengl, T., 2007. A Practical Guide to Geostatistical Mapping of Environmental Variables. Scientific and Technical Research Series1018-5593143Official Publications of the European Communities. ISBN 978-92-79-06904-8.

Hengl, T., Heuvelink, G.B.M., Rossiter, D.G., 2007. About regression-kriging: from equations to case studies. In: Computers & Geosciences. vol. 33. Elsevier, pp. 1301–1315.

Inna, P., Alksei, M., 2016. Statistical methods of ecological modeling. Res. J. Appl. Sci. 1815-932X11 (6), 321–326.

Jangwon, S., Sung-Min, K., Yosoon, C., 2017. An overview of GIS-based modeling and assessment of mining-induced hazards: soil, water, and Forest. Int. J. Environ. Res. Public Health 14 (1463), 1–12.

Lillesand, T.M., Kiefer, R.W., Chipman, J.W., 2009. Remote Sensing and Image Interpretation, fifth ed. John Wiley and Sons, Inc., New York, p. 753.

Longley, P.A., Goodchild, M.F., Maguire, D.J., Rhind, D.W., 2011. Geographic Information System and Science. vol. 2. John Wiley & Sons, Ltd., England, p. 487.

Nas, B., 2009. Geostatistical approach to assessment of spatial distribution of groundwater quality. Pol. J. Environ. Stud. 18 (6), 1073–1082.

Setianto, A., Triandini, T., 2013. Comparison of kriging and inverse distance weighted (IDW) interpolation methods in lineament extraction and analysis. J. SE Asian Appl. Geol. 2086-51045 (1), 21–29.

Singh, V.P., Fiorentino, M., 1996. Geographical Information Systems in Hydrology. Springer-Science & Business Media, Dordrecht, . ISBN 978-90-481-4751-9p. 439.

Chapter 3

An Introduction to Various Spatial Analysis Techniques

C.R. Paramasivam[*,1] and S. Venkatramanan[*,†,‡,§]

*Department of Geology, Alagappa University, Karaikudi, India †Department of Earth and Environmental Sciences, Pukyong National University, Busan, South Korea ‡Department for Management of Science and Technology Development, Ton Duc Thang University, Ho Chi Minh City, Vietnam §Faculty of Applied Sciences, Ton Duc Thang University, Ho Chi Minh City, Vietnam

[1]Corresponding author: e-mail: pusivam@gmail.com

Chapter Outline

3.1 INTRODUCTION

Spatial analysis can be done using various techniques with the aid of statistics and geographical information systems (GIS). A GIS facilitates attribute interaction with geographical data in order to enhance interpretation accuracy and prediction of spatial analysis (Gupta, 2005). The spatial analysis that is involved in GIS can build geographical data and the resulting information will be more informative than unorganized collected data. According to the requirement of end user, a suitable geospatial technique is chosen to be implemented with GIS. This selection of the geospatial technique will define the classification and method of analysis to be used (Burrough, 2001).

The word "analysis" used alone refers to data querying and data manipulation. Whereas spatial analysis refers to statistical analysis based on patterns and underlying processes. It is a kind of geographical analysis that elucidates patterns of personal characteristics and spatial appearance in terms of geostatistics and geometrics, which are known as location analysis. It involves statistical and manipulation techniques, which could be attributed to a specific geographic database (Cucala et al., 2018; Burrough, 2001).

Suppose the assigned GIS task is to record sampling stations chosen in a selected study site with different patterns, then by implementing spatial techniques appropriate results can be obtained (Burrough, 2001). These results further show the sample location's characteristics, such as dispersed or clustered. Spatial information relates to the position, area, shape, and size of objects on Earth and this information is stored as coordinates and topology (Cucala et al., 2018; Fischer et al., 1997; Gupta, 2005).

The sampling stations were observed for only the area of interest in the entire domain. This area is derived applying quantitative and statistical techniques on the spatial attributes of GIS database (Fig. 3.1).

The spatial analysis can be refined and made interactive, i.e., transformation, manipulation of maps, and applied simple mathematical facts (Bourgault and Marcotte, 1991). The spatial data can be derived from large databases providing detailed information and trends (Higgs et al., 1998). For example, multivariable or factor analysis allows changes in variables.

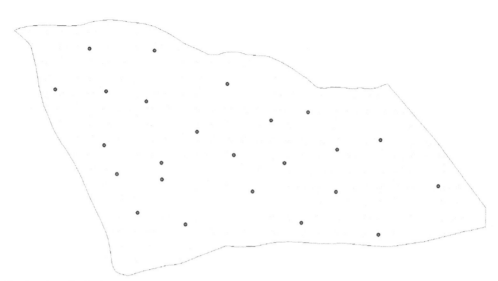

FIG. 3.1 Sampling locations distribution map.

FIG. 3.2 Components of spatial data.

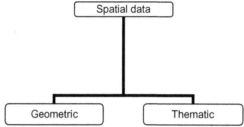

The principle components of which could be data correlation with eigen values. This chapter aims to impart a basic method by which spatial data could be analyzed.

A GIS database computes spatial location, distribution, and relationship. Fundamentally, spatial analysis is a set of methods producing refined results with spatial correlation. A spatial link is observed between geometric and thematic data and attributes in the data components are identified (Fig. 3.2). Nowadays all GIS software has modules designed to handle spatial data and positions are connected with other features and details either spatial or nonspatial characters (Burrough, 2001).

The range of methods deployed for spatial analysis varies with respect to the type of the data model used. Measurement of length, perimeter and area of the features is a very common requirement in spatial analysis (Parasiewicz et al., 2018; Clark and Evans, 1954). However different methods are used to make measurements based on the type of data used i.e. vector or raster. Invariably, the measurements will not be exact, as digitized feature on map may not be entirely similar to the features on the ground, and moreover in the case of raster, the features are approximated using a grid cell representation (Oliver and Webster, 2007).

Many methods can be linked with GIS software, and the most applicable methods are discussed in this chapter, such as inverse distance weighted, natural neighbor inverse distance weighted, spline, kriging, and topo to raster methods. The suite of analyses should be incorporated into a GIS package, ensuring that a user can still intervene to choose the most appropriate form of analysis (Cucala et al., 2018; Fischer et al., 1997).

3.2 TYPES OF SPATIAL ANALYSIS

There are different types of spatial analysis, but in all types of spatial analysis, locations are very important. Generally, spatial analysis is "a group of methods whose results change when the locations of the objects being analyzed change." For example, calculating the number of locations in a particular domain is spatial analysis because the output is directly based on the locations.

3.2.1   Inverse Distance Weighting

Inverse distance weighting (IDW) is one of the interpolation methods that are considered to be simple to use. Here the locations used have identified values or other unidentified locations have calculated ones. This method is used to forecast unidentified values for any geographical location data. For example, precipitation, height, depth, concentrations of chemical parameters, pollution levels, and so on.

IDW is the simplest interpolation method. A neighborhood for the interpolated point is identified and a weighted average is taken within this neighborhood. The weights observed show it as a decreasing function of distance. However, the user can control the mathematical form of the weighting function and the size of the neighborhood (expressed as a radius or a number of points) and other options are available (Setianto and Triandini, 2013).

IDW interpolator presumes each input point to be locally influenced, which diminishes with distance. It assigns greater weights to the points closer to the processing cell than to those further away. A specified number of points or all points within a specified radius can be used to determine the output value of each location. The presumption being that the variable mapped decreases in influence with its distance from the sampled location.

Maximum and minimum values can be obtained through IDW moving average interpolator in variable data. This seems to highlight a possible new value for the reading points within the area (Setianto and Triandini, 2013). An example is shown in Fig. 3.3.

Further, having found the grid node values and data points, the radius can be identified. A few or all of the data points could be involved in the interpolation process. The chosen data points that are nearer the node points influence the value to be computed (Fig. 3.4).

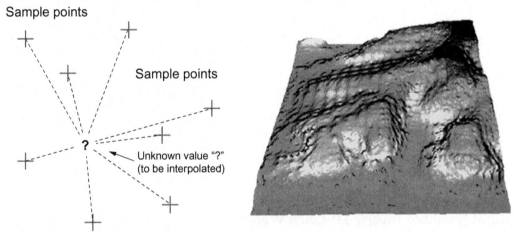

FIG. 3.3 Moving average inverse distance weighting interpolation. *(IDW Interpolation; Courtesy: QGIS.)*

FIG. 3.4 Radius search in IDW interpolation.

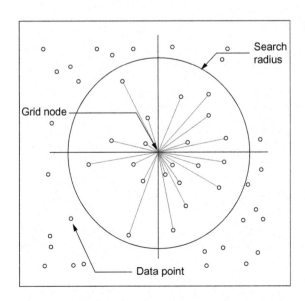

FIG. 3.5   NNIDW interpolation. *(Natural IDW: Courtesy: ESRI.)*

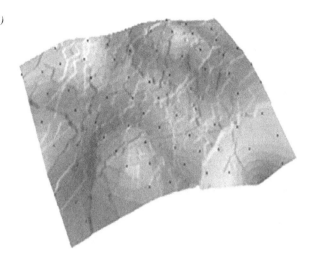

FIG. 3.6   An example of spline generated surface. *(Spline: Courtesy: ESRI.)*

3.2.2   Natural Neighbor Inverse Distance Weighted

Natural neighbor IDW (NNIDW) interpolation can be effectively implemented for interpolation and extrapolation methods dealing with huge data sets. This technique works on cluster scatter points and identifies datasets. An example of NNIDW is shown in Fig. 3.5.

This method follows IDW interpolation except that the points and weighted distance are interpolated to form a triangulation that chooses only the closest nodes. Therefore it can be adapted to all environments wherein the sample data are distributed with uneven density.

3.2.3   Spline

In this technique the values are calculated using a mathematical function that minimizes the overall surface curvature (Fig. 3.6). The resultant point shows a smooth surface passing exactly through the given input points. This method can be adapted to gently varying surfaces like elevation, water table heights, or pollution.

3.2.4   Spline Interpolations

The spatial interpolation method finds the data in a continuous area and forecasts unknown points using the observed data to fill in data that is missing or cannot be obtained (Fig. 3.7). Errors may be expected in cases of uneven distribution. In such cases, the tension spline interpolation (TSPLINE) technique can be applied.

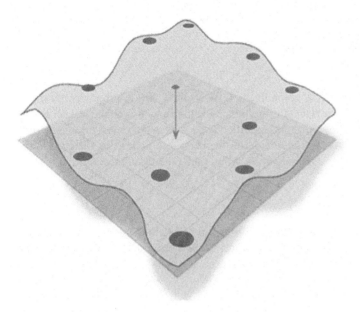

FIG. 3.7   An example of spline interpolation.

FIG. 3.8   A sample formulation of kriging. *(Kriging: Courtesy: ESRI.)*

Generally, spline interpolations can be applied to a large number of data interpolations. The results obtained through this method are closely aligned to those of kriging methods. The use of TSPLINE also means that covariance function structure estimation can be avoided (Xiao et al., 2016).

3.2.5   Kriging

Kriging is a geostatistical interpolation technique that considers both the distance and the degree of variation between known data points when estimating values in unknown areas (Fig. 3.8). Kriging is a multistep process. It starts with the exploratory statistical analysis of the data, variogram modeling, creation of the surface, and (optionally) exploration of a variance surface. The kriging method will be more successful when spatially correlated distance or directional bias is present in the data. It is widely applied in soil science and geology.

Kriging is similar to IDW in that it assigns weights to the surrounding measured values in deriving a prediction for an unmeasured location. The general formula for both interpolators involves the calculation of a weighted sum of the data (Eq. 3.1):

$$\hat{Z}(s_0) = \sum_{i=1}^{N} \lambda_i Z(s_i) \tag{3.1}$$

where:

$Z(s)$ = the measured value at the ith location.

λ = an unknown weight for the measured value at the ith location.

s = the prediction location.

N = the number of measured values.

The weight, λ, depends upon the distance of the prediction location in the IDW method. But in kriging method, the weights are based upon the distance between the measured points and the prediction location and also on the overall spatial arrangement of the measured points. The spatial autocorrelation must be quantified to be used in the spatial arrangement of weights. Hence in ordinary kriging the weight, λ, depends on a fitted model to the measured points, the distance to the prediction location, and the spatial relationships among the measured values around the prediction location.

3.3 KRIGING TYPES

Geostatistical interpolation consists of ordinary kriging interpolation, simple kriging interpolation, universal kriging interpolation, indicator kriging, probability kriging, disjunctive kriging, and topo to raster (Hengl et al., 2007; Setianto and Triandini, 2013).

3.3.1 Ordinary Kriging

Ordinary kriging assumes the model (Eq. 3.2):

$$Z(s) = \mu + \varepsilon(s) \tag{3.2}$$

where μ is an unknown constant. The assumption of a constant mean should be reasonable. Scientifically this assumption is to be rejected. However, as a simple prediction method, it has remarkable flexibility. It uses semivariogram/covariance and transformations, removes trends, and allows for measurement error.

3.3.2 Simple Kriging

In Eq. (3.3), simple kriging assumes the model:

$$Z(s) = \mu + \varepsilon(s) \tag{3.3}$$

where μ is a known constant.

This method is similar to ordinary kriging, which uses semivariogram/covariance and transformations and allows for measurement error. Here, all the parameters and covariant should be known.

3.3.3 Universal Kriging

Universal kriging assumes the model shown in Eq. (3.4):

$$Z(s) = \mu(s) + \varepsilon(s) \tag{3.4}$$

where $\mu(s)$ is some deterministic function.

Universal kriging can also use semivariogram/covariance and transformations and allows for measurement error, but the parameters and covariant will be unknown.

3.3.4 Indicator Kriging

Indicator kriging assumes the model shown in Eq. (3.5):

$$I(s) = \mu + \varepsilon(s) \tag{3.5}$$

where μ is an unknown constant and $I(s)$ is a binary variable. The threshold of the continuous data is used to create binary data. In simple terms, the observed data are considered to be 0 or 1. Using binary variables indicator kriging follows ordinary kriging. Like the other methods, indicator kriging uses either semivariogram or covariance.

3.3.5 Probability Kriging

Probability kriging assumes the model shown in the following equations (Eqs. 3.6, 3.7):

$$I(s) = I(Z(s) > c) = \mu + \varepsilon(s) \tag{3.6}$$

$$Z(s) = \mu + \varepsilon(s) \tag{3.7}$$

where μ and μ are unknown constants and $I(s)$ is a binary variable created by using a threshold indicator, $I(Z(s) > c)$. Now there could be two types of random errors, $\varepsilon(s)$ and $\varepsilon(s)$, indicating that there is an autocorrelation for each and a cross-correlation between them. Probability kriging follows indicator kriging, but it uses cokriging to overcome the errors (Hengl et al., 2007; Setianto and Triandini, 2013; Oliver and Webster, 2007).

Probability kriging can use semivariogram or covariance, cross-covariance, and transformations, but does not allow measurement error to occur (Bourgault and Marcotte, 1991).

3.3.6 Disjunctive Kriging

Disjunctive kriging assumes the model:

$$f(Z(s)) = \mu + \varepsilon(s) \tag{3.8}$$

where μ is an unknown constant and $f(Z(s))$ is an arbitrary function of $Z(s)$. It tries to do more than ordinary kriging with increased costs. The bivariate normality assumption and approximations are applied to the functions $f(Z(s))$. These assumptions and solutions obtained from this interpolation method are more complex. Semivariogram or covariance and transformations are adapted but prohibit the measurement of errors.

3.3.7 Topo Grid/Topo to Raster

This interpolation technique extracts the elevation surfaces from different types of input data (elevation points, contour lines, stream lines, and lake polygons). An iterative finite difference interpolation technique is used in this method. By optimizing this method, the computational efficiency of local interpolation methods, such as IDW interpolation can be achieved without losing the surface continuity of global interpolation methods, such as kriging and spline (Gupta, 2005).

Basically, the topo grid interpolation technique is a discretized thin-plate spline technique wherein the roughness penalty has been modified in order to allow the fitted DEM to capture abrupt changes in terrain, such as streams, ridges, and cliffs. It is specifically designed to work intelligently with contour inputs (Augusto Filho et al., 2016; Wahba, 1990).

The topo to raster interpolation method can be derived from existing elevation points, contours, stream lines, and lake polygons. An example is shown in Fig. 3.9.

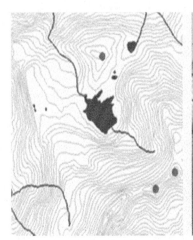 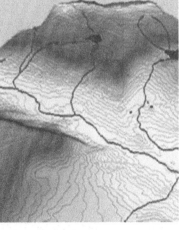

FIG. 3.9 Topo to raster interpolation. *(Topo to raster: Courtesy: ESRI.)*

3.4 SUMMARY

The association of spatial data with GIS is the most effective way to realize and visualize direction changes in the Earth's surface characteristics through the use of maps or statistical calculation. The visualized surface demonstrates x, y, and z positions along with topographic 3D projections. The spatial approach using different methods can provide a smooth map of variogram and spatial query even though these techniques are relatively undeveloped. The gathering of GIS domain functionality could solve the problem of edge effects, particularly in the smoothing of irregular data. A broad view on GIS analysis can be operated in terms of the spatial phenomenon and characteristics of the Earth's surface. Hence it can be employed by decision-making authorities as a tool for regional planning and development. The data analysis domain put together using GIS functions aims to obtain spatial relationship, patterns, and trends. Spatial analysis can generate new information about the Earth's features.

REFERENCES

Augusto Filho, O., Soares, W., Fernandez, C.I., 2016. Mapping of the water table levels of unconfined aquifers using two interpolation methods. J. Geogr. Inf. Syst. (8), 480–494.

Bourgault, G., Marcotte, D., 1991. Multivariable variogram and its application to the linear model of coregionalization. Math. Geol. 23 (7), 899–900.

Burrough, P.A., 2001. GIS and geostatistics: essential patners for spatial analysis. In: Environmental and Ecological Statistics. vol. 8. Kluwer Academic Publishers, pp. 361–377.

Clark, P.J., Evans, F.C., 1954. Distances to nearest neighbor as a measure of spatial relationships in populations. Ecology 35 (4), 445–453.

Cucala, L., Genin, M., Occelli, F., Soula, J., 2018. A multivariate nonparametric scan statistic for spatial data. Spatial Stat. 29, 1–14.

Fischer, M.M., Gopal, S., Staufer, P., Steinnocher, K., 1997. Evaluation of neural pattern classifiers for a remote sensing application. Geogr. Syst. 4 (2), 195–226.

Gupta, R.P., 2005. Remote Sensing Geology, second ed. Springer International Edition, p. 627.

Hengl, T., Heuvelink, G.B.M., Rossiter, D.G., 2007. About regression-kriging: from equations to case studies. Comput. Geosci. 33, 1301–1315.

Higgs, G., Senior, M.L., Williams, H.C.W.L., 1998. Spatial and temporal variation of mortality and deprivation 1: widening health inequalities. Environ. Plan. 30, 1661–1682.

Oliver, M.A., Webster, R., 2007. Kriging: a method of interpolation for geographical information systems. Int. J. Geogr. Inf. Syst. 4 (3), 313–332.

Parasiewicz, P., Prus, P., Suska, K., Marcinkowski, P., 2018. $E = mc^2$ of environmental flows: a conceptual framework for establishing a fish-biological foundation for a regionally applicable environmental low-flow formula. Water J. MDPI Switzerland 10 (1501), 1–19.

Setianto, A., Triandini, T., 2013. Comparison of kriging and inverse distance weighted (IDW) interpolation methods in lineament extraction and analysis. J. Southeast Asian Appl. Geol. 5 (1), 21–29.

Wahba, G., 1990. Spline Models for Observational Data. Society for Industrial and Applied Mathematics (SIAM), Philadelphia, PA, p. 169.

Xiao, Y., Gu, X., Yin, S., Shao, J., Cui, Y., Zhang, Q., Niu, Y., 2016. Geostatistical interpolation model selection based on ArcGIS and spatio-temporal variability analysis of groundwater level in piedmont plains, Northwest China. Springer Plus 5 (425), 1–15.

Section B

Types of Geospatial and Geostatistical Techniques

Chapter 4

Supplement of Missing Data in Groundwater-Level Variations of Peak Type Using Geostatistical Methods

Sang Yong Chung*, S. Venkatramanan*,†,‡,§, Hussam Eldin Elzain*, S. Selvam¶ and M.V. Prasanna‖

*Department of Earth and Environmental Sciences, Pukyong National University, Busan, South Korea †Department for Management of Science and Technology Development, Ton Duc Thang University, Ho Chi Minh City, Vietnam ‡Faculty of Applied Sciences, Ton Duc Thang University, Ho Chi Minh City, Vietnam §Department of Geology, Alagappa University, Karaikudi, India ¶Department of Geology, V.O. Chidambaram College, Thoothukudi, India ‖Department of Applied Geology, Faculty of Engineering and Science, Curtin University Malaysia, Miri, Malaysia

Chapter Outline

4.1 INTRODUCTION

Many groundwater monitoring wells are installed for the effective conservation and management of groundwater. Groundwater level and quality are measured from automatic recorders in the monitoring wells. However, groundwater-level data are often missed because of power outage or digital-sensor problems. The missing data need to be interpolated using proper statistical methods, because they degrade the continuity of monitored data.

The use of geostatistics is necessary for the reproduction of missing data, because groundwater-levels show various changes and irregularities. Kriging can interpolate the missing intervals with the minimum errors in the case of small fluctuation in water-level data. Chung et al. (2001) reproduced missing data in a sinuous-type long-term groundwater-level with few errors using ordinary kriging.

Examples of kriging applications include the estimation of aquifer parameters (Loaiciga et al., 1996), the analysis of groundwater flow (Jensen et al., 1996), and the estimation of groundwater level and hydraulic gradient (Philip and Kitanidis, 1989). Recently kriging has been used for the optimization of groundwater-level observation networks (Theodossiou and Latinopoulos, 2006), the risk assessment of nitrate contamination (Hu et al., 2005), and the evaluation of arsenic-contamination potential (Liu et al., 2004).

In the case of irregularly fluctuating water-level data, kriging has limitations in reproducing the variability of water-level data. Conditional simulation can produce the irregular fluctuations with relatively few statistical errors (Journel and Huijbregts, 1978; Chiles and Delfiner, 1999). Chung and Wheatcraft (1993) showed that conditional simulation was far superior to kriging in the estimation of two-dimensional hydraulic conductivity distributions using hydraulic conductivity data of the Borden site in Canada. Conditional simulation is widely used for groundwater-flow modeling and contaminant-transport modeling through the simulation of hydraulic conductivity distribution (Gómez-Hernández et al., 1997; Capilla et al., 1997; Hendricks Franssen et al., 2003).

4.2 GROUNDWATER LEVEL DATA WITH A PEAK TYPE VARIATION

In this study, the groundwater-level data of alluvium and bedrock in the Daegu Bisan National Groundwater Monitoring Well, Korea were used to compare kriging with conditional simulation for the interpolation of missing data. Fig. 4.1 shows the variations of two groundwater-level data sources from the Daegu Bisan National Groundwater Monitoring Well. They have very irregular fluctuations and there are several missing intervals of groundwater level. Table 4.1 shows the general statistical values for the groundwater levels of alluvium and bedrock.

FIG. 4.1 Missing interval in groundwater levels of a groundwater monitoring well.

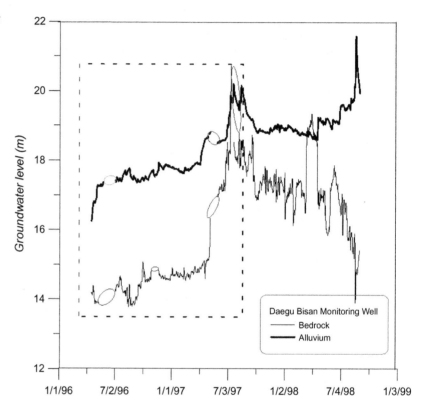

TABLE 4.1 General Statistics for Groundwater Levels of a Groundwater Monitoring Well

	Statistical Values	
Statistics	Alluvium	Bedrock
No. of data	383	409
Mean	17.77	15.43
Median	17.72	14.72
Standard deviation	0.48	1.55
Variance	0.23	2.40
Skewness	0.51	1.16
Kurtosis	1.02	−0.09
Minimum value	16.24	13.77
Maximum value	19.31	20.74

4.3 GEOSTATISTICAL METHODS

4.3.1 Kriging

Kriging is a local estimation technique of the best linear unbiased estimator (BLUE) for the unknown values of spatial and temporal variables. Kriging is expresses as:

$$Z_K^* = \sum_{i=1}^{n} \lambda_i Z_i \tag{4.1}$$

where Z_K^* is an estimate by kriging, λ_i is a weight for Z_i, and Z_i is a variable. The weight is determined to ensure that the estimator is unbiased and that the estimation variance is minimal (Journel and Huijbregts, 1978).

The unbiased condition of kriging is:

$$E\{Z_V - Z_K^*\} = 0 \tag{4.2}$$

where Z_V is an actual value and Z_K^* is an estimated value.

The sum of weights is:

$$\sum_{i=1}^{n} \lambda_i = 1.0 \tag{4.3}$$

The estimation variance of kriging variance is:

$$\sigma_K^2 = E\left\{ [Z_V - Z_K^*]^2 \right\} = \overline{C}(V, V) + \mu - \sum_{i=1}^{n} \lambda_i \overline{C}(v_i, V) \tag{4.4}$$

where $\overline{C}(V, V)$ represents the covariances between sample variables, μ is Langrange parameter, and $\overline{C}(v_i, V)$ represents the covariances between the sample variable and the estimates.

Various kinds of kriging have been developed to be suitable for the characteristics of used data, i.e., ordinary kriging for stationary data, universal kriging for nonstationary data, cokriging for a group of correlated data, etc. In this study, ordinary kriging was used to produce the graphs of groundwater-level data.

4.3.2 Variogram

The spatial dependence between sample data is necessary for the determination of kriging weights. The measure of the spatial dependence is the semivariogram expressed as:

$$\gamma(h) = \frac{1}{2N(h)} \sum_{i=1}^{N(h)} [Z(x_i) - Z(x_i + h)]^2 \tag{4.5}$$

where $Z(x_i)$ and $Z(x_i + h)$ are observed variables at sampling point x_i and $x_i + h$, and $N(h)$ is the number of pairs of samples separated by the lag h. An experimental semivariogram needs to be fitted to a theoretical semivariogram model for the kriging interpolation.

The covariances between sample data are obtained from the following relation:

$$C(h) = sill - \gamma(h) \tag{4.6}$$

where $C(h)$ is covariance, and $\gamma(h)$ is semivariogram.

4.3.3 Conditional Simulation

The principles of conditional simulation are expressed as:

$$Z_{SC}^*(x) = Z_{OK}^*(x) + \left[Z_S(x) - Z_{SK}^*(x) \right] \tag{4.7}$$

where $Z_{SC}^*(x)$ is a conditional simulation and $Z_{OK}^*(x)$ is a kriged value at a point x.

$Z_S(x)$ is a nonconditional realization at a point x and $Z_{SK}^*(x)$ is a kriged value of a nonconditional realization $Z_S(x)$.

Turning band method (TBM; Journel, 1974) was used for a nonconditional realization $z_s(x)$ at a point x. TBM turns multidimensional simulations into several independent one-dimensional simulations for reasonable computer costs. $z_s(x)$ is a realization of three-dimensional random function $Z_S(x) = Z_s(u, v, w)$ which is a second-order stationary, and has a zero expectation and a covariance of $C(h) = E\{Z_s(x)Z_s(x+h)\}$.

The equation of TBM is expressed as:

$$z_s(x) = \frac{1}{\sqrt{(N)}} \sum_{i=1}^{N} z_i(x) \tag{4.8}$$

where $z_s(x)$ is multidimentional nonconditional realization at a point x, $z_i(x)$ is one-dimensional nonconditional realization, and N is the number of turning band lines ($N = 15$ in three-dimensional realization). TBM was developed by Matheron (1973) and practically applied by Journel (1974).

4.4 COMPARISON OF INTERPOLATION CAPABILITY

4.4.1 Statistical Validation Test

Some statistical errors were used for the accuracy validation between kriging and conditional simulation.

Mean error (ME):

$$ME = \frac{1}{N} \sum_{i=1}^{n} [Z(x) - Z^*]_i \tag{4.9}$$

Standard deviation of error (SDE):

$$VE = \frac{1}{N-1} \sum_{i=1}^{N} (Error - ME)^2 \tag{4.10}$$

$$SDE = \sqrt{VE} \tag{4.11}$$

Square root-mean-squared errors (SRMSE):

$$MSE = \frac{1}{N} \sum_{i-1}^{m} [Z(x) - Z^*(x)]_i^2 \tag{4.12}$$

$$SRMSE = \sqrt{MSE} \tag{4.13}$$

4.4.2 Interpolation of Artificial Missing Data

Alluvial groundwater-level data collected by the Daegu Bisan National Groundwater Monitoring Well were used for a comparison of the interpolation capability of kriging with that of conditional simulation. Experimental data were sampled from November 1, 1997 to April 15, 1998. Four kinds of missing intervals were arbitrary chosen from these data. Ten missing segments of data were identified from January 1, 1998 to January 10, 1998; 20 from January 1, 1998 to January 20; 40 from January 1, 1998 to February 19; and 60 from January 1, 1998 to March 10, respectively.

Fig. 4.2 shows the results of interpolation for the missing data (*black* color) using kriging and conditional simulation. The distributions of data interpolated by kriging (*red* color) are almost linear in shape, but those interpolated by conditional simulation (*blue* color) show similar fluctuations as the original distributions.

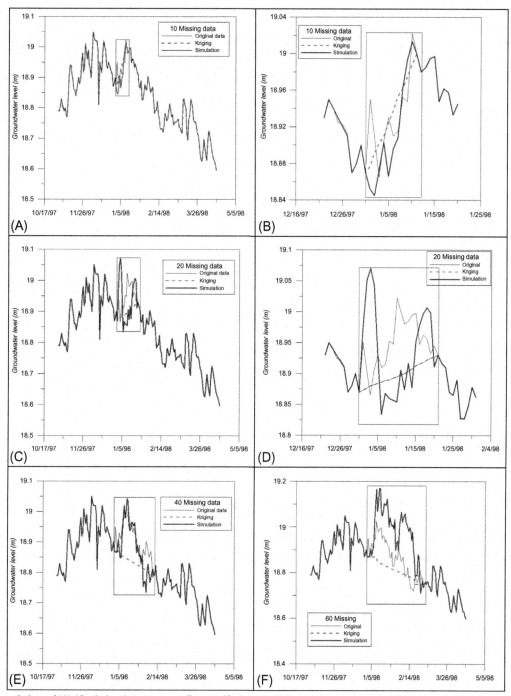

FIG. 4.2 Interpolations of (A) 10 missing data segments, (B) magnified graph of 10 missing data segments, (C) 20 missing data segments, (D) magnified graph of 20 missing data segments, (E) 40 missing data segments, and (F) 60 missing data segments.

Table 4.2 shows the statistical errors resulting from the interpolations of four kinds of missing data. The SRMSEs of kriging are smaller than conditional simulation for the time periods with 10, 20, and 60 missing data segments, but the SRMSE of conditional simulation is smaller than kriging for the time period with 40 missing data segments. Table 4.3 is the number of smaller deviations from original data. Kriging has more numbers than conditional simulation for the time period with 20, and 60 missing data segments, but conditional simulation has more numbers than kriging for the time period with 40 missing data segments. Thus conditional simulation is far superior to kriging for the interpolation of the 40 missing data segments.

TABLE 4.2 Statistics of Errors for Four Kinds of Missing Data

Statistics		10 Missing Data Segments	20 Missing Data Segments	40 Missing Data Segments	60 Missing Data Segments
Mean error (ME)	Kriging	0.002	−0.051	−0.0774	−0.052
	Conditional simulation	−0.019	−0.012	−0.0258	0.108
Standard deviation of error (SDE)	Kriging	0.0354	0.0355	0.0448	0.0538
	Conditional simulation	0.0429	0.0981	0.0503	0.0935
Square root of mean square error (SRMSE)	Kriging	0.0336	0.0620	0.0892	0.0748
	Conditional simulation	0.0450	0.0964	0.0560	0.1424

TABLE 4.3 Number of Smaller Deviations From Original Data

	Kriging		Conditional Simulation	
Data Segments	No. of Data Segments	Ratio (%)	No. of Data Segments	Ratio (%)
10 Missing	5	50	5	50
20 Missing	13	65	7	35
40 Missing	17	42.5	23	57.5
60 Missing	40	66.7	20	33.3

4.5 INTERPOLATION OF ACTUAL MISSING DATA

Kriging and conditional simulation were used for the interpolation of groundwater-level data actually missing from that collected by the Daegu Bisan National Groundwater Monitoring Well with peak-type variations.

4.5.1 Application of Kriging

Fig. 4.3 shows the results of interpolation by kriging of missing data for the groundwater-levels in alluvium and bedrock. Kriging reproduced the missing data with minimum errors. However, the distributions of reproduced data in alluvium and bedrock were nearly linear and didn't show the fluctuations of groundwater-levels.

4.5.2 Application of Conditional Simulation

Fig. 4.4 shows the results of interpolation by conditional simulation of missing data for groundwater-levels in alluvium and bedrock. Conditional simulation reproduced the missing data of alluvium and bedrock with reasonable fluctuations.

4.5.3 Cross Validation Test

Cross validation test was developed by Davis (1987) to examine the suitability of a specified vaiogram or covariance model for the given data.

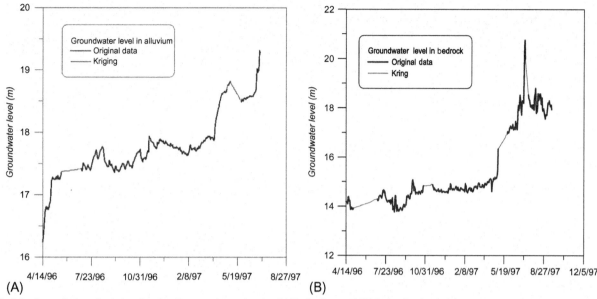

FIG. 4.3   Interpolation of missing data for the groundwater levels of (A) alluvium and (B) bedrock using kriging.

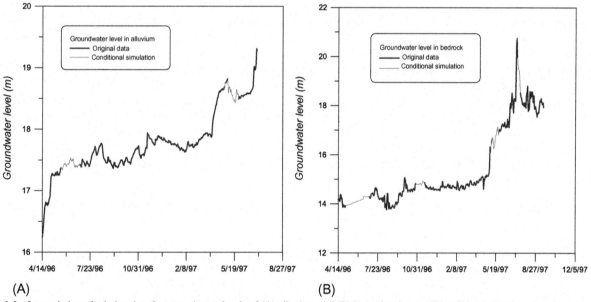

FIG. 4.4   Interpolation of missing data for groundwater levels of (A) alluvium and (B) bedrock using conditional simulation.

The test is using a reduced error (RE) which is defined by the error $(Z(x) - Z^*(x))$ divided by the square root of kriging variance $\left(\sqrt{\sigma_K^2}\right)$ of sample data:

$$RE = \frac{Z(x) - Z^*(x)}{\sqrt{\sigma_K^2}} \tag{4.14}$$

Kriging variance is calculated from kriging, but cannot be calculated from conditional simulation. Thus, mean error (ME), standard deviation of error (SDE) and Square Root of Mean Square Error (SRMSE) were used for the cross validation test between kriging and conditional simulation.

TABLE 4.4 Statistics of Errors for the Cross-Validation Test of Original Data

Statistics		Alluvium	Bedrock
Mean error (ME)	Kriging	0.00003	−0.00032
	Conditional Simulation	0.00028	−0.00064
Standard deviation of error (SDE)	Kriging	0.02199	0.10755
	Conditional Simulation	0.05171	0.23271
Square root of mean square error (SRMSE)	Kriging	0.02198	0.10742
	Conditional Simulation	0.05164	0.23245

Table 4.4 shows the statistical errors of cross-validation test. Statistical errors of kriging were smaller than conditional simulation, but the fluctuations of groundwater-levels couldn't be generated by kriging.

4.6 CONCLUSIONS

Alluvial groundwater-level data from the Daegu Bisan National Groundwater Monitoring Well were used for the comparison of the interpolation capabilities of kriging with those of conditional simulation. Experimental data were sampled from November 1, 1997 to April 15, 1998. Four kinds of missing intervals, i.e., 10, 20, 40, and 60 missing data segments, were arbitrarily chosen. The distributions of the missing data interpolated by kriging was almost linear in shape, although the statistical errors of kriging were smaller than those of conditional simulation. However, the data interpolated by conditional simulation showed the similar fluctuations to the original distributions, even though conditional simulation made larger statistical errors than kriging.

Groundwater-level data reproduced by kriging at the actual intervals of a groundwater monitoring well were completely consistent with the original data, but the missing data formed a nearly linear shape because kriging produced estimates with minimum errors. On the other hand, conditional simulation produced very similar fluctuations to the original distributions, although its estimates contained larger statistical errors than the kriging estimates. Therefore conditional simulation is widely used for the reproduction of irregular fluctuations such as groundwater level, hydraulic conductivity, and tidal fluctuation.

REFERENCES

Capilla, J.E., Gómez-Hernández, J.J., Sahuquillo, A., 1997. Stochastic simulation of transmissivity fields conditional to both transmissivity and piezometric data, 2. Demonstration on a synthetic aquifer. J. Hydrol. 203, 175–188.

Chiles, J.-P., Delfiner, P., 1999. Geostatistics: Modeling Spatial Uncertainty. John Wiley & Sons, Inc. 695 p.

Chung, S.Y., Wheatcraft, S.W., 1993. Application of fractal interpolation functions for modeling hydraulic conductivity distributions. J. Geol. Soc. Korea 29, 168–175.

Chung, S.Y., Shim, B.W., Kang, D.H., Kim, B.W., Park, H.Y., Won, J.H., Kim, G.B., 2001. Interpolation of missing groundwater level data using kriging at a national groundwater monitoring well. J. Geol. Soc. Korea 37, 421–430.

Davis, B.M., 1987. Uses and abuses of cross-validation in geostatistics. Math. Geol. 19, 241–248.

Gómez-Hernández, J.J., Sahuquillo, A., Capilla, J.E., 1997. Stochastic simulation of transmissivity fields conditional to both transmissivity and piezometric data, part 1, theory. J. Hydrol. 204, 162–174.

Hendricks Franssen, H.J., Gómez-Hernández, J.J., Sahuquillo, A., 2003. Coupled inverse modeling of groundwater flow and mass transport and the worth of concentration data. J. Hydrol. 281, 281–295.

Hu, K., Huang, Y., Li, H., Li, B., Chen, D., White, R.E., 2005. Spatial variability of shallow groundwater level, electrical conductivity and nitrate concentration, and risk assessment of nitrate contamination in North China Plain. Environ. Int. 31, 896–903.

Jensen, J.L., Corbett, P.W.M., Pickup, G.E., Ringrose, P.S., 1996. Permeability semivariograms, geological structure, and flow performance. Math. Geol. 28 (4), 419–436.

Journel, A.G., 1974. Geostatistics for conditional simulation of ore bodies. Econ. Geol. 69, 673–687.

Journel, A.G., Huijbregts, C.J., 1978. Mining Geostatistics. Academic Press. 600 p.

Liu, C.W., Jang, C.S., Liao, C.M., 2004. Evaluation of arsenic contamination potential using indicator kriging in the Yun-Lin aquifer (Taiwan). Sci. Total Environ. 321, 173–188.

Loaiciga, H.A., Leipnik, R.B., Hudak, P.F., Marino, M.A., 1996. 1-, 2-, and 3-Dimensional effective conductivity of aquifers. Math. Geol. 28 (5), 563–584.

Matheron, G., 1973. The intrinsic random functions and their applications. Adv. Appl. Probab. 5.

Philip, R.D., Kitanidis, P.A., 1989. Geostatistical estimation of hydraulic gradients. Ground Water 27 (6), 855–865.

Theodossiou, N., Latinopoulos, P., 2006. Evaluation and optimization of groundwater observation networks using the kriging methodology. Environ. Model. Softw. 21, 991–1000.

Chapter 5

A Comparative Study of Spatial Interpolation Technique (IDW and Kriging) for Determining Groundwater Quality

Prafull Singh and Pradipika Verma

Amity Institute of Geoinformatics and Remote Sensing, Amity University, Noida, India

Chapter Outline

5.1 INTRODUCTION

Surface and groundwater resources are an important source of water for many purposes. Due to climate change and anthropogenic pressure, like increasing populations, industrialization, urbanization, and complex land-use practices, which are very common in the developing countries like India, the quality and quantity of water resources are severely deteriorating (Singh et al., 2013, 2015, 2016). It is very important to study the spatial and temporal distribution of pollutants in water bodies to develop appropriate management plans and map pollutant concentrations. Periodic water-quality monitoring and conservation of water resources are very important (Sun et al., 1992). Many studies have been reported on water-quality mapping and monitoring using a geographical information system (GIS) and statistical tools. GIS plays an important role in groundwater studies especially for site suitability analysis, managing site inventory data, estimating the vulnerability of groundwater to pollution, groundwater movement modeling, modeling of transport, monitoring the leaching of solutes, and integrating groundwater-quality assessment models using spatial data to create spatial decision systems (Engel and Navulur, 1999). There are many studies on the analysis of data carried out by combining GIS and statistical methods (Levallois et al., 1998).

Geostatistics can be considered as a collection of numerical techniques that deal with the characterization of spatial attributes, employing primarily random models in a manner similar to the way in which time series analysis characterizes temporal data (Olea, 1999). Generally, geostatistics is used as a management and decision tool. Researchers all over the world have been using the concept of geostatistics and their applications for mapping pollutants and measuring the spatial structure of the geochemical components of groundwater (Bierkens and Burrough, 1993a,b; Belkhiri and Narany, 2015).

Geostatistics is a technique for estimating the values of properties (at unsampled places) that vary in space from more or less sparse sample data (Oliver and Webster, 1991). Geostatistics evolved from the work carried out by the mining engineer D.G. Krige during the early 1950s. These ideas were formalized and extended by the mathematician G. Matheron.

With the help of geostatistics one can measure values that are unknown, produce a map, and improve sampling by validating sampling strategies. It became a branch of science in the 1960s when it was used in the mining industry, and it was later extended to many other fields, including geomorphology, geology, hydrology, and geography. The technique of geostatistics is mainly deployable in predicting the values associated with spatial or spatio-temporal phenomena. It is a class of statistics that takes into account the spatial, and sometimes spatial and temporal, coordinates of the data within the analysis.

Tools in geostatistics can describe spatial patterns of a given sample and interpolate the unknown values. Geostatistics techniques could also be used in the measurement of uncertainty and to obtain information on possible values for locations other than the ones whose values are interpolated. The unique feature of geostatistics is the use of regionalized variables; these variables are completely deterministic and in the range of random variables. Phenomena with geographic distribution are well explained by regionalized variables such as the elevation of the ground's surface. During sampling it is not possible to take every reading from every location, therefore the values of unknown data at particular locations are collected and these samples are used to measure the value of unknown points. The size, orientation, shape, and spatial arrangement of the measured points affects the ability to predict the values at unmeasured locations. The prediction of these unknown values on the basis of known sample points is termed interpolation. The term interpolation can be defined as the estimation of values for attributes at locations that are not sampled from measurements at point locations within the same area or region. Interpolation is basically applied to convert the data from observations taken from a few points to the continuous fields, so that the spatial pattern of other entities can be analyzed from the sampled spatial patterns obtained from these measurements. Thus spatial interpolation is a method for obtaining surface data from point data or point observation. In this method points of known values are used to predict the values of unknown points forming the surface. The principle behind the spatial interpolation is based on Tobler's first law of geography, "Everything in space is related to every other thing, but points close together are more likely to be similar than the points which are far apart." There are two types of interpolation methods that are popularly used:

a. Nongeostatistical interpolation methods
b. Geostatistical interpolation methods.

5.1.1 Nongeostatistical Interpolation Methods

Nongeostatistical interpolation methods include various techniques such as:

Inverse distance weighting (IDW), which is a nongeostatistical interpolation method. In IDW it is assumed that a point having an unknown value is affected more by the control points near to it than those that are far away. The original data are generally placed on a regular grid or sometimes irregularly distributed over an area and interpolations are made on a denser regular grid for generating a map. A linear interpolator is its simplest form. In this method the computation of weights is carried out from a linear function of the distance between sets of point data and points to be predicted. This method has no inbuilt means for verification of the accuracy of predictions. Therefore the quality of the map can only be analyzed by considering validated sample points.

Radial basis functions (RBF) methods are a series of exact interpolation techniques that stipulate that the surface must go through each measured sample value. There are five different radial basis functions: thin plate spline, spine with tension, completely regularized spline, multiquadric function, and inverse multiquadric function. Spline is another nongeostatistical method that uses mathematical function for the estimation of values and minimizes overall surface curvature (Li and Heap, 2008). It consists of polynomials, with each polynomial of degree p being local rather than global. Surfaces or pieces of line described by the polynomials (i.e., they are fitted to a small number of data points exactly) are fitted together so that they join smoothly (Burrough and McDonnell, 1998). For each predicted point, a local trend surface is fitted on a polynomial surface using the nearby samples (Venables and Ripley, 2002). Thin plate splines, also known as "laplacian smoothing splines," were developed by Wahba and Wendelberger in 1980 for climate data. This method provides a measure of spatial accuracy (Hutchinson, 1995; Wahba and Wendelberger, 1980).

Trend surface analysis is a special case of linear regression model, which uses geographical coordinates to predict the values of the primary variables. It separates data into local variations and regional trends (Collins and Bolstad, 1996).

The global polynomial method of interpolation is a deterministic but inexact method used for surfaces that change slowly and gradually. A plane is a special case. A plane can be fitted between the sample points to support our assumption on the overriding trend. We can then find out the unknown height from the value on the plane for the assumed location. The plane may lie above certain points and below others.

Minimizing error is the goal of interpolation. The error can be measured by subtracting each calculated point from its assumed value on the plane, after which it needs to be squared and then added. The resultant sum is called a "least squares" fit. In first-order global polynomial interpolation, this process acts as the theoretical base.

Local polynomial interpolation is used when the area slopes, levels off, and then slopes again. A flat plane fitted in the study site gives poor assumption for the unmeasured values, but if we are allowed to fit many smaller overlapping planes, and then use the center of each plane as the assumption for each location in the study area, the resulting surface will be more flexible and accurate. This is the main concept behind the local polynomial interpolation.

5.1.2 Geostatistical Interpolation Methods

These techniques are applied when there is irregular attribute variation and the density of sample may be such that simple interpolation methods may produce unreliable predictions. These methods give probabilistic estimates of the quality of interpolation. Not only to the sample numbers and sampling patterns, but also to the assumptions made using the geostatistical methods, which are included in the built-in structure of the variable. This spatial structure is characterized by spatial autocorrelation and the semivariogram. Using the semivariogram, kriging is used for interpolation of the unknown points.

5.1.2.1 Semivariogram

This is a basic tool for the analysis of spatial structure. "The semivariogram is a mathematical description of the relationship (structure) between the variance of pairs of observations (data points) and the distance separating these observations (h)" (Mabit and Bernard, 2007). A semivariogram describes the between-population variance within a distance class (*y*-axis) according to the geographical distance between pairs of populations (*x*-axis). The fitted curve minimizes the variance of the errors. The variogram model is used to define the weights of the kriging function.

Semivariogram is described mathematically as:

$$\gamma(h) = \frac{1}{2} \sum_{i=1}^{N} (Z(x+h) - Z(x))^2$$

where function $\gamma(h)$ is semivariogram and $Z(x)$ is random variable.

In semivariogram model, sill, nugget, and range are the parameters. The most important part of a semivariogram is its shape near the origin until the range, as the closest points/or group of pairs are given more weight in the interpolation process. Fig. 5.1 shows graphical representation of sill, nugget, and range.

Nugget (Co): Nugget represents unresolved, subgrid scale variation or measurement error and is seen on the semivariogram as the intercept of the semivariogram.

Sill (Co + C1): The value of the semivariogram as the lag (*h*) goes to infinity; it is equal to the total variance of the dataset.

Range (a): The distance where the model first flattens out is known as the range.

The semivariogram properties, the sill, range, and nugget, can provide insights into which model will fit the best (Cressie, 1993; Burrough and McDonnell, 1998). The most common models are linear, spherical, and exponential. Various types of variogram models are shown in Fig. 5.2.

Commonly used semivariogram models are given below:

Spherical model: The spherical model is the most commonly used model:

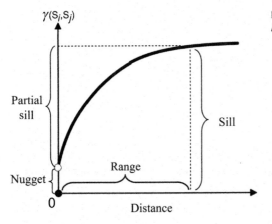

FIG. 5.1 Semivariogram parameters, i.e., sill, nugget, and range. *(Source: http://planet. botany.uwc.ac.za/.)*

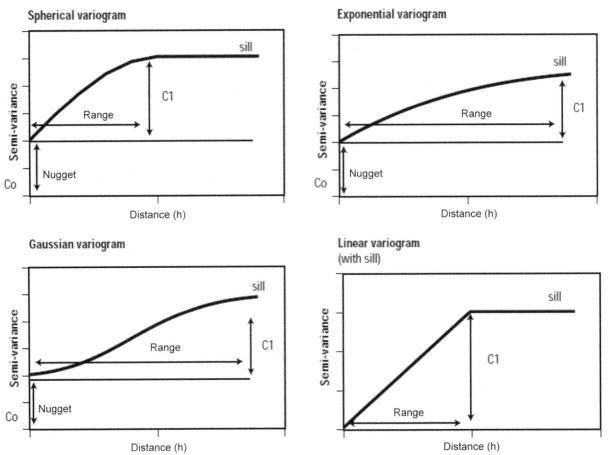

FIG. 5.2 Showing various types of variogram models. (Source: *Hartkamp, A.D., de Beurs, K., Stein, A., White, J., 1999. Interpolation Techniques for Climate Variables. Geographic Information Systems Series 99-01. International Maize and Wheat Improvement Center (CIMMYT), Mexico. ISSN:1405-7484.)*

$$\gamma(h) = \left\{ c\left[1.5\left(\frac{h}{a}\right) - \frac{1}{2}\left(\frac{h}{a}\right)^3 \right] \right\}; \quad \text{for}|h| < a$$
$$= C \qquad\qquad\qquad ;\text{for } |h| > a$$

where "C" is sill and "a" is range.

Exponential model: The tangent at the origin intersects the sill at a point with an abscissa "a":

$$\gamma(h) = C\left[1 - \exp\left(\frac{h}{a}\right) \right]$$

where "C" and "a" are sill and range, respectively.

Gaussian model: It represents extremely continuous phenomena.

$$\gamma(h) = C\left[1 - \exp\left[-\left(\frac{h}{a}\right)^3 \right] \right]$$

where "C" and "a" stand for sill and range, respectively.

Finally, the best variogram model (spherical, linear, etc.) and its parameters (nugget, sill, range, etc.) have to be determined in order to validate the modeling of the spatial autocorrelation through the variogram's parameter optimization.

5.1.2.2 Kriging

Kriging is an eminent geostatistical interpolation method. Kriging as a geostatistics method has been in use since the 1960s. Kriging is an estimation procedure that uses known values and semivariogram to determine unknown values. It was named after D.G. Krige from South Africa. The procedures involved in kriging incorporate measures of error and uncertainty when determining estimations. "The approach is based on the use a random field, and a number of assumptions such as stationarity and spatial ergodicity, so as the reduce the needed information to a so-called variogram that can be estimated from the available measurements" (Loquin and Dubois, 2010). Based on the semivariogram, optimal weights are assigned to known values in order to calculate unknown ones. The variogram changes with distance and the weights depend on the known sample distribution. Kriging assumes that the spatial variation is neither totally random nor deterministic, i.e., it deals with the regionalized variable. "Kriging interpolation through variography provides an optimal interpolation estimate from observed values and their spatial relationships" (Olea, 1991; Wackernagel, 1995). "Kriging uses nearby points weighted by distance from the interpolate location and the degree of autocorrelation or spatial structure for those distances, and calculates optimum weights at each sampling distance" (Isaaks and Srivastava, 1990).

Physical measurements may be inaccurate, as uncertainties are present in the process that affect the validity of the resulting interpolation. Statistically the validity of the variogram clashes with the measured values. The bigger the area of study, the larger the amount of data, and the more valid is the variogram. In contrast, the validity of local analysis will be less accurate as the amount of data available will be smaller:

$$Z(s) = \mu(s) + \varepsilon'(s)$$

This is a general equation of kriging where $Z(s)$ is the variable of interest, decomposed into a deterministic trend $\mu(s)$ and a random, autocorrelated errors $\varepsilon'(s)$. The symbol simply indicates the location (containing x and y coordinates).

Types of kriging are given in the following section:

- *Ordinary kriging (OK)*: this is a standardized version of kriging. "Ordinary kriging is an estimation technique known as the Best Linear Unbaised Estimator (BLUE) that has the great advantage of using the semivariogram information" (Cressie, 1993). Here the predictions are based on the model:

$$Z(s) = \mu + \varepsilon'(s)$$

where T is an unknown constant and $\varepsilon'(s)$ is the spatially correlated stochastic part of the variation.

- *Simple kriging*: this is the most basic form of kriging, which assumes that the measured values are realizations of a stationary random function with a constant mean μ. Simple kriging is:

$$Z(s) = \mu + \varepsilon'(s)$$

where μ is a known constant.

- *Universal kriging*: this does not require prior knowledge of the mean, but does require a model for trend surface. Its algorithm can generate the trend model by fitting a polynomial function to the local data. It assumes the model:

$$Z(s) = \mu(s) + \varepsilon'(s)$$

where $\mu(s)$ is some deterministic function.

- *Disjunctive kriging*: this assumes the model:

$$F(Z(s)) = \mu_1 + \varepsilon'(s)$$

where μ_1 is an unknown constant and $f(Z(s))$ is an arbitrary function of $Z(s)$. Disjunctive kriging requires the bivariate normality assumption and approximations to the functions $f_i(Z(s_i))$.

- *Indicator kriging*: this assumes the model:

$$I(s) = \mu + \varepsilon'(s)$$

where μ is an unknown constant and $I(s)$ is a binary variable.
"These methods can be used to produce the following surfaces:

- Maps of kriging predicted values
- Maps of kriging standard errors associated with predicted values
- Maps of probability, indicating whether or not a predefined critical level was exceeded
- Maps of quantiles for a predetermined probability level

The exceptions to this are indicator and probability kriging, which produce the following:

- Maps of probabiltity, indicating whether or not a predefined critical level was exceeded
- Maps of standard errors of indicators" (Johnston, 2001).

5.1.3 Use of Geostatistics for the Study of Groundwater Quality

Geostatistics is a very important technique when it comes to the determination of groundwater quality and the distance covered by contaminants/pollutants. Among groundwater-quality parameters, nitrate is one of the most important pollutants. High concentrations of nitrate in groundwater can affect public health. Nitrate contamination in groundwater is a very common global phenomena and it has been reported by many researchers (Hemant, 2013). In India, nitrate represents one of the most common groundwater contaminants in rural and urban areas. The presence of nitrate has been reported in several areas in Tamil Nadu, Orissa, Maharashtra, Bihar, Karnataka, Gujarat, Madhya Pradesh, Uttar Pradesh, Rajasthan, and other parts of India (CGWB Report, 2010). Both naturally occurring and anthropogenic factors are responsible for the formation of nitrate. Chemical reactions occurring in the Earth's atmosphere result in the mixing of atmospheric nitrogen with rainwater lead to the formation of nitrate and ammonium ions.

Nitrate is a very common nitrogenous compound formed due to natural processes but anthropogenic interventions are responsible for increasing nitrate concentrations in groundwater. Anthropogenic sources of nitrate responsible for the contamination of groundwater include septic tanks, fertilizer used for irrigation purposes, domestic animals in residential areas, and nitrate leaching from stored manure for agricultural purposes. Nitrogen concentrations more than the permissible limit in potable water leads to infant methaemoglobinaenia (blue-baby syndrome), gastric cancer, goiter, metabolic disorder, birth malformations, hypertension, and livestock poisoning.

The use of geostatistics for mapping groundwater quality has been analyzed in several studies. Rizzo and Mouser (2000) used geostatistics in their study for analyzing groundwater quality (Mouser, 2004). Nazari-zade et al. (2006) used geostatistics to study the spatial variability of groundwater quality in the Balarood Plain. Their results showed that a spherical model is the best model for fitting an experimental variogram of EC, Cl, and SO_4 variables. In another similar study, analysis of the spatial distribution of groundwater quality using geostatistics was carried out by Mehrjardi et al. (2008). IDW, kriging, and cokriging were used for the interpolation of groundwater quality, and kriging and cokriging were found to be superior to IDW. Other similar studies have been done by Goovaerts et al. (2005), who compared the performances of multi-Gaussian and indicator kriging for modeling probabilistically the spatial distribution of arsenic concentrations in groundwater. Various studies (Ahmadi and Sedghamiz, 2007) have found that kriging is a beneficial and capable tool for detecting those critical regions that need more attention to achieve sustainable use of groundwater. Another study done by Adhikary et al. (2010) used a geostatistical approach for the preparation of thematic maps of groundwater-quality parameters such as bicarbonate, calcium, chloride, electrical conductivity (EC), magnesium, nitrate, sodium, and sulfate, with concentrations equal to or greater than their respective groundwater pollution cutoff values. Their study showed that a spherical model was the best fit for groundwater-quality assessment.

5.2 GEOGRAPHIC SETUP AND HYDROGEOLOGICAL CHARACTERISTICS

The case study is focused on Lucknow City, which is situated in the middle Ganga Plain, Uttar Pradesh and is one of the most rapidly growing cities of Central India. The city covers approximate $429.50 \, km^2$. Its boundary lies between the latitude $26°45'0''$ N and $26°55'0''$ N and longitude $80°50'0''$ E and $81°5'0''$ E (Fig. 5.3). The city has faced fast and continuous urban expansion over the last few decades. A maximum area is covered by settlement or impervious surface in comparison to other surface features (forest/plantation, agricultural area and open land/waste land), which are relatively more pervious in nature. Lucknow is a part of the Ganga basin with flat alluvial terrain; elevation ranges from 103 to 130 m above mean sea level (AMSL) with south-east slope in general. Groundwater occurs in the pore spaces of the unconsolidated alluvium sediments in the zone of saturation under phreatic and semiconfined conditions. In deeper aquifer it occurs under semiconfined to confined conditions. Quaternary sediments have been divided into older and newer alluvium. The older alluvium is comprised of gray-to-brown colored silt clay and sand with or without kankar of middle-to-late Pleistocene period. The newer alluvium overlies the older alluvium and has been subdivided into terrace alluvium and channel alluvium and they belong to the Holocene period. The newer alluvium comprises of light gray silt, clay, and fine-to-medium and coarse-grained gray sand, which is micaceous in nature. The subsurface of the Lucknow area is essentially composed of unconsolidated alluvial sediments produced by weathering and the erosion of the Himalaya region. The area is mainly fed by the Gomati River and only two main canals, the Sarda canal and Nawab Gaziuddin Haider canal, along with a few minor distributaries are present.

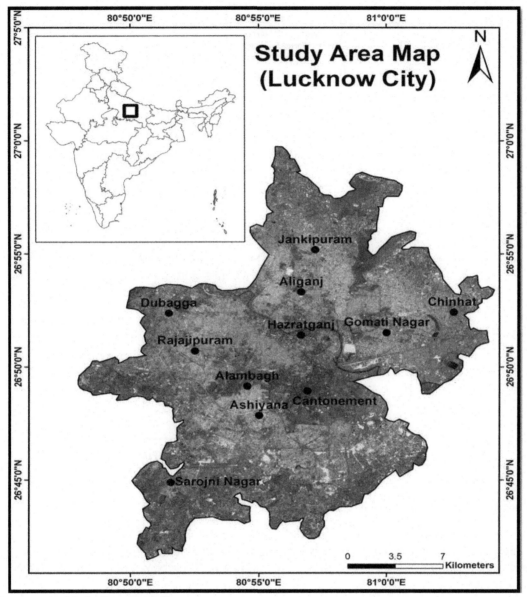

FIG. 5.3   Location map of the Lucknow City, India.

The Gomati River is characterized by sluggish flow throughout the year, except during monsoon season when heavy rainfall causes a manifold increase in the runoff. There are 26 nalas that drain into the Gomati River between Gaughat and Pipraghat, 11 of which, such as Gaughat nala, Pata nala, Sarkata nala, etc., are located on the right bank, and 12 of which, such as Nadwa nala, Khadra nala, Mahanagar nala, Kukrail nala, etc., are located on the left bank. In Lucknow, the development of groundwater resources is due to presence of thick Quaternary deposits forming a multitier aquifer system. The unconsolidated near-surface alluvial sediments of the Ganga Plain are generally potential water-bearing materials. The availability of groundwater in these alluvial deposits is controlled by the relative thickness of sand and mixed-clay horizons. Sand layers make the potential aquifer in the area and its presence increasing with depth of aquifer formation.

5.3 DATA AND METHODS USED

The standard research methodology has been applied for the preparation of a database and the generation of a predicted interpolated groundwater-quality map of Lucknow City using kriging and IDW techniques in a GIS environment. The detailed methodology followed in the present work has been shown in a flow chart (Fig. 5.4). The standard water-sampling

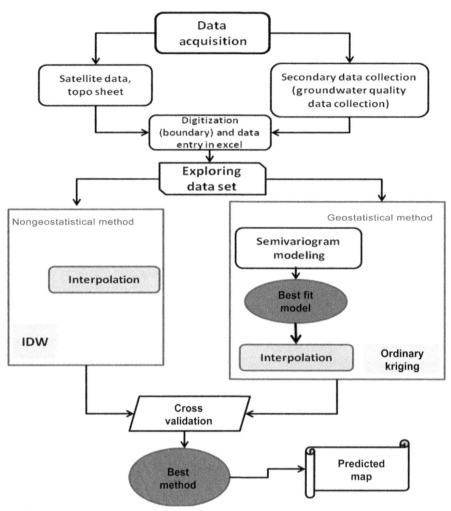

FIG. 5.4 Methodology followed in the present work.

techniques were applied for the collection of groundwater samples, which includes the collection of water samples from a bore well and hand pump in the study area. Preprocessing includes a GPS survey, georeferencing, digitization of the boundary, and generation of a GIS database.

5.3.1 Exploring the Dataset

The first step in geostatistical analysis is to understand and explore the dataset. This step is used for closely examining the data and its trends through histogram and QQ plot, which helps in comparing the distribution of the data with standard normal distribution. A histogram is useful to understand whether the data has a normal distribution.

Semivariogram cloud: Semivariogram plots the semivariance value of point pairs, i.e., the difference squared between the values of each pair of locations on "*y*" axis versus lag distance (distance separating each pair) on "*x*" axis. It helps to observe the spatial autocorrelation between the calculated sample points. In spatial autocorrelation it is presumed that the points that lie close to one another are comparatively more alike. The semivariogram cloud helps to study this relationship.

5.3.2 Interpolation Methods

The interpolation methods used in the present work have been discussed in the previous sections. The concentration of nitrate in the groundwater of the study area has been taken for applying and compares the interpolation techniques.

5.3.3 Semivariogram Modeling

In the next step, semivariogram modeling has been applied to show the spatial relationship. The experimental semivariogram is calculated by averaging the difference squared of the z-values over all pairs of observations with the specified separation distance and direction. It is plotted as a two-dimensional graph (Higgins et al., 2005). A theoretical variogram model is then fitted to the experimental semivariogram by adjusting the three key parameters namely sill, nugget, and range, and the efficacy of fit is calculated. The semivariogram properties—sill, range, and nugget—can provide insights into which model will fit the best (Cressie, 1993; Burrough and McDonnell, 1998). The most common models are linear, spherical, exponential, and Gaussian model. From the fitted semivariogram model, various properties of the data are determined: the sill, the range, and the nugget.

5.3.4 Cross-Validation

This gives an idea about how well the model predicts the unknown values. For all points cross-validation sequentially omits a point, predicts its value using the rest of the data, and then compares the measured and predicted values. The objective of cross-validation is to make an informed decision about which model provides the most accurate predictions. For a model that provides accurate predictions, the standardized mean error should be close to zero, and the root-mean-square error and average standard error should be as small as possible. The error in the root-mean-square standardized error should be close to 1. When the average estimated prediction standard errors are close to the root-mean-square prediction errors from cross-validation, then we can be confident that the prediction standard errors are appropriate (ESRI, 2001).

The difference between the estimated value Z and the corresponding measured value Z_i is the experimental error:

$$\varepsilon_i = Z(s_i)Zi(s_i)$$

The cross-validation statistics of mean error (ME)

$$ME = \frac{1}{n}\sum_{i=1}^{n}\varepsilon$$

Root-mean-square error (RMSE)

$$RMSE = \sqrt{\frac{1}{n}\sum_{i=1}^{n}\varepsilon_i^2}$$

where ε_i is the prediction standard error for location s_i.

ME should be close to zero. Root-mean-square error and average standard error are indices that represent the efficacy of prediction, and root-mean-square standardized error compares the error variance with a theoretical variance, such as kriging variance, and it should be close to unity (Diodato, 2005).

5.3.5 Prediction Map

After completing the cross-validation step, an output map was calculated using each of the methods. It helps to find unknown values by generating contours and surfaces using the two variables.

5.4 RESULTS AND DISCUSSION

In the present study, first an exploratory analysis of the existing data pertaining to the study area was performed, followed by a semivariogram analysis, and thereafter a comparison of the interpolation techniques (e.g., IDW and ordinary kriging). The best interpolation technique for groundwater-quality mapping was used for output generation and prediction mapping. The generation concentration and statistics for nitrate concentrations in the groundwater (e.g., mean, minimum, maximum, and standard deviation) are given in Table 5.1. The histogram of the variable comprised right skewed data and did not follow a well-shaped curve; however, the log transformation of the variable followed a close-to-normal curve.

TABLE 5.1 Statistical Information About Nitrate Concentrations in Groundwater

Variable Names	No. of Points	Min	Max	Mean	Std. Dev.
Nitrate	49	0.3	155.76	22.13	32.47

5.4.1 Quantile-Quantile Plots

A good dataset's QQ PLOT should show a line falling close to normal. The nitrate concentrations in the groundwater do not demonstrate this relation; however, their log transformations do fall close to the normal line. This indicates that the data are not normally distributed and the resultant data are showing log normal; therefore the log transformation has given better results (Fig. 5.5).

The semivariogram shows the spatial dependence of groundwater nitrate concentration points. The semivariogram fitted for nitrate concentration of the study area and graphical representation provides a picture of the spatial correlation of the data points with their neighbors (Fig. 5.6). Kriging and IDW techniques for groundwater-quality mapping of the

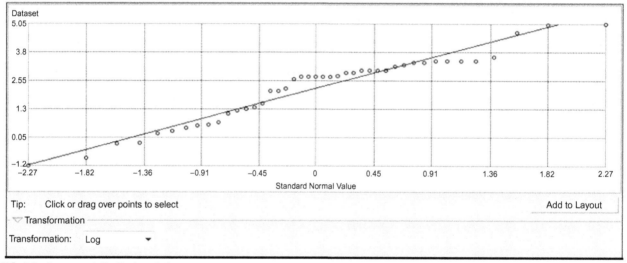

FIG. 5.5 Log normal distribution of nitrate.

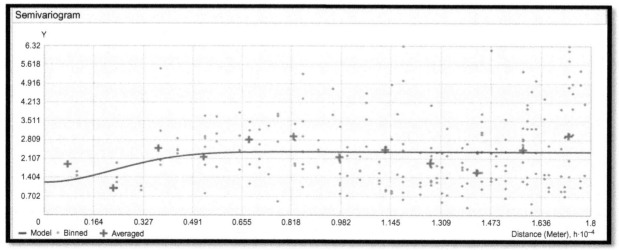

FIG. 5.6 Semivariogram of nitrate concentrations.

TABLE 5.2 Comparison of Interpolation Techniques

Interpolation Techniques	Predicted Errors Root-Mean-Square
Inverse distance weighting (IDW)	40.64
Kriging (ordinary kriging)	37.78

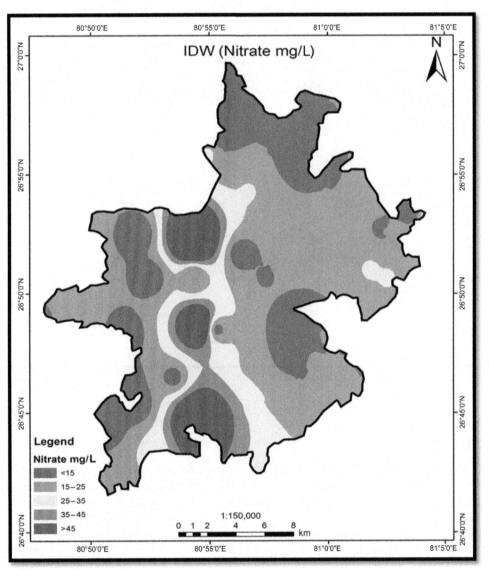

FIG. 5.7 Predicted map showing inverse distance weighting for nitrate dataset.

Lucknow area were performed and kriging was found to be better than the IDW for groundwater-quality mapping. RMSE was used for assessing the various techniques' performances (Table 5.2). The smallest RMSE indicates accurate predictions (Mohammad et al., 2010). The predicted nitrate-concentration map of the study area using IDW and kriging is shown in Figs. 5.7 and 5.8.

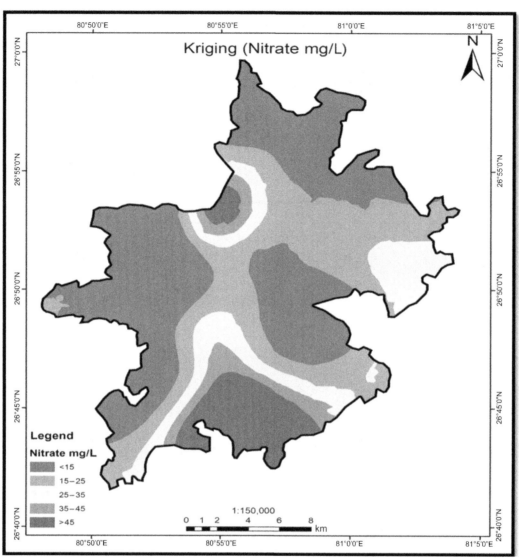

FIG. 5.8 Predicted map showing ordinary kriging for nitrate dataset.

5.5 CONCLUSION

The observation from the analysis of 49 groundwater samples of the study area for nitrate-concentration analysis comprises important thematic information using advance interpolation techniques, which show the areas at highest risk of high nitrate concentrations. It was also suggested from the comparative assessment of two important interpolation techniques in the area that the kriging method gives the better result for nitrate-concentration prediction for the study area. From this case study it can be concluded that geostatistical techniques can provide good-quality spatial distribution mapping of groundwater-quality parameters and help to predict the location of unknown sample points from the measured or collected sample points.

ACKNOWLEDGMENTS

The first author is grateful to the Science and Engineering Research Board (SERB), Department of Science and Technology, Government of India, for providing the necessary funding support under the FastTrack Young Scientist Scheme (Grant No. SR/FTP/ES-83/2013) to carry out the present research work. Authors are also thankful to Amity University for providing the necessary infrastructure to carry out this work.

REFERENCES

Adhikary, P.P., Chandrasekharan, H., Chakraborty, D., Kamble, K., 2010. Assessment of groundwater pollution in west Delhi, India using geostatistical approach. Environ. Monit. Assess 167, 599–615.

Ahmadi, S.H., Sedghamiz, A., 2007. Geostatistical analysis of spatial and temporal variations of groundwater level. Environ. Monit. Assess. 129 (1), 277–294.

Belkhiri, L., Narany, T.S., 2015. Using multivariate statistical analysis, geostatistical techniques and structural equation modeling to identify spatial variability of groundwater quality. Water Resour. Manage. 29 (6), 2073–2089.

Bierkens, M.F.P., Burrough, P.A., 1993a. The indicator approach to categorical soil data. I Theory. J. Soil Sci. 44, 361–368.

Bierkens, M.F.P., Burrough, P.A., 1993b. The indicator approach to categorical soil data. II Application to mapping and land use suitability analysis. J. Soil Sci. 44, 369–381.

Burrough, P.A., McDonnell, R.A., 1998. Principles of Geographical Information Systems. Oxford University Press, New York 333 p.

CGWB Report, 2010. Groundwater Quality in Shallow Aquifers of India, CGWB, Ministry of Water Resources. Government of India, Faridabad.

Collins, F.C., Bolstad, P.V., 1996. A comparison of spatial interpolation techniques in temperature estimation. In: Proceedings of the 3rd International Conference/Workshop on Integrating GIS and Environmental Modeling. National Center for Geographic Information and Analysis, Santa Barbara, Santa Fe, NM, Santa Barbara, CA.

Cressie, N., 1993. Statistics for Spatial Data, Revised Ed. Wiley, New York.

Diodato, N., 2005. The influence of topographic variables on the spatial variability of precipitation over small regions of complex terrain. Int. J. Climatol. 25, 351–363.

Engel, B.A., Navulur, K.C.S., 1999. The role of geographical information systems in groundwater engineering. In: Delleur, J.W. (Ed.), The Handbook of Groundwater Engineering. In: vol. 21. CRC, Boca Raton, pp. 1–16.

ESRI, 2001. Using ArcGIS Geostatistical Analyst. ESRI Press, Redlands, CA.

Goovaerts, P., AvRuskin, G., Meliker, J., Slotnick, M., Jacquez, G., Nriagu, J., 2005. Geostatistical modeling of the spatial variability of arsenic in groundwater of Southeast Michigan. Water Resour. Res. 41 (7), W07013.

Hemant, W.K., 2013. Scenario of nitrate contamination in groundwater: its causes and prevention. Int. J. Chem Tech Res. 5 (4), 1921–1926.

Higgins, N.A., Burge, F., Charnock, T.W., Teale, P., 2005. Statistical estimation and characterization techniques for use during accident response. Oxfordshire, 101 pp.

Hutchinson, M.F., 1995. Interpolating mean rainfall using thin plate smoothing splines. Int. J. Geograph. Informat. Sys. 9, 385–403.

Isaaks, E.H., Srivastava, R.M., 1990. An Introduction to Applied Geostatistics. Oxford University Press, USA.

Johnston, K., Environmental Systems Research Institute (Redlands C), 2001. Using Arcgis Geostatistical Analyst [Internet]. Environmental Systems Research Institute. Available from: http://books.google.com/books?id=kFZyQgAACAAJ.

Levallois, P., Thériault, M., Rouffignat, J., Tessier, S., Landry, R., Ayotte, P., 1998. Groundwater contamination by nitrates associated with intensive potato culture in Québec. Sci. Total Environ. 217, 91–101.

Li, J., Heap, A.D., 2008. A review of spatial interpolation methods for environmental scientists. In: Geoscience Australia Record 23.p. 137.

Loquin, K., Dubois, D., 2010. Kriging with ill-known variogram and data [Internet]. In: Deshpande, A., Hunter, A. (Eds.), Scalable Uncertainty Management. Springer Berlin Heidelberg, Berlin, Heidelberg, pp. 219–235.

Mabit, L., Bernard, C., 2007. Assessment of spatial distribution of fallout radionuclides through geostatistics concept. J. Environ. Radioact. 97 (2–3), 206–219.

Mehrjardi, R.T., Jahromi, M.Z., Mahmodi, S., Heidari, A., 2008. Spatial distribution of groundwater quality with geostatistics (case study: Yazd-Ardakan plain). World Appl. Sci. J. 4 (1), 09–17.

Mohammad, Z.-M., Ruhollah, T.-M., Ali, A., 2010. Evaluation of geostatistical techniques for mapping spatial distribution of soil pH, salinity and plant cover affected by environmental factors in Southern Iran. Not. Sci. Biol. [Internet]. Available from: http://notulaebiologicae.ro/nsb/article/viewArticle/4997.

Mouser, P.J., 2004. Evaluation of geostatistics for combined hydrochemistry and microbial community fingerprinting at a waste disposal site [Internet]. In: World Water Congress 2004, Salt Lake City, UT, p. 106.

Nazari-zade, F., Arshadiyan, F., Zand-vakily, K., 2006. Study of spatial variability of Groundwater quality of Balarood Plain in Khuzestan province. In: Proceedings of the First Congress of Optimized Exploitation from Water Source of Karoon and Zayanderood Plain. Shahrekord University, Shahrekord, pp. 1236–1240.

Olea, R.A., 1991. Geostatistical Glossary and Multilingual Dictionary. International Association for Mathematical Geology Studies in Mathematical Geology. No. 3, 177p. Oxford University Press, New York.

Olea, R.A., 1999. Geostatistics for Engineers and Earth Scientists. Kluwer Academic Publishers, United States.

Oliver, M.A., Webster, R., 1991. How geostatistics can help you. Soil Use Manag. 7, 206–217.

Rizzo, D.M., Mouser, J.M., 2000. Evaluation of geostatistics for combined hydrochemistry and microbial community fingerprinting at a waste Disposal Site. pp. 1–11.

Singh, S.K., Srivastava, P.K., Pandey, A.C., 2013. Fluoride contamination mapping of groundwater in northern India integrated with geochemical indicators and GIS. Water Sci. Technol. 13 (6), 1513–1523.

Singh, S.K., Srivastava, P.K., Singh, D., Han, D., Gautam, S.K., Pandey, A.C., 2015. Modelling groundwater quality over a humid subtropical region using numerical indices, earth observation datasets and X-ray diffraction technique: a case study of Allahabad district, India. Environ. Geochem. Health 37, 157–180.

Singh, S.K., Singh, P., Gautam, S.K., 2016. Appraisal of urban lake water quality through numerical index, multivariate statistics and earth observation data sets. Int. J. Environ. Sci. Technol. 13, 445–456.

Sun, H., Bergstrom, J.C., Dorfman, J.H., 1992. Estimating the benefits of groundwater contamination control. South. J. Agric. Econ. 24, 63.

Venables, W.N., Ripley, B.D., 2002. Modern Applied Statistics with S. Springer, New York, pp. 271–300.

Wackernagel, H., 1995. Multivariate Geostatistics. Springer-Verlag, 256 pp.

FURTHER READING

Artkamp, A.D., 1999. Interpolation Techniques for Climate Variables. CIMMYT.

Baalousha, H., 2010. Assessment of a groundwater quality monitoring network using vulnerability mapping and geostatistics: a case study from Heretaunga Plains, New Zealand. Agric. Water Manage. 97 (2), 240–246.

Goovaerts, P., 1997. Geostatistics for Natural Resources Evaluation. Oxford University Press, USA.

Krishna, A.K., Satanarayanan, M., Govil, P.K., 2009. Assessment of heavy metal pollution in water using multivariate statistical techniques in an industrial area: a case study from Patancheru, Medak District, Andhra Pradesh, India. J. Hazard. Mater. 167 (1), 366–373.

Marko, K., Al-Amri, N.S., Elfeki, A.M.M., 2014. Geostatistical analysis using GIS for mapping groundwater quality: case study in the recharge area of Wadi Usfan, Western Saudi Arabia. Arab. J. Geosci. 7 (12), 5239–5252.

Schwartz, F.W., Zhang, H., 2002. Fundamentals of Ground Water, first ed. Wiley.

Chapter 6

Methods for Assessing the Groundwater Quality

R. Rajesh*, L. Elango† and K. Brindha‡

*Department of Civil Engineering, Indian Institute of Science, Bangalore, India †Department of Geology, Anna University, Chennai, India ‡Hydrogeology Group, Institute of Geological Sciences, Freie Universität Berlin, Berlin, Germany

Chapter Outline

6.1 INTRODUCTION

Groundwater is a vital resource for drinking and irrigation in rural areas especially in arid and semiarid provinces. The deficiency of hygienic drinking water is harmful to the health of the people in many developing countries (Nash and McCall, 1995). The quality of water as a world resource is diminishing rapidly due to the substantial increases in industrialization and urbanization and the expansion of agricultural activities. Degradation of groundwater quality is mainly categorized by the hydrogeochemical characteristics that result from the multifaceted interactions of geology, hydrogeology, topography, drainage systems, hydrometeorology, and anthropogenic activities, which lead to groundwater deterioration (Kim et al., 2005). Groundwater deterioration not only affects water quality but also impedes economic development and social wealth (Milovanovic, 2007). Groundwater-quality studies have demonstrated that for the many countries it has noticeably deteriorated in recent years (Magesh and Chandrasekar, 2013; Selvam et al., 2014; Masoud, 2014; Li et al., 2013; Chen and Feng, 2013; Moosavirad et al., 2013; Jeong, 2001; Elhatip et al., 2003; Lee et al., 2003; Varol and Davraz, 2014). The chemical composition of groundwater plays a significant role in the categorizing and evaluating of water quality. Hydrogeochemical studies of groundwater provide a better understanding of the probable variations in quality. Water quality is assessed using different methods. Some of the common methods used to study the quality of water include hydrogeochemical methods (Rajesh et al., 2012; Rajesh, 2014; Chung et al., 2015), and remote sensing with GIS techniques (Brindha and Elango, 2012; Rajaveni et al., 2014). Fuzzy logic and geostatistical analysis are relatively new techniques in the assessment of water quality (Sajil Kumar et al., 2013; Adhikary et al., 2012).

Water quality index (WQI) is an effective tool to assess the state of an ecosystem, and this method is based on a group of physicochemical and biological characteristics of water samples (Namibian, 2007; Simoes et al., 2008). However, WQI is a

GIS and Geostatistical Techniques for Groundwater Science. https://doi.org/10.1016/B978-0-12-815413-7.00006-7

relatively simple approach to assessing the complex influence of overall deterioration, and it provides little evidence in terms of the sources of deterioration. Selvam et al. (2014) evaluate the WQI method using a geographic information system (GIS) for groundwater resources around the coastal city of Tuticorin. The study explained that the higher values occurring in the southwest portion of premonsoon period and southeast-southwest portion during postmonsoon period are mainly due to the leaching of ions and anthropogenic activities. Magesh and Chandrasekar (2013) determined that mineral dissolution and anthropogenic influences resulted in high concentrations of ions in groundwater; using WQI and GIS techniques they identified spatial variation in the groundwater quality in the Viruthunagar district, Tamil Nadu. GIS are evolving to identify and explore the sources of groundwater contamination as well as to propose mitigation measures. GIS is proving to be a powerful tool for formulating solutions for groundwater resource-management problems such as determining water quality with respect to the inherent features of the study site including geology, landuse, identifying groundwater potential zones, etc.

Multivariate statistical analysis is a numerical and independent method of groundwater classification permitting the grouping of groundwater samples and the creation of correlations between chemical parameters and groundwater samples. The combined use of multivariate geostatistical methods may also be helpful in the hydrogeochemical assessment of aquifers using spatial variation of the major factors inducing groundwater quality (Ceron et al., 2000; Guler and Thyne, 2004; Andrade et al., 2008; Dassi, 2011; Masoud, 2014; Bouzourra et al., 2014).

The present study was carried out in a portion of Nalgonda Region, southern India. This area is an intensively irrigated region where people depend upon groundwater for their everyday usage. The granitic terrain of Archean age, which is a feature in most parts of central-southern India, has an arid climate and substantial irrigation activity (Rajesh et al., 2012). The Nalgonda district in southern India is well known for having high concentrations of fluoride in its groundwater (Rao et al., 1993). Brindha and Elango (2010) and Brindha et al. (2011) identified the source of the high concentrations of fluoride. Nitrates were found to be higher in this area due to the leaching of animal waste (Brindha and Elango, 2010). Causes for the presence of bromide in groundwater were determined by Brindha and Elango (2010). Rajesh et al. (2012) and Rajesh (2014) identified the influence of hydrogeochemical processes on temporal changes in groundwater quality. Brindha and Elango (2010) assessed the groundwater quality based on minor ions (fluoride, bromide, nitrate) using GIS techniques. Rajaveni et al. (2014) evaluated the drainage density groundwater fluctuations and weathering of dykes does not affect the groundwater flow path in shallow unconfined aquifer. All these studies are concentrated on hydrogeochemical processes, minor ion concentrations, and drainage density related to water-level fluctuations and thus there is a gap in the knowledge about the overall groundwater quality in this area. To accomplish sustainability in groundwater resources it is important to draw on a combined method for the identification of the problem to select a combined groundwater-management approach. The present study aims to fill the gaps in previous studies of the area and to evaluate the groundwater quality for drinking and irrigation purposes by using GIS, WQI, and multivariate statistical analysis in shallow aquifers of granitic terrain in a part of southern India.

6.2 STUDY AREA

The study area is located at a distance of about 85 km east-southeast from Hyderabad, Andhra Pradesh, India. It covers an area of about 724 km^2 in a portion of the Nalgonda region, southern India (Fig. 6.1A). The study area boundary has been delineated with a watershed boundary as far as possible. The southeast side of the study area is surrounded by the Nagarjuna Sagar Reservoir and the southern side of the area is bounded by the Pedda Vagu River. A portion of the northern boundary is confined by Gudipalli Vagu River (Rajesh et al., 2012; Rajesh, 2014). This area lies under a tropical region, which is characterized by an arid to semiarid climate. Summer is typically from April to June, when temperatures range from a maximum of 44°C during the day to a minimum of 28°C at night. Winter is from December to February, when the maximum daytime temperature is around 35°C and the minimum temperature is about 20°C at night. The average rainfall is around 600 mm/year. Most of the rainfall occurs during the southwest monsoon period from July to September. The agricultural activity is practiced depending on the climatic conditions and the availability of water sources. Paddy is the principle crop grown in this area while other crops include sweet lime, castor, cotton, grams, and groundnut (Rajesh et al., 2012; Rajesh, 2014). Based on the Central Ground Water Board report (CGWB, 2007) 57.20% of the agricultural area is irrigated using groundwater and 38.63% is irrigated by surface-water resources.

6.2.1 Geology and Hydrogeology

The topography map of the area is derived from the shuttle radar topography mission (SRTM) data. The topography of the area comprised an undulating terrain that has a maximum elevation of 348 m in the northwest and a minimum elevation of 170 m in the east. In general, the ground-surface slopes in an southeastern direction. There are several small hillocks in this

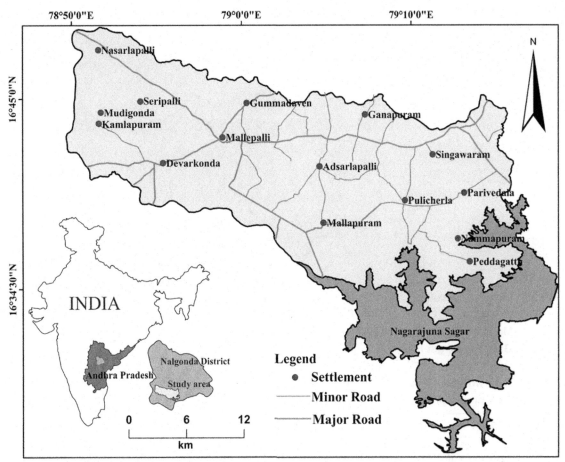

FIG. 6.1 Location of the study area.

area with heights ranging from 100 to 200 m. The surface runoff creates the dendritic to subdendritic drainage patterns in this area (Rajesh et al., 2012; Rajesh, 2014). The geological map was prepared after GSI (Geological Survey of India, 1995). The basement rocks consist of granite and granitic gneiss (Fig. 6.2). The rocks are generally medium- to coarse-grained. These rocks are crisscrossed by numerous dolerite dykes and quartz veins. The granitic rocks are intensely weathered and the thickness of weathered zones range from 4 to 15 m. The geological formation of this area dips at an angle of about 3 to 5 degrees toward the southeast. Srisailam formation is the youngest formation of the Cuddapah supergroup, overlies the granite formation with a distinct unconformity in the southeastern part of the Peddagattu and Lambapur area, and is mainly arenaceous consisting of pebbly-gritty quartzite shale with dolomitic limestone intercalated with shale quartzite and massive quartzite (Rajesh et al., 2012; Rajesh, 2014).

The total area is characterized by four distinct layers of top soil, weathered rock, moderately weathered rock, and fractured rock, which acts as an unconfined aquifer in this area. Pore spaces are developed in the weathered portions and form potential water-bearing zones. There are a number of wells in this area that supply water for domestic and agricultural purposes. The depth of the dug wells ranges from 1.45 to 20 m below ground level. Most of the wells contain groundwater from the weathered and fractured zone. The diameter of the dug wells ranges from 2 to 5 m; bore wells are generally 15 cm in diameter and the depth is greater than 10 m (Rajesh et al., 2012; Rajesh, 2014).

6.3 METHODOLOGY

6.3.1 Groundwater Sampling and Analysis

A preliminary well inventory survey was carried out of 250 wells. Based on this survey, 45 wells were chosen and collected water samples (Fig. 6.2) during January 2010 in clean polyethylene bottles of 500 mL capacity. These bottles were rinsed two to three times with the samples before collection. In the bore wells, the water samples were collected after pumping the water for a sufficient time so as to ensure the collection of formation water. In the case of open wells, water samples were

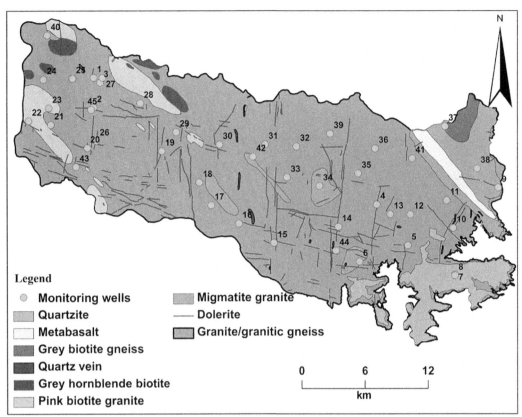

FIG. 6.2 Geology map of the area.

collected 30 cm below the water level using a depth sampler. The groundwater level was measured using Solinist. The pH and electrical conductivity (EC) were measured using Eutech portable digital meters. Collected groundwater samples were transported to the laboratory and filtered using 0.45 μm millipore filter paper. Major cation (calcium [Ca], magnesium [Mg], sodium [Na], and potassium [K]) and anion (chloride [Cl] and sulfate [SO_4^{-2}]) concentrations in groundwater were determined using Metrohm 861 advanced compact ion chromatograph and using the appropriate standards (Rajesh et al., 2012; Rajesh, 2014). Blanks and standards were run simultaneously during the measurement to ensure accuracy in the results. The concentrations of carbonate and bicarbonate were determined by titrating against H_2SO_4 as per the standard method (APHA, 1995). The quality of the analysis was confirmed by standardization using blank, spike, and also duplicate samples. As an additional assurance of accuracy, the chemical analysis was verified by calculating the ion-balance error, which was generally within 5% (Rajesh et al., 2012; Rajesh, 2014).

6.3.2 Estimation of the Water Quality Index

WQI is defined as a rating technique that demonstrates the composite influence of individual water-quality parameters on the overall quality of water for human consumption (Mitra, 1998). For this study, 10 water-quality parameters were selected. The parameters used to develop a WQI depend on the purpose for which the water is used. Parameters were selected according to the availability of data as well as their relative importance in defining water quality for human consumption. The parameters for this purpose follow the WHO guidelines (WHO, 2004). WQI is calculated by assigning weights to the measured parameters based on their relative importance. The maximum weight 5 was assigned to parameters like Na, Ca, K, bicarbonate (HCO_3^-), and total dissolved solids (TDS); 3 was assigned to Cl and Mg because these parameters are very closely related with groundwater deterioration. The minimum weight 1 was given to sulfate, since it doesn't contribute to groundwater deterioration (Sajil Kumar et al., 2013). In the second step, the relative weight (W_i) is computed with the following equation:

$$W_i = w_i \Big/ \sum_{i=1}^{n} w_i \qquad (6.1)$$

TABLE 6.1 Relative Weight of Chemical Parameters Based on WHO (2004)

Water Quality Parameters	Unit	WHO (2004)	Weight (w_i)		
		Desirable Limits	Permissible limits	Relative Weight	$W_i = w^i / \sum_{i=1}^{n} w_i$
pH		6.5	8.5	–	–
EC	mg/L	–	200	–	–
TDS	mg/L	500	1500	5	0.156
Na	mg/L	–	200	5	0.156
K	mg/L	–	100	5	0.156
Ca	mg/L	75	200	5	0.156
Mg	mg/L	30	150	3	0.093
HCO$_3$	mg/L	–	300	5	0.156
Cl	mg/L	200	600	3	0.093
SO$_4$	mg/L	200	400	1	0.031
				$\sum w_i = 32$	$\sum w_i = 1$

where W_i is the relative weight, w_i is the weight of each parameter, and n is the number of parameters. The calculated weighting factors of each parameter are given in Table 6.1.

In third step, a quality rating scale (q_i) for each parameter is assigned by dividing its concentration in each water sample by its respective standard according to the guidelines laid down in the WHO (2004), and the result is multiplied by 100:

$$q_i = (C_i/S_i) \times 100 \tag{6.2}$$

For calculating *WQI*, *SI* is firstly determined for each chemical parameter with the following equation:

$$SI_i = W_i \times q_i \tag{6.3}$$

$$WQI = \sum SI_i \tag{6.4}$$

where SI_i is the subindex of ith parameter, q_i is the rating based on concentration of ith parameter, and n is the number of parameters. The calculated *WQI* values are used to categorize the groundwater quality as excellent, good, marginal, poor, and unsuitable for drinking.

6.3.3 GIS Analysis

The base map of study area was digitized from the survey of India (1995) toposheet using ArcGIS 10.2 software. The precise locations of sampling points were determined in the field using TRIMBLE Geoexplorer 3 and the exact longitudes and latitudes of the sampling points are imported using a GIS platform. The spatial distribution for groundwater-quality parameters like hardness, pH, TDS, Na, Ca, Mg, K, HCO$_3$, SO$_4$, and Cl were done with the help of ArcGIS 10.2 software.

6.3.4 Spline Spatial Analysis

Spline interpolation technique was used for spatial variation and the parameter values are classified according to WHO (2004) standards for drinking water. Evaluations of spline interpolation comprise the weighted averaged values of surrounding sample points.

6.3.5 Grouping Analysis

Grouping analysis tool is one of the methods used for mapping clusters. It performs a classification procedure that tries to find natural clusters in sample data. Given a number of groups to create, it will look for a solution where all the features within each group are as similar as possible, and all the groups themselves are as different as possible. Feature similarity is based on a set of attributes specified for the analysis of field parameters and to be incorporated into the spatial properties. When spatial constraints are specified, the algorithm employs a connectivity graph (minimum spanning tree) to find natural groupings. The grouping analysis tool uses a K means algorithm. It evaluates the optimal number of group parameters and the grouping analysis tool assesses the effectiveness of dividing features into anywhere between 2 and 15 groups. Grouping effectiveness is measured using the Calinski-Harabasz pseudo F-statistic, which is a ratio reflecting within-group similarity and between-group differences:

$$\left(\left(R^2/_n - (c-1)\right)\right)/\left(\left((1-R^2)/(n-n-c)\right)\right) \tag{6.5}$$

where

$$R^2 = \frac{SST - SSE}{SST} \tag{6.6}$$

SST is a reflection of between-group difference and SSE reflects within-group similarity:

$$SST = \sum_{i=1}^{n_c}\sum_{j=1}^{n_i}\sum_{k=1}^{n_v}\left(V_{ij}^k - \overline{V^k}\right)^2 \tag{6.7}$$

$$SSE = \sum_{i=1}^{n_c}\sum_{j=1}^{n_i}\sum_{k=1}^{n_v}\left(V_{ij}^k - \overline{V^k}\right)^2 \tag{6.8}$$

where

n = the number of features,
n_i = the number of features in group i,
n_c = the number of classes (groups),
n_v = the number of variables used to group features,
V_{ij}^k = the value of the kth variable of the jth feature in the ith group,
$\overline{V^k}$ = the mean value of the kth variable,
V_i^k = the mean value of the kth variable in group i.

Therefore the grouping analysis method is used to prepare a spatial grouping map of groundwater types, USSL classification, and Wilcox Classification in the study area.

6.3.6 Statistical Analysis

In this study, cluster, correlation matrix, and factor analysis were used to identify the spatial variation, assess groundwater quality, and identify the sources of deterioration using STATISTICA version 8 (StatSoft, 2007). Cluster analysis is used for grouping objects (cases) into classes (clusters) on the basis of similarities within a class and dissimilarities between different classes, and also plays an important role in interpreting the data and indicating patterns. Cluster analysis needs the sample data to be standardized, because each component of water quality has a different unit. Standardization of data ensures that each variable has the same influence in the analysis. Data were standardized to the Z score ($m = 0$ and $S = 1$) by applying the following equation:

$$Z = \frac{X - m}{S} \tag{6.9}$$

where Z is the standardized value, X is the value of sample data, m is the mean, and S is the standard deviation. Factor analysis is a chemometric technique to minimize the loss of information based on correlations between several variables. This method converts data matrix into a square, symmetric matrix that expresses the degrees of interrelationships either between variables or between the objects on which these variables are measured (Davis, 2002). The square matrix is obtained by pre- or postmultiplying the data matrix by its transpose. This method extracts the eigenvalues and eigenvectors from the covariance matrix of variables. Dalton and Upchurch (1978) have revealed that factor scores can be related to the intensity of the chemical process described by each factor; extreme negative numbers (<-1.0) reflect areas essentially unaffected by the process and positive scores ($>+1.0$) reflect areas most affected. Near-zero scores approximate areas affected to an average degree by the chemical process of that particular factor. The determination of the factor analysis of the hydrogeochemical data is to explain the observed relationship in simple terms expressed as a new set of variables called factors (Venkatramanan et al., 2014; Singaraja et al., 2014).

6.4 RESULTS AND DISCUSSION

6.4.1 Water Quality for Drinking Purpose

The physicochemical characteristics of the groundwater samples were statistically evaluated, and the results, such as minimum, maximum, average, and standard deviation parameters, are given in Table 6.2. The quality of groundwater is very significant to the evaluation of suitability for drinking and irrigation purposes. The spatial variation of water-quality parameters, such as EC, TDS, Na, K, Ca, Mg, Cl, SO_4, and HCO_3 values, are based on WHO standards (2004). These are categorized by desirable, permissible, and not permissible. The pH of the study area is 7 to 8. This indicates an alkaline nature and that the samples are under WHO permissible limits. At the northwestern and southwestern portion of the study area, pH (Fig. 6.3) values exceed the permissible limits.

The alkalinity may rely on the existence of bicarbonate ions, which are formed by the free association of CO_2 with water to form carbonic acid, which influence the pH of the groundwater (Azeez et al., 2000).

The EC varies from 509 to 3907 μS/cm of the study area. Higher values of EC (Fig. 6.4) are identified in the southeast and southwest, which fall under the category of not permissible according to WHO standards. To define the suitability of groundwater for different purposes it is necessary to categorize the groundwater on its hydrochemical characteristics based on its EC values (Handa, 1969), which are represented in Table 6.3.

A TDS map (Fig. 6.5A) shows that the southeast and southwest portion are affected with high values, that they fall under not permissible limits based on WHO standards, and that values vary from 318 to 2442 mg/L. High TDS values are a result of the leaching of salts from the surrounding soil, which decreases water quality. This may cause gastrointestinal irritation in human beings and can have a laxative effect (WHO, 2004). To determine the suitability of groundwater for different purposes, it is necessary to categorize the groundwater depending upon its hydrochemical characteristics based on the TDS values (Davis and DeWiest, 1966; Freeze and Cherrey, 1979), which are represented in Fig. 6.5B and C, respectively.

The total hardness of water is influenced by the concentration of ions such as Ca and Mg. The hardness of water is affected by the dissolved Ca and, to a lesser extent, Mg. It is usually expressed as the equivalent quantity of $CaCO_3$ (WHO, 2004). TH varies from 142 to 920 mg/L, and Fig. 6.6 represents the southeast portion of the study area, which exceeds the limits of WHO standards. The TH in groundwater is estimated by the following equation (Todd, 1959):

$$TH\,(mg/L) = 2.497Ca^{2+} + 4.115Mg^{2+} \tag{6.10}$$

TABLE 6.2 Statistical Water-Quality Parameters of Groundwater in the Study Area

Water-Quality Parameters	Unit	Minimum	Maximum	Mean	SD
pH		7.15	8.12	7.50	0.23
EC	μS/cm	509.30	3907.25	1079.26	552.54
TDS	mg/L	318.31	2442.03	674.54	345.33
Na	mg/L	41.03	316.15	111.60	60.11
K	mg/L	1.58	245.48	20.32	43.88
Ca	mg/L	33.31	452.75	89.00	62.65
Mg	mg/L	16.10	86.40	35.21	12.34
HCO_3	mg/L	186.12	488.46	292.13	73.33
Cl	mg/L	24.00	613.00	115.76	100.18
SO_4	mg/L	15.92	317.90	59.42	44.91
TH	mg/L	142.05	920.11	256.68	115.49

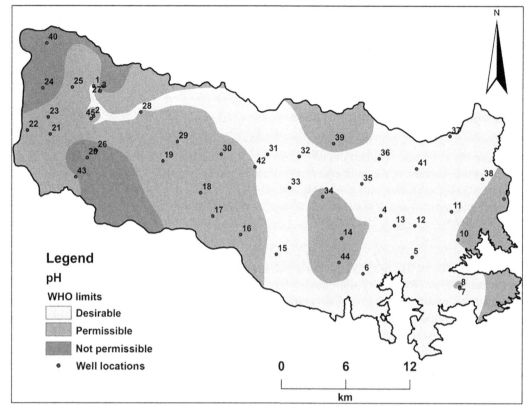

FIG. 6.3 Spatial variation map of pH.

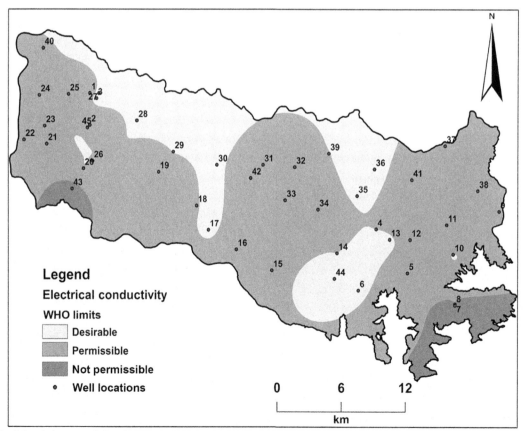

FIG. 6.4 Spatial variation map of electrical conductivity.

TABLE 6.3 Handa (1969) Classified the Water Based on Electrical Conductivity

EC (μS/cm)	Salinity Index	Percentage of Samples
0–250	Low	–
251–750	Medium	20
751–2250	High	76
2251–6000	Very high	4
6001–10,000	Extensive high	–
10,001–20,000	Weakly brine	–
20,001–50,000	Moderately brine	–
50,001–100,000	Highly brine	–
>100,000	Extremely high brine	–

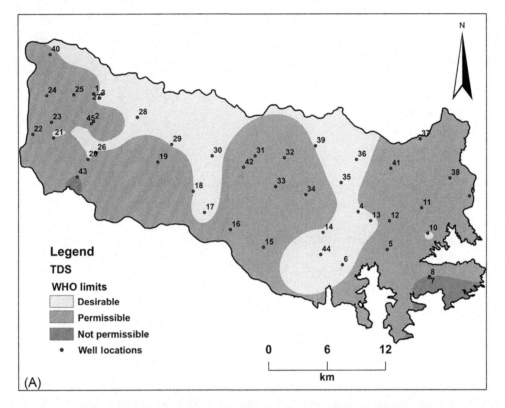

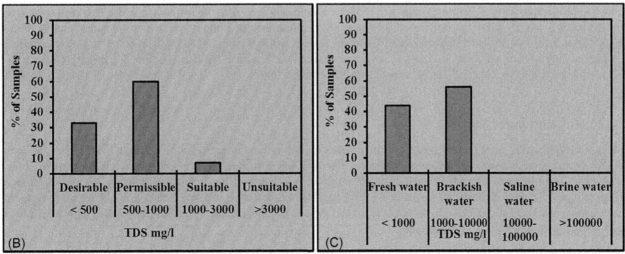

FIG. 6.5 (A) Spatial variation map of total dissolved solids, (B) classification of groundwater based on total dissolved solids (Davis and DeWiest, 1966), and (C) Freeze and Cherrey (1979) classification on TDS.

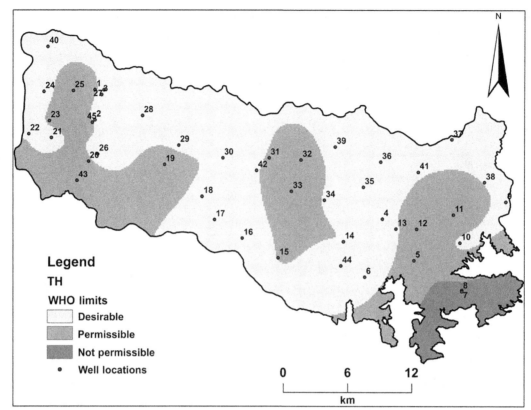

FIG. 6.6   Spatial variation map of total hardness (TH).

TABLE 6.4 Sawyer and McCarthy (1967) Classified the Groundwater Based on Hardness

Total Hardness as CaCO$_3$	Water Class	Percentage of Samples
<75	Soft	–
74–150	Moderately hard	3
150–300	Hard	84
>300	Very hard	13

The category of the groundwater of the study area based on hardness (Sawyer and McCarthy, 1967) has been represented in Table 6.4. Moderately hard levels were found in 2% of the samples, 84% were found to be hard, and 13% were found to be very hard; this is mainly due to the study area soil types, i.e., chromusters and Rhodustalfs, which contain of high levels of Ca and Mg and this leaches into the groundwater during the rainfall (Rajesh et al., 2012). High levels of hardness may affect the water-supply system; account for excessive soap consumption; and cause calcification of arteries, urinary concretions, diseases of the kidney and bladder, and stomach disorders (CPCB, 2008).

6.4.2 Groundwater Types

The groundwater types show the effects of the chemical reactions that occur between the minerals within the lithologic framework and in the groundwater. Hydrogeochemical diagrams aim to facilitate the interpretation of evolutionary trends, particularly in groundwater systems, when they are interpreted in conjunction with distribution maps and hydrogeo-chemical types. An overall characterization of the hydrogeochemical data is possible if the hydrogeochemical types of water are known, generally referred to as water type, and expressed using various plots, such as Durov (1948), trilinear Piper (1944), and Schoeller (1965) diagrams.

The analytical data obtained from the groundwater samples are plotted on a Piper trilinear diagram to understand the hydrogeochemical composition in the study area. The hydrogeochemical evolution can be understood from the Piper plot, which has been divided into six categories: Ca-HCO$_3$, Na-Cl, mixed Ca-Na-HCO$_3$, mixed Ca-Mg-Cl, Ca-Cl, and Na-HCO$_3$ type. Based on these categories the results were obtained from the Piper plot and these represent the spatial grouping map (Fig. 6.7) showing the six groundwater types present in the study area. The cation chemistry shows that the alkalies (Na$^+$ and K$^+$) exceed the earth alkalines (Ca^{2+} and Mg^{2+}) and that strong acids exceed weak acids. In the case of anions, strong acid shows dominance over weak acid, and HCO$_3$$^-$ and Cl$^-$ have almost equal influence to Na$^+$, indicating that the groundwater chemistry of this area is controlled by the hydrogeochemical processes within the geology.

6.4.3 Major Ions Chemistry

The controlling factor of groundwater in the order of dominance of cations in the study area is Ca^{2+} > Na$^+$ > Mg^{2+} > K$^+$, whereas the anions is Cl$^-$ > HCO$_3$$^-$ > SO$_4$$^{-2}$ (Rajesh et al., 2012; Rajesh, 2014). Ca concentration varies from 33 to 453 mg/L and Fig. 6.8A shows the southeast portion of the study area has high concentrations: 62% of the samples fall under the desirable limit, 35% are within permissible limits, and 3% are not permissible according to WHO standards. The high concentrations of Ca lead to the development of kidney or bladder stones in the human body (Mckee and Wolf, 1963). The concentrations of Na vary from 41 to 316 mg/L, and Fig. 6.8B shows that levels are high in the northeast-northwest and southeast-southwest areas. Based on the WHO standards, 89% of the samples fall under permissible limit and 11% of the samples exceed the limits. High Na levels are harmful and cause cardiac, renal, and circulatory diseases (Laubusch and McCammon, 1955).

Mg concentrations vary from 16 mg/L to 86 mg/L: 30% of the samples fall under the desirable limit, 62% of the samples are within permissible limits, and 8% of samples exceed the permissible limits of the WHO standard. Fig. 6.8C shows that the southeast-southwest and northwest portions of the study area have high concentrations. A high concentration of Mg causes scouring diseases in livestock (Mckee and Wolf, 1963). K concentrations vary from 2 to 245 mg/L: 82% of the samples fall within permissible limits and 18% of the samples fall are not permissible. Fig. 6.8D shows that the southeast-southwest and northeast portions of the study area have high levels.

Cl concentrations vary from 24 to 623 mg/L: 93% of the samples fall within desirable limits and 7% of the samples contain levels not permissible as per WHO standard. The southeast-southwest portion of the study area (Fig. 6.8E) have high concentrations. High concentrations of Cl may be due to Cl-bearing rocks such as sodalite chloroapatite,

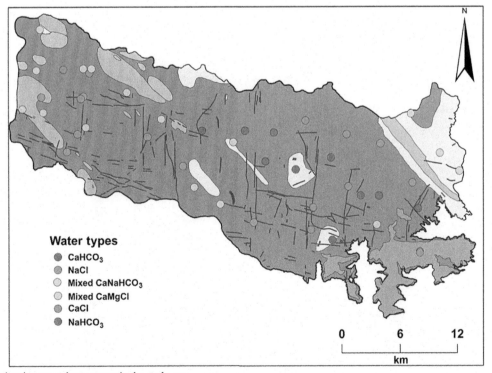

FIG. 6.7 Map showing groundwater types in the study area.

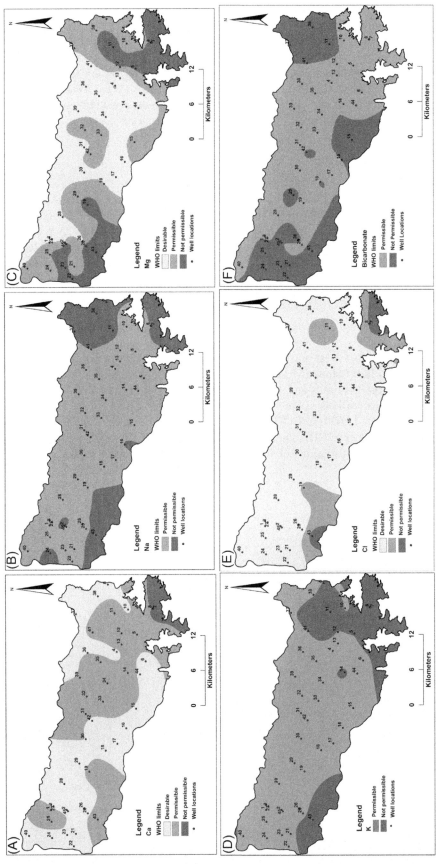

FIG. 6.8 (A–F) Map showing spatial variation of Ca, Na, Mg, K, Cl, and HCO$_3$.

although these are very minor constituents of igneous and metaphoric rocks and contribute very little to the total amount of groundwater (Karanth, 1987). High levels may be injurious to health and affect the heart and kidneys. Taste, indigestion, corrosion and palatability are also affected (CPCB, 2008).

Bicarbonate concentrations vary from 186 to 486 mg/L: 52% of the samples fall within permissible limits and 48% of samples fall without permissible limits. The bicarbonate map (Fig. 6.8F) shows that the northeast-northwest, southeast-southwest, and small patches in the central portion of the study area have high levels. The groundwater used for irrigation undergoes evaporation leading to an increase in the concentration of the ions. This evaporation-enriched irrigated water enters the groundwater zone as recharge and is then pumped away for irrigation. This pumped groundwater is used for irrigation and its further evaporation from the irrigated area leads to an increase in the concentration of salts, especially carbonates, in the soils (Rajesh et al., 2012; Rajesh, 2014). Sulfate concentrations vary from 16 to 318 mg/L: 98% of the samples fall within desirable levels and 2% of samples fall under permissible levels.

6.4.4 Water Quality Index

WQI is used to simplify the reporting of complex water quality data. This science-based communication tool is used in the testing of multivariable water data against numeric water-quality guidelines and/or objectives to produce a single unitless number that represents the overall water quality. WQI has been used to rank the overall water quality in spatial comparisons of sites. The application of the WQI was found to be a good tool to assess complete water quality as it relates to international water-quality guidelines. This paper assesses water quality for the ecosystem initiatives and employs WQI, which provides a convenient means for summarizing complex water-quality data that can be easily understood by the public, water distributors, planners, managers, and policy makers. The WQI method was used to evaluate spatial changes in water quality (Sajil kumar et al. 2013; Varol and Davraz, 2014).

The WQI of the study area varies from 51 to 639. A spatial distribution map of WQI of the groundwater in the study area is presented in Fig. 6.9A. A higher value is noted for the northeast and southeast that follows the trend of EC and major ions. The WQI was used to interpret the groundwater quality parameters, thus the results indicate the significant changes in the quality of water. According to WQI classifications, 1% of the samples are categorized as excellent and represent an area (Fig. 6.9B) of 15 km^2, 78% of the samples are good and represent an area of 443 km^2, 13% of the samples are marginal and represent an area of 105 km^2, 4% of the samples are poor and represent an area of 80 km^2, and 4% of the samples are unsuitable for drinking purposes and represent an area is 80 km^2.

6.4.5 Groundwater Quality for Irrigation Purposes

A evaluation of the suitability of the groundwater for irrigation is based on the estimation of Na content, which is compared to the total cations in the aquifer system. High concentrations of Na in water makes it unsuitable for irrigation activities because the Na$^+$ ion is involved in cation-exchange processes, which tend to affect the ability of soils to sustain crop productivity. The Na$^+$ ion adsorbs onto a cation exchange, affecting the ability of soil aggregates to disperse, thus reducing soil permeability (Tijani, 1994; Yidana et al., 2010). The irrigational suitability of groundwater in the study area was evaluated based on FAO guidelines by EC, SAR, RSC, USSL classification, Na%, and Wilcox diagram.

Relation between Electrical Conductivity With Sodium Adsorption Ratio (SAR)

The total content of soluble salts, such as Na, Ca, and Mg, and its relative proportion affects the suitability of groundwater for irrigation. The EC and Na contents are significant in categorizing irrigation water. According to Richards (1954) the irrigation water is categorized into four groups: low (EC <250 μS/cm), medium (250–750 μS/cm), high (750–2250 μS/cm), and very high (2250–5000 μS/cm) salinity. High EC in water leads to the formation of saline soil, whereas high Na content in water causes alkaline soil.

Based on EC values, 20% of samples have medium, 76% of samples have high, and 4% of samples have very high salinity for irrigation (Table 6.5). The common parameters used that need to be considered for determining groundwater suitability for irrigation are salinity, Na%, and SAR. Irrigation water holding a high quantity of Na will increase the passage of Na content into the soil, which affects the soil permeability and texture making the soil hard to plow and cultivate (Trivedy and Goel, 1984; Nagarajan et al., 2010). SAR and EC proportionally can be used to evaluate irrigation water quality. The SAR recommended by the salinity laboratory of the US Department of Agriculture (Wilcox, 1955) is calculated using Eq. (6.11):

$$SAR = rNa^+ / \left[r\left(Ca^{2+} + rMg^{2+}\right)/2 \right]^{1/2} \tag{6.11}$$

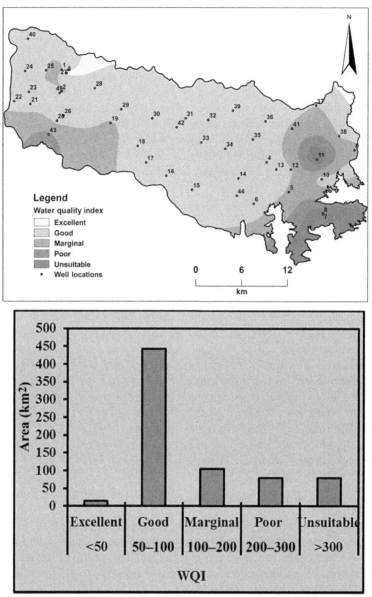

FIG. 6.9   (A) Integrated water-quality index map and (B) water-quality index map covering the study area.

TABLE 6.5 Richards (1954) Classified the Irrigation Water Based on Electrical Conductivity

EC (μS/cm)	Water Quality	Percentage of Samples
<250	Low	–
250–750	Medium	20
750–2250	High	76
2250–5000	Very high	4

In the study area the SAR value varies from 1 to 7 mg/L, and >18 shows that groundwater is unsuitable for irrigation purposes (Sahinci, 1991), indicating that all of samples are suitable for irrigation purposes. A meticulous analysis with respect to the irrigation suitability of the groundwater was executed by plotting the data on the USSL (1954) classification diagram and its results represent the spatial grouping map (Fig. 6.10): 20% of the samples fall within the C2S1 field

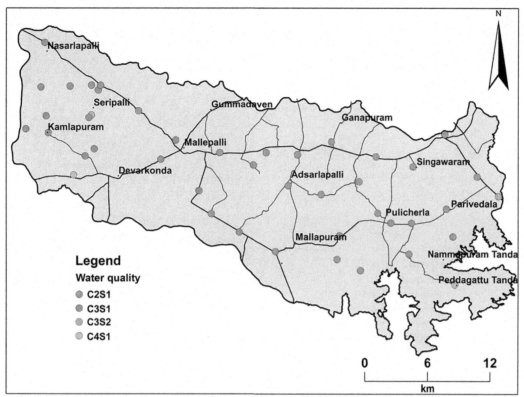

FIG. 6.10 Map showing the spatial distribution of the water quality for irrigation use (USSL, 1954).

TABLE 6.6 Irrigation Water Quality of Groundwater Based on Na%

Na%	Water Class	Percentage of Samples
<20	Excellent	–
20–40	Good	24
40–60	Permissible	51
60–80	Doubtful	25
>80	Unsuitable	–

and 73% within C3S1, which represent the medium-to-high salinity and low alkalinity of the water used for irrigation; and 4% of the samples fall within the C3S2 field (Table 6.4) and 4% within C4S1, which represent the very-high salinity and low-Na hazard, it affects the irrigation purpose during dry periods. Based on this map the groundwater of the study area is suitable for use in irrigation.

Sodium Percentage (Na%)

The percent of Na content is a signifying parameter that is used to evaluate the suitability of the groundwater for irrigation purposes (Wilcox, 1955). Na mixed with carbonate can lead to the formation of alkaline soils, whereas the Na mixed with chloride forms a saline soil. Na percentage (Na%) is calculated using Eq. (6.12):

$$Na\% = Na^+ + K^+ \times 100/(Ca^{2+} + Mg^{2+} + Na^+ + K^+) \tag{6.12}$$

Na% in groundwater was estimated in the study area (Table 6.6). Groundwater samples are plotted into EC and Na% in the Wilcox diagram and the results represent the spatial grouping analysis method (Fig. 6.11) 64% of the samples fall into the excellent-to-good, 9% into the good-to-permissible, 24% into permissible-to-doubtful, and 3% into doubtful-to-unsuitable category. Na concentrations play a vital role in the evaluation of groundwater quality for irrigation because Na affects the

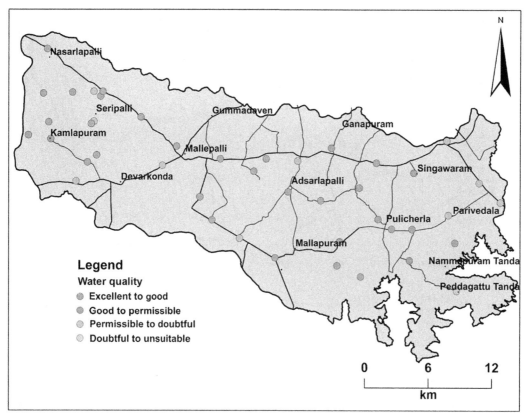

FIG. 6.11 Map showing the spatial distribution of the water quality for irrigation use (Wilcox, 1955).

texture as well as the permeability of soil (Tijani, 1994; Gholami et al., 2013; Belkhiri and Mouni, 2012; Dar et al., 2012; Nagarajan et al., 2010). High Na% is due to the dissolution of minerals, weathering of rock formations, and the accumulation of chemical fertilizers in irrigation waters (Subba Rao, 2002).

6.4.6 Residual Sodium Carbonate

Residual sodium carbonate (RSC) is a significant parameter used to evaluate the suitability of irrigation water (Ragunath, 1987) and can be calculated using Eq. (6.13):

$$RSC = \left(CO_3^{-2} + HCO_3^{-}\right) - \left(Ca^{2+} + Mg^{2+}\right) \tag{6.13}$$

Based on Lloyd and Heathcote (1985), irrigation water is categorized as suitable (<1.25), marginal (1.25–2.5), and not suitable (>2.5). According to RSC 89% of the samples are suitable, 7% are marginal, and 4% are unsuitable. Overall, the residual Na carbonate is suitable for irrigation in the study area.

6.4.7 Permeability Index

The permeability index (PI) is also a valuable method to identify the suitability of groundwater for irrigation. Irrigation purpose of groundwater was determined the effects of Na, Ca, Mg and HCO_3 concentration in soil permeability after long term (Nagaraju et al., 2006). WHO (1989) categorizes the permeability index as Class I ($>75\%$), Class II (25%–75%), and Class III ($<25\%$), respectively. The PI varies from 7% to 34% in the study area: 58% of the water samples were Class II, Class I, and Class II and categorized as good for irrigation with 75% or more of maximum permeability (Varol and Davraz, 2014); and 42% of the water samples were Class III, which is unsuitable with 25% of maximum permeability.

6.4.8 Magnesium Hazard

In general, Ca and Mg in groundwater is in equilibrium. Szaboles and Darab (1964) proposed a Mg hazard (MH) for evaluating the suitability of water quality for irrigation. Mg damages the structure of soil, while water that contains more Na reflects high salinity. Usually high Mg values affect the interchangeable Na in irrigated soils. In an equilibrium state additional Mg can affect the soil quality by making it alkaline in nature, which affects crop yields. MH is categorized in terms of the Mg ratio (MR). This is a ratio of Mg concentration is composition of Ca and Mg concentration of ions. It is calculated by using the Eq. (6.14):

$$MR = \left[Mg^{2+} / \left(Ca^{2+} Mg^{2+} \right) \right] \times 100 \tag{6.14}$$

where all ionic concentrations are expressed in milli equivalents per liter. If the MR exceeds the value of 50, the water is judged as harmful and unsuitable for irrigation because it undesirably affects the crop yields. In the study area the MR varies from 32 to 79 meq/L: 24% of the samples are suitable for irrigation and 76% of the samples exceed the value of 50 and are unsuitable for irrigation.

6.5 MULTIVARIATE STATISTICAL ANALYSIS

6.5.1 Correlation Matrix

The correlation matrix is used to find the correlation between the physicochemical properties of groundwater, it exposes the source of solutes and the processes that produce the observed water's constitution (Azaza et al., 2011). A high correlation coefficient (close to 1 or 1) means a good positive relationship between two variables and a value around zero means no relationship at a significant level of $P < 0.05$. More specifically, the parameters of $r > 0.7$ are strongly correlated while an r value 0.5 and 0.7 represents a moderate correlation (Manish et al., 2006).

The correlation matrices were prepared using the following parameters: pH, EC, TDS, TH, and major ions (Table 6.7). pH indicated a negative correlation with all parameters. EC indicated a high positive correlation with TDS, TH, Na, K, Ca, Mg, Cl, and SO$_4$. TDS indicated a high positive correlation with TH, K, Ca, Cl, and SO$_4$, and a moderate correlation with Na and Mg. TH indicated a high positive correlation with K, Ca, Mg, Cl, and SO$_4$. Some groups indicated a strong-to-moderate correlation, $r > 0.7$, therefore Na-K-Mg-Cl-SO$_4$-HCO$_3$, K-Ca-Mg-Cl-SO$_4$, Ca-Mg-Cl-SO$_4$, Mg-Cl-SO$_4$, and Cl-SO$_4$. This indicates that the variations in ionic concentration are chiefly the result of the weathering of rock formation and the dissolution of minerals in the study area. Similar observations were made in a few other regions of south India (Srinivasa Rao et al., 1997; Manish et al., 2006; Singaraja et al., 2014).

TABLE 6.7 Pearson's Correlation Matrix of Physicochemical Parameters

	pH	EC	TDS	TH	Na	K	Ca	Mg	Cl	SO$_4$	HCO$_3$
pH	1.00										
EC	0.15	1.00									
TDS	0.15	**1.00**	1.00								
TH	0.20	**0.90**	**0.90**	1.00							
Na	0.11	**0.77**	**0.77**	0.47	1.00						
K	0.17	**0.84**	**0.84**	**0.82**	0.51	1.00					
Ca	0.30	**0.82**	**0.82**	**0.93**	0.34	**0.80**	1.00				
Mg	0.02	**0.78**	**0.78**	**0.84**	**0.55**	**0.64**	**0.59**	1.00			
Cl	0.15	**0.97**	**0.97**	**0.89**	**0.72**	**0.83**	**0.85**	**0.72**	1.00		
SO$_4$	0.26	**0.89**	**0.89**	**0.89**	0.54	**0.88**	**0.91**	**0.63**	**0.90**	1.00	
HCO$_3$	0.20	0.28	0.28	0.08	**0.65**	0.09	−0.14	0.41	0.13	0.00	1.00

The bold value indicates the correlation coefficient of >0.60 at significant 0.05 levels.

6.5.2 Cluster Analysis

Cluster analysis (CA) was carried out using groundwater samples from some classes. The similarity between objects was measured by squared Euclidean distances, and Ward's method of divisive hierarchical clustering method was used for CA (Kim et al., 2014). Fig. 6.12 characterizes a dendrogram based on the CA for the groundwater samples. CA combined the groundwater samples of the study area into three types of cluster groups. Group I consisted of pH, Na, K, Ca, Mg, Cl, and SO_4; Group II consisted of TH and HCO_3; and Group III consisted of EC and TDS. The CA groups are mainly derived from the incongruent dissolution of minerals from rock and the high Ca and Mg in soil is mainly chromusters and Rhodustalfs types of soils occurs and it leaches into the groundwater during the rainfall (Rajesh et al., 2012).

6.5.3 Factor Analysis

Factor analysis is one of the most significant statistical methods for the interpretation of the hydrogeochemistry of groundwater (Subba Rao, 2002). Factor analysis leads to the determination of the basic independent dimensions of variables. Weathering and anthropogenic activities are the two main processes responsible for changing the hydrogeochemical composition of the groundwater (Jeong, 2001). The variables for factor analysis were pH, EC, TDS, TH, Na, K, Ca, Mg, Cl, HCO_3, and SO_4 (Table 6.8). In general, the factor will be connected to the largest eigenvalue and will describe the greatest amount of variance in the data set. Liu et al. (2003) categorizes factor loadings as strong, moderate, and weak according to the total values of >0.75, 0.75–0.50, and 0.50–0.30, respectively. These categories were used to classify the data in this study (Table 6.5).

Factor 1 of the groundwater in the study area is categorized by the strong loads between EC, TDS, TH, K, Ca, Cl and SO4 and moderate loads between Na and Mg (Table 6.8). Factor 1 explains the natural hydrogeochemical characteristics of groundwater by water-rock interaction, it is mainly derived from weathering of rocks and dissolution of minerals. Strong loads of EC and TDS in the groundwater are due to the intercalation quartzite and shale found in the southeastern portion of the study area. Strong loads of K are due to the disintegration of K-bearing minerals like K-feldspar such as orthoclase and microcline; most of the area is covered by granite. The high load of Ca is due to pyroxene, amphibolite, and feldspar and the leaching of soils such as chromusters and Rhodustalfs (Rajesh et al., 2012; Rajesh, 2014). Cl mainly comes from feldspathoid sodalite and phosphate mineral apatite (Hem, 1970), and also Cl-containing fertilizers. Strong loads of SO_4 are found in the feldspathoid group of igneous rocks (Hem, 1985) and also in the oxidation states of organic matter in the sulfur cycle, which in turn are a source of energy to split bacteria (Mckee and Wolf, 1963). A moderate load of Na occurs as the soda lime feldspars with sodium-rich plagioclase feldspar (Albite) and sodium-rich silicates (nephline and sodalite). Na-rich irrigation water allows the displacement of exchangeable cations of Ca and Mg from the clay minerals in the soil and the replacement of cations by sodium (Matthess, 1982). High Mg values are due to the presence of pyroxene and amphibole minerals in the study area.

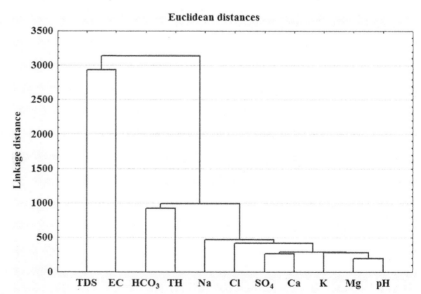

FIG. 6.12 Dendrogram of Q-mode of hierarchical cluster analysis of groundwater samples of the study area.

TABLE 6.8 Results of the Factor Analysis of Physicochemical Parameters

	F1	F2
pH	0.31	**0.53**
EC	**0.94**	0.31
TDS	**0.94**	0.31
TH	**0.96**	0.06
Na	**0.54**	**0.74**
K	**0.89**	0.07
Ca	**0.94**	0.19
Mg	**0.73**	0.43
Cl	**0.95**	0.20
SO$_4$	**0.96**	0.02
HCO$_3$	0.02	**0.89**
Eigen value	7.41	1.78
% variance	67.35	16.19
Cumulative %	67.35	83.53

Factor 2 of the groundwater in the study area is categorized by a strong load of HCO$_3$ and moderate loads of Na, pH (Table 6.6). It indicates the weathering and dissolution of carbonate minerals from the soil. A strong load of HCO$_3$ ions with alkali and alkaline earth metals enables the interpretation of weathering sources. The source of HCO$_3$ derives from the dissolution of carbonate minerals from the soil reacting with carbonic acid and the degradation of organic matter from the bacteria. Strong pH values indicate that the major ion concentration is controlled by pH and is mainly derived from the interaction of the rocks in the study area.

6.6 CONCLUSIONS

Groundwater quality and its suitability for drinking and irrigation purposes were evaluated in the study area. The hydrogeochemical studies of the groundwater in the granitic terrain of southern India delivered the following conclusions:

– A total of 45 groundwater samples were collected and analyzed for various physicochemical parameters. The pH 4% of samples exceeds the permissible limits set by WHO (2004). Based on the EC and TDS groundwater was above the permissible limits, and this is contributed by the dissolution of minerals from the geological formation. One-third of total hardness exceeds the permissible limit, explaining the hard to very hard groundwater in the study area.

– GIS-based spatial analysis using a spline interpolation technique was used to represent the spatial variation of pH, EC, TH, and major ions such as Na, K, Ca, Mg, Cl, SO$_4$, and HCO$_3$. The higher concentrations mostly appear in the southeast-southwest portion of the study area, which indicates that the intercalation of quartzite and shale formation increases concentrations.

– Groundwater was classified into types to aid in the understanding of the hydrogeochemical characteristics: Ca-HCO$_3$, Na-Cl, mixed Ca-Na-HCO$_3$, mixed Ca-Mg-Cl, Ca-Cl, and Na-HCO$_3$ are the dominant water types in the study area. This is due to the water-rock interaction and anthropogenic contamination.

– WQI was used to evaluate the groundwater quality and its suitability for drinking purposes. WQI classification shows excellent quality in a 15 km^2 area representing 1% of the samples, good quality for a 443 km^2 area representing 78% of the samples, marginal quality for a 105 km^2 area representing 13% of the samples, poor quality for an 80 km^2 area representing 4% of the samples, and unsuitable for drinking purposes for an 80 km^2 area representing 4% of the samples. This shows the effect of the leaching of ions, weathering of rocks, and anthropogenic activities.

– FAO classification for EC FAO guidelines of EC, SAR, residual sodium bicarbonate, PI, MH, USSL classification, Na%, and Wilcox diagram. The SAR values demonstrate that all samples are suitable for irrigation. The Na% explains that 64% of the samples fall into the excellent-to-good, 9% into the good-to-permissible, 24% into the permissible-to-doubtful, and 3% into the doubtful-to-unsuitable category. The PI value of the study area shows that 58% of samples are good for irrigation and 42% of the samples are unsuitable for irrigation. MR values demonstrate that 24% of the samples are suitable for irrigation and that 76% of the samples are unsuitable for irrigation.

- Multistatistical analyses were used for interpreting the characteristics of groundwater quality. A correlation matrix reveals the EC indicated a high positive correlation with TDS, TH, Na, K, Ca, Mg, Cl, and SO_4. TDS indicated a high positive correlation with TH, K, Ca, Cl, and SO_4, and a moderate correlation with Na and Mg. TH indicated a high positive correlation with K, Ca, Mg, Cl, and SO_4. Some groups indicated a strong-to-moderate correlation, $r>0.7$, therefore Na-K-Mg-Cl-SO_4-HCO_3, K-Ca-Mg-Cl-SO_4, Ca-Mg-Cl-SO_4, Mg-Cl-SO_4, and Cl-SO_4. Cluster analysis classified the groundwater samples into three types of clusters group: Group I consists of pH, Na, K, Ca, Mg, Cl, and SO_4 derives from the weathering of minerals; Group II contains TH and HCO_3 leaching of high levels of Ca and Mg into the groundwater during the rainfall; and Group III includes EC and TDS derives from the dissolution of minerals.
- There are two factors related to the formation of groundwater quality in the study area. Factor 1 characterizes the natural hydrogeochemical characteristics of groundwater through the water-rock interactions, which can be described by the dissolution of minerals in rocks. Factor 2 indicates the weathering and dissolution of carbonate minerals from the soil.
- The mechanism of groundwater chemistry of the study area is mainly controlled by geogenic sources, rather than anthropogenic source.

ACKNOWLEDGMENTS

The authors would like to thank the Board of Research in Nuclear Sciences, Department of Atomic Energy, and Government of India (Grant No. 2007/36/35) for their financial support. The Department of Science and Technology's Funds for Improvement in Science and Technology scheme (Grant No. SR/FST/ESI-106/2010) and University Grants Commission's Special Assistance Programme (Grant No. UGC DRS II F.550/10/DRS/ 2007 (SAP-1)) are also acknowledged as the analytical facilities created from these funds were used to carry out part of this work.

REFERENCES

Adhikary, P.P., Dash, C.J., Chandrasekharan, H., Rajput, T.B.S., Dubey, S.K., 2012. Evaluation of groundwater quality for irrigation and drinking using GIS and geostatistics in a peri-urban area of Delhi, India. Arab. J. Geosci. 5, 1423–1434.

Andrade, E.M., Araujo, H., Palácio, Q., Souzab, I.H., Leño, R.A., Guerreiro, M.J., 2008. Land use effects in groundwater composition of an alluvial aquifer (Trussu River, Brazil) by multivariate techniques. Environ. Res. 106 (2), 170–177.

APHA, 1995. Standard Methods for the Examination of Water and Waste Water.

Azaza, F.H., Ketata, M., Bouhlila, R., Gueddari, M., Riberio, L., 2011. Hydrogeochemical characteristics and assessment of drinking water quality in Zeuss-Koutine aquifer, southeastern Tunisia. Environ. Monit. Assess. 174, 283–298.

Azeez, P.A., Nadarajan, N.R., Mittal, D.D., 2000. The impact of monsoonal wetland on groundwater chemistry. Pollut. Res. 19 (2), 249–255.

Belkhiri, L., Mouni, L., 2012. Hydrochemical analysis and evaluation of groundwater quality in El Eulma area, Algeria. Appl. Water Sci. https://doi.org/ 10.1007/s13201-012-0033-6.

Bouzourra, H., Bouhlila, R., Elango, L., Slama, F., Ouslati, N., 2014. Characterization of mechanisms and processes of groundwater salinization in irrigated coastal area using statistics, GIS, and hydrogeochemical investigations. Environ. Sci. Pollut. Res. Int. https://doi.org/10.1007/s11356-014-3428-0.

Brindha, K., Elango, L., 2010. Study on bromide in groundwater in parts of Nalgonda District, Andhra Pradesh. Earth Sci. India 3 (1), 73–80.

Brindha, K., Elango, L., 2012. Groundwater quality zonation in a shallow weathered rock aquifer using GIS. Geo-Spat. Inf. Sci. 15 (2), 95–104.

Brindha, K., Rajesh, R., Murugan, R., Elango, L., 2011. Fluoride contamination in groundwater in parts of Nalgonda District, Andhra Pradesh, India. Environ. Monit. Assess. 172, 481–492.

Central Ground Water Board, 2007. Ground Water Information Nalgonda District, Andhra Pradesh. Publishing CGWB. http://cgwbgovin/DistrictProfile/ AP/Nalgonda.pdf.

Ceron, J.C., Jimenez-Espinosa, R., Pulido-Bosch, A., 2000. Numerical analysis of hydrochemical data: a case study (Alto Guadalentín, southeast Spain). Appl. Geochem. 15 (7), 1053–1067.

Chen, L., Feng, Q., 2013. Geostatistical analysis of temporal and spatial variations in groundwater levels and quality in the Minqin oasis, Northwest China. Environ. Earth Sci. 70, 1367–1378.

Chung, S.Y., Venkatramanan, S., Park, N., Rajesh, R., Ramkumar, T., Kim, B.W., 2015. An assessment of selected hydrochemical parameter trend of Nakdong River water in South Korea; using time series analyses and PCA. Environ. Monit. Assess. 187, 4192.

CPCB, 2008. Guideline for Water Quality Management. Central Pollution Control Board, Parivesh Bhawan.

Dalton, M.G., Upchurch, S.B., 1978. Interpretation of hydrochemical facies by factor analysis. Groundwater 16, 228–233.

Dar, I.A., Sankar, K., Dar, M.A., 2012. Groundwater quality evaluation of Mamundiyar River Basin, India using ARCGIS 9.2 platform continental. Environ. Sci. 6 (1), 1–24.

Dassi, L., 2011. Investigation by multivariate analysis of groundwater composition in a multilayer aquifer system from North Africa: a multi-tracer approach. Appl. Geochem. 26 (8), 1386–1398.

Davis, J.C., 2002. Statistics and Data Analysis in Geology, third ed. Wiley, New York.

Davis, S.N., DeWiest, R.J., 1966. Hydrogeology. Wiley, New York.

Durov, S.A., 1948. Natural waters sand graphic representation of their composition. Dokl. Akad. Nauk SSSR 59, 87–90.

Elhatip, H., Afsin, M., Kuscu, L., Dirik, K., Kurmac, A., Kavurmac, M., 2003. Influences of human activities and agriculture on groundwater quality of Kayseri-Incesu-Dokuzpinar springs central Anatolian part of Turkey. Environ. Geol. 44, 490–494.

Freeze, A.R., Cherrey, J.A., 1979. Groundwater. Prentice-Hall, New Jersey.

Geological Survey of India, 1995. Geology and Minerals Map of Nalgonda District, Andhra Pradesh, India.

Gholami, A., Shahinzadeh, N., Afrous, A., Papan, P., 2013. An assessment of groundwater quality for agricultural use (case study: Loor plain, Khouzestan, Iran). Int. J. Farm. Allied Sci. 2-21, 890–894.

Guler, C., Thyne, G.D., 2004. Hydrologic and geologic factors controlling surface and groundwater chemistry in Indian Wells-Owens Valley area, southeastern California, USA. J. Hydrol. 285, 177–198.

Handa, B.K., 1969. Description and classification of media for hydro-geochemical investigations. In: Symposium on Ground Water Studies in Arid and Semiarid Regions, Roorkee.

Hem, J.D., 1970. Study and interpretation of the chemical characteristics of natural water. U.S. Geological Survey, Water Supply Paper 1473, 363.

Hem, J.D., 1985. Study and Interpretation of Chemical Characteristics of Natural Water. U.S. Geological Survey. (Water Supply Paper No. 2254).

Jeong, C.H., 2001. Mineral-water interaction and hydrogeo-chemistry in the sank wang mine area, Korea. Geochem. J. 35, 1–12.

Karanth, K.R., 1987. Ground Water Assessment, Development and Management. Tata McGraw-Hill, New Delhi, p. 720.

Kim, J., Kim, R., Lee, J., Cheong, T., Yum, B., Chang, H., 2005. Multivariate statistical analysis to identify the major factors governing groundwater quality in the coastal area of Kimje, South Korea. Hydrol. Process. 19, 1261–1276.

Kim, T.H., Chung, S.Y., Park, N., Hamm, S.Y., Lee, S.Y., Kim, B.W., 2014. Combined analyses of chemometrics and kriging for identifying groundwater contamination sources and origins at the Masan coastal area in Korea. Environ. Earth Sci. 67 (5), 1373–1388.

Laubusch, E.J., McCammon, C.S., 1955. Water as a sodium source and its relation to sodium restriction therapy patient response. Am. J. Public Health 45, 1337–1338.

Lee, S.M., Min, K.D., Woo, N.C., Kim, Y.J., Ahn, C.H., 2003. Statistical models for the assessment of nitrate contamination in urban groundwater using GIS. Environ. Geol. 44, 210–221.

Li, P., Wu, J., Qian, H., 2013. Assessment of groundwater quality for irrigation purposes and identification of hydrogeochemical evolution mechanisms in Pengyang County, China. Environ. Earth Sci. 69 (7), 2211–2225.

Liu, C.W., Lin, K.H., Kuo, Y.M., 2003. Application of factor analysis in the assessment of groundwater quality in a blackfoot disease area in Taiwan. Sci. Total Environ. 313 (1–3), 77–89.

Lloyd, J.W., Heathcote, J.A., 1985. Natural Inorganic Hydrochemistry in Relation to Groundwater. Clarendon, Oxford, p. 294.

Magesh, N.S., Chandrasekar, N., 2013. Evaluation of spatial variations in groundwater quality by WQI and GIS technique: a case study of Virudunagar District, Tamil Nadu, India. Arab. J. Geosci. 6, 1883–1898.

Manish, K., Ramanathan, A., Rao, M.S., Kumar, B., 2006. Identification and evaluation of hydrogeochemical processes in the ground-water environment of Delhi, India. J. Environ. Geol. 50, 1025–1039.

Masoud, A., 2014. Groundwater quality assessment of the shallow aquifers west of the Nile Delta (Egypt) using multivariate statistical and geostatistical techniques. J. Afric. Earth Sci. 95, 123–137.

Matthess, G., 1982. The Properties of Groundwater. Wiley, New York, p. 498.

Mckee, J.E., Wolf, H.W., 1963. Water Quality Criteria, California State Water Quality Control Board, Pub. No.: 3-A.

Milovanovic, M., 2007. Water quality assessment and determination of pollution sources along the Axios/Vardar River, Southeast Europe. Desalination 213, 159–173.

Mitra, B.K., 1998. Spatial and temporal variation of ground water quality in sand dune area of Aomori Prefecture in Japan. (Paper Number 062023). In: 2006 ASAE Annual Meeting.

Moosavirad, S.M., Janardhana, M.R., Khairy, H., 2013. Impact of anthropogenic activities on the chemistry and quality of groundwater: a case study from a terrain near Zarand City, Kerman Province, SE Iran. Environ. Earth Sci. 69 (7), 2451–2467.

Nagarajan, R., Rajmohan, N., Mahendran, U., Senthamilkumar, S., 2010. Evaluation of groundwater quality and its suitability for drinking and agricultural use in Thanjavur City, Tamil Nadu, India. Environ. Monit. Assess. 171, 289–308.

Nagaraju, A., Suresh, S., Killham, K., Hudson-Edwards, K., 2006. Hydrogeochemistry of waters of Mangampeta barite mining area, Cuddapah Basin, Andhra Pradesh, India. Turk. J. Eng. Environ. Sci. 30, 203–219.

Namibian, M., 2007. A new water quality index for environmental contamination contributed by mineral processing: a case study of Amang (tin tailing) processing activity. J. Appl. Sci. 7, 2977–2987.

Nash, H., McCall, G.J.H. (Eds.), 1995. Groundwater quality. 17th Special Report. Chapman and Hall, London.

Piper, A.M., 1944. A graphic procedure in the chemical interpretation of water analysis. Trans. Am. Geophys. Union 25, 914–923.

Ragunath, H.M., 1987. Groundwater. Wiley, New Delhi, p. 563.

Rajaveni, S.P., Brindha, K., Rajesh, R., Elango, L., 2014. Spatial and temporal variation of groundwater level and its relation to drainage and intrusive rocks in a part of Nalgonda District, Andhra Pradesh, India. J. Indian Soc. Remote Sens. https://doi.org/10.1007/s12524-013-0328-6.

Rajesh, R., 2014. Hydrogeology and Hydrogeochemical Characterisation of Groundwater of a Part of Nalgonda District, Andhra Pradesh India. (Ph.D. Thesis). Anna University, Department of Geology. http://shodhganga.inflibnet.ac.in/handle/10603/15062.

Rajesh, R., Brindha, K., Murugan, R., Elango, L., 2012. Influence of hydrogeochemical processes on temporal changes in groundwater quality in a part of Nalgonda District, Andhra Pradesh, India. Environ. Earth Sci. 65, 1203–1213.

Rao, N.V.R., Rao, K.S., Schuiling, R.D., 1993. Fluorine distribution in waters of Nalgonda District, Andhra Pradesh, India. Environ. Geol. 21, 84–89.

Richards, L.A., 1954. Diagnosis and Improvement of Saline and Alkali Soils. US Department of Agriculture Handbook 60, Washington.

Sahinci, A., 1991. Geochemistry of Natural Waters. Reform Publications, Sec 2, p. 33.

Sajil Kumar, P.J., Elango, L., James, E.J., 2013. Assessment of hydrochemistry and groundwater quality in the coastal area of South Chennai, India. Arab. J. Geosci. 7 (7), 2641–2653. https://doi.org/10.1007/s12517-013-0940-3.

Sawyer, C., McCarthy, P., 1967. Chemical and Sanitary Engineering, second ed. McGraw-Hill, New York.

Schoeller, H., 1965. Qualitative evaluation of groundwater resource. In: Methods and Techniques of Ground-Water Investigation and Development. UNESCO, pp. 54–83.

Selvam, S., Manimaran, G., Sivasubramanian, P., Balasubramanian, N., Seshunarayana, T., 2014. GIS-based evaluation of water quality index of groundwater resources around Tuticorin coastal city, South India. Environ. Earth Sci. 7 (6), 2847–2867.

Simoes, F.S., Moreira, A.B., Bisinoti, M.C., Gimenez, S.M.N., Yabe, M.J.S., 2008. Water quality index as a simple indicator of aquaculture effects on aquatic bodies. Ecol. Indic. 8, 476–484.

Singaraja, C., Chidambaram, S., Anandhan, P., Prasanna, M.V., Thivya, C., Thilagavathi, R., 2014. A study on the status of saltwater intrusion in the coastal hard rock aquifer of South India. Environ. Dev. Sustain. https://doi.org/10.1007/s10668-014-9554-5.

Srinivasa Rao, Y., Krishna Reddy, T.V., Nayudu, P.T., 1997. Groundwater quality in the Niva river basin, Chittoor District, Andhra Pradesh, India. J. Environ. Geol. 32, 56–63.

StatSoft, 2007. STATISTICA Version 8. StatSoft, OK, USA.

Subba Rao, N., 2002. Geochemistry of groundwater in parts of Guntur District, Andhra Pradesh, India. Environ. Geol. 41, 552–562.

Szaboles, I., Darab, C., 1964. The influence of irrigation water of high sodium carbonate content of soils. In: Proceedings of 8th International Congress of ISSS, Trans, IIpp. 803–812.

Tijani, M.N., 1994. Hydrochemical assessment of groundwater in Moro area, Kwara State, Nigeria. Environ. Geol. 24, 194–202.

Todd, D.K., 1959. Groundwater Hydrology. Wiley, New York, p. 336.

Trivedy, R.K., Goel, P.K., 1984. Chemical and Biological Methods for Water Pollution Studies. Environmental Publication, Karad.

United States Salinity Laboratory Staff, 1954. Diagnosis and Improvement of Saline and Alkali Soils. U.S. Department of Agriculture Handbook 60, Washington, DC. 160 p.

Varol, S., Davraz, A., 2014. Evaluation of the groundwater quality with WQI (water quality index) and multivariate analysis: a case study of the Tefenni plain (Burdur/Turkey). Environ. Earth Sci.. https://doi.org/10.1007/s12665-014-3531-z.

Venkatramanan, S., Chung, S.Y., Ramkumar, T., Gnanachandrasamy, G., Vasudevan, S., 2014. Application of GIS and hydrogeochemistry of groundwater pollution status of Nagapattinam District of Tamil Nadu, India. Environ. Earth Sci. 73, 4429–4442.

Guidelines for drinking water quality. WHO, (Ed.), 2004. Recommendations. third ed. In: vol. 1. WHO, Geneva, p. 515.

WHO, 1989. Health guidelines for the use of wastewater in agriculture and aquaculture. Report of a WHO Scientific Group – Technical Report Series 778. WHO, Geneva, p. 74.

Wilcox, L.V., 1955. Classification and Use of Irrigation Water. US Geol. Dep. Agric. Arc 969, p. 19.

Yidana, S.M., Banoeng-Yakubo, B., Akabzaa, T.M., 2010. Analysis of groundwater quality using multivariate and spatial analyses in the Keta basin Ghana. J. Afr. Earth Sci. 58 (2), 220–234.

Chapter 7

Spatial Prediction of Groundwater Level Using Models Based on Fuzzy Logic and Geostatistical Methods

Ata Allah Nadiri*, Rahman Khatibi†, Fatemeh Vahedi‡ and Keyvan Naderi‡

*Department of Earth Sciences, Faculty of Natural Sciences, University of Tabriz, Tabriz, Iran †GTEV-ReX Limited, Swindon, United Kingdom ‡Department of Earth Sciences, University of Tabriz, Tabriz, Iran

7.1 INTRODUCTION

The application of modeling techniques based on fuzzy logic (FL) in the spatial analysis of groundwater levels is investigated as an alternative to the techniques based on traditional geostatistical approaches, such as kriging and co-kriging. The techniques already applied to the spatial analysis of groundwater levels are well established and there are a number of comprehensive reviews (ASCE, 1990; Barca and Passarella, 2008; Cay and Uyan, 2009; Finke et al., 2004; Gundogdu and Guney, 2007; Ma et al., 1999; Ta'any et al., 2009). Kriging techniques are advanced geostatistical algorithms used in geostatistical spatial analysis to generate water-level data from a sample of scattered datapoints with z-values. These techniques have limitations and when a set of available data does not comply with their underlying assumptions, the reliability of their results will be undermined. Such limitations were also true of first-generation modeling techniques based on fuzzy logic due to the use of prescribed fuzzy rules and therefore the outputs from this approach were dependent on the prescribed rules. However, recent advances in these techniques have created a true bottom-up and data-driven capability such that the rules are derived from the data available and are not prescribed. In this way, the role of assumptions on the generated groundwater levels can be minimal. This study therefore investigated the application of Sugeno fuzzy logic (SFL), developed by Takagi and Sugeno (1985) and Mamdani fuzzy logic (MFL), developed by Mamdani and Assilian (1975).

Applications of FL-based models for hydrology are wide ranging and include those by Burrough (1989), Metternicht (2001), Bogardi et al. (2003), and Kholghi and Hosseini (2009). In particular, geostatistics has been used widely to assess water-quality problems and examples include Chowdhury et al. (2010), Ko et al. (2009), and Sen (2009). Pozdnyakova and Zhang (1999) applied the geostatistical methods of kriging and cokriging to estimate sodium adsorption ratios for an agricultural field. Goovaerts et al. (2005) compared the performances of multi-Gaussian and indicator kriging for probabilistic modeling of the spatial distribution of arsenic concentrations in the groundwater of southeast Michigan. Other studies have assessed the spatial distribution of groundwater quality using geostatistical approaches (Eslami et al., 2013; Mehrjardi et al., 2008; Nas, 2009; Zehtabiana et al., 2013).

Some of the findings from the above research activities are that kriging techniques perform better than other interpolation techniques, like weighted moving average (WMA) and inverse distance weighting (IDW), in characterizing spatial variability. Geostatistical methods provide reliable results when data are: (i) normally distributed and (ii) their mean and

GIS and Geostatistical Techniques for Groundwater Science. https://doi.org/10.1016/B978-0-12-815413-7.00007-9

variance are stationary and therefore do not vary significantly in space. Significant deviations from normality and stationarity can cause problems in terms of accurate interpolations and therefore this makes a research case for investigating the application of FL with bottom-up data-driven capabilities.

Recent generations of models based on FL do not suffer from these limitations and they are even robust, as they cope with data that are subject to ambiguous boundaries and gradual transitions between the defined sets (see Kadkhodaie-Ilkhchi and Amini, 2009). As such, the popularity of FL-based models has recently amplified with applications to wider hydrogeological problems including: (i) water-quality classification (Barbieri et al., 2001; Guler et al., 2002; Guler and Thyne, 2004; Kord and Moghaddam, 2014); (ii) spatial prediction of groundwater quality (Kumar et al., 2010; Samson et al., 2010); (iii) groundwater pollution distribution (Kiurski-Milošević et al., 2015; Muhammetoglu and Yardimci, 2006; Tutmez and Hatipoglu, 2010); and (iv) spatial analysis of groundwater vulnerability and risk (Ozbek and Pinder, 1998; Uricchio et al., 2004; Gemitzi et al., 2006; Fijani et al., 2013; Nadiri et al., 2014).

To examine the capability of FL models, some researchers have compared the results of FL models with those of geostatistics. Tutmez and Hatipoglu (2010) compared the performances of kriging and fuzzy computing in spatial interpolation of nitrate concentration in groundwater. The results and performance evaluations indicated that FL-based models perform better than the kriging models. Kord and Moghaddam (2014) implemented spatial analysis of the aquifer potable groundwater for the Ardabil plain using FL and kriging; they report that the fuzzy model provides better results in comparison with the kriging.

Groundwater level is a key parameter in aquifer management and maintaining groundwater quality. Modeling groundwater levels is now a capability that involves a diverse range of techniques, but their validations require monitoring groundwater levels. In spite of the key role of monitoring of groundwater levels, the available data are often sparse due to time and budgetary constraints and observation wells are often provided sparingly. This problem is unlikely to be resolved at present time and therefore interpolation techniques are likely to remain topical for some time.

There are some studies on the interpolation techniques used in spatial generation of water tables. Kurtulus et al. (2011) used ordinary kriging (OK), and adaptive neuro fuzzy-based inference system (ANFIS) to assess hydraulic head distribution in an aquifer unit. Both methods produced realistic results, even if OK performed slightly better than ANFIS. Kholghi and Hosseini (2009) investigated the efficiency of OK and ANFIS to generate data on groundwater levels in an unconfined aquifer and reported ANFIS to be more efficient than OK. Hamad (2009) carried out geostatistical analysis of groundwater levels including spatial interpolation by OK using semivariogram and covariance. Varouchakis and Hristopoulos (2013) compared the interpolation performance of OK, universal kriging (UK), and disjunctive kriging (DK) with traditional techniques, such as IDW and minimum curvature (MC) for groundwater levels in a sparsely gauged basin.

This chapter presents an investigation using the SFL and MFL and compares their performances against the widely used OK and co-kriging (CK) techniques to generate data on groundwater levels in the Meshginshahr Plain of the Ardabil Province of northwest Iran. The results of these four different techniques (SFL, MFL, OK, and CK) are compared with one another to assess their performance using the data for the study area.

7.2 METHODOLOGY

7.2.1 Fuzzy Logic

This chapter investigates the performance of SFL and MFL in predicting groundwater levels at the location of observation wells distributed throughout a study area against the baseline of geostatistical tools. Both SFL and MFL are based on fuzzy logic, which, since its introduction by Zadeh (1965), has evolved from a prescribed rules-based artificial intelligence (AI) technique to an AI modeling approach capable of learning the rules from site-specific data, and this is now a successful modeling strategy that is applied widely, as outlined above. Fuzzy sets are distinguished from ordinary sets in terms of partial membership functions, which allows the technique to cope with vague data; hence the term fuzzy (see Nadiri et al., 2018a,b). Fuzzy set theory is now a mathematical tool used to handle uncertainties arising from the vagueness of some data sources.

One of the applications of fuzzy set theory is in modeling, which transforms input data as membership functions related to output data through the definition of fuzzy rules and this chapter introduces its application as an alternative to geostatistics in generating data on the water table of an aquifer. The classic basis of a fuzzy system comprises three basic components: (i) *fuzzification*: this defines the membership function to transform any given data into the degree of membership in a fuzzy set, where the value of the membership function ranges from 0 to 1 (*no membership* is denoted by 0; *full membership* by 1; *partial membership* by a value in between); (ii) *fuzzy rules or inference engine*: this operates on an "if" → "then" principle, where "if" corresponds to a vector of explanatory or input variables, and "then" corresponds to consequences

(Hamamin and Nadiri, 2018; Nadiri et al., 2018c); and (iii) *defuzzification*: this extracts the model prediction by filtering the input values through the constructed fuzzy rule (Coppola et al., 2002).

Inference engines in the early generation of FL applications were prescriptive but the emergence of learning approaches and data-clustering techniques created a bottom-up data-driven capability, in which rules are learned from the data and this chapter uses two such techniques: SFL introduced by Takagi and Sugeno (1985) and MFL introduced by Mamdani and Assilian (1975). SFL and MFL differ in the way they treat the output membership functions; where SFL either employs linear or constant functions, as in the subtractive clustering (SC) method (see Chiu, 1994), MFL employs the fuzzy c-means (FCM) clustering method (Bezdec, 1981; Lee, 2004; Newton et al., 1992). As argued by Nadiri et al. (2017a,b), the number of rules is a problem of diminishing returns, such that too many rules will increase computational costs without necessarily adding significantly to accuracy, and, conversely, too few rules will make a model run efficient but will very likely undermine accuracy.

7.2.2   Geostatistical Techniques

This chapter also uses OK and CK techniques and so the focus is on geostatistical tools generating data on groundwater levels (water table) through the use of interpolation techniques. Spatial interpolation techniques estimate the statistical structure inherent in the available measured data using statistical properties, although this is normally limited to the first two moments of the variable, i.e., the mean and the covariance or the semivariogram (ASCE, 1990). Geostatistical interpolation techniques (e.g., kriging) account for the spatial configuration of the sample points around the prediction location by using both the mathematical and statistical properties of the measured data and quantifying the spatial autocorrelation among the measured points.

Kriging is widely used in geology, hydrology, environmental monitoring, and other fields to interpolate spatial data and it is a geostatistical term for optimum linear prediction of spatial processes (Danielsson et al., 1998; Stein, 1999; Prakash and Singh, 2000; Gloaguen et al., 2001; Razack and Lasm, 2006; Idrysy and Smedt, 2007; Lamsal et al., 2009). The basic assumption in kriging is that the data are stationary within their stochastic process, although some methods additionally require that the data are normally distributed (Nas, 2009). The various versions of kriging available include: simple kriging, OK, UK, block kriging, CK, and DK. Of these, OK is in wide use, as a reliable estimation method (Yamamoto, 2000). The main tool in geostatistics is the semivariogram, which expresses the spatial dependence between neighboring observations. The semivariogram, $\gamma_{(h)}$, is defined as one-half the variance of the difference between the attribute values at all points separated by h as follows:

$$\gamma_{(h)} = \frac{1}{2N_{(h)}} \sum_{i=1}^{N(h)} [Z(x_i) - Z(x_i + h)]^2 \tag{7.1}$$

where $Z(x)$ indicates the magnitude of the variable and $N_{(h)}$ is the total number of pairs of attributes that are separated by a distance h. The kriging technique is an exact interpolation estimator used to find the best linear unbiased estimate, which must have minimum variance of the estimation error. Detailed discussions of kriging methods and their descriptions can be found in Goovaerts (1997). In kriging, the estimated value, Z^*, at any point X_0 is given as follows:

$$\gamma_{(h)} = \frac{1}{2N_{(h)}} \sum_{i=1}^{N(h)} [Z(x_i) - Z(x_i + h)]^2 \tag{7.2}$$

where λ_i, is the weight for the known value z at location x_i.

CK is another technique employed in the chapter but notably there are little differences between CK and kriging techniques. CK considers a secondary variable and makes use of the cross-correlation between main and secondary variables. The general equation of CK estimator is:

$$Z^*(X_i) = \sum_{i=1}^{n} \lambda_i . Z(X_i) . \sum_{i=1}^{n} \lambda_k . y(X_k) \tag{7.3}$$

where λ_i is the weight related to the Z variable, λ_k is the weight of the secondary variable, $Z^*(X_i)$ is estimated value for X_i, $Z(X_i)$ is the value of observed main variable, and $y(X_k)$ is the observed value of the secondary variable.

The accuracy of all of the models in this study was evaluated using R^2 and normalized root mean square error (NRMSE) between the measured and predicted values as follows:

$$R^2 = \frac{\left(\sum_{i=1}^{n} X_i - \overline{X}_i\right) \cdot \left(\sum_{i=1}^{n} \hat{X}_i - \overline{\hat{X}}_i\right)}{\sqrt{\sum_{i=1}^{n} X_i^2 \hat{X}_i^2}} \tag{7.4}$$

where n is the total number of data, X_i is the observed and $\hat{X}_i$ is the calculated groundwater level, and $\overline{X}_i$ and $\overline{\hat{X}}_i$ are the mean of the observed and the calculated data, respectively.

As an error measurement of the NRMSE is used, it is expressed as:

$$\text{NRMSE} = \sqrt{\frac{\sum_{i=1}^{n} \left(X_i - \hat{X}_i\right)^2}{\sum_{i=1}^{n} X_i^2}} \tag{7.5}$$

7.3 DATA AVAILABILITY

7.3.1 Study Area

The Meshginshahr Plain is approximately $704 \, \text{km}^2$ and is located in the Province of Ardabil, Northwest Iran (Fig. 7.1). The aquifer of this plain is unconfined. Mount Sabalan (widely known as Savalan), with a height of 4814 m AMSL, is the highest point in the region. Based on Emberger (1930), the prevailing climate in this area is semiarid-cold. The average annual

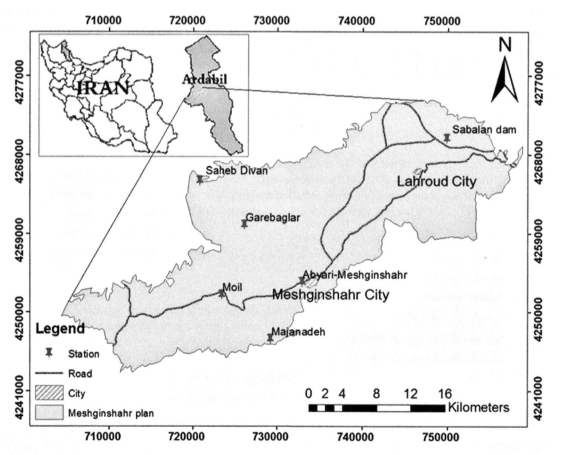

FIG. 7.1 Location of study area.

temperature is 11.66°C (according to the stations at the Meshginshahr Water Supply Office, Sabalan Dam, and Sahebdivan, for the years 2009–2012) and the average annual rainfall is 289 mm (based on measurements at the stations of the office of Meshginshahr Water Supply, Qarabaghlar, Majnade, Moil, on the plain, for the years 2001–12).

The mains cities around the plain are Meshginshahr (also known as Khiyov) and Lahrud (also known as Lari). Agriculture is the main economic activity on the plain and represents the main demands on the groundwater. This underlines the role of efficient management of these resources creating a focus on managing information on groundwater levels.

7.3.2   Data Analysis

The data from 23 observation wells on the Meshginshahr Plain were used for spatial prediction of groundwater levels in the study area. The data of 18 observation wells were used to calibrate the fuzzy models in the training phase, and the remaining five observation wells were used to test the fuzzy models in the testing phase; see Fig. 7.2 for the layout of these wells. For both fuzzy models the inputs comprised Universal Transverse Mercator X and Y coordinates of UTMx, UTMy; digital elevation model (DEM); and groundwater levels at the observation wells in September 2012. The output was the groundwater level for the present month at the observation wells.

Groundwater levels at the observation wells were used to also generate spatial distribution of groundwater levels by OK. In the CK technique the parameter that had the highest correlation coefficient (0.98) with the primary variable was selected as an auxiliary variable. In this study, DEM was selected as an auxiliary input variable (Fig. 7.3). Input variables in CK were groundwater levels and DEM at the coordinate point of each observation well. The same training and testing datasets were used for all methods for a fair comparison.

7.4   RESULTS

In this study four techniques (SFL, MFL, OK, and CK) were used to investigate the spatial analysis for predicting groundwater levels in the Meshginshahr Plain. The results are presented in this section.

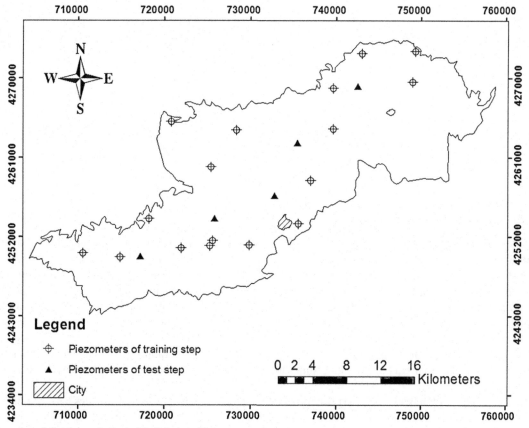

FIG. 7.2 Location of the observed sites in Meshginshahr Plain.

FIG. 7.3 Relation between groundwater level and digital elevation model (DEM) (used DEM as second variable in cokriging).

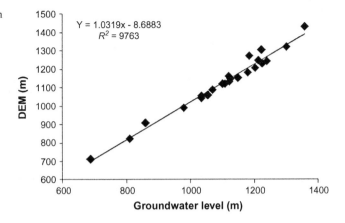

FIG. 7.4 Schematic diagram for SFL with four inputs (UTMx, UTMy, DEM, and GWL) and one output (groundwater levels).

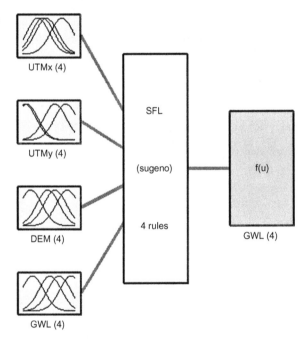

System SFL: 4 inputs, 1 outputs, 4 rules

7.4.1 Performances of Fuzzy Models

To investigate the performance of SFL and MFL, these techniques were implemented using the following steps: (i) Gaussian membership functions for inputs were used and data were classified by clustering methods—SC is used in the SFL model and FCM in the MFL model to develop fuzzy rules; (ii) data clustering methods minimized NRMSE to identify the optimum cluster radius and the number of rules; (iii) the defuzzification method used centroid calculation to produce crisp output; and (iv) the groundwater level in each observation well was identified using the least squares optimization technique for both models. Fig. 7.4 illustrates schematically the implementation of the SFL model from input data to the generation of groundwater levels following the four rules.

SFL Results: Fig. 7.5 shows the results for SFL, in which the optimum cluster radius is 0.6 m and it generates four fuzzy if-then rules and establishes four Gaussian membership functions for the SFL model. The scale of NRMSE in Fig. 7.5 is logarithmic for clarity.

MFL Results: For the MFL model, the FCM clustering method rendered six clusters and six rules based on the minimum average NRMSE and its value at the optimum cluster radius is found to be 0.022 m. Notably, SFL and MFL were carried out using the same data sets with the identical input and output variables.

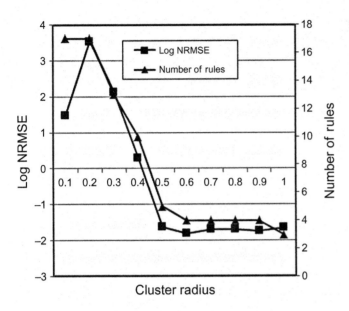

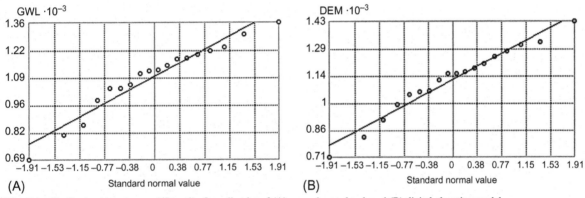

FIG. 7.6 Data distribution histogram and Quantile-Quantile-plot of (A) groundwater level; and (B) digital elevation model.

7.4.2 Performances of Geostatistical Techniques

OK and CK are used as baseline techniques for the prediction of spatial analysis of groundwater levels and therefore their results will serve as the basis for benchmarking. The suitability of the data is studied in terms of Quantile-Quantile (QQ)-plots comprising standardized normal value against the groundwater level or elevation values at the observation stations and these are displayed in Fig. 7.6. These results do not suggest any significant trend and the data are likely to be normally distributed. They are therefore suitable for the application of OK and CK.

The empirical semivariogram provides information on the spatial autocorrelation of datasets. However, it does not provide information for all possible directions and distances. For this reason and to ensure that kriging predictions have positive kriging variances, it is necessary to fit a model to the empirical semivariogram (Johnston et al., 2001). To identify the parameters with the least error, various techniques were tried and their predicted values in the semivariogram method were compared with the real values. Results in Table 7.1 show the stable type curve gives the least RMSE and hence it was selected for both OK and CK.

The subsequent semivariogram model of stable type is expressed as:

$$\gamma(h, \theta) = \theta_s \left[1 - \exp\left(-3 \left(\frac{\|h\|}{\theta_r} \right)^{\theta_e} \right) \right] \text{ for all } h, \tag{7.6}$$

where h is the distance of pairs of attributes $0 \leq \theta_e \leq 2$, $\theta_s \geq 0$ is the partial sill parameter, and $\theta_r \geq 0$ is the range parameter (Johnston et al., 2001). These semivariograms based on Eq. (7.6) are shown in Fig. 7.7.

TABLE 7.1 RMSE Values for Each of Semivariogram Techniques

| | RMSE | |
Models	Kriging	Co-kriging
Circular	62.04	78.92
Spherical	62.04	78.92
Tetraspherical	62.04	46.12
Pentaspherical	62.04	43.89
Exponential	62.04	42.45
Gaussian	62.04	78.92
Rational quadratic	62.04	44.66
Hole effect	62.04	52.16
K-bessel	62.04	32.85
J-bessel	62.04	50.93
Stable	59.24	32.38

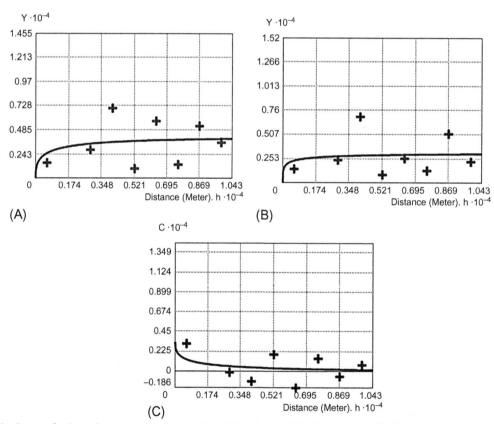

FIG. 7.7 Semivariograms for the study area—variogram γ on the y-axis) against spatial variations x-axis. (A) Variogram of groundwater levels (GWL) (ordinary kriging [OK] and co-kriging [CK]), (B) variogram of digital elevation model (DEM)—CK only, (C) cross variogram of GWL-DEM (CK).

Low nugget effect and consequently low ratio of nugget variance to sill can generally be used to classify the spatial dependence (Cambardella et al., 1994). A variable is considered to have strong spatial dependence if the ratio is less than 0.25, a moderate spatial dependence if the ratio is between 0.25 and 0.75, and if greater than 75%, the variable shows only weak spatial dependence (Liu et al., 2006). Based on Table 7.2, it is concluded that both groundwater level and DEM are highly spatially correlated.

TABLE 7.2 Parameters of Groundwater Level Semivariograms

Method	Nugget	Sill	Range	Nugget/Sill	R^2
Ordinary kriging	0	3176	10,132	0	.87
Co-kriging	0	3306	10,132	0	.96

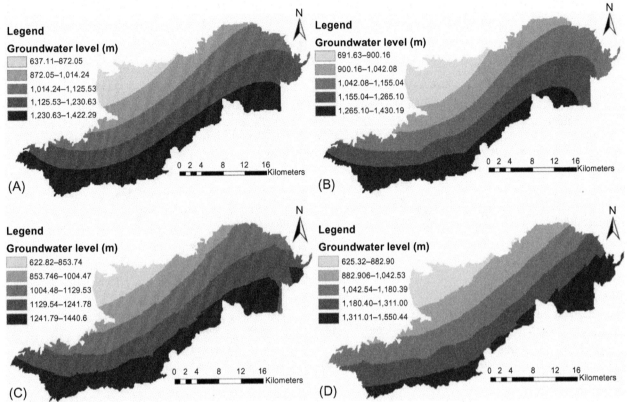

FIG. 7.8 Spatial distribution of groundwater level by (A) Sugeno fuzzy logic, (B) Mamdani fuzzy logic, (C) ordinary kriging, and (D) co-kriging.

7.5 DISCUSSION

Groundwater level maps were generated by SFL and MFL using a rule-base derived from the data and those by OK and CK using stable semivariogram, as presented by Eq. (7.6). The maps are displayed in Fig. 7.8A–D, respectively. Their performances are demonstrated by the scatter diagrams in Fig. 7.9 and the values of the performances of NRMSE and R^2 for both training and testing phases are presented in Table 7.3.

The performance of the four techniques studied in the chapter are investigated visually using scatter diagrams, as displayed in Fig. 7.9, and by comparing their performance measures, as presented in Table 7.3. Fig. 7.9 shows that the scatter in the fuzzy models are significantly reduced when these are compared with the performances of the geostatistical techniques.

The performance measures presented in Table 7.3 are used as the quantitative basis for the intercomparison of the four techniques under investigation (SFL, MFL, OK and CK). Table 7.3 shows that the performance of SFL is better than the other three techniques and these may be ranked as: SFL, MFL, CK, and OK. The improvements by SFL and MFL are significant, particularly as their implementations do not impose any restrictive assumptions. In comparing the performances of OK with CK, the results in Table 7.3 suggest that the use of the secondary variable by CK improves its performance in comparison with OK.

Attention is drawn to the role of AI techniques in that any new knowledge captured from the applications are bottom-up and data-driven and, as such, these techniques involve learning from local data often without little restrictive assumptions. Hence the findings reported in the chapter are true for the study area with respect to the tools used and some variations are expected if other modeling strategies are formulated. Also, the findings are driven by the measured data and do not apply to different sites. However, the methodology is generic and allow for broader applications.

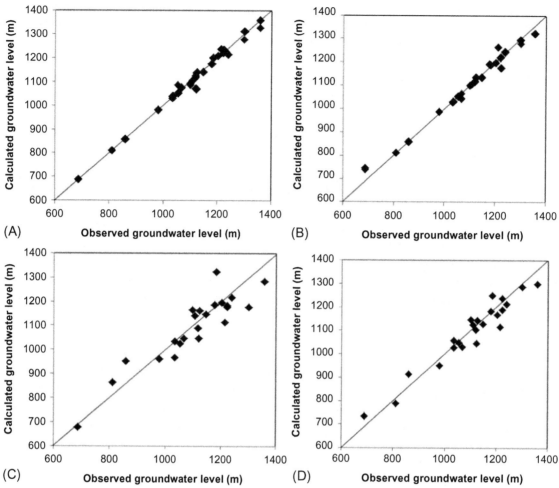

FIG. 7.9 Scatter diagram of the models investigated for generating groundwater level: (A) Sugeno fuzzy logic, (B) Mamdani fuzzy logic, (C) ordinary kriging, and (D) co-kriging.

TABLE 7.3 Comparison of Performance of Developed Models in Training and Testing Steps

Model	Training Phase		Testing Phase	
	NRMSE	R^2	NRMSE	R^2
SFL	0.008	.99	0.023	.90
MFL	0.015	.99	0.028	.80
OK	0.012	.98	0.056	.31
CK	0.022	.98	0.053	.42

7.6 CONCLUSION

The performance of the four modeling techniques were studied in this chapter to seek improvements in the spatial analysis of groundwater levels to predict future values using past groundwater levels at the observation stations. Data requirements are limited to the coordinates of the observation wells, a digital elevation map of the study area, and past observed groundwater levels at the observation wells. The investigation focused on the intercomparison of two state-of-the-art AI methods (SFL and MFL) with two geostatistical methods (OK and CK). The better performing model was shown to be SFL in terms of its performance metrics compared with MFL, CK, and OK. Notably, MFL demonstrated performances that were close to those of SFL, but CK performed significantly better than OK owing to its use of an auxiliary variable.

REFERENCES

ASCE Task Committee, 1990. Review of geostatistics in geohydrology. Part II: Applications. J. Hydraul. Eng. 116, 633–658.

Barbieri, P., Adami, G., Favretto, A., Lutman, A., Avoscan, W., Reisenhofer, E., 2001. Robust cluster analysis for detecting physico-chemical typologies of freshwater from wells of the plain of Friuli (northeastern Italy). Anal. Chim. Acta 440, 161–170.

Barca, E., Passarella, G., 2008. Spatial evaluation of the risk of groundwater quality degradation. A comparison between disjunctive kriging and geostatistical simulation. Environ. Monit. Assess. 137, 261–273.

Bezdec, J.C., 1981. Pattern Recognition with Fuzzy Objective Function Algorithms. Plenum Press, New York.

Bogardi, I., Bardossy, A., Duckstein, L., Pongracz, R., 2003. Fuzzy logic in hydrology and water resources. In: Demicco, R.V., Klir, G.J. (Eds.), Fuzzy Logic in Geology. Academic Press, New York.

Burrough, P.A., 1989. Fuzzy mathematical methods for soil survey and land evaluation. J. Soil Sci. 40, 477–492.

Cambardella, C.A., Moorman, T.B., Novak, J.M., Parkin, T.B., Karlen, D.L., Turco, R.F., Konopka, A.E., 1994. Field scale variability of soil properties in Central Iowa soils. Soil Sci. Soc. Am. J. 58, 1501–1511.

Cay, T., Uyan, M., 2009. Spatial and temporal groundwater level variation geostatistical modeling in the city of Konya, Turkey. Water Res. 81 (12), 2460–2470.

Chiu, S.L., 1994. Fuzzy model identification based on cluster estimation. J. Intell. Fuzzy Syst. 2 (3), 267–278.

Chowdhury, M., Alouani, A., Hossain, F., 2010. Comparison of ordinary kriging and artificial neural network for spatial mapping of arsenic contamination of groundwater. Stoch. Environ. Res. Risk Assess. 24, 1–7.

Coppola, E.A., Duckstein, L., Davis, D., 2002. Fuzzy rule-based methodology for estimating monthly groundwater recharge in a temperate watershed. J. Hydrol. Eng. (4), 326–335.

Danielsson, A., Carman, R., Rahm, L., Aigars, J., 1998. Spatial estimation of nutrient distributions in the Gulf of Riga sediments using Cokriging. ECSS 46, 713–722.

Emberger, L., 1930. La vegetation de la region mediterraneenne. Essai d'une classification des groupments vegetaux. Rev. Gen. Bot. 42, 641–662.

Eslami, H., Dastorani, J., Javadi, M., Chamheidar, H., 2013. Geostatistical evaluation of ground water quality distribution with GIS (case study: Mianab-Shoushtar plain). Bull. Env. Pharmacol. Life Sci. 3 (1), 78–82.

Fijani, E., Nadiri, A.A., Moghaddam, A.A., Tsai, F., Dixon, B., 2013. Optimization of DRASTIC method by supervised Committee machine artificial intelligence to assess groundwater vulnerability for Maragheh-Bonab plain aquifer, Iran. J. Hydrol. 530, 89–100.

Finke, P.A., Brus, D.J., Bierkens, M.F.P., Hoogland, T., Knotters, M., Vries, F., 2004. Mapping groundwater dynamics using multiple sources of exhaustive high-resolution data. Geoderma 123, 23–39.

Gemitzi, A., Petalas, C., Tsihrintzis, V.A., Pisinaras, V., 2006. Assessment of groundwater vulnerability to pollution: a combination of GIS, fuzzy logic and decision-making techniques. Environ. Geol. 49 (5), 653–673.

Gloaguen, E., Chouteau, M., Marcotte, D., Chapuis, R., 2001. Estimation of hydraulic conductivity of an unconfined aquifer using cokriging of GPR and hydrostratigraphic data. J. Appl. Geophys. 47, 135–152.

Goovaerts, P., 1997. Geostatistics for Natural Resourcesevaluation. Oxford University Press, New York.

Goovaerts, P., AvRuskin, G., Meliker, J., Slotnick, M., Jacquez, G., Nriagu, J., 2005. Geostatistical modeling of the spatial variability of arsenic in groundwater of southeast Michigan. Water Resour. Res. 41, 1–19.

Guler, C., Thyne, G.D., 2004. Delineation of hydrochemical facies distribution in a regional groundwater system by means of fuzzy c-means clustering. Water Resour. Res. 40, 1–11.

Guler, C., Thyne, G.D., McCray, J.E., Turner, A.K., 2002. Evaluation of graphical and multivariate statistical methods for classification of water chemistry data. Hydrogeol. J. 10, 455–474.

Gundogdu, K., Guney, I., 2007. Spatial analysis of groundwater levels using universal kriging. J. Earth Syst. Sci. 116, 49–55.

Hamad, S., 2009. Geostatistical analysis of groundwater levels in the south Al Jabal Al Akhdar area using GIS. In: GIS Ostrava Symposium. Ostrava, Czech Republic, 25–28 January, Ostrava.

Hamamin, D.F., Nadiri, A.A., 2018. Supervised committee fuzzy logic model to assess groundwater intrinsic vulnerability in multiple aquifer systems. Arab. J. Geosci. 11 (8), 176.

Idrysy, E.H.E., Smedt, F.D., 2007. A comparative study of hydraulic conductivity estimations using geostatistics. Hydrogeol. J. 15, 459–470.

Johnston, K., Hoef, J.M.V., Krivoruchko, K., Lucas, N., 2001. Using ArcGIS Geostatistical Analyst. GIS by ESRI, Redlands.

Kadkhodaie-Ilkhchi, A., Amini, A., 2009. A fuzzy logic approach to estimating hydraulic flow units from well log data: a case study from the Ahwaz oilfield, South Iran. J. Petrol. Geol. 32 (1), 67–78.

Kholghi, M., Hosseini, S.M., 2009. Comparison of groundwater level estimation using neuro-fuzzy and ordinary kriging. Environ. Model. Assess. 14, 729–737.

Kiurski-Milošević, J.Ž., Miloradov, M.B.V., Ralević, N.M., 2015. Fuzzy model for determination and assessment of groundwater quality in the city of Zrenjanin, Serbia. HEM IND 69 (1), 17–28.

Ko, K.-S., Lee, J.-S., Kim, J.-G., Lee, J., 2009. Assessments of natural and anthropogenic controls on the spatial distribution of stream water quality in southeastern Korea. Geosci. J. 13 (2), 191–200.

Kord, M., Moghaddam, A.A., 2014. Spatial analysis of Ardabil plain aquifer potable groundwater using fuzzy logic. J. King Saud Univ. Sci. 26, 129–140.

Kumar, N.V., Mathew, S., Swaminathan, G., 2010. Multifactorial fuzzy approach for the assessment of groundwater quality. J Water Resour. Prot. 2, 597–608.

Kurtulus, B., Flipo, N., Goblet, P., Vilain, G., Tournebize, J., Tallec, G., 2011. Hydraulic head interpolation in an aquifer unit using ANFIS and ordinary kriging. Comput. Intell. 343, 265–276.

Lamsal, S., Bliss, C.M., Graetz, D.A., 2009. Geospatial mapping of soil nitrate-nitrogen distribution under a mixed-land use system. Pedosphere 19 (4), 434–445.

Lee, K.H., 2004. First Course on Fuzzy Theory and Applications. SpringerVerlag, Berlin.

Liu, D., Wang, Z., Zhang, B., Song, K., Li, X., Li, J., Li, F., Duan, H., 2006. Spatial distribution of soil organic carbon and analysis of related factors in croplands of the black soil region, northeast China. Agric. Ecosyst. Environ. 113, 73–81.

Ma, T.S., Sophocleous, M., Yu, Y.S., 1999. Geostatistical applications in groundwater modeling in south-central Kansas. J. Hydrol. Eng. 4 (1), 57–64.

Mamdani, E.H., Assilian, S., 1975. An experiment in linguistic synthesis with a fuzzy logic controller. Int. J. Man Mach Stud. 7, 1–13.

Mehrjardi, R.T., Jahromi, M.Z., Mahmodi, S., Heidari, A., 2008. Spatial distribution of groundwater quality with geostatistics, case study: Yazd-Ardakan plain. World Appl. Sci. J. 4 (1), 9–17.

Metternicht, G., 2001. Assessing temporal and spatial changes of salinity using fuzzy logic, remote sensing and GIS. Foundations of an expert system. Ecol. Model. 144, 163–179.

Muhammetoglu, A., Yardimci, A., 2006. A fuzzy logic approach to assess groundwater pollution levels below agricultural fields. Environ. Monit. Assess. 118, 337–354.

Nadiri, A.A., Chitsazan, N., Tsai, F.T.C., Moghaddam, A.A., 2014. Bayesian artificial intelligence model averaging for hydraulic conductivity estimation. J. Hydrol. Eng. 19, 520–532.

Nadiri, A.A., Gharekhani, M., Khatibi, R., Sadeghfam, S., Asgari Moghaddam, A., 2017a. Groundwater vulnerability indices conditioned by supervised intelligence committee machine (SICM). Sci. Total Environ. 574, 691–706.

Nadiri, A.A., Gharekhani, M., Khatibi, R., Asgari Moghaddam, A., 2017b. Assessment of groundwater vulnerability using supervised committee to combine fuzzy logic models. Environ. Sci. Pollut. Res. 24 (9), 8562–8577.

Nadiri, A.A., Taheri, Z., Khatibi, R., Barzegari, G., Dideban, K., 2018a. Introducing a new framework for mapping subsidence vulnerability indices (SVIs): ALPRIFT. Sci. Total Environ. 628, 1043–1057.

Nadiri, A.A., Shokri, S., Tsai, F.T-C., Moghaddam, A.A., 2018b. Prediction of effluent quality parameters of a wastewater treatment plant using a supervised committee fuzzy logic model. J. Clean. Prod. 180, 539–549.

Nadiri, A.A., Asadi, S., Babaizadeh, H., Naderi, K., 2018c. Hybrid fuzzy model to predict strength and optimum compositions of natural Alumina-Silica-based geopolymers. Comput. Concrete. 21 (1), 103–110.

Nas, B., 2009. Geostatistical approach to assessment of spatial distribution of groundwater quality. Poblish J. Environ. Stud. 18 (6), 1073–1082.

Newton, S.C., Pemmaraju, S., Mitra, S., 1992. Adaptive fuzzy leader clustering of complex data sets in pattern recognition. IEEE Trans. Neural Netw. 3 (5), 794–800.

Ozbek, M.M., Pinder, G.F., 1998. A fuzzy logic approach to health risk based design of groundwater remediation. Comp. Met. Water Res. 12 (1), 115–122.

Pozdnyakova, L., Zhang, R., 1999. Geostatistical analyses of soil salinity in a large field. Precis. Agric. 1, 153–165.

Prakash, M.R., Singh, V.S., 2000. Network design for groundwater monitoring: a case study. Environ. Geol. 39 (6), 628–632.

Razack, M., Lasm, T., 2006. Geostatistical estimation of the transmissivity in a highly fractured metamorphic and crystalline aquifer (Man-Danane region, Western Ivory Coast). J. Hydrol. 325, 164–178.

Samson, M., Swaminathan, G., Kumar, N.V., 2010. Assessing groundwater quality for potability using a fuzzy logic and GIS – a case study of Tiruchirappalli city– India. Comput. Model. 14 (2), 58–68.

Sen, Z., 2009. Fuzzy groundwater classification rule derivation from quality maps. Water Qual. Expo. Health 1, 115–122.

Stein, M.L., 1999. Interpolation of Spatial Data: Some Theory for Kriging. Springer Verlag, New York.

Ta'any, R., Tahboub, A., Saffarini, G., 2009. Geostatistical analysis of spatiotemporal variability of groundwater level fluctuations in Amman-Zarqa basin, Jordan: a case study. Environ. Geol. 57 (3), 525–535.

Takagi, H., Sugeno, M., 1985. Fuzzy identification of systems and its application to modeling and control. IEEE SMC 5, 116–132.

Tutmez, B., Hatipoglu, Z., 2010. Comparing two data driven interpolation methods for modeling nitrate distribution in aquifer. Ecol. Inform. 5, 311–315.

Uricchio, V.F., Giordano, R., Lopez, N., 2004. A fuzzy knowledge-based decision support system for groundwater pollution risk evaluation. J. Environ. Manag. 73, 189–197.

Varouchakis, E.A., Hristopulos, D.T., 2013. Comparison of stochastic and deterministic methods for mapping groundwater level spatial variability in sparsely monitored basins. Environ. Monit. Assess. 185, 1–19.

Yamamoto, J.K., 2000. An alternative measure of the reliability of ordinary kriging estimates. Math. Geol. 32 (4), 489–509.

Zadeh, L.A., 1965. Fuzzy sets. Inf. Control. 8 (3), 338–353.

Zehtabiana, G.R., Asgarib, H.M., Tahmouresc, M., 2013. Assessment of spatial structure of groundwater quality variables based on the geostatistical simulation. Desert 17, 215–224.

FURTHER READING

Nadiri, A.A., Fijani, E., Tsai, F.T.C., Moghaddam, A.A., 2013. Supervised committee machine with artificial intelligence for prediction of fluoride concentration. J. Hydroinf. 15 (4), 1474–1490.

Chapter 8

Remote Sensing and Fuzzy Logic Approach for Artificial Recharge Studies in Hard Rock Terrain of South India

C.K. Muthumaniraja*, S. Anbazhagan[†], A. Jothibasu[†] and M. Chinnamuthu[†]

*Groundwater Division, Tamil Nadu Water Supply and Drainage Board (TWAD), Erode, India [†] Centre for Geoinformatics and Planetary Studies, Periyar University, Salem, India

Chapter Outline

8.1 INTRODUCTION

Groundwater is a vital natural resource as a major source of water for society, industries, and agricultural purposes. Groundwater resources can be improved by surface water resources in the form of winter and spring floods. Surface storage projects require huge investment and become filled with silt and unusable after a few years. However, water storage in aquifers protect and improve water quality. Therefore artificial recharge planning is an optimal solution to water crises in most areas. The purpose of artificial recharge is to refill the groundwater aquifer and to reuse that water, which ensures higher-quality water at a later date. Feeding water is provided using facilities designed for the purpose or sometimes through changing the natural conditions of the region (Bize et al., 1972). The replenishment of groundwater aquifers through artificial recharge has been carried out in various parts of the world for the last 6 decades; however, its importance was only realized in India about 4 decades ago (Karanth, 1963).

In recent years, remote sensing and geographical information system (GIS) techniques have been successfully applied in artificial recharge studies (Anbazhagan and Ramasamy, 1993). Suitable sites for artificial recharge were identified by analysis of geological, geomorphological, subsurface geological, and water-level fluctuation data through thematic as well as statistical modeling (Anbazhagan and Ramasamy, 1997, 2002). In the second phase, various methods of artificial recharge were identified based on different controlling terrain parameters. In the priority areas the following were recommended: desiltation of existing tanks; flooding and dendritic furrowing; and the creation of percolation ponds, en-echelon dams, injection wells, and subsurface dams (Ramasamy and Anbazhagan, 1994). Remote sensing provides spatially extensive, multitemporal, and cost-effective data that comprise a very useful tool in identifying hydrogeological processes (Meijerink, 1996; Rango and Shalaby, 1998; Hoffmann, 2005; Tweed et al., 2007). Indirect analysis of some observable geological structures, geomorphology, and their hydrologic characteristics using remote sensing enables the targeting of groundwater (Rai et al., 2005; Becker, 2006). GIS is a powerful tool for the integration and analysis of multithematic layers and in delineating potential areas of groundwater (Saraf and Choudhury, 1998; Chowdhury et al., 2009; Machiwal et al., 2011; Hammouri et al., 2012; Fashae et al., 2014; Anbazhagan and Jothibasu, 2016).

Dadrasi (2008) compared the fuzzy algorithm with other conceptual models compatible with GIS to locate the flood spread zone. The research was done in six cities of Khorasan-Razavi province, Iran. The results indicate that the fuzzy logic

GIS and Geostatistical Techniques for Groundwater Science. https://doi.org/10.1016/B978-0-12-815413-7.00008-0

sum operator had the highest efficiency. Balachandar et al. (2010) and Ravi Shankar and Mohan (2005) have used GIS for the identification of site-specific artificial-recharge techniques in different parts of India. Ghayoumian et al. (2007) applied fuzzy-logic GIS techniques to determine the most suitable areas for artificial groundwater recharge in a coastal aquifer in the Gavbandi Drainage Basin. Nouri (2003) adopted the same GIS platform fuzzy techniques to carry out their study. Fuzzy logic as part of the GIS technique has been widely used to estimate groundwater quality and for transport modeling of groundwater flow (Boughriba et al., 2010). In addition, the fuzzy logic method has been widely employed in groundwater-recharge mapping for site selections. However, the integration of remote sensing, GIS, and fuzzy logic has not been applied for the delineation of artificial groundwater recharge maps. The main objective of this study is to develop a GIS model based on fuzzy logic methods for locating suitable sites for artificial groundwater recharge in the Thalaivasal block, Tamil Nadu.

8.2 STUDY AREA

The study area of Thalaivasal block is located in the Salem district, in the southern part of India. Based on extensive groundwater development, the block is categorized as a "dark block" in the state of Tamil Nadu. This means that the groundwater development reaches >70% of available resources (CGWB, 2009). The block is bounded by Kalrayan hills in the north, Chinna Salem block in the east, Veppanthattai block in the south, and Gangavalli and Attur blocks (Salem district) in the west. It lies between 78°39′00″ to 78°50′40″ eastern longitudes and 11°23′00″ to 11°42′00″ northern latitudes. The total area of block is 400 km^2 and bifurcated into 41 panchayat villages (Fig. 8.1).

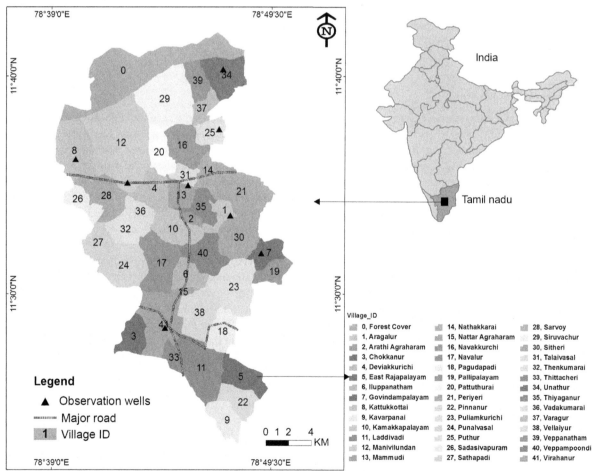

FIG. 8.1 Location of the study area shows the village boundary and location of observation well.

The Thalaivasal block is known for aggressive agricultural practices, such as tapioca and sugarcane cultivation. A semiarid climate prevails, with the annual average rainfall reaching 700 mm, which is much lower than the state average rainfall of 980 mm. The block is mostly covered by pediments and pediplain with extensive coverage of black soil. Presently, agricultural practices are mostly oriented based on the availability of groundwater resources. Gneisses are the major rock types covered in the northern and southern areas. In total, an area of approximately 300 km^2 (70%) is covered by gneissic rocks comprises of granitoid gneiss, granitic gneiss, biotite gneiss, and hornblende biotite gneiss. The charnokitic groups of rocks occupy about 100 km^2 of the central part of the block. The general trend of the geologic formations is a northeast to southwest direction. The gneissic rocks are mostly foliated and weathered when compared to charnockites. The joints and fracture-controlled lineaments are mostly permeable and support groundwater accumulation and exploration. At present, both dug wells and bore wells in the block provide the groundwater supplies for drinking and irrigation. Vasishta nadhi and Sweta nadhi are the two major rivers that flow into the area.

8.3 METHODOLOGY

In the present study, to delineate suitable sites for artificial recharge, the Survey of India (SOI) toposheet, IRS P6 LISS III, SRTM-DEM, and water level data have been utilized. Satellite data and ArcGIS 9.3 software have been used for the preparation of thematic layers. The detailed methodology adopted in the study is presented schematically in Fig. 8.2.

8.3.1 Fuzzy Logic and Membership Functions

Fuzzy logic is commonly used in spatial planning in order to allow the spatial objects on a map to be treated as members of a set. In a classic case, which is sometimes called "crisp," an object either belongs to a set or not. However, in fuzzy set theory a candidate object can take on membership values between 0 and 1, which reflects a degree of membership (Zadeh, 1965; Bonham-Carter, 1996). The benefit of fuzzy logic is that a new analysis, or a change in the rules or the criteria is not required, which saves time and effort. In fuzzy systems, values are indicated by a number (called a truth value) in the range from 0 to 1, where 0.0 represents absolute falseness and 1.0 represents absolute truth. While this range evokes the idea of probability, fuzzy logic and fuzzy sets operate quite differently from probability.

Using the integration of two or more parameters with fuzzy membership functions for the same set, the different operators can be employed to combine the membership values together. It was found that five operators were useful for combining exploration datasets, namely the fuzzy AND, fuzzy OR, fuzzy algebraic product, fuzzy algebraic sum, and fuzzy gamma operator (Tangestani, 2001). The fuzzy operators are as follows:

Fuzzy AND: This is equivalent to a Boolean AND (logical intersection) operation on classical set values:

$$\mu \text{combination} = \text{MIN}\,(\mu_A, \mu_B, \ldots\ldots\ldots\mu_N) \tag{8.1}$$

Fuzzy OR: This is equivalent to a Boolean OR (logical union) on classical set values:

$$\mu \text{combination} = \text{MAX}\,(\mu_A, \mu_B, \ldots\ldots\ldots\mu_N) \tag{8.2}$$

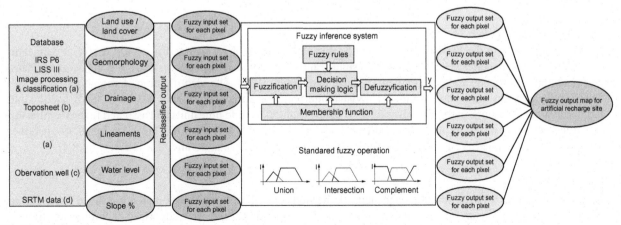

FIG. 8.2 Methodology adopted in the present study.

Fuzzy algebraic product: The combined membership function is defined as:

$$\mu\text{combination} = \prod_{i=1}^{n} \mu i \tag{8.3}$$

where μ_i is the fuzzy membership function for the Ith map, $i = 1, 2, 3 \ldots$, n maps are to be combined. The combined fuzzy membership values tend to be very small with this operator, due to the effect of multiplying several numbers <1. Nevertheless, all the contributing membership values have an effect on the result, unlike the fuzzy AND or fuzzy OR operators.

Fuzzy algebraic sum: This operator is complementary to the fuzzy product, being defined as:

$$\mu\text{combination} = 1 - \prod_{i=1}^{n} (1 - \mu i) \tag{8.4}$$

The result is always larger (or equal to) the largest contributing fuzzy membership value. The effect is therefore "increasive." The "increasive" effect of combining several favorable pieces of evidence is automatically limited by the maximum value of 1.0. Fuzzy algebraic product is an algebraic product, but fuzzy algebraic sum is not an algebraic summation.

Gamma operation: This is defined in terms of the fuzzy algebraic product and the fuzzy algebraic sum by the representation:

$$\mu\text{combination} = (\text{FAS})^{\gamma} * (\text{FAP})^{1-\gamma} \tag{8.5}$$

where γ is a parameter chosen in the range (0, 1). When γ is 1 the combination is the same as the fuzzy algebraic sum, and when γ is 0 the combination is equal to the fuzzy algebraic product. Judicious choice of the γ produces output values that ensure a flexible compromise between the "increasing" tendencies of the fuzzy algebraic sum and the "decreasing" effects of the fuzzy algebraic product where γ is a parameter chosen in the range (0, 1).

8.3.2 Fuzzy Membership Function and Ranking Assigned to Thematic Classes

The fuzzy membership values were assigned to the selected six thematic layers based on their classification on the respective of groundwater recharge conditions. The weights are assigned to the recharge conditioning factors from 0 to 1 (Delft, 2000).

8.3.2.1 Land Use/Cover

Land use/land cover (LULC) is an important factor in the selection of artificial recharge sites for groundwater recharge. Information about LULC is necessary to help quantify the water budget. It affects evapotranspiration, runoff, and recharge of the groundwater system. Eight LULC classes were interpreted from IRS P6 LISS III satellite data (Fig. 8.3). These classes were further classified into highly suitable, suitable, moderately suitable, least suitable, and unsuitable based on recharge conditions and then fuzzy membership values were assigned to barren/fallow land (0.58), barren rocky (0.31), built-up land (0.38), dry crop land (0.86), forest cover (0.45), lake (1.00), river (0.98) and wet crop land (0.91) (Table 8.1).

8.3.2.2 Geomorphology

Geomorphology deals with the various landforms and structural features of an area and it is very useful in delineating groundwater recharge categories. Mapping activities significantly contribute in deciphering areas of groundwater recharge and their potential for groundwater development (Singhal and Gupta, 1999). The identification of geomorphological features in the study area was done using an on-screen visual interpretation approach utilizing IRS P6 LISS III satellite data. The geomorphological features in the Thalaivasal block can be categorized into nine units and assigned fuzzy membership, such as buried pediment moderate (0.96), buried pediment shallow (0.95), duri crust (0.31), pediment (0.96), pediment (block cotton soil) (0.89), reserve forest (0.33), settlement (0.32), tank (0.98), and valley fill (1.00) (Fig. 8.4).

8.3.2.3 Drainage

SOI toposheets were utilized to digitize drainage maps for the study area. Thalaivasal block region is drained to Swetha and Vasista nathi with many stream orders from the source to the mouth and these are categorized according to Strahler's method. The Strahler's (1952) system is: the order of the stream is 1 if a stream has no contributing tributaries. Two streams

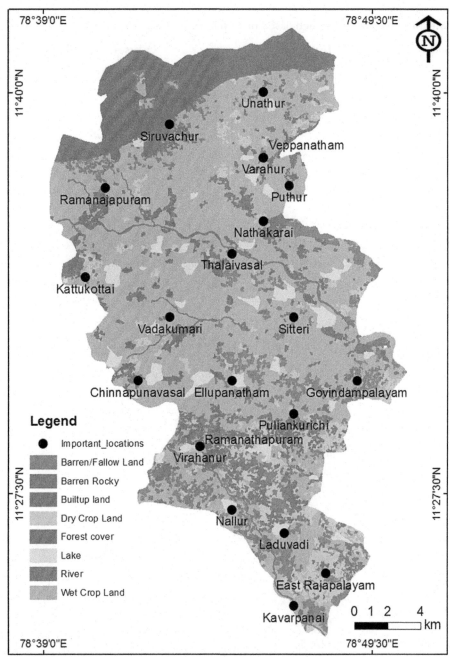

FIG. 8.3 Land use/land cover interpreted from IRS P6 LISS III satellite data.

with same order i unite to give a stream the order $i+1$ and if streams of different orders unite, the new stream retains the order of the highest order stream. Based on the Strahler's stream order system, Thalaivasal block has 5th order streams and these were assigned to fuzzy membership as I order (0.52), II order (0.96), III order (0.98), IV order (1.00), and V order (1.00) (Fig. 8.5).

8.3.2.4 Lineament Density

Lineament is another important parameter considered for artificial recharge structures. A lineament is a linear feature in a landscape that is an expression of an underlying geological structure such as a fault, fracture, or joint. Lineaments are generally referred to in the analysis of remote sensing of fractures or structures. The lineaments are interpreted using IRS P6

TABLE 8.1 Fuzzy Membership Values Assigned Based on Importance of Artificial Recharge

Sl. No.	Themes	Features	Classes	Fuzzy Membership
1	Land use/Land cover	Forest cover	Least suitable	0.45
		Wet crop land	Highly suitable	0.91
		Barren/Fallow land	Moderately suitable	0.58
		Dry crop land	Highly suitable	0.86
		Barren rocky	Unsuitable	0.31
		Builtup land	Unsuitable	0.38
		Lake	Highly suitable	1
		River	Highly suitable	0.98
2	Geomorphology	Burried pediment moderate	Highly suitable	0.96
		Burried pediment shallow	Highly suitable	0.95
		Duri crust	Unsuitable	0.31
		Pediment	Highly suitable	0.96
		Tank	Highly suitable	0.98
		Pediment (Black cotton soil)	Highly suitable	0.89
		Valley fill	Highly suitable	1
		Settlement	Unsuitable	0.32
		Reserve Forest	Unsuitable	0.33
3	Drainage	I Stream order	Moderately suitable	0.52
		II Stream order	Highly suitable	0.96
		III Stream order	Highly suitable	0.98
		IV Stream order	Highly suitable	1
		V Stream order	Highly suitable	1
4	Lineament density	<0.06 km/sq.km (Low)	Least suitable	0.45
		0.06–0.13 km/sq.km (Low)	Moderately suitable	0.58
		0.13–0.20 km/sq.km (Moderate)	Suitable	0.72
		0.20–0.27 km/sq.km (High)	Highly suitable	0.85
		0.27–0.34 km/sq.km (High)	Highly suitable	1
5	Water level (m, bgl)	1.00–5.12	Unsuitable	0.35
		5.12–7.51	Suitable	0.65
		7.51–10.64	Suitable	0.82
		10.64–16.19	Highly suitable	0.95
		16.19–22.29	Highly suitable	1
6	Slope %	Leveled (<1%)	Highly suitable	1
		Gently (1%–5%)	Highly suitable	0.98
		Moderate (5%–25%)	Highly suitable	0.95
		Steep (25%–75%)	Least suitable	0.45
		Very Steep (>75%)	Unsuitable	0.28

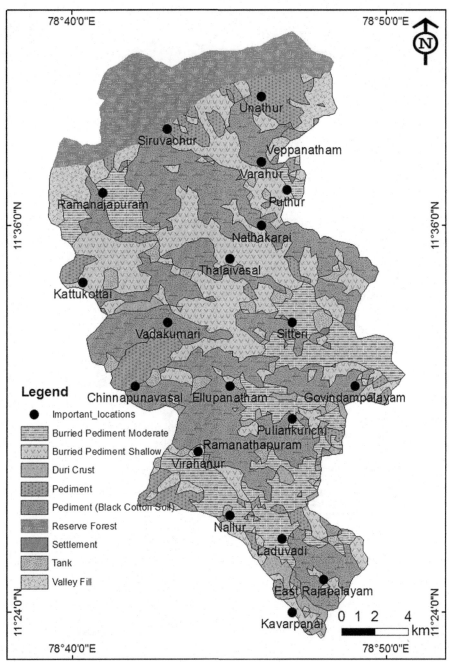

FIG. 8.4 Geomorphological features of the study area.

LISS III satellite data in the Thalaivasal block. In the present study, lineament-length density (L_d) was used, which represents the total length of lineaments in a unit area:

$$L_d = \frac{\sum_{i=1}^{i=n} L_i}{A}$$

(8.6)

where $\sum_{i=1}^{i=n} L_i$ represents the total length of lineaments (L) and A is the unit area (L^2). A high L_d value infers high secondary porosity, thus indicating a zone with high levels of groundwater recharge. The lineament density classified into very high, high, moderate, low and very low. These classes are assigned to fuzzy membership: <0.06 km/ km^2 (0.45), 0.06–0.13 km/ km^2 (0.58), 0.13–0.20 km/km^2 (0.72), 0.20–0.27 km/km^2 (0.85), and 0.27–0.34 km/km^2 (1.00) (Fig. 8.6).

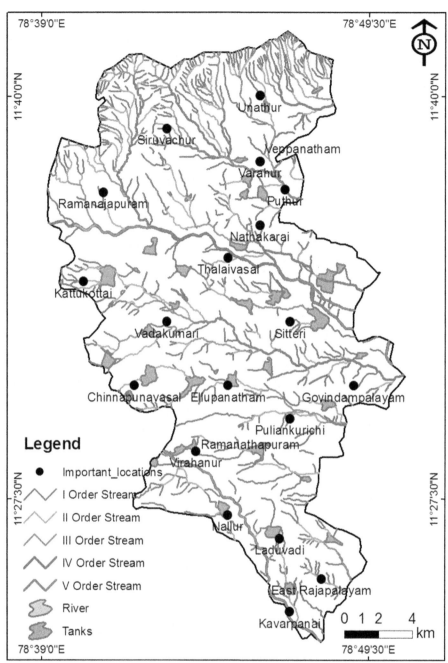

FIG. 8.5 Drainage pattern in the study area.

8.3.2.5 Water Level

Water level is an important indicator for groundwater recharge. In the post-monsoon period of 2012 water-level data were collected from observation wells (Fig. 8.1) and was utilized in and the preparation of a water level map using inverse distance weighting interpolation (Fig. 8.7). In the post-monsoon season the water level generally varies from 1 to 22.29 m below ground level (bgl). The depth to the water level is divided into five zones then assigned to fuzzy logic such as 1.00–5.12 (0.35), 5.12–7.51 (0.65), 7.51–10.64 (0.82), 10.64–16.19 (0.95), and 16.19–22.29 (1.00). The deeper water level condition occurs in the northwestern and southern part of the study area. The central part of the image shows moderate water levels and gradually increases in the direction of northern west and southern corners, and reaches up to 22 m.

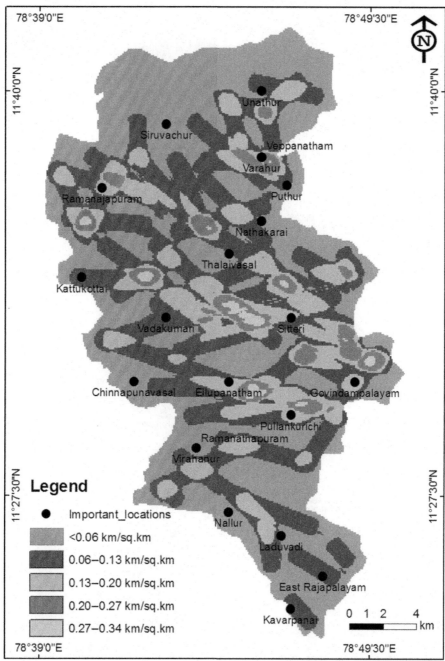

FIG. 8.6 Lineament densities in Thalaivasal block.

8.3.2.6 Slope

In many studies related to groundwater flow and storage, the slope is often ignored; especially in areas with less mountainous terrain (Al Saud, 2010). Rainfall is the main source of groundwater recharge in both tropic and subtropic regions. The slope gradient directly influences the infiltration of rainfall. Larger slopes produce a smaller recharge because water flows rapidly down a steep slope during rainfall, so it does not have sufficient time to infiltrate the surface and recharge the saturated zone. The slope analysis function in the GIS is used to assess the slope variation in the study area using data obtained from the digital terrain model (DTM) from the SRTM database for Thalaivasal block:

$$SL = 100\, x\, \frac{\sqrt{DX^2 + DY^2}}{\text{Pixel Size}} \tag{8.7}$$

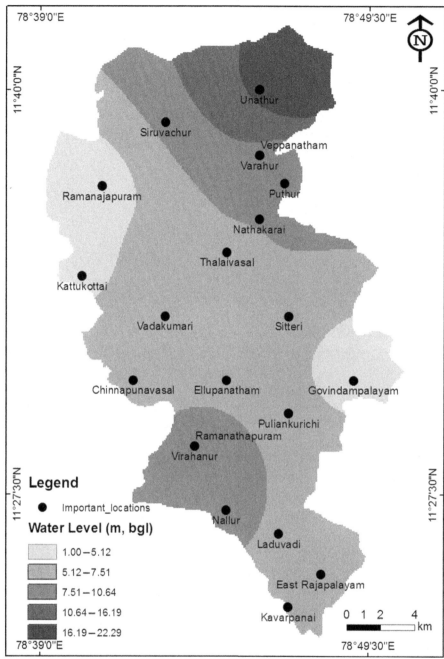

FIG. 8.7 Spatial distribution of water level in the study area.

where DX and DY are the filtered digital elevation model (DEM) values with the horizontal and vertical gradient filters, and pixel size is the height value of the DEM. Based on the slope percentage the study area is level to gently sloping. The Thalavasal block was classified into five slope percentage categories and assigned to fuzzy membership such as leveled (<1%) (1.00), gently (1%–5%) (0.98), moderate (5%–25%) (0.95), steep (25%–75%) (0.45), and very steep (>75%) (0.28) zones (Fig. 8.8).

8.4 RESULTS AND DISCUSSION

The six groundwater-recharge conditioning factors were imported into a GIS environment and classes of each factor were reclassified and standardized and an add or combine step was performed. In the add or combine step fuzzy logic explores the interaction of the possibility of the phenomenon belonging to multiple sets, as opposed to weighted overlay and weighted

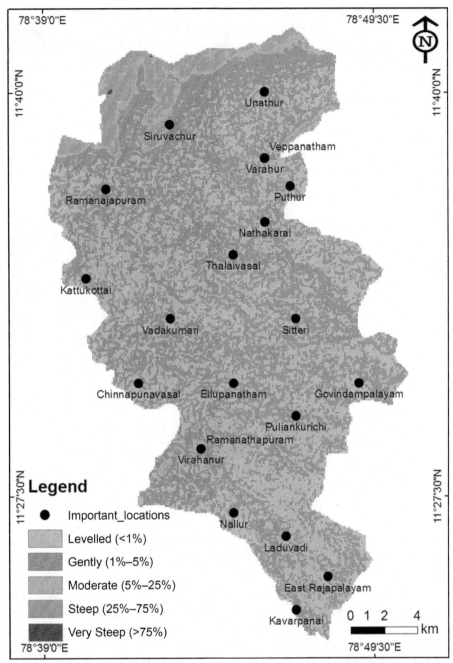

FIG. 8.8 Slope percentage generated from SRTM data.

sum (Das et al., 1997). An et al. (1994) argue that the five operators are useful for combining exploration datasets, namely the fuzzy AND, fuzzy OR, fuzzy algebraic product, fuzzy algebraic sum, and fuzzy gamma operator. We performed a multiplicity operator to combine fuzzy membership values using six groundwater-recharge factor maps. More specifically, we used a fuzzy algebraic product and fuzzy algebraic sum tools to map groundwater recharge (*r*) based on the following equation and it shows the upper and lower limits of the weights assigned to groundwater recharge and provides a broad idea of the artificial recharge of the study area (Fig. 8.9):

$$\text{Product} = r1*r2*r3*r4*r5*r6 \tag{8.8}$$

$$\text{WGR} = \text{sum} = 1 - ((1-r1)*(1-r2)*(1-r3)*(1-r4)*(1-r5)*(1-r6)) \tag{8.9}$$

where *r*1, *r*2 … *r*6 are fuzzy membership values for each conditioning factor class.

FIG. 8.9 Fuzzy membership function of the study area.

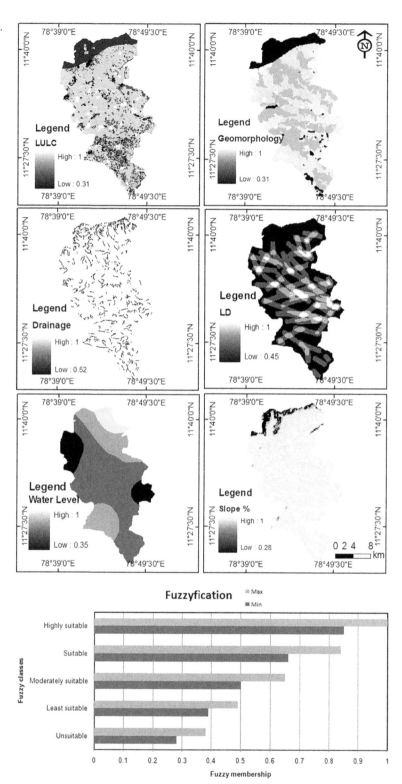

The groundwater recharge zones has been classified into highly suitable, suitable, moderately suitable, least suitable, and unsuitable. The final artificial groundwater-recharge map is shown in Fig. 8.10. The highly suitable category ranges from 0.84 to 1.00 and represents 7.72% of the total area. The range of this zone is also >0.85, again with the same concept of membership. The weight value for the suitable category is 0.65 to 0.84 and this category covers 18.62% of the total area.

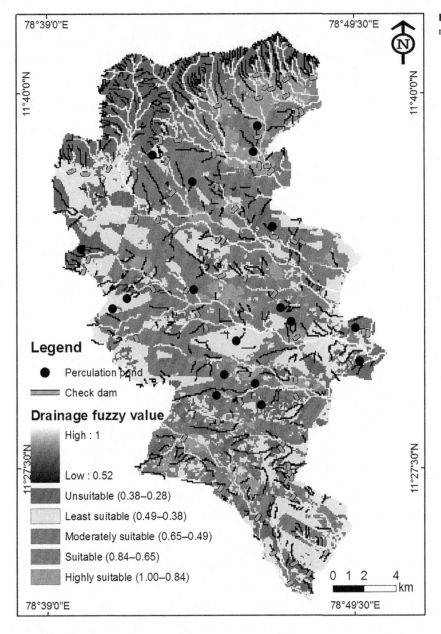

FIG. 8.10 Fuzzy sites suitability for artificial recharge in Thalaivasal block.

The suitable category is just above the gamma value taken for the overlay analysis. The moderately suitable category has a fuzzy number ranging between 0.49 and 0.65 and it represents 23.87% of the total area. This zone's category has a fuzzy number in the range of the gamma value so it is considered as moderate. It can also be inferred from Table 8.1 that the fuzzy number for the least suitable category varies from 0.38 to 0.49 and this category represents 25.87% of the total area. This particular range of the recharge zone has a fuzzy membership value only slightly lower than the gamma value. The last category for the groundwater recharge zone is unsuitable and this has an extremely low fuzzy number, i.e., between 0.28 and 0.38 and this category covers 23.92% of the total area. Based on the fuzzy logic artificial recharge zones and the selected six themes, 41 panchayat villages in the Thalaivasal block were recommended to manage the aquifer for a sustainable environment (Table 8.2A–D). Based on the Table 8.2A–D the village wise recommendation were evaluated for artificial recharge structures.

TABLE 8.2 Village Wise Data Base for Aquifer Management

Village ID	Name of Village	Land Use/Land Cover	Geomorphology	Drainage	Lineament/Fracture	Water Level (m)	Recharge Condition (m)	Artificial Recharge Suitability	Aquifer Management
(A)									
1	Aragalur	Barren Fallow land	Pediment (Black Cotton Soil)	Stream Order-II, Tank-Nil	Low	6	6–7	Least	Percolation pond & Desiltation of tanks
		Wet Crop land	Buried Pediment Shallow		High	9	10–12	High	
2	Arathi Agraharam	Barren Fallow land	Pediment (Black Cotton Soil)	Stream order-Nil, Big Tank-I (West)	Low	6	6–8	Least	Desiltation of existing tank
		Lake	Pediment		Low	1	1–2	Moderate	
		Wet Crop land	Buried Pediment Shallow		High	7	9–12	High	
3	Chokkanur	Lake	Pediment	Stream Order-I,II, Medium Tank-I (Northwest)	Low	1	0.5–1	Moderate Moderate	Check dams, desiltation of tanks
		River	Valley Fill		High	8	10–16	High	
4	Deviyakurichi	Barren Fallow land	Pediment (Black Cotton Soil)	Stream Order-I, II Md Tank-I,	Low	6	7–9	Least	Check dams, Desilting of existing tank.
		Lake	Pediment		Low	1	1–2.5	Moderate	
		Wet Crop land	Buried Pediment Shallow		High	7	10–13	High	
5	East Rajapalyam	Barren Fallow land	Pediment (Black Cotton Soil)	Stream Order-I,II, III, Medium Tank-I	Low	5	5–8	Least	check dams.
		Lake	Pediment		Low	1	1–2	Moderate	
		Wet Crop land	Buried Pediment Shallow		High	6	10–13	High	
6	Iluppanatham	Barren Fallow land	Pediment (Black Cotton Soil)	Stream Order-I,II Medium Tank-I	Low	5	6–7	Least	Percolation pond & desilting of tanks
		Dry crop land	Buried Pediment Moderate		High	22	10–14	Moderate	
7	Govindam-palyam	Barren Fallow land	Pediment (Black Cotton Soil)	Stream Order-I,II Small Tank-I	Low	5	6–7	Least	Check dams
		Lake	Pediment		Low	1	1.5–2	Moderate	
		Dry crop land	Buried Pediment Moderate	Check Dam	Moderate	4	10–16	Moderate	

No.	Village	Land use	Geomorphology	Drainage		No.	Range		Recommendation
8	Kattukottai	Barren Fallow land	Pediment (Black Cotton Soil)	Stream Order-I to IV, Tank-Nil	Low	5	5–7	Least	Check dams & Desilting tanks
		Barren rocky	Structural hill		–	–	–	–	
		River	Valley Fill,		High	8	10–16	High	
		Dry crop land	Buried Pediment Moderate			10	10–17	Moderate	
9	Kavarapanai	Barren Fallow land	Pediment (Black Cotton Soil)	Stream Order-I,II Medium Tank-I	Low	6	6–8	Least	Check dams & Desilting tanks
		Lake	Pediment		Low	1	1–2	Moderate	
		Wet Crop land	Buried Pediment Shallow		High	9	10–12	High	
		Barren rocky	Duri crust		Low	7	1–4	Least	
10	Kamakkapalayam	Barren Fallow land	Pediment (Black Cotton Soil)	Stream Order-I,II,III, Small Tank-I	Low	5	5–7	Least	Percolation ponds & desilting tanks
		Wet Crop land	Buried Pediment Shallow		High	8	6–9	High	
(B)									
11	Laduvadi	Barren Fallow land	Pediment (Black Cotton Soil)	Stream Order-I to IV, Medium Tank-I	Low	5	6–7	Least	Percolation ponds & desilting tanks
		Dry crop land	Buried Pediment Moderate		Moderate	7	10–16	Moderate	
		Barren rocky	Duri crust		Low	6	GL-5	Least	
12	Manivilundan	Barren Fallow land	Pediment (Black Cotton Soil)	Stream Order-I to IV, A big Reservoir located at the foot hill Valley	Low	4	6–7	Least	Series of check dams
		Barren rocky	Structural hill		–	–	–	–	
		River	Valley Fill,		High	15	9–18	High	
		Dry crop land	Buried Pediment Moderate			8	10–16	Moderate	
		Wet Crop land	Buried Pediment Shallow		High	7	10–12	High	
13	Mummudi	Barren Fallow land	Pediment (Black Cotton Soil)	Stream Order-I,II, Tank-Nil	Low	5	7–8	Least	checkdams across Vasishtanadhi
		Lake	Pediment		Low	1	1.5–2	Moderate	
		Dry crop land	Buried Pediment Moderate		Moderate	9	10–16	Moderate	

Continued

TABLE 8.2 Village Wise Data Base for Aquifer Management—cont'd

Village ID	Name of Village	Land Use/Land Cover	Geomorphology	Drainage	Lineament/Fracture	Water Level (m)	Recharge Condition (m)	Artificial Recharge Suitability	Aquifer Management
14	Nathakarai	Barren Fallow land	Pediment (Black Cotton Soil)	Stream Order-I,II, Tank-Nil	Low	5	4–7	Least	percolation pond
		Lake	Pediment		Low	1	1–2		
		Wet Crop land	Buried Pediment Shallow		High	6	8–9	High	
15	Nattur Agraharam	Barren Fallow land	Pediment (Black Cotton Soil)	Stream Order-I,II, Tank-Nil	Low	5	5–7	Least	checkdams
		Lake	Pediment		Low	1	0.5–1	Moderate	
		Dry crop land	Buried Pediment Moderate			9	10–16	Moderate	
		River	Valley Fill		High	5	9–12	High	
16	Navakurichi	Barren Fallow land	Pediment (Black Cotton Soil)	Stream Order-I,II Medium Tank-2	Low	6	6–8	Least	checkdams Desiltation of existing tanks
		Wet Crop land	Buried Pediment Shallow		High	7	8–9	High	
17	Navalur	Barren Fallow land	Pediment (Black Cotton Soil)	Stream Order-I,II Small tank-1	Low	1	1.5–2	Moderate	checkdams Desiltation of existing tanks
		Barren Fallow land	Pediment (Black Cotton Soil)		Low	6	4–7	Least	
		Lake	Pediment		Low	2	1–3	Moderate	Moderate
18	Pagadapadi	Barren Fallow land	Pediment (Black Cotton Soil)	Stream Order-I,II Small Tank-1	Low	6	5–7	Least	checkdams Desiltation of existing tanks
		Dry crop land	Buried Pediment Moderate		Moderate	8	10–15	Moderate	
19	Pallipalayam	Barren Fallow land	Pediment (Black Cotton Soil)	Stream Order-I,II, III Tank Nil	Low	5	4–8	Least	percolation pond
		Lake	Pediment		Low	1	1–2	Moderate	
		Wet Crop land	Buried Pediment Shallow,		High	6	9–10	High	
		Dry crop land	Buried Pediment Moderate		Moderate	8	11–14	Moderate	
20	Pattuthurai	Barren Fallow land	Pediment (Black Cotton Soil)	Stream Order-III, IV Small Tank-I	Low	6	3–8	Least	percolation pond
		Lake	Pediment		Low	1	1–2	Moderate	
		Wet Crop land	Buried Pediment Shallow		High	7	10–12	High	

(C)

No.	Village	Land use	Geomorphology	Drainage					Remarks
21	Periaeri	Barren Fallow land	Pediment (Black Cotton Soil)	Stream Order-I,II, III, Big Tank-I	Low	6	4–8	Least	Desiltation of existing tank check dams
		Lake	Pediment		Low	1	1–2.5	Moderate	
		Wet crop land	Buried Pediment Shallow		High	7	10–12	High	
22	Pinnanur	Barren Fallow land	Pediment (Black Cotton Soil)	Stream Order-I,II Small Reservoir check dam	Low	6	3–7	Least	check dam
		Lake	Pediment		Low	1	1.5–2	Moderate	
		Wet Crop land	Buried Pediment Shallow		High	7	9–11	High	
23	Puliankurichi	Barren Fallow land	Pediment (Black Cotton Soil)	Stream Order-I,II, III Medium Tank-I	Low	6	6–8	Least	Check dam
		Lake	Pediment		Low	1	1–2.5		
		Wet Crop land	Buried Pediment Shallow,		High	7	10–12	High	
		Dry crop land	Buried Pediment Moderate		Moderate	8	12–15	Moderate	
24	Punalvasal	Barren Fallow land	Pediment (Black Cotton Soil)	Stream Order-I,II Big Tank-I Small Tank-I	Low	5	4–7	Least	check dam
		Wet Crop land	Buried Pediment Shallow		High	8	9–12	High	
25	Puthur	Barren Fallow land	Pediment (Black Cotton Soil)	Stream Order-III Medium Tank-I	Low	6	6–7	Least	check dams
		Wet Crop land	Buried Pediment Shallow		High	7	10–12	High	
		Dry crop land	Buried Pediment Moderate		Moderate	8	12–15	Moderate	
26	Sadhasiva puram	Barren Fallow land	Pediment (Black Cotton Soil)	Stream Order-I,II Small Tank-I check dam	Low	6	7–8	Least	part of the village
		Lake	Pediment		Low	2	2–2.5	Moderate	
		Wet Crop land	Buried Pediment Shallow		High	7	10–14	High	

Continued

TABLE 8.2 Village Wise Data Base for Aquifer Management—cont'd

Village ID	Name of Village	Land Use/Land Cover	Geomorphology	Drainage	Lineament/Fracture	Water Level (m)	Recharge Condition (m)	Artificial Recharge Suitability	Aquifer Management
27	Sathapadi	Barren Fallow land	Pediment (Black Cotton Soil)	Stream Order-I,II Big Tank-I	Low	6	2–8	Least	Desilting of existing tank.
		Lake	Pediment		Low	1	1.5–2	Moderate	
28	Sarvoy	Barren Fallow land	Pediment (Black Cotton Soil)	Stream Order-I,II, III Big Tank-I, Small Tank-I	Low	5	3–7	Least	Desilting of existing tank
		Wet crop land	Buried Pediment Shallow		High	9	9–14	High	
29	Siruvachur	Barren Fallow land	Pediment (Black Cotton Soil)	Stream Order-I to IV, Small Tank-I	Low	5	3–5	Least	Check dams
		Barren rocky	Structural hill		–	–	–	–	
		Lake	Pediment	Cascade of check dam	Low	1	1–2	Moderate	
30	Sitheri	Barren Fallow land	Pediment (Black Cotton Soil)	Stream Order-I,II, III Big Tank-I	Low	6	4–8	Least	Checkdams
		Lake	Pediment		Low	2	2–2.5	Moderate	
		Dry crop land	Buried Pediment Moderate		Moderate	9	14–16	Moderate	
(D)									
31	Thalaivasal	Barren Fallow land	Pediment (Black Cotton Soil)	Stream Order-III Medium Tank-I	Low	6	4–7	Least	Check dam across Vasishta Nadhi
		Wet Crop land	Buried Pediment Shallow		High	8	10–13	High	
32	Thenkumari	Barren Fallow land	Pediment (Black Cotton Soil)	Stream Order-I,II Medium Tank-I	Low	6	3–8	Least	Desiltation of existing tanks
		Lake	Pediment		Low	1	1–2	Moderate	
33	Thittacheri	Wet Crop land	Buried Pediment Shallow	Stream Order-I,II Medium Tank-I	High	7	9–12	High	Desilting of existing tank
		Dry crop land	Buried Pediment Moderate		Moderate	8	10–16	Moderate	

No.	Village	Land use	Geomorphology	Drainage					Recommendation
34	Unathur	Barren rocky	Structural hill	Stream Order-I to IV Small Tank-I	–	–	–	–	seven Check dams
		Lake	Pediment		Low	1	1.5–2	High	
		River	Valley Fill		High	8	10–18	Least	
35	Thiyaganur	Barren Fallow land	Pediment (Black Cotton Soil)	Stream Order-I,II Medium Tank-I	Low	6	4–7	Least	check dam
		Wet Crop land	Buried Pediment Shallow		High	10	9–12	High	
36	Vadakumari	Barren Fallow land	Pediment (Black Cotton Soil)	Stream Order-I,II Small Tank-I	Low	7	3–8	Least	check dam
		Lake	Pediment		Low	1	1.5–2	Moderate	
		Barren rocky	Duricrust		Low	4	1–3	Least	
37	Varagur	Barren Fallow land	Pediment (Black Cotton Soil)	Stream Order-I &III, Tank-Nil	Low	6	3–7	Least	check dam & percolation pond
		Lake	Pediment		Low	2	2–2.5	Moderate	
		Wet Crop land	Buried Pediment Shallow		High	7	10–12	High	
38	Vellaiyur	Barren Fallow land	Pediment (Black Cotton Soil)	Stream Order-I,II, III Big Tank-I	Low	5	2–7	Least	percolation pond
		Lake	Pediment		Low	1	1–2	Moderate	
		Wet Crop land	Buried Pediment Shallow		High	12	10–14	High	
		Barren rocky	Duri crust		Low	8	GL-5	Least	
39	Veppanatham	Barren Fallow land	Pediment (Black Cotton Soil)	Stream Order-I,II Small Tank-I	Low	5	2–8	Least	check dams & one percolation pond
		Barren rocky	Structural hill		–	–	–	–	
		Lake	Pediment		Low	1	1–2.5	High	
		Wet Crop land	Buried Pediment Shallow		High	7	10–13	High	
40	Veppampondi	Barren Fallow land	Pediment (Black Cotton Soil)	Stream Order-I,II Medium Tank-I	Low	4	3–7	Least	percolation pond
		Wet Crop land	Buried Pediment Shallow		High	6	GL-5	Moderate	
		Dry crop land	Buried Pediment Moderate		Moderate	15	14–16	Moderate	

Continued

TABLE 8.2 Village Wise Data Base for Aquifer Management—cont'd

Village ID	Name of Village	Land Use/ Land Cover	Geomorphology	Drainage	Lineament/ Fracture	Water Level (m)	Recharge Condition (m)	Artificial Recharge Suitability	Aquifer Management
41	Virahanur	Barren Fallow land	Pediment (Black Cotton Soil)	Stream Order-I,II Big Tank-I	Low	5	4–8	Least	Check dams
		Lake	Pediment		Low	1	1–2	Moderate	
		Dry crop land	Buried Pediment Moderate,		Moderate	15	12–16	Moderate	
		River	Valley Fill		High	14	12–22	High	

8.5 CONCLUSION

The present study is important for the sustainable use of the groundwater resource through enhancing groundwater recharge by proper management. The results revealed that the application of remote sensing and GIS techniques helped groundwater exploration by narrowing down the target areas for conducting detailed hydrogeological surveys on the ground. The results indicate that the most effective groundwater recharge zone is located in the Thalaivasal block. In this region the gravelly stratum and the concentration of drainage also helps the stream flow to recharge the groundwater system.

ACKNOWLEDGMENT

The authors thank State Groundwater Department, Tamil Nadu for providing water level data.

REFERENCES

Al Saud, M., 2010. Mapping potential areas for groundwater storage in Wadi Aurnah basin, western Arabian peninsula, using remote sensing and geographic information system techniques. Hydrogeol. J. 18, 1481–1495.

An, P., Moon, W., Bonham-Carter, G., 1994. Uncertainty management in integration of exploration data using the belief function. Nonrenew. Resour. 3, 60–71.

Anbazhagan, S., Jothibasu, A., 2016. Geoinformatics in groundwater potential mapping and sustainable development: a case study from southern India. Hydrol. Sci. J. 61 (6), 1109–1123.

Anbazhagan, S., Ramasamy, S.M., 1993. Role of remote sensing in geomorphic analysis for water harvesting structures. In: Proceedings of National Seminar and Water Harvesting, pp. 208–217.

Anbazhagan, S., Ramasamy, S.M., 1997. Geophysical resistivity survey and potential site selection for artificial recharge, India. In: Marinos, P.G. et al., (Eds.), International Symposium on Engineering Geology and Environment, 23–25 June, AthensIn: vol. 2, pp. 1169–1173.

Anbazhagan, S., Ramasamy, S.M., 2002. Remote sensing based artificial recharge studies – a case study from Precambrian Terrain, India. J. Manage. Aquif. Recharge Sustain., 553–556.

Balachandar, D., Alaguraja, P., Sundaraj, P., Rutharvelmurthy, K., Kumaraswamy, K., 2010. Application of remote sensing and GIS for artificial recharge zone in Sivaganga District, Tamilnadu, India. Int. J. Geomat. Geosci. 1 (1), 84–97.

Becker, M.W., 2006. Potential for satellite remote sensing of groundwater. Groundwater 44 (2), 306–318.

Bize, J., Bourguet, L., Lemoine, J., 1972. Artificial Recharge of Groundwater. Masson, Paris, p. 199.

Bonham-Carter, G.F., 1996. Geographic Information Systems for Geosciences Modeling with GIS. Pergamon, Love Printing Service Ltd, Oxford, p. 398.

Boughriba, M., Barkaoui, A., Zarhloule, Y., Lahmer, Z., Verdoya, M., 2010. Groundwater vulnerability and risk mapping ofthe Angad transboundary aquifer using DRASTIC index method in GIS environment. Arab. J. Geosci. 3, 207–220.

Central Ground Water Board-India, 2009. Manual on Artificial Recharge of Groundwater. Ministry of Water Resource Board, Government of India.

Chowdhury, A., Jha, M.K., Chowdary, V.M., Mal, B.C., 2009. Integrated remote sensing and GIS-based approach for assessing groundwater potential in West Medinipur district, West Bengal, India. Int. J. Remote Sens. 30 (1), 231–250.

Dadrasi, A., 2008. Comparing fuzzy logic model with other concept models of compatible GIS development in locating of flood spread suitable areas with application of satellite sensor data ETM. In: Proceedings of Conference of Geomatics, 12–13 May. National Cartographic center, Tehran (in Persian).

Das, S., Behra, S.C., Kar, A., Narendra, P., Guha, N.S., 1997. Hydrogeomorphological mapping in groundwater exploration using remotely sensed data – case study in Keonjhar district in Orissa. J. Indian Soc. Remote Sens. 25, 47–259.

Delft, 2000. A Project Report on Use of Artificial Neural Network and Fuzzy Logic for Integrated Water Management: Review of Applications. IHE Delft Hydro Informatics.

Fashae, O.A., Tijani, M.N., Talabi, A.O., Adedeji, O.I., 2014. Delineation of groundwater potential zones in the crystalline basement terrain of SW-Nigeria: an integrated GIS and remote sensing approach. Appl. Water Sci. 4 (1), 19–38.

Ghayoumian, J., MohseniSaravi, M., Feiznia, S., Nouri, B., Malekian, A., 2007. Application of GIS techniques to determine areas most suitable for artificial groundwater recharge in a coastal aquifer in southern Iran. J. Asian Earth Sci. 30 (2), 364–374.

Hammouri, N., El-Naqa, A., Barakat, M., 2012. An integrated approach to groundwater exploration using remote sensing and geographic information system. J. Water Resour. Prot. 4, 717–724.

Hoffmann, J., 2005. The future of satellite remote sensing in hydrogeology. Hydrogeol. J. 13 (1), 247–250.

Karanth, K.R., 1963. Groundwater Assessment, Development and Management. Tata McGrawHill Publishing Company Limited, New Delhi, p. 576.

Machiwal, D., Jha, M.K., Mal, B.C., 2011. Assessment of groundwater potential in a semiarid region of India using remote sensing, GIS and MCDM techniques. Water Resour. Manag. 25 (5), 1359–1386.

Meijerink, A.M.J., 1996. Remote sensing applications to hydrology: groundwater. Hydrol. Sci. J. 41 (4), 549–561.

Nouri, B., 2003. Identification of suitable sites for groundwater artificial recharges using remote sensing and GIS in Gavbandi watershed. Thesis (MSc), Tehran University.

Rai, B., Tiwar, A., Dubey, V.S., 2005. Identification of groundwater prospective zones by using remote sensing and geoelectrical methods in Jharia and Raniganj coalfields, Dhanbad district, Jharkhand state. J. Earth Syst. Sci. 114 (5), 515–522.

Ramasamy, S.M., Anbazhagan, S., 1994. Remote sensing for artificial recharge of groundwater. Bull. Natl. Nat. Resour. Manage. Syst. Bangalore 18, 35–37.

Rango, A., Shalaby, A.I., 1998. Operational applications of remote sensing in hydrology: success, prospects and problems. Hydrol. Sci. J. 43 (6), 947–968.

Ravi Shankar, M.N., Mohan, G., 2005. A GIS based hydrogeomorphic approach for identification of sitespecific artificial-recharge techniques in the Deccan Volcanic Province (DVP). J. Earth Syst. Sci. 114 (5), 505–514.

Saraf, A.K., Choudhury, P.R., 1998. Integrated remote sensing and GIS for groundwater exploration and identification of artificial recharge sites. Int. J. Remote Sens. 19 (10), 1825–1841.

Singhal, B.B.S., Gupta, R.P., 1999. Applied Hydrogeology of Fractured Rocks. Kluwer Academic Publishers, Dordrecht.

Strahler, A.N., 1952. Quantitative analysis of watershed geomorphology. Trans. Am. Geophys. Union 38, 913–920.

Tangestani, M.H., 2001. Landslide Susceptibility Mapping Using the Gamma Operation in a GIS, Iran. .

Tweed, S.O., Leblanc, M., Webb, J.A., Lubczynski, M.W., 2007. Remote sensing and GIS for mapping groundwater recharge and discharge areas in salinity prone catchments, southeastern Australia. Hydrogeol. J. 15, 75e96.

Zadeh, L., 1965. Fuzzy sets. Inf. Control 8, 338–353.

FURTHER READING

Anbazhagan, S., 1994. Fracture pattern study for ground water exploration in part of Dharmapuri dt. Tamil Nadu. Bhu-Jal News New Delhi 8 (2), 8–12.

Manap, M.A., Sulaiman, W.N.A., Ramli, M.F., Pradhan, B., Surip, N., 2013. A knowledge-driven GIS modeling technique for groundwater potential mapping at the Upper Langat Basin, Malaysia. Arab. J. Geosci. 6 (5), 1621–1637. https://doi.org/10.1007/s12517-011-0469-2.

Morovati, M., et al., 2008. Artificial recharge of aquifer is suitable solution for increasing the water level of groundwater. J. Hum. Environ. 17 (18), 68–76.

Chapter 9

Groundwater Quality Assessment Using Multivariate Statistical Methods for Chavara Aquifer System, Kerala, India

R. Anand Krishnan, Jamal Ansari, M. Sundararajan, Cincy John and P.M. Saharuba

CSIR—National Institute for Interdisciplinary Science and Technology, Council Of Scientific and Industrial Research, Thiruvananthapuram, India

Chapter Outline

9.1 INTRODUCTION

Basically, groundwater is an integral part of the environment and, hence, it is a major life-sustaining resource for all living beings (Krishnakumar et al., 2013). There are many processes involved during the movement of groundwater from recharge to discharge areas that include precipitation, mixing, ion exchange, redox condition, leaching, and dissolution. These geochemical changes mainly depend on the chemical, physical, and biological properties of the host rocks, along with temperature and climatic changes (Thivya et al., 2013). Mineral weathering and anthropogenic processes are the two major processes that control the hydrogeochemical characteristics of groundwater. Several studies have highlighted the role of weathering in groundwater chemistry and regulating the concentration of dissolved ions in groundwater (Cloutier et al., 2008; Yidana et al., 2008; Banoeng-Yakubo et al., 2009); groundwater contaminants have been carried out in various parts of the world (Min et al., 2003; Pérez del Villar et al., 2003; Chae et al., 2004).

9.1.1 Study Area

Chavara is located in the Kollam District and lies between 8°59′43.44″ latitude and 76°32′14.145″ longitude. It has tropical climate characterized by an average monthly temperature of about 30°C. The major activities are agriculture, fishing, aquaculture, etc. The major water bodies are the Kallada River, Ashtamudi Lake, and TS canal. The study area comprises an area of 12.56 km².

9.1.2 Land Use/Land Cover

The buffer zone for the study was an area within a 2 km radius. The scope of services included literature survey, field studies, impact assessment, and preparation of the Environmental Impact Assessment (EIA) document. The various areas of study included baseline data generation for water quality, water quantity, the soil characteristics, the geology of the area, and a hydrological survey.

GIS and Geostatistical Techniques for Groundwater Science. https://doi.org/10.1016/B978-0-12-815413-7.00009-2

9.2 MATERIALS AND METHODS

Groundwater samples were collected randomly from wells in 99 locations of the study area during February 2014 (Fig. 9.1 and Table 9.1). Water was pumped for some time to ensure that there was complete homogeneity. Then the extracted water samples were collected in precleaned polythene bottles and analyzed for various water-quality parameters as per standard procedure. pH, electrical conductivity (EC), and total dissolved solids (TDS) were measured in the field itself using a water-analysis kit (Deep Vision-191). Na^+ and K^+ were analyzed using the spectrometric method. Ca^{2+} and Cl were determined by the volumetric titration method. Nitrate, phosphate, and fluoride were determined using a spectropho-tometer (Shimadzu UV-1800). Descriptive statistics were carried out using the Statistic Package for Social Sciences (SPSS) package (Khan, 2011). Spatial distributions of geochemical parameters were prepared on Arc-GIS software using the inverse distanced weighted (IDW) interpolation method (Lee et al., 2006).

9.3 RESULTS

The minimum, maximum, mean, and standard deviation of physicochemical parameters of groundwater quality, such as pH, EC, TDS, and major elements (Ca^{2+}, Mg^{2+}, Na^+, K^+, Cl^-, HCO_3^-, NO_3^-, PO_4^{3-}, and F^-), are presented (Krishnakumar et al., 2013) (Tables 9.1 and 9.2). Among these, pH is a very important factor. In the studied samples, the pH values ranged from 2.62 to 7.91, indicating that the groundwater was of acidic nature with little alkalinity due to its bicarbonate form. The EC value of groundwater varied from 94.7 to 33,100 µS/cm, with an average of 1137 µS/cm. A higher EC value was noticed in the north western part of study area. The TDS value ranged from 46 to 9614 mg/L (Fig 9.2). According to Davis and DeWiest (1966), TDS is classified into four different categories based on drinking and irrigation suitability, and this classification showed that 87% of the samples were within the permissible limit for

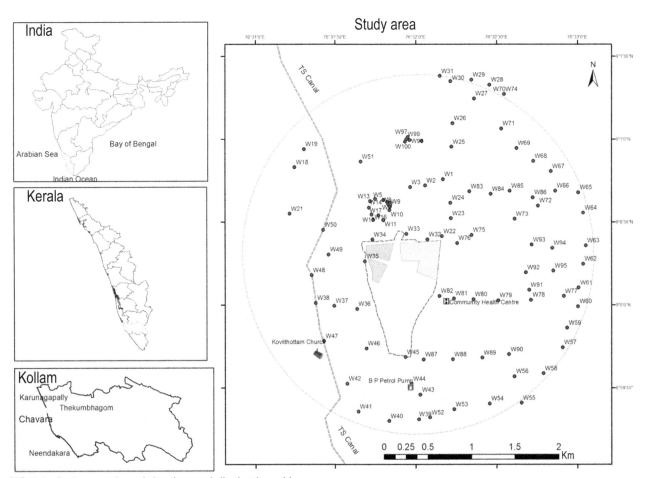

FIG. 9.1 Study area and sample location map indicating the positions.

TABLE 9.1 Chemical Composition of Groundwater Samples Collected From the Study Area

Point ID	pH	TDS (mg/L)	Cl (mg/L)	Iron (mg/L)	EC (μS/cm)	HCO_3^- (mg/L)	Na^+ (mg/L)	K^+ (mg/L)	PO_4^{3-} (mg/L)	Mg^{2+} (mg/L)	Ca^{2+} (mg/L)
W1	5.24	125	35.5	0.06	256.2	24	41.2	5.8	0.001	3.141	9.619
W2	5.27	75	23.075	4.5	153.7	24	8.7	2.1	0.002	5.05	5.611
W3	6.89	801	227.2	0.1	1631	180	362.8	12.9	0.007	25.94	40.08
W4	5.27	206	53.25	0	419	30	61.8	9.2	0.065	8.091	11.222
W5	6.08	280	71	0.11	571	60	73.5	4.5	0.015	9.784	32.865
W6	5.67	228	40.825	0.03	464	50	51.1	10.5	0.022	9.63	30.460
W7	6.62	145	35.5	0.97	296	64	48.5	3.8	0.05	5.235	14.428
W8	6.83	190	42.6	0.73	388	94	41.5	7.3	0	6.99	33.667
W9	6.62	207	60.35	0.003	422	66	54.1	3.4	0	5.7	36.072
W10	5.77	138	37.275	0.002	280.6	36	45.7	1.7	0.005	3.962	12.825
W11	5.39	268	60.35	0.03	547	16	84.7	5.1	0.014	6.674	19.238
W12	5.65	166	53.25	0	339	38	53.9	6.1	0	5.473	13.627
W13	5.81	102	23.43	0.014	207.9	40	17.5	0	0.008	5.101	18.436
W14	5.9	67	21.3	0.54	136.6	20	13.4	0	0.012	2.764	9.619
W15	4.81	176	66.03	0.006	360	4	44.3	3	0	6.266	21.643
W16	6.73	176	24.14	0.024	359	116	9.1	6	0.015	4.101	55.310
W17	4.62	100	22.01	0.135	204	4	22.2	5.1	0.013	2.198	9.619
W18	7.64	7164	3301.5	0.005	14650	218	2216	30	0.173	253.3	324.648
W19	6.91	1399	482.8	0	2856	150	355	2.5	0.221	29.96	168.336
W21	7.13	143	60.35	0.052	292.3	50	31.65	1.87	0.171	9.032	16.032
W22	7.13	272	36.92	0.06	554	182	23.36	3.75	0.011	14.95	77.755
W23	6.35	84	23.43	0	171.2	40	8.09	1.91	0.198	1.883	17.635
W24	6.71	239	54.67	0.104	488	100	28.08	2.88	0.273	8.098	56.112
W25	6.6	154	31.24	0	314	98	17.79	1.83	0.441	6.527	32.86
W26	7.18	99	27.69	0.037	199.9	40	10.63	2.4	0.173	2.616	20.04
W27	5.91	126	42.6	0.03	257.2	30	28.51	1.19	0.509	7.635	12.024
W28	5.81	132	39.05	0	270.2	30	29.91	3.82	0.223	5.68	16.032
W29	5.65	163	49.7	0.085	333	12	60.25	2.97	0.03	4.231	4.809
W30	7.86	198	12.07	0.051	405	100	21.43	12.17	0.087	5.636	48.096
W31	7.91	434	51.12	0.054	886	196	115.71	16.98	0.027	6.162	44.889
W32	3.33	1514	674.5	6.01	33100	0	70.23	4.6	0.024	46.63	258.11
W33	7.22	3497	1597.5	0.49	7130	54	115.71	29.25	0.028	147.6	801.6
W34	7.66	507	248.5	0.51	1035	66	22.97	3.03	0	24.63	119.43
W35	2.62	5974	1952.5	63.82	12190	0	115.71	10.05	0.019	48.53	801.6
W36	7.56	583	213	0.099	1190	246	115.71	6.17	0	4.508	24.048
W37	7.9	1491	568	5.8	3060	160	115.71	12.3	0.192	4.69	0
W38	7.71	305	35.5	0.67	625	198	41.17	9.54	1.297	10.28	68.9376
W39	6.9	267	42.6	0.063	546	122	38.54	6.02	0	7.266	72.144

Continued

TABLE 9.1 Chemical Composition of Groundwater Samples Collected From the Study Area—cont'd

Point ID	pH	TDS (mg/L)	Cl (mg/L)	Iron (mg/L)	EC (μS/cm)	HCO_3^- (mg/L)	Na^+ (mg/L)	K^+ (mg/L)	PO_4^{3-} (mg/L)	Mg^{2+} (mg/L)	Ca^{2+} (mg/L)
W40	7.17	190	28.4	0.095	388	122	27.78	7.82	0	13.51	43.2864
W41	7.05	308	42.6	0.102	611	198	43.04	12.44	0	16.76	72.144
W42	6.12	146	42.6	0.091	298.4	64	27.29	0	0	7.267	24.84
W43	6.3	122	35.5	0.243	248.6	32	27.06	0	0	4.253	23.24
W44	7.04	203	42.6	0.138	414	120	18.96	0	0.227	7.267	62.524
W45	6.79	230	42.6	0.122	470	118	40.16	7.58	0.11	7.138	52.104
W46	7.27	473	106.5	0.155	965	126	63.26	62.93	0.329	15.18	85.7712
W47	6.92	304	71	0.246	620	198	57.65	9.33	0.23	17.84	68.9376
W48	6.41	156	49.7	1.301	319	24	43.49	0	0.043	8.303	22.4448
W49	7.13	793	248.5	0	614	104	87.8	0	0.012	34.2	84.168
W50	7.07	313	31.95	0.03	37	210	16.1	0	0.071	12.27	95.3904
W51	7.72	535	63.9	0	1092	238	70.4	17.12	0.103	57.95	53.7072
W52	6.61	307	49.7	0	628	50	10.58	0	0.353	10.47	59.3184
W53	6.21	240	35.5	0	489	86	13.51	0	0.011	14.2	27.25
W54	7.28	348	56.8	0	705	112	21.35	0	0.022	12.58	74.548
W55	5.88	224	49.7	0	496	32	34.73	0	0.001	3.832	16.833
W56	5.77	207	63.9	0.022	422	22	92.2	0.68	0.94	3.5	8.016
W57	5.71	131	31.95	0	266.7	30	13.45	0	0.002	5.063	18.436
W58	5.9	182	31.95	1.968	372	58	0	0	0	8.854	44.889
W59	6.27	153	31.95	0	313	38	66.1	28.15	0.034	4.223	13.627
W60	3.29	334	88.75	0.493	683	0	99.84	6.78	0.033	10.3	16.8336
W61	5.68	108	28.4	0	223	22	38.65	8.37	0.02	5.131	12.024
W62	6.22	884	17.75	0	171	24	18.77	27.49	0.027	3.78	12.024
W63	5.25	63	17.75	0	129.3	16	20.67	9.53	0.004	2.929	7.2144
W64	6.31	157	53.25	0	319	34	50.65	1.35	0	8.601	25.651
W65	6.18	97	17.75	0	198.2	18	31.94	0.15	0	3.285	16.032
W66	6.13	76	28.4	0	156.3	24	33.26	0.97	0.009	1.985	11.2224
W67	5.49	99	28.4	0	203	22	45.31	2.18	0.012	3.728	12.024
W68	5.47	148	17.75	0.028	303	16	40.86	9.73	0.057	5.901	11.2224
W69	5.54	127	39.05	0.102	259.6	30	34.56	3.15	0.02	6.503	13.6272
W70	5.96	183	46.15	0.077	373	60	43.26	3.4	0.027	9.23	28.056
W71	5.94	221	49.7	0.046	450	28	71.11	2.69	0.055	9.064	14.4288
W72	6.33	195	74.55	0.032	397	72	37.04	9.22	0.026	9.141	37.6752
W73	5.67	109	39.05	0.034	223.4	24	40.65	1.28	0.019	2.459	8.8176
W74	5.7	187	46.15	0.032	381	58	59.38	5.33	0.014	5.343	19.2384
W75	6.42	50	49.7	0.235	102.3	20	11.88	1.22	0.004	2.53	4.8096
W76	6.21	195	17.75	0.05	398	100	55.08	4.57	0.007	6.888	32.064
W77	5.77	129	39.05	0.025	263.7	44	14.82	3.49	0.005	7.892	28.8576

TABLE 9.1 Chemical Composition of Groundwater Samples Collected From the Study Area—cont'd

Point ID	pH	TDS (mg/L)	Cl (mg/L)	Iron (mg/L)	EC (µS/cm)	HCO_3^- (mg/L)	Na^+ (mg/L)	K^+ (mg/L)	PO_4^{3-} (mg/L)	Mg^{2+} (mg/L)	Ca^{2+} (mg/L)
W78	5.9	75	21.3	0	151.6	40	8.15	0	0.006	0	15.2304
W79	6.38	267	42.6	0.402	545	184	43.12	0	0.002	3.141	53.7072
W80	7.11	181	17.75	0	370	170	5.97	0	0.024	14.79	64.9296
W81	7.5	315	35.5	0	642	214	42.57	0	0.034	12.94	70.5408
W82	7.13	293	42.6	0.345	599	168	21.24	0	0.036	10.77	80.9616
W83	6.93	46	21.3	0	94.7	18	8.85	0	0.014	2.509	3.2064
W84	5.32	148	46.15	0.34	302	10	34.07	0	0.004	6.666	12.024
W85	5.01	92	42.6	0	187.7	16	35.48	0	0.018	1.74	4.809
W86	6.24	176	35.5	0.344	360	100	17	0	0.033	7.76	40.88
W87	7.24	200	31.95	0	408	138	24.92	0	0.008	5.683	35.270
W88	5.99	408	78.1	0.284	834	234	71.9	7.46	0.019	11.79	88.176
W89	5.65	140	35.5	0.853	286	26	42.73	1.55	0.08	5.17	12.02
W90	5.75	137	24.85	1.192	280	28	29.93	1.47	0.01	6.378	20.04
W91	7.08	388	46.15	0.521	794	228	61.06	6.46	0.177	16.11	96.192
W92	5.65	133	46.15	0.192	270	24	47.87	0	0.014	3.213	12.024
W93	6.22	125	35.5	0.246	265	40	41.66	0	0.019	2.417	12.024
W94	6.38	195	35.5	0.122	398	76	42.01	4.65	0.023	14.44	32.064
W95	5.82	159	24.85	0.365	327	32	52.97	0	0.039	3.092	20.04
W96	5.63	58	17.75	0.009	188	26	9.37	0	0.26	3.17	8.016
W97	5.61	155	24.85	2.344	317	26	60.98	1.44	0.057	6.122	12.02
W98	6.32	146	17.75	0.082	297	76	37.34	0	0.034	3.715	40.08
W99	5.59	199	56.8	0.487	242	24	43.4	0	0.018	4.117	12.02
W100	6.02	108	31.95	0.246	222	46	21.1	0	0.024	2.814	24.04

TABLE 9.2 Minimum, Maximum, Mean, and Standard Deviation of Physicochemical Parameters

S. No	Parameters	Min	Max	Mean	Std. Deviation
1	pH	2.62	7.91	6	0.938
2	TDS (mg/L)	46	7164	416.8	985.19
3	Cl^- (mg/L)	12.07	19525	296.4	415.63
4	Iron	0	63.82	1.002	6.45
5	EC (µS/cm)	94.7	33100	1137.07	16597.35
6	HCO_3^-	0	246	75.717	67.41
7	Sodium	0	2216	70.12	224.03
8	Potassium	0	62.93	5.269	8.67
9	Phosphate	0	1.297	0.081	0.181
10	Magnesium	0	253.3	12.96	29.64
11	Calcium	0	801.6	54.47	117.52

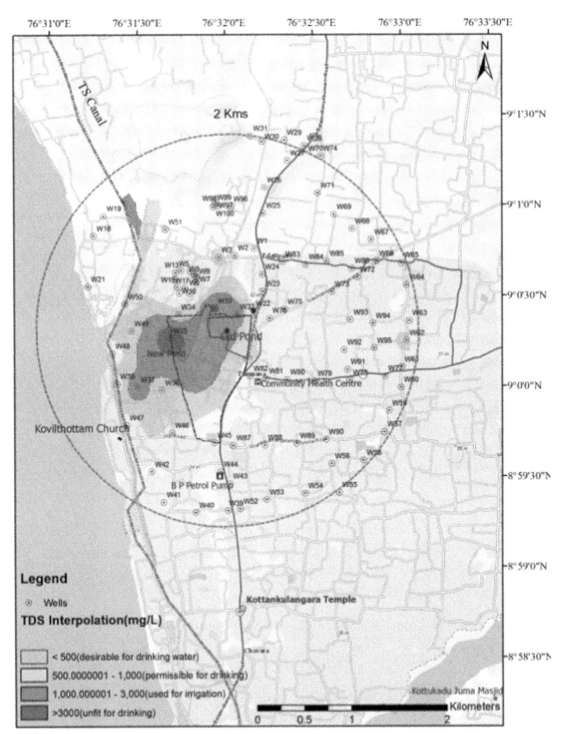

FIG. 9.2 Spatial distribution of TDS based on the Davis and DeWiest method (1966).

drinking (Table 9.3). However, 3.03% of samples were unfit for drinking as well as irrigation. Freeze and Cherry (1979) also classified the TDS, based on fresh, brackish, saline, and bryne water types, and this classification showed that about 88% of the groundwater came under the fresh category and the remaining 11% was of the brackish water type. The concentration of iron in the observed samples ranged from 0 to 63.82 mg/L with six samples exceeding the limit. The maximum value of 63.82 mg/L showed the high contamination of iron in the samples. The sodium concentration in the samples ranged from 0 to 2216 mg/L. The BIS (1998) guideline shows the maximum permissible limit for sodium in drinking water to be 200 mg/L. In most of the

TABLE 9.3 Groundwater Quality Based on TDS (Davis and DeWiest)

Parameter	Range	Classification	Sample Numbers	No. of Samples	% of Samples
Davis and DeWiest (1966)					
TDS (mg/L)	<500	Desirable for drinking water	1, 2, 4, 5, 6, 7, 8, 9, 10, 11, 12, 13, 14, 15, 16, 17, 21, 22, 23, 24, 25, 26, 27, 28, 29, 30, 31, 38, 39, 40, 41, 42, 43, 44, 45, 46, 47, 48, 50, 52, 53, 54, 55, 56, 57, 58, 59, 60, 61, 63, 64, 65, 66, 67, 68, 69, 70, 71, 72, 73, 74, 75, 76, 77, 78, 79, 80, 81, 82, 83, 84, 85, 86, 87, 88, 89, 90, 91, 92, 93, 94, 95, 96, 97, 98, 99, 100	87	87.88
	500–1000	Permissible for drinking water	3, 34, 36, 49, 51, 62	6	6.06
	1000–3000	Useful for irrigation	19, 32, 37	3	3.03
	>3000	Unfit for drinking and irrigation	18, 33, 35	3	3.03
Freeze and Cherry (1979)					
TDS (mg/L)	<1000	Freshwater	1, 2, 3, 4, 5, 6, 7, 8, 9, 10, 11, 12, 13, 14, 15, 16, 17, 21, 22, 23, 24, 25, 26, 27, 28, 29, 30, 31, 34, 36, 38, 39, 40, 41, 42, 43, 44, 45, 46, 47, 48, 49, 50, 51, 52, 53, 54, 55, 56, 57, 58, 59, 60, 61, 62, 63, 64, 65, 66, 67, 68, 69, 70, 71, 72, 73, 74, 75, 76, 77, 78, 79, 80, 81, 82, 83, 84, 85, 86, 87, 88, 89, 90, 91, 92, 93, 94, 95, 96, 97, 98, 99, 100	88	88.88
	1000–10,000	Brackish water	18, 19, 32, 33, 35, 37	11	11.11
	10,000–100,000	Saline water	Nil		
	>100,000	Brine water	Nil		

samples, Na^+ concentration was within the permissible limits and only two samples exceeded the permissible limits. The magnesium concentration ranged from 0 to 253 mg/L and Mg^{2+} concentrations in the studied samples showed that four exceeded the allowable limits. The calcium value of groundwater samples ranged from 0 to 801 mg/L, with four samples exceeding the allowable limits.

Among the anionic concentrations, chloride plays a predominant role in the studied groundwater samples (Krishnakumar et al., 2013). The chloride value ranged from 12 to 19525 mg/L, with 6% of samples exceeding the permissible limits (250 mg/L). Phosphate concentration ranged from 0.1 to 1.2 mg/L; a higher concentration of PO_4^{3-} is indicative of pollution and the major sources are domestic sewage and agriculture wastes.

9.3.1 Groundwater Quality Based on TDS

Using the Davis and DeWiest method of classification a quantity of TDS <500 is desirable for drinking water, 500–1000 is permissible for drinking, 1000–3000 is allowable for irrigation, and >3000 is unfit for drinking. TDS is higher along the eastern side of the TS Canal.

In the Freeze and Cherry method of classification the values of TDS can be grouped into two classes <1000 and >1000, and the allowable levels are the same as in the Davis and DeWiest method.

According to the Davis and DeWiest method about 3%–6% of the total samples are unfit for drinking and 3% of the total samples were useful for irrigation purposes. Using the Freeze and Cherry method it was found that 11% was not suitable for drinking.

9.4 DISCUSSION

9.4.1 Saline Water Intrusion

The Cl^-/HCO_3^- ratio can be classified into three groups: <0.5 for no intrusion, 0.5–6.6 for moderate intrusion, and >6.6 for high intrusion. It was observed that 28.52% of the samples had no intrusion, 62.2% had moderate intrusion, and 9.18% had high intrusion. About 50% of the analyzed samples showed strong intrusion.

9.4.2 Factor Analysis

Factor analysis is very useful to interpret groundwater-quality data and to understand specific hydrogeochemical processes. Spatial distribution of TDS was based on Davis and DeWiest method (1966) (Fig. 9.3). R-mode factor analysis is the most widely used statistical method. Hence, varimax with Kaiser normalization rotated factor loading matrix was performed (Krishnakumar et al., 2013). In the analysis three principal components were extracted accounting for 75% of the variance, which shows the major effective controlling agents in the groundwater (Tables 9.4 and 9.5). Factor 1 accounted for about 37% of the variance, which is positively contributed by EC, TDS, calcium, magnesium, sodium. TDS have loadings of 0.845, and control the overall mineralization, while Cl^- has a loading of 0.905. This is derived from saline water intrusion, salt pan deposits, or agricultural return flowing into the groundwater (Ramkumar et al., 2010).

The loadings are higher for TDS, chloride, sodium, and magnesium. The higher loadings for sodium indicate the high saline-water intrusion in the area. The alkalinity values are comparatively less for component one and negative for component two, which indicates the absence of bicarbonates in the groundwater. The bicarbonate ion has no effect on the variation of water chemistry when analyzed. Iron has moderate loadings in the first component but increases considerably in the second component. This indicates the high presence of iron. The loadings of phosphate are comparatively low (Fig. 9.4).

9.4.3 Cluster Analysis

Cluster analysis is another data-reduction method that is used to classify entities with similar properties (Pathak et al., 2008). Through hierarchical cluster analysis (HCA) the associations among the stations were obtained using Ward's method with Euclidean distance as a similarity measure with dendrogram plots (Krishnakumar et al., 2013) (Fig. 9.7). Physicochemical parameter concentrations were used as variables to show the spatial heterogeneity among the stations as a result of sequence and their relationship in the level of contamination (Krishnakumar et al., 2013). There were two major cluster groups obtained from the sampling. Group I, clustered with a limited number of stations, interprets the spatial similarity in the physicochemical composition as influenced by human interference and the extent of pollutants from the salt pan and aquaculture activity (Krishnakumar et al., 2013). Group I is typically represented by the 31st sample, which is moderately polluted. Group II (a) is represented by the 18th, 34th, and 32nd sample sites (Table 9.6) where there is maximum contamination, a fact that is supported by the spatial distribution of the results. Group II (b) is the sample list with the least contamination. Here the spatial analysis diagrams are classified into two categories (Fig. 9.5). The first category consists of elements that are very high in concentration and the second category those that are low in concentration. The first category consists of calcium, chloride, and bicarbonate and the second potassium, phosphate, and sodium.

Iron is the major contaminant; from Fig. 9.6 it can be seen that the concentration of iron is highest toward the eastern part and lowest in the western part of the TS canal.

9.4.4 Water Quality Index

WQI provides the overall results of water quality for drinking purpose. It is classified into five: excellent (<50), good (51–100), poor (101–200), very poor (201–300). Based on the performance of water-quality parameters, maximum, and minimum weight was assigned (Krishnakumar et al., 2013). A maximum weight of 5 was assigned for the parameters TDS and chloride, because they play a vital role in groundwater quality (Vasanthavigar et al., 2010). Other parameters, such as bicarbonate, sodium, potassium, calcium, and magnesium were assigned the weights 1–4 depending on their importance in water-quality evaluation. After assigning the weight for each parameter, the relative weight was calculated (Table 9.7).

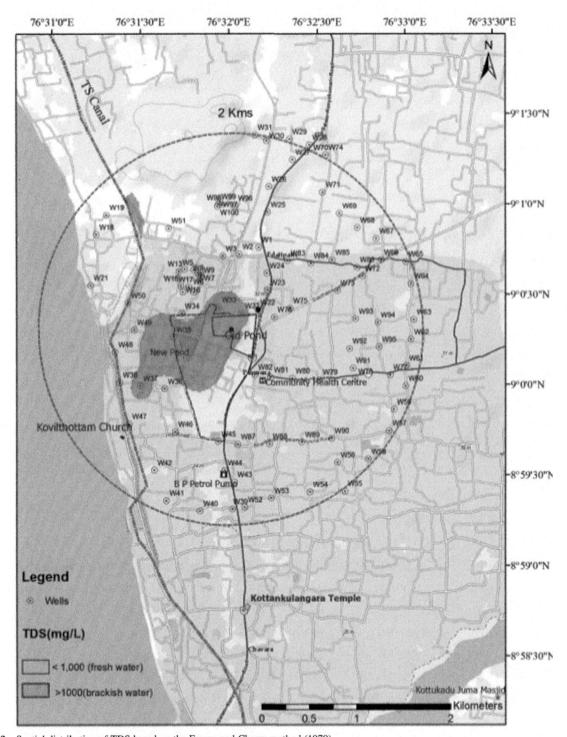

FIG. 9.3 Spatial distribution of TDS based on the Freeze and Cherry method (1979).

Finally, a quality rating (q_i) is assigned in each parameter based on the concentration of the sample (C_i) divided by the drinking water quality standard (S_i) using the following equation:

$$Q_i = (C_i/S_i) \times 100$$

WQI is computed based on the following equation:

$$SI_i = W_i/q_i$$

TABLE 9.4 Rotated Component Matrix

	__Rotated Component Matrix__		
	Component		
	1	2	3
pH	.112	−.434	.784
TDS	.845	.497	.076
Cl	.905	.385	.044
Iron	.119	.893	−.126
EC	.576	.450	−.165
Alkalinity	.229	−.173	.774
Sodium	.906	−.126	.099
Potassium	.426	.107	.391
Phosphate	−.166	.258	.628
Magnesium	.949	.100	.152
Calcium	.546	.712	.098

Extraction method: principal component analysis. Rotation method: varimax with Kaiser normalization.
Rotation converged in seven iterations.

TABLE 9.5 Variance Calculation Using PCA

	__Total Variance Explained__					
	Initial Eigen Values			Rotation Sums of Squared Loadings		
Component	Total	% of Variance	Cumulative %	Total	% of Variance	Cumulative (%)
1	4.967	45.155	45.155	4.171	37.920	37.920
2	2.236	20.325	65.480	2.225	20.224	58.143
3	1.048	9.526	75.006	1.855	16.862	75.006
4	.882	8.014	83.019			
5	.748	6.800	89.820			
6	.537	4.884	94.704			
7	.356	3.240	97.943			
8	.181	1.642	99.585			
9	.027	.242	99.827			
10	.014	.132	99.959			
11	.005	.041	100.000			

Extraction method: principal component analysis (PCA).

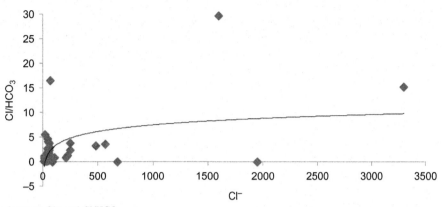

FIG. 9.4 Relationship between Cl$^-$ and Cl/HCO$_3$.

TABLE 9.6 Different Water Groups in Chavara Region Using Cluster Analysis

Cluster	Sample Number
Group I	31
Group II (a)	32, 34, 18
Group II (b)	26 68 27 1 92 42 56 76 10 88 91 89 7 41 20 24 97 67 83 47 63 12 28 94 96 58 15 98 13 25 99 60 72 64 66 17 84 74 82 14 62 2 77 22 65 95 9 71 69 73 8 93 75 39 86 43 29 16 85 87 79 23 44 52 4 55 6 70 54 87 90 30 37 40 80 46 49 81 21 78 38 5 11 51 53 59 61

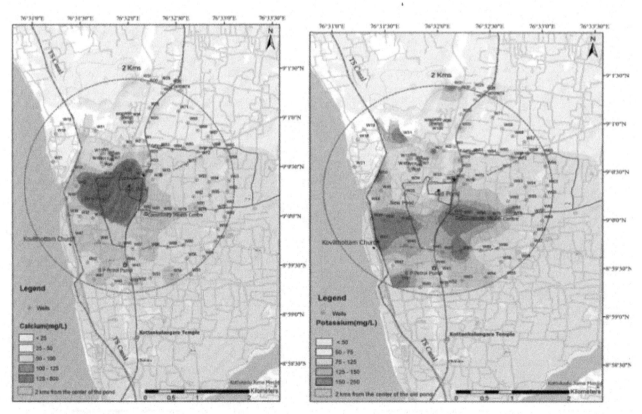

FIG. 9.5 Spatial distribution of physicochemical parameters.   *(Continued)*

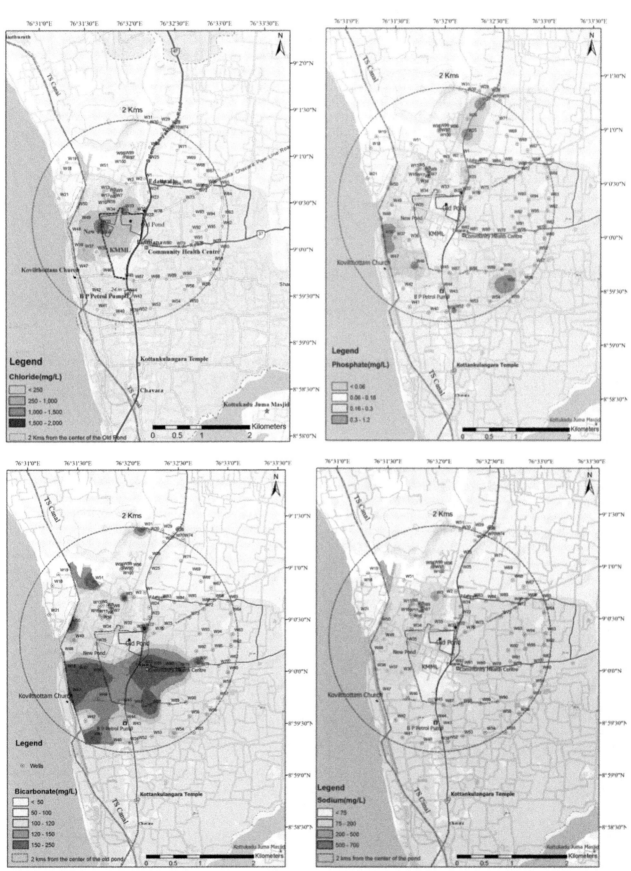

FIG. 9.5, CONT'D

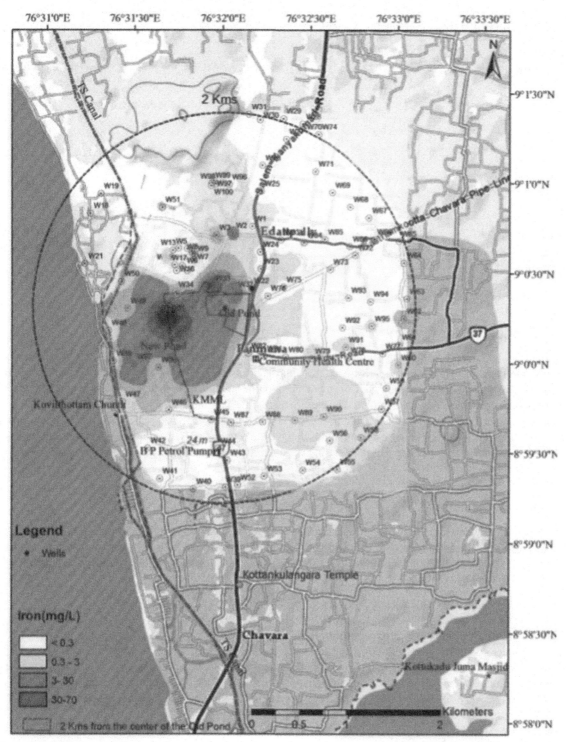

FIG. 9.6 Spatial distribution of iron.

$$\mathrm{WQI} = \sum SI_i$$

where SI_i is the subindex of ith parameter (Krishnakumar et al., 2013).

Therefore the water quality rating of the studied samples obtained ranges from 29 to 393. The total SI value is high for iron and sodium (Fig. 9.7).

TABLE 9.7 Water Quality Index and Relative Weight

Chemical Parameters	BIS 10500 1991	Weight (W_i)	Relative Weight (W)
pH	6.5	5	0.17241
Alkalinity	200	1	0.0344
Iron	0.3	5	0.1724
TDS	500	4	0.1379
Chloride	250	3	0.10344
Magnesium	30	2	0.06896
Calcium	200	3	0.034
Sodium	200	2	0.0686
Potassium	0.1	4	0.1379

9.4.5 Irrigation Water Quality

The irrigation suitability was based on the salinity Na% measured using the sodium absorption ratio (SAR). Excess salinity reduces the osmotic activity of plants and thus interferes with the absorption of water and nutrients from the soil. Salinity and toxicity generally needed to be considered in the evaluation for the suitable quality of the groundwater for irrigation. EC is a good measure for the salinity hazard to crops. The eastern side of the TS Canal is highly affected (Fig. 9.8). Conductivity is itself not a human or aquatic health concern, but because it is easily measured, it can serve as an indicator of other water-quality problems. Water with high mineral content tends to have higher conductivity, which is a general indication of high dissolved-solid concentrations in water. Therefore using conductivity measurements we can easily identify water-quality problems (Table 9.8).

9.4.6 Sodium Percentage

Sodium percentage is widely used to assess the groundwater quality, because a higher level of sodium in irrigation water may increase the exchange of sodium with irrigated soil and affect soil permeability and structure and create toxic conditions for plants (Bangar et al., 2008; Durfer and Backer, 1964; Todd, 1980). Hence, Na% is assessed based on Raghunath (1987) using the following equation:

$$Na\% = (Na + K) \times 100/(Ca + Mg + Na + K)$$

Based on the relative proportions of cation concentration, 20% of samples were within permissible limits, 10.2% were excellent, 17% were good, 31% were doubtful, and 6% were unsuitable for irrigation (Fig. 9.9). Excess sodium in water produces the undesirable effects of changing soil properties and reducing soil permeability. Hence, the assessment of sodium concentration is necessary when considering suitability for irrigation. The sample locations fall along the coastal regions and Na might have been acquired from the seawater through saline intrusion (Chidambaram et al., 2007).

9.4.7 Sodium Absorption Ratio

SAR has been extensively used as an indicator of sodium hazard in irrigation water (Gholami and Srikantaswamy, 2009). Richards (1954) classified water-quality based on the following equation:

$$SAR = Na^+ / \left(\left(Ca^+ + Mg^{2+} \right)^{1/2} / 2 \right) meq/L$$

According to Richard's classification, the computed SAR value showed that 35% of samples were excellent, 29% were good, and 13% were doubtful for irrigation. High concentrations of sodium ion in irrigation water have a tendency to

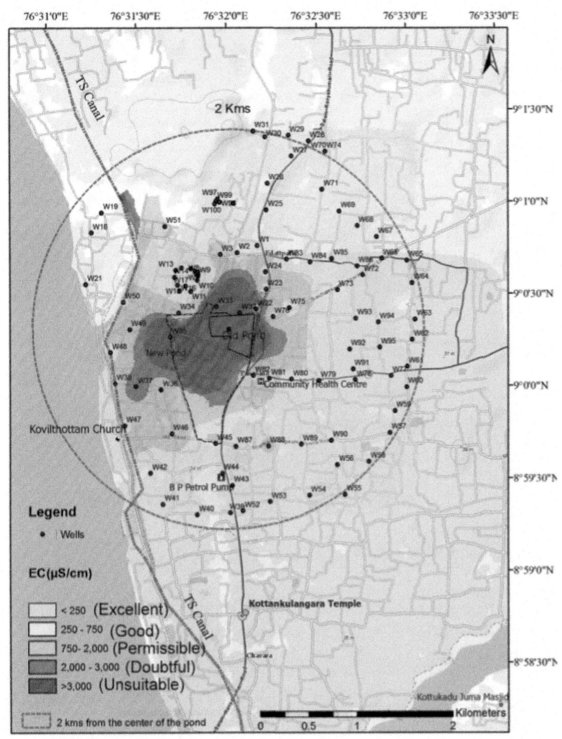

FIG. 9.7 Spatial distribution of suitability of groundwater for irrigation based on EC.

be easily absorbed into the clay particles due to the exchange of calcium and magnesium ions. These exchange processes decrease the soil permeability and reduce internal drainage.

From Richard's classification of SAR (Table 9.9) it was found that about 13% was doubtful and 9% was unsuitable for irrigation. In the case of Na%, about 31% was doubtful and 6% was unsuitable for irrigation.

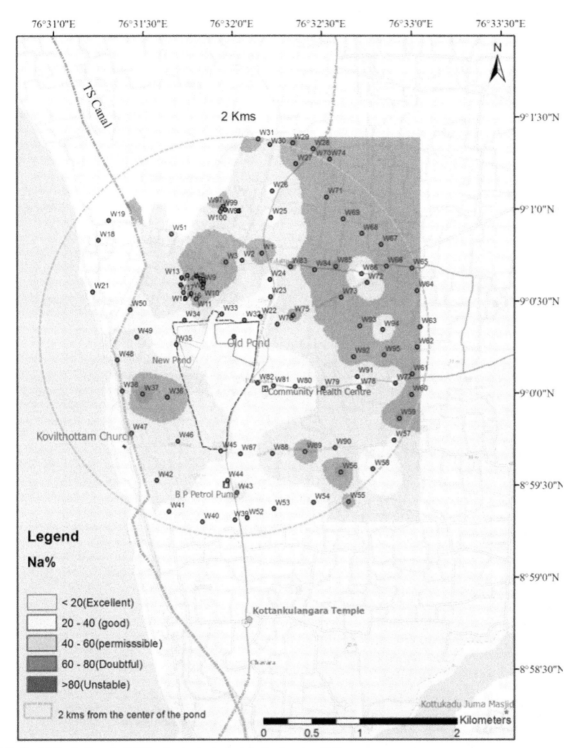

FIG. 9.8 Spatial distribution of suitability of groundwater for irrigation based on Na%.

TABLE 9.8 Irrigation Quality of Groundwater Based on EC

Parameter	Range	Classification	Sample Numbers	No. of Samples	% of Samples
Raghunath (1987) EC (µS/cm)	<250	Excellent	W2, W3, W14, W17, W26, W43, W61, W62, W63, W65, W66, W67, W75, W78, W83, W85, W96, W99, W100	19	19.387
	250–750	Good	W1, W4, W5, W6, W7, W8, W9, W10, W11, W12, W15, W16, W22, W24, W25, W27, W28, W29, W30, W31, W38, W39, W40, W41, W42, W44, W45, W47, W50, W52, W53, W54, W55, W56, W58, W59, W60, W64, W68, W69, W70, W71, W72, W74, W76, W79, W80, W81, W82, W84, W86, W87, W88, W89, W90, W92, W93, W94, W95, W97, W98	61	62
	750–2000	Permeable	W31, W34, W36, W49, W51, W88, W91	7	7.142
	2000–3000	Doubtful	W19	1	1.0204
	>3000	Unsuitable	W18, W32, W33, W35, W37	5	5.102

TABLE 9.9 Irrigation Quality of Groundwater Based on SAR and Na%

			No.	%
Richards (1954) SAR	Excellent	W2, W13, W14, W16, W23, W24, W25, W26, W30, W32, W33, W34, W35, W38, W39, W40, W41, W42, W44, W50, W53, W57, W58, W62, W75, W77, W78, W80, W81, W82, W83, W86, W87, W96, W100	35	35.71
	Good	W6, W8, W15, W17, W21, W27, W28, W43, W45, W46, W47, W48, W49, W51, W55, W63, W64, W65, W69, W70, W72, W76, W79, W84, W88, W90, W91, W94, W98	29	29.51
	Doubtful	W1, W5, W7, W10, W12, W61, W67, W68, W71, W73, W89, W92, W93	13	13.265
	Unsuitable	W3, W4, W11, W18, W19, W36, W37, W56, W60	9	9.183
Raghunath (1987) Na%	Excellent	W32, W33, W34, W35, W50, W52, W54, W58, W80, W83	10	10.2
	Good	W16, W22, W23, W24, W25, W26, W30, W38, W39, W40, W44, W53, W57, W78, W81, W87, W91	17	17.34
	Permissible	W2, W8, W9, W13, W14, W21, W42, W43, W45, W46, W49, W51, W72, W79, W88, W90, W94, W96, W98, W100	20	20.40
	Doubtful	W5, W6, W7, W10, W11, W12, W15, W17, W18, W19, W27, W28, W31, W55, W60, W61, W62, W63, W64, W66, W67, W69, W71, W73, W75, W76, W84, W85, W95, W97, W99	31	31.63
	Unsuitable	W3, W29, W36, W37, W56, W59	6	6.122

9.5 CONCLUSION

The present study of groundwater samples in Chavara shows that the water quality of the studied samples exceeded the Bureau of Indian Standards (BIS) limits in some areas, which may affect human health. From the TDS results and comparing the other parameters, we can see that of the total 99 samples analyzed, 9% of the samples were affected by saline-water intrusion. Industrial effluents and sewage wastes increased the affected area to $0.73 \, km^2$. From the method used to estimate SAR, EC, and Na% it was seen that around 63% of the samples were suitable for irrigation purposes. It is obvious that Chavara aquifer is moderately polluted and needs a long-term plan to restore groundwater quality. From the cluster and factor analysis, it can be seen that the eastern part of the TS Canal is affected. The study on the water quality for drinking showed that about 85% of the samples were good for drinking. To ensure a good management of the resource, proper water-extraction structures and sincere maintenance of the surface water resources are needed.

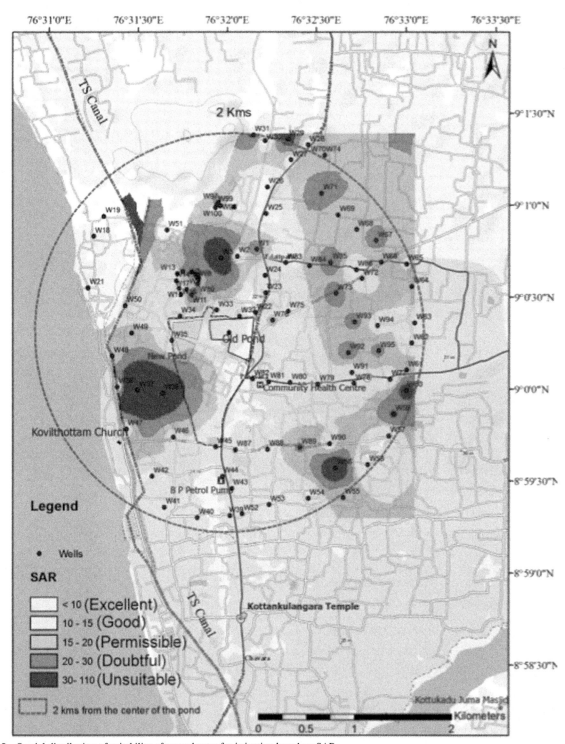

FIG. 9.9 Spatial distribution of suitability of groundwater for irrigation based on SAR.

ACKNOWLEDGMENTS

The authors would like to thank Dr. Ajit Haridas, Dr. K.P. Prathish, and, Dr. Rugmini Sukumar of CSIR-NIIST for their invaluable review comments and suggestions. The financial support was provided by The Director, Council of Scientific and Industrial research (CSIR) National Institute for Interdisciplinary Science and Technology (NIIST), Thiruvananthapuram.

REFERENCES

Bangar, K.S., Tiwari, S.C., Vermaandu, S.K., Khandkar, U.R., 2008. Quality of groundwater used for irrigation in Ujjain district of Madhya Pradesh, India. J. Environ. Sci. Eng. 50 (3), 179–186.

Banoeng-Yakubo, B., Yidana, S.M., Nti, E., 2009. Hydrochemical modeling of groundwater using multivariate statistical methods—the Volta Region, Ghana. KSCE J. Civ. Eng. 13 (1), 55–63.

Chae, G.T., Kim, K., Yun, S.T., Kim, K.H., Kim, S.O., Choi, B.Y., Kim, H.S., Rhee, C.W., 2004. Hydrogeochemistry of alluvial groundwaters in an agricultural area: an implication for groundwater contamination susceptibility. Chemosphere 55, 369–378.

Chidambaram, S., Vijayakumar, V., Srinivasamoorthy, K., Anandhan, P., Prasanna, M.V., Vasudevan, S., 2007. A study on variation in ionic composition of aqueous system in different lithounits around Perambalur region, Tamil Nadu. J. Geol. Soc. India 70 (6), 1061–1069.

Cloutier, V., Lefebvre, R., Therrien, R., Savard, M.M., 2008. Multivariate statistical analysis of geochemical data as indicative of the hydrogeochemical evolution of groundwater in a sedimentary rock aquifer system. J. Hydrol. 353, 294–313.

Davis, S.N., DeWiest, R.J.M., 1966. Hydrogeology. Wiley, New York.

Durfer, C.M., Backer, E., 1964. Public Water Supplies of the Three Largest Cities in the U.S. US Geological Survey Water Supply Paper No. 1812. p. 364.

Freeze, R.A., Cherry, J.A., 1979. Groundwater. Prentice-Hall, New Jersey, p. 604.

Gholami, S., Srikantaswamy, S., 2009. Analysis of agricultural impact on the Cauvery river water around KRS dam. World Appl. Sci. J6 (8), 1157–1169.

Khan, T.A., 2011. Multivariate analysis of hydrochemical data of the groundwater in parts of Karwan–Sengar sub-basin, Central Ganga basin, India. Global NEST J. 13 (3), 229–236.

Krishnakumar, P., Lakshumanan, C., Kishore, V.P., Sundararajan, M., Santhiya, G., Chidambaram, S., 2013. Assessment of groundwater quality in and around Vedaraniyam, South India. Environ. Earth Sci. 1866-6280. https://doi.org/10.1007/s12665-013-2626-2.

Lee, C.S.L., Li, X.D., Shi, W.Z., Cheung, S.C.N., Thornton, I., 2006. Metal contamination in urban, suburban, and country park soils of Hong Kong: a study based on GIS and multivariate statistics. Sci. Total Environ. 356, 45–61.

Min, J.H., Yun, S.T., Kim, K., Kim, H.S., Kim, D.J., 2003. Geologic controls on the chemical behavior of nitrate inriverside alluvial aquifers, Korea. Hydrol. Process. 17, 1197–1211.

Pathak, J.K., Alam, M., Sharma, S., 2008. Interpretation of groundwater quality using multivariate statistical technique in Moradabad city, Western Uttar Pradesh State, India. E-J. Chem. 5 (3), 607–619.

Pérez del Villar, L., Reyes, E., Delgado, A., Nunez, R., Pelayo, M., Cózar, J.S., 2003. Argillization processes at the El Berrocal analogue granitic system (Spain): mineralogy, isotopic study and implications for the performance assessment of radwaste geological disposal. Chem. Geol. 193 (3–4), 273–293.

Raghunath, H.M., 1987. Groundwater, second ed. New Age International Publications, New Delhi.

Ramkumar, T., Venkatramanan, S., Mary, I.A., Tamilselvi, M., Ramesh, G., 2010. Hydrogeochemical quality of groundwater in Vedaraniyam town, Tamil Nadu, India. Res. J. Environ. Earth Sci. 2 (1), 44–48.

Richards, L.A., 1954. Diagnosis and Improvement of Saline and Alkali Soils. Agricultural HandbookUS Laboratory Staff, US Department of Agriculture, p. 60.

Thivya, C., Chidambaram, S., Singaraja, C., Thilagavathi, R., Prasanna, M.V., Anandhan, P., Jainab, I., 2013. A study on the significance of lithology in groundwater quality of Madurai district, Tamil Nadu (India). Environ. Dev. Sustain. https://doi.org/10.1007/s10668-013-9439-z.

Todd, D.K., 1980. Groundwater Hydrology, second ed. John Wiley and Sons, New York, p. 535.

Vasanthavigar, M., Srinivasamoorthy, K., Vijayaragavan, K., Rajiv Ganthi, R., Chidambaram, S., Anandhan, P., Manivannan, R., Vasudevan, S., 2010. Application of water quality index for groundwaterquality assessment: Thirumanimuttar sub-basin, Tamilnadu, India. Environ. Monit. Assess. 171, 595–609. https://doi.org/10.1007/s10661-009-1302-1.

Yidana, S.M., Ophori, D., Banoeng-Yakubo, B., 2008. Hydrochemical evaluation of the Volta Basin: the Afram plains area. J. Environ. Manage. 88, 697–707.

Chapter 10

Overview, Current Status, and Future Prospect of Stochastic Time Series Modeling in Subsurface Hydrology

Priyanka Sharma*,a, Deepesh Machiwal† and Madan Kumar Jha‡

*SWE Department, College of Technology and Engineering, Maharana Pratap University of Agriculture and Technology, Udaipur, India †ICAR-Central Arid Zone Research Institute, Regional Research Station, Bhuj, India ‡AgFE Department, Indian Institute of Technology Kharagpur, Kharagpur, India

Chapter Outline

10.1 INTRODUCTION

Groundwater is a major source of freshwater supply used to meet domestic, agricultural, industrial, and other water demands throughout the world (Li et al., 2013). Demand for this precious natural resource to support ecosystems is significantly high in arid and semiarid regions where rainfall is highly variable and considerably lower than the evaporation rate. Despite the fact that groundwater plays a crucial role in sustaining livelihoods and preserving the economy of many nations across the globe, this hidden resource suffers from a negligence of its adequate management (Sophocleous, 2010). In recent times, it has been reported that groundwater levels are being depleted and groundwater quality is deteriorating in both developed and developing countries (Konikow and Kendy, 2005; Fogg and LaBolle, 2006; Rodell et al., 2009; Bhanja et al., 2017). In contrast to its depletion in some areas, rising groundwater levels in other arid and semiarid regions are also poses a serious threat to ecosystems (Konukcu et al., 2006; Guganesharajah et al., 2007; Ritzema et al., 2008; Han et al., 2011). Therefore there is an urgent need for the efficient and effective management of this subsurface resource in order to ensure its long-term sustainability, which is a challenging issue in a scenario of burgeoning populations and climate change.

In the literature a variety of tools and techniques have been suggested for the sustainable management of groundwater resources. In groundwater management, the forecasting of future groundwater levels and chemical concentrations is of the utmost importance. In previous studies researchers predicted future groundwater scenarios mostly through simulation modeling tools such as analytical and/or numerical groundwater flow and transport models (e.g., Wu et al., 2003). However, these models are based on certain simplifying assumptions while describing groundwater flow and transport processes

---

a. *Present address*: School of Agriculture, Lovely Professional University, Phagwara, India.

GIS and Geostatistical Techniques for Groundwater Science. https://doi.org/10.1016/B978-0-12-815413-7.00010-9

through mathematical equations. The model assumptions are about the direction of flow, geometry of the aquifer, the heterogeneity or anisotropy of sediments or bedrock in the aquifer, the contaminant transport mechanisms, and chemical reactions. Thus a model is an approximation of real-world physical and natural systems, which may or may not depict the exact physical processes occurring within the systems due to the involvement of uncertainties.

Stochastic time series modeling offers a convenient and valuable tool for the identification of trends, development and simulation of hydrologic and/or climatic models, and forecasting of hydrologic and hydrogeologic time series under future climate-change/variable conditions (Salas, 1993). Stochastic time series modeling can be applied to all sorts of data because of their statistical background as long as the behavior of the subsurface system to be modeled is linear or can be transformed to linear. Stochastic time series models are very useful for the modeling of the systems that cannot be easily described by physical principles and properties. Thus the stochastic modeling technique may be advantageous over the physically-based flow and transport modeling tools in subsurface hydrology. However, there have been only a few applications of stochastic time series modeling in subsurface hydrology, and mainly studies have been restricted to the identification of trends in subsurface hydrologic variables (Machiwal and Jha, 2006; Panda et al., 2007; Shamsudduha et al., 2009). Hence, there exists great scope to widen the domain of stochastic modeling to encompass subsurface hydrology and hydrogeology in addition to surface hydrology.

This chapter aims at highlighting the current status and prospects for stochastic time series modeling in subsurface hydrology. It provides an overview of time series analysis, explains its components, and briefly discusses the hypotheses involved in stochastic modeling. It then details the step-by-step procedure for applying stochastic time series modeling.

10.2 TIME SERIES ANALYSIS

A time series is a sequence of numerical values of any variable arranged in chronological order of their occurrence. A time series may be discrete or continuous, and it can be of univariate, bivariate, or multivariate type. The aim of time series analysis is to detect and describe quantitatively each of the generating processes underlying a given sequence of observations (Shahin et al., 1993). Time series analysis helps describe behavior of a given time series in terms of time and to relate observations to some structural rules of behaviors. One of the main objectives of time series analysis is to develop a model that contains well-defined components each explaining a portion of the decomposed time series. The components of the time series are either deterministic and/or stochastic. The deterministic components are identified, detected, quantified, and then removed from the time series, and then the residuals are used for developing stochastic models. Thus in time series modeling, past observations of a time series are carefully collected and rigorously studied to develop an appropriate model to describe the inherent structure of the series. Once validated, the developed model is used to forecast the future values of the series.

10.2.1 Components of Time Series

Time series can be divided into two major components: (i) deterministic and (ii) stochastic. A time series may consist of only deterministic events, only stochastic events, or a combination of the two (Machiwal and Jha, 2012). A hydrological time series x_t can be represented by a decomposition model of the additive type as shown in the following equation (Kottegoda, 1980):

$$x_t = T_t + P_t + S_t \tag{10.1}$$

where T_t = trend component, P_t = periodic component, S_t = stochastic component, and t = discrete values of time 1, 2, 3, ..., N. The first two components represent specific deterministic features and contain no element of randomness.

10.2.2 Stochastic Modeling

In subsurface hydrology, the main limitation for the researchers is the restricted quantum of available data, which makes representing a real picture of the subsurface hydrologic processes difficult. Most hydrologic time series studies aimed at understanding and quantitatively describing the *population*, as well as the stochastic process that generates it, are based on a limited number of *samples*. Similarly, a hydrologic time series can be predicted and/or simulated based on historical data using statistical tools, e.g., probabilistic or stochastic models. This practice of dealing with a subsurface hydrologic time series is known as "stochastic modeling," and the parameters associated with statistical and probabilistic terms are called "stochastic parameters."

10.3 CHECKING HYPOTHESES PRIOR TO STOCHASTIC MODELING

There are certain hypotheses about the hydrologic time series that should be confirmed as true prior to stochastic modeling. These hypotheses include for the following: (i) the hydrologic time series should follow the normal probability distribution, (ii) time series should possess stationarity, (iii) time series should be homogenous, (iv) time series should be free from trends, and (v) time series should be nonperiodic. Details about the techniques used to check these hypotheses of the subsurface hydrologic time series are mentioned in the following sub-sections.

10.3.1 Checking Normality of Time Series

The hypothesis of presence of normality is crucial for ensuring the reliability of the parametric statistical tests. Since 1990, advances in computer technology have enabled the effective application of statistical analyses. However, some of the researchers ignored the normality assumption before applying parametric tests to hydrologic time series (Adeloye and Montaseri, 2002). A variety of methods are available to test whether a time series describes the statistical normal distribution or not. The methods of normality testing can be categorized into two methods, graphical and statistical, as shown in Fig. 10.1. The graphical method is a qualitative method that is based on the visual comparison of the actual shape of the data distribution and normal probability distribution. It does not quantify the difference between the normal distribution and the sample distribution and thus it does not test whether the nonnormality is significant, if present. On the other hand, statistical normality tests have some potential problems. On testing the normality of a hydrologic time series with a small sample size, the statistical tests generally have little chance of rejecting the null hypothesis that the data come from a normal distribution. As a result most small-sample time series easily pass normality tests. In contrast, even a minor departure from a normal distribution may be found statistically significant in large sample-size time series, although it should be mentioned that small deviations from a normal distribution may not alter the results of a parametric test. The best practice is to evaluate the normality of the hydrologic time series by using both graphical and adequate statistical tests (Machiwal and Jha, 2012).

10.3.2 Detecting Homogeneity of Time Series

The hypothesis of homogeneity is considered true when a sample taken from any portion of the complete hydrologic time series represents the same statistical population with a time invariant mean. Hence, the significance of changes in the mean value of a time series is detected by employing statistical homogeneity tests. Three widely used homogeneity tests include the von Neumann test, Cumulative Deviations test, and Bayesian test, and their details can be found in Buishand (1982). In addition, there are few multiple comparison tests, e.g., Bartlett, Dunnett, Link-Wallace, Hartley, and Tukey tests, which have been popular in geotechnical studies (Phoon et al., 2003) but have not gained a wide popularity in hydrology. Various homogeneity tests are enlisted in Fig. 10.1, and their detailed procedures can be found in Kanji (2001), and Machiwal and Jha (2012).

10.3.3 Examining Stationarity of Time Series

Stationarity of a hydrologic time series may be tested by applying two general approaches: (i) parametric, and (ii) nonparametric. In literature, parametric tests are usually preferred by the researchers such as economists working in time domain, who make certain hypotheses about the nature of their data. However, nonparametric tests are favored by the researchers such as electrical engineers working in frequency domain, who consider the system as a black box without making any hypotheses about the nature of system. In the field of hydrology, both parametric and nonparametric tests have been used. The main advantage of the nonparametric tests is nondependency on the hypothesis that the *population* is normally distributed (Bethea and Rhinehart, 1991). On contrary, the nonparametric tests are less robust than parametric tests. Methods used to test stationarity of a time series are grouped in Fig. 10.1.

10.3.4 Evaluating the Presence of Trends

Similar to tests for stationarity, there exist parametric and nonparametric tests for evaluating trends in hydrologic time series. The parametric tests are reported to be more robust than the nonparametric tests but the former require fulfillment of normality and independence conditions, which are rarely met in most of the hydrological datasets (Machiwal and Jha, 2012). Perhaps, this may be the reason that the nonparametric tests have sought wide applications for identifying trends in hydrology (e.g., Machiwal and Jha, 2015a). Methods for detecting trends in hydrologic time series have been shown in

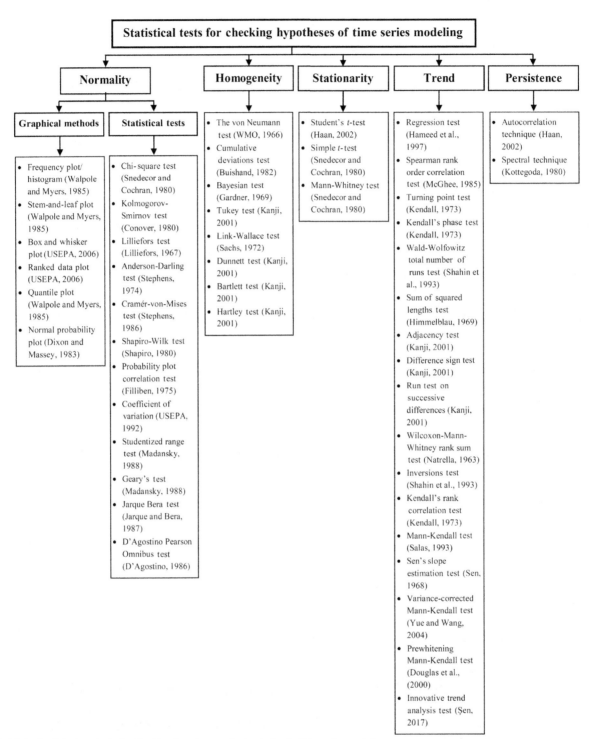

FIG. 10.1   Statistical methods for checking hypotheses prior to stochastic time series modeling.

Fig. 10.1. It is revealed that among various available nonparametric tests, two tests, i.e., Kendall's Rank Correlation (KRC) and Mann-Kendall (M-K) tests, are customarily applied for detecting trends in hydrologic time series.

10.3.4.1   The Effect of the Presence of Serial Correlation on the Mann-Kendall Test

It is reported by Yue et al. (2002) that the performance of the Mann-Kendall (M-K) test is affected by the presence of a serial correlation in the time series. Hence, trend-free prewhitening of the original time series is proposed prior to applying the

M-K test to avoid the effect of serial correlation on the robustness of the test (Douglas et al., 2000). The work of Yue et al. (2002) was further extended by Yue and Wang (2002), who emphasized that prewhitening is not suitable for removing the influence of serial correlation on the M-K test when trend does exist in a time series. A few recent studies (e.g., Sang et al., 2014) have further highlighted the fact that a prewhitening approach cannot really improve trend identification using the M-K test, but also may cause the wrong results to be produced. Therefore some studies suggested variance correction (VC) to the original data before applying the M-K test to eliminate the influence of serial correlation on the test (Hamed and Rao, 1998; Yue et al., 2002; Yue and Wang, 2004).

10.3.5 Testing Persistence in Time Series

Persistence is sometimes treated as periodicity, although there is a slight difference between these two terms. In some studies, persistence is also considered as randomness (McMohan and Mein, 1986). Hence, statistical tests used to identify the randomness of a hydrologic time series can also be used for testing persistence (Machiwal and Jha, 2006). Persistence in a time series can be detected by following two approaches: (i) time domain (autocorrelation technique), and (ii) frequency domain (spectral technique) (Fig. 10.1). In the literature the time domain approach is preferred to the frequency domain approach (Schwankl et al., 2000). The autocorrelation technique has also been used for examining periodicity in hydrologic time series.

10.4 FRAMEWORK FOR STOCHASTIC MODELING

Once all necessary hypotheses of a time series are tested, and the series is found normal, homogenous, stationary, free from trends, and without any periodicity and persistence, then the time series can be used for stochastic modeling. The procedure for applying stochastic modeling is discussed in detail in the following section, and a flowchart demonstrating the step-by-step procedure is shown in Fig. 10.2.

10.4.1 Making Time Series Ready for Stochastic Modeling

10.4.1.1 Detrending the Series

When a trend is found in a time series it is removed by detrending, which is a statistical or mathematical approach for removing trends from the time series. It is performed by fitting a linear time trend to the data and then subtracting the estimated trend from the series (Zhang and Qi, 2005). Detrending of a time series is carried out as shown in the following equation (Montgomery et al., 2015):

$$x_t - T_t = P_t + S_t \tag{10.2}$$

10.4.1.2 Standardizing the Series

The periodicity can be effectively removed by standardizing the original data series, and thereby constructing a transformed and stationary time series for further stochastic model development. The standardization of the series is done by subtracting mean from the variable and then dividing the residuals by standard deviation of the original series as shown in the following expression (Kottegoda, 1980):

$$y_t = \frac{x_t - \bar{x}_t}{\sigma_i} \tag{10.3}$$

where y_t = standardized or stationary time series in the mean and variance, $\bar{x}_t$ = sample mean of the series, and σ_i = standard deviations of the series.

10.4.1.3 Transforming the Series

There are a few specific types of transformations, which can be applied to transform a time series into stationary, and making free from trends and periodicity by stabilizing the variance of the data.

FIG. 10.2 Flowchart illustrating step-by-step procedure of stochastic modeling.

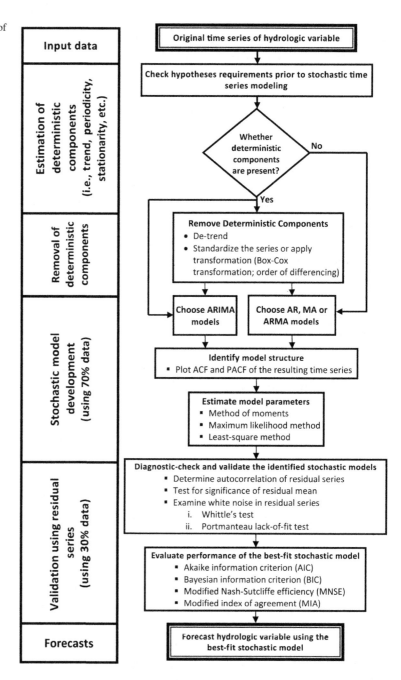

Box-Cox Transformation

In stochastic modeling after removing the trend and/or periodic components the residual series is assumed to be independent, homoscedastic (i.e., variance is a constant), and usually normally distributed. However, if the constant variance and normality assumptions are not true, they are often reasonably well-satisfied when the observations x_t are transformed by a Box-Cox transformation (Box and Cox, 1964; Hipel et al., 1977). The Box-Cox transformation is also known as the power transformation and is given by following formula (Hipel et al., 1977):

$$x_t^{\lambda} = \lambda^{-1} \left[(x_t + const)^{\lambda} - 1 \right] \lambda \neq 0 \tag{10.4}$$

$$x_t^{\lambda} = \ln(x_t + const)\lambda \approx 0 \tag{10.5}$$

where "*const*" = a constant term added to standardized data to make the complex data set positive and λ = another constant whose value is varied to get the desired transformation.

The values of "*const*" and λ are varied till the coefficient of skewness of the transformed series x_t becomes near zero. If $\lambda = 1$, there is no transformation. Typical values of λ used with time series data are $\lambda = 0.5$ (a square root transformation), $\lambda = 0$ (logarithmic transformation), $\lambda = -0.5$ (reciprocal square root transformation), and $\lambda = -1$ (inverse transformation) (Montgomery et al., 2015). The logarithmic transformations are commonly used for reducing right skewness, and this transformation is often useful for series that must be greater than zero and that grow exponentially. Similarly, inverse or reciprocal transformation can be used for nonzero data.

Order of Differencing

Differencing is a very useful method for removing trends and cycles present in the time series. The dth differencing operator is applied to an original time series x_t to create a new series z_t whose value at time t is the difference between x_{t+d} and x_t. Differencing is generally used to convert the nonstationary time series into a stationary time series (Montgomery et al., 2015). First order differencing, x_{t_1} is defined in the following expression:

$$x_{t_1} = x_t - x_{t-1} \tag{10.6}$$

Second order differencing, x_{t_2} is defined in the following expression:

$$x_{t_2} = x_{t_1} - x_{(t-1)_2} \tag{10.7}$$

and so on.

Differencing can allow the trend component to change through time. The first difference accounts for a trend that impacts the change in the mean of the time series, the second difference accounts for changes in the slope of the time series, and so forth. Usually, one or two differences are all that is required in practice to remove an underlying trend in the data (Montgomery et al., 2015).

10.4.2 Choosing Stochastic Model

In this section, few important classes of time series models are discussed.

10.4.2.1 Autoregressive Model

An autoregressive (AR) model of the order p assumes that a linear stochastic process can be represented such that the present value, x_t, of the process is a function of the previous values, $x_{t-1}, x_{t-2}, x_{t-3}$, etc., of the same process, and the model can be defined by the following expression (Shahin et al., 1993):

$$x_t = \mu + \alpha_1(x_{t-1} - \mu) + \alpha_2(x_{t-2} - \mu) + \ldots + \alpha_p\left(x_{t-p} - \mu\right) + \epsilon_t \tag{10.8}$$

where μ = mean of the series; $\alpha_1, \alpha_2, \ldots, \alpha_p$ = coefficients of the model; and ϵ_t = purely random process component.

10.4.2.2 Moving Average Model

We assume that ϵ_t is a purely random process with zero mean and variance σ_ϵ^2, then the process is said to be a moving average process of order q or MA(q) process that can be defined by the following expression (Shahin et al., 1993):

$$x_t = \mu + \epsilon_t + \theta_1\epsilon_{t-1} + \theta_2\epsilon_{t-2} + \cdots + \theta_{t-q}\left(\theta_q \neq 0\right) \tag{10.9}$$

where $\theta_1, \theta_2, \ldots, \theta_q$ = coefficients of model parameters.

10.4.2.3 Autoregressive Moving Average Model

A mixed autoregressive moving average (ARMA) model of order (p,q) may be expressed as (Shahin et al., 1993):

$$\begin{aligned} x_t = {} & \mu + \alpha_1(x_{t-1} - \mu) + \alpha_2(x_{t-2} - \mu) + \cdots + \alpha_p\left(x_{t-p} - \mu\right) + \epsilon_t \\ & + \theta_1\epsilon_{t-1} + \theta_2\epsilon_{t-2} + \cdots + \theta_{t-q} \end{aligned} \tag{10.10}$$

10.4.2.4 Autoregressive Integrated Moving Average Model

In a real-world time series of hydrology and hydrogeology most of the variables are nonstationary and, hence, the AR, MA, and ARMA models are not applicable to nonstationary processes. It is observed that sometimes the differences between successive values of a nonstationary time series form a stationary time series. Therefore a backward shift operator can be used to convert a nonstationary process into a stationary one by taking second-order differences. Briefly, it can be stated that if $w_t = \nabla^d x_t$ is stationary, where ∇^d represents the dth difference of the process x_t, then w_t is described by a stationary ARMA model of order (p,q). This can also be termed as autoregressive integrated moving average (ARIMA) process of order (p,d,q).

A stochastic ARIMA model can be given by following expression (Shahin et al., 1993):

$$x_t = w_t + B w_t + B^2 w_t + \cdots \tag{10.11}$$

where B = backward shift operator.

An ARIMA model can be generalized in many ways by considering the different terms in the models. The resulting models range from simple autoregressive (AR) models ($p=0$), moving average (MA) models ($q=0$), ARMA models ($d=0$) to ARIMA models. The ARIMA model is termed as a nonstationary model since it takes into account the nonstationarity of the data in the model. If the seasonal terms are absent and the differencing term "d" is set equal to zero then the model is called a stationary autoregressive moving average (ARMA) [identified by (p,q)].

10.4.3 Identifying the Model Structure

In the model structure identification step, orders of the model are determined. Detailed guidelines for identification of possible model structure of ARMA models can be found in standard textbooks (e.g., Hipel and Mcleod, 1994; Machiwal and Jha, 2012). A general idea about the model structure can be obtained by considering the following four criteria.

10.4.3.1 Plot of the Original Series

A visual inspection of the time plot of the original time series can reveal one or more of the time series characteristics, i.e., seasonality, trends, persistence, long-term cycles, and/or extreme values or outliers.

10.4.3.2 Plot of Standardized Series

The time plot of the standardized time series can be visually examined to get information as to whether the time series is stationary or not.

10.4.3.3 Plot of Autocorrelation Function

The autocorrelation function (ACF) can be used to detect recurrence or periodicity in a time series. The ACF measures the correlation of a variable x_t at time t with x_{t+k} shifted by some time delay or lag k. This covariance is called the autocovariance at lag k and is defined by (Box et al., 2008):

$$\gamma_k = \text{cov}[x_t, x_{t+k}] = E[(x_t - \bar{x}_t)(x_{t+k} - \bar{x}_t)] \tag{10.12}$$

Similarly, the autocorrelation at lag k is:

$$\rho_k = \frac{\sum[(x_t - \bar{x}_t)(x_{t+k} - \bar{x}_t)]}{\sqrt{\sum\left[(x_t - \bar{x}_t)^2\right]\sum\left[(x_{t+k} - \bar{x}_t)^2\right]}} \tag{10.13}$$

$$= \frac{\sum[(x_t - \bar{x}_t)(x_{t+k} - \bar{x}_t)]}{\sigma_x^2} \tag{10.14}$$

where ρ_k = autocorrelation function with time lag k and k = time lag between the correlated pairs (x_t, x_{t+k}). Since, for a stationary process when $k=0$, the variance of the time series $\sigma_x^2 = \gamma_0$. Thus the autocorrelation at lag k, that is, the correlation between x_t and x_{t+k}, is:

$$\rho_k = \frac{\gamma_k}{\gamma_0} \tag{10.15}$$

10.4.3.4 Plot of Partial Autocorrelation Function

In choosing the appropriate structure of the stochastic model, the partial autocorrelation function (PACF) plays an important role in identifying the extent of the lag in an autoregressive model. By plotting the PACFs one could determine the appropriate lags p in a time series x_t (Box and Jenkins, 1970). The PACF at lag k, denoted by $\alpha(k)$, is the autocorrelation between x_t and x_{t+k} with the linear dependence of x_{t+k} through to x_{t+k-1} removed; equivalently, it is the autocorrelation between x_t and x_{t+k} that is not accounted for lags 1 to $k-1$, inclusive:

$$\alpha(1) = Cor(x_t, x_{t+1}) \tag{10.16}$$

$$\alpha(k) = Cor\big(x_{t+k} - P_{t+k}(x_{t+k}), x_t - P_{t,k}(x_t)\big), \quad \text{for } k \geq 2 \tag{10.17}$$

where $P_{t+k}(x) =$ the projection of x onto the space spanned by $x_{t+1}, ..., x_{t+k-1}$; $\alpha(t) =$ PACF at lag t; and $\alpha(k) =$ PACF at lag k.

The PACF plots are commonly used for identifying the order of an autoregressive model (Box et al., 2008). The PACF of an AR(p) process is zero at lag $p+1$ and greater. If the sample ACF plot indicates that an AR model may be appropriate, then the sample PACF plot is examined to help identify the order (p). One looks for the point on the plot where the PACF for all higher lags are essentially zero. Placing on the plot an indication of the sampling uncertainty of the sample PACF is helpful for this purpose: this is usually constructed on the basis that the true value of the PACF, at any given positive lag, is zero. The entire procedure is briefly described here. An approximate test that a given partial correlation is zero (at a 5% significance level) is given by comparing the sample PACF values against the critical region with upper and lower limits given by $\pm 1.96/\sqrt{n}$, where n is the record length (number of points) of the time-series being analyzed. This approximation relies on the assumption that the record length is moderately large (say $n > 30$) and that the underlying process has a finite second moment. The general attributes of all these useful identification functions are summarized in Table 10.1.

10.4.4   Estimating Model Parameters

The structure identification process leads to a tentative formulation of the model and then model parameters can be estimated by the statistical analysis of the time series. Three fundamental methods are used for the estimation of model parameters: (i) the method of moments, (ii) the method of least-squares, and (iii) the method of maximum likelihood, which are briefly discussed in the following section.

10.4.4.1 Method of Moments

The method of moments is one of the easiest methods for estimating the parameters of the chosen stochastic models, although it is not the most efficient. Using this method unknown parameters can be estimated by matching theoretical

TABLE 10.1 Guidelines to Identify Model Parameters on the Basis of ACF and PACF (Machiwal and Jha, 2012)

S. No.	Model Parameter	Characteristics of ACF	Characteristics of PACF
1	One autoregressive (p)	Exponential decay	Spike at lag-1, no correlations for other lags
2	Two autoregressive (p)	A sine-wave shape pattern or a set of exponential decays	Spikes at lags 1 and 2, no correlation for other lags
3	One moving average (q)	Spike at lag-1, no correlation for other lags	Damps out exponentially
4	Two moving average (q)	Spikes at lags-1 and 2, no correlation for other	A sine-wave shape pattern or a set of exponential decays lags
5	One autoregressive (p) and one moving average (q)	Exponential decay starting at lag-1	Exponential decay starting at lag-1

moments with the appropriate sample moments. Let us consider sample time series $x_1, x_2 \ldots x_n$. The nth sample moment is defined as follows:

$$M_n = \left(\frac{1}{n}\right) \sum_{i=1}^{n} x_i^n \tag{10.18}$$

10.4.4.2 Method of Least-Squares

The method of moments sometimes gives unsatisfactory estimates for model parameters, in which case the method of least-squares can be adopted. Using this method the parameters are estimated in two steps: (i) a preliminary or initial estimation of model parameters is made, and (ii) an exact estimate of the parameters is obtained. Let us consider that the following model is fitted to the time series $x_1, \ldots, x_n$:

$$x_t = f(x_{t-1}, x_{t-2}, \ldots, \omega_1, \omega_2, \ldots, \omega_m) + a_t \tag{10.19}$$

where $\omega_1, \omega_2, \ldots, \omega_m$ = model parameters and a_t = residual series.

In the least-squares method, the following condition must be minimized:

$$\sum_{t=1}^{N} (x_t - \hat{x}_t)^2 = \sum_{t=1}^{N} (x_t - f(x_{t-1}, x_{t-2}, \ldots, \hat{\omega}_1, \hat{\omega}_2, \ldots, \hat{\omega}_m))^2 \tag{10.20}$$

In this method, all partial derivatives with respect to the estimated values of the parameters, $\hat{\omega}_1, \hat{\omega}_2, \ldots, \hat{\omega}_m$, are equated to zero. Therefore it can be defined as the following:

$$\frac{\partial \sum_{t=1}^{N} (x_t - \hat{x}_t)^2}{\partial \hat{\omega}_1} = 0, \ldots, \frac{\partial \sum_{t=1}^{N} (x_t - \hat{x}_t)^2}{\partial \hat{\omega}_m} = 0 \tag{10.21}$$

These equations are solved simultaneously to obtain estimates of the model parameters.

10.4.4.3 Method of Maximum Likelihood

The values of the parameters that maximize the likelihood function, or equivalently the log-likelihood function, are called maximum likelihood estimates (Box et al., 2008). In this method the likelihood values are calculated for each of the candidate models. The model with highest likelihood value is selected. The general form of log-likelihood value for the ith model for a Gaussian process is given by (Kashyap and Rao, 1976):

$$L_i = \ln \left[P\left(z, \hat{\phi}_i\right) \right] - n_i \tag{10.22}$$

The specific likelihood function in this general class can be written as (Kashyap and Rao, 1976):

$$L_i = -\frac{N}{2} \ln (\sigma_i) - n_i \tag{10.23}$$

where L_i = likelihood value; P = probability density function; z = vector of historical series; $\hat{\phi}_i$ = vector of parameters and residual variance $(\theta_1, \theta_2, \ldots; \hat{\phi}_1, \hat{\phi}_2, \ldots; \sigma_i)$; σ_i = residual variance; n_i = number of parameters; and N = number of data values.

10.4.5 Diagnostic Checking and Validating Identified Models

The whole time series is divided into two parts containing the initial 70% and last 30% of the dataset. The first part with 70% of the data is utilized for developing the stochastic model by following a model structure identification step. The remaining 30% of the dataset is utilized for the verification or validation of the identified and developed models. The identified stochastic model is run using the time series data for the validation period and the values are predicted for the same period. Then a residual series for the selected model is constructed for the validation period. The residuals are the differences between the observed and model-predicted values. The selected model is diagnostically checked using three criteria to ascertain whether the assumptions used in building the model are valid or not. The three criteria are: (i) the residual series should be normally distributed; (ii) no significant periodicity should be present in the residual series; and (iii) the residual series should be uncorrelated. The independent model diagnostic-checking is accomplished through careful analysis of the residual series, the histogram of the standardized residual series, sample correlation, and a diagnosis test (Ljung and Box, 1978). Once tested and successfully verified, the developed stochastic model can be applied for further forecasting of the hydrologic variable. For validating the selected model, the following tests can be adopted to examine whether the assumptions used in the building of the model are valid or not. All the validation tests are applied to the residual series only. The residual series is generated using the following expression (Mujumdar and Kumar, 1990):

$$a_t = x_t - \sum_{j=1}^{p} \emptyset_j x_{t-1} - \sum_{j=1}^{q} \theta_j a_{t-1} + C \tag{10.24}$$

where p = number of AR parameters; $\emptyset_j$ = jth AR parameter; q = number of MA parameters; θ_j = jth MA parameter, and C = constant.

10.4.5.1   Determining Autocorrelations in Residual Series

The ACFs of the residual series are calculated with their confidence limits (upper and lower limits). Autocorrelograms of residual a_t for all the selected models are constructed to adjudge the mutual dependency. If the autocorrelograms of residual a_t are found to appear within the corresponding limits, the residual series obtained from the model is considered as mutually independent, as they are not significantly different from zero.

10.4.5.2   Testing Significance of Residual Mean

The purpose of this test is to examine the validity of the assumption that the series x_t has the zero mean. For this purpose, a statistic, $\eta(x)$, is defined as:

$$\eta(x) = \left(N^{1/2} \times \bar{a}_t \right) / \hat{\rho}^{1/2} \tag{10.25}$$

where $\bar{a}_t$ = estimate of the residual mean and $\hat{\rho}$ = estimate of the residual variance.

The statistic $\eta(x)$ is approximately distributed by $t(\alpha, N-1)$, where α is the significance level of the test. If the value of $\eta(x) \leq t(\alpha, N-1)$, then mean of the residual series is not significantly different from zero and, hence, the residual series passes the validity test.

10.4.5.3   Examining White Noise in Residual Series

An important assumption in the stochastic time series modeling studies is that the residual series a_t must necessarily be a white noise sequence (or that the residual series is uncorrelated). The white noise or absence of correlation in the residual series may be tested by using two tests: (i) Whittle's test, and (ii) Portmanteau lack-of-fit test, which are briefly described in the following subsections.

Whittle's Test

This test involves the construction of the covariance matrix (Whittle, 1952). The covariance ρ_k at lag k of the residual series a_t is estimated by:

$$\rho_k = \frac{1}{(N-k)} \sum_{j=k-1}^{N} a(j)a(j-k) \tag{10.26}$$

where k = 0, 1, 2, ..., k_{max}.

The value of k_{max} is normally chosen as 15% of the sample size, i.e., k_{max} = 0.15N. The covariance matrix τ_{n1} is then constructed as:

$$\tau_{n1} = \begin{bmatrix} \rho_0 & \rho_1 & \rho_2 & \cdots & \rho_{k_{max}} \\ \rho_1 & \rho_0 & \rho_2 & \cdots & \rho_{k_{max}-1} \\ \rho_2 & \rho_1 & \rho_0 & \cdots & \rho_{k_{max}-2} \\ \vdots & \vdots & \vdots & \ddots & \vdots \\ \rho_{k_{max}} & \rho_{k_{max}-1} & \rho_{k_{max}-2} & \cdots & \rho_0 \end{bmatrix} \tag{10.27}$$

This is a square symmetric matrix of size $n_1 = k_{\max}$. The statistic $\eta(a)$ is defined as:

$$\eta(a) = \left(\frac{N}{n_1 - 1}\right)\left(\frac{\hat{\rho}_0}{\hat{\rho}_1} - 1\right) \tag{10.28}$$

where $\hat{\rho}_0 = $ lag zero correlation $(=1)$ and:

$$\hat{\rho}_0 = \frac{\det \tau_{n1}}{\det \tau_{n1-1}} \tag{10.29}$$

where $det\tau_{n1}$ and $det\tau_{n1-1} = $ determinants of matrices τ_{n1} and τ_{n1-1}, respectively. The statistic $\eta(a)$ is distributed approximately as $F_\alpha(n_1, N - n_1)$. If $\eta(a) \leq F_\alpha(n_1, N - n_1)$ then the residual series is considered as uncorrelated.

Portmanteau Lack-of-Fit Test

The Portmanteau lack-of-fit test is used for checking the adequacy of stochastic models. The statistic for the Portmanteau lack-of-fit test is given by (Bhattacharya and Burman, 2016):

$$Q = N \sum_{k=1}^{L} \rho_k^2(a) \tag{10.30}$$

where $\rho_k(a) = $ estimated autocorrelations of the residuals a_t and $L = $ maximum lag considered. The static Q is approximately described by Chi-square (χ^2) distribution with L-p-q degree of freedom. The value of L is taken up to 10%–30% of the sample size N, and the test is performed for different values of L. The adequacy of the ARMA model for x_t may be checked by comparing $\chi^2(L$-p-$q)$ of a given significance level. If $Q < \chi^2(L$-p-$q)$, series a_t is considered as an independent series and, hence, the stochastic models are inadequate. The Portmanteau lack-of-fit test is reported to be inferior to the Whittle's test and, hence, the latter is recommended to examine white noise in the residual series (Kashyap and Rao, 1976). However, the Portmanteau test is preferred in hydrological studies.

10.4.6 Performance Evaluation for Selecting the Best-Fit Model

In order to create a better balance among generalization ability, parsimony, and training speed, the use of two additional performance-indicator tests, i.e., the Akaike information criterion (AIC) and the Bayesian information criterion (BIC), is suggested (Akaike, 1974; Rissanen, 1978). In addition, the comparative performance of the stochastic models can also be evaluated using correlation-based or error-based criteria. However, correlation-based criteria, such as correlation coefficient and coefficient of determination, are limited by their oversensitivity to outliers and insensitivity to the additive and proportional differences between the observations and regression-based predictions (Moore, 1991). Therefore two error-based criteria, namely modified Nash-Sutcliffe efficiency (MNSE) (Legates and McCabe, 1999), and modified index of agreement (MIA) (Willmott et al., 1985), can be employed to ensure a more efficient evaluation. The four evaluation criteria are briefly summarized in the following section.

10.4.6.1 Akaike and the Bayesian Information Criteria

The goal is to minimize AIC to obtain a network with best generalization. Although root mean square error (RMSE) statistics are expected to progressively improve as more parameters are added to the model, the AIC and BIC statistics penalize the model for having more parameters, and therefore tend to result in more parsimonious models (Hsu et al., 1995). Model selection is performed by looking for the minimum BIC value. It turns out that the final form of this criterion is rather similar to that of AIC but one can see that the penalty due to the number of model parameters is multiplied by the "natural logarithmic" value of the RMSE. As a consequence BIC leans more than AIC toward lower-dimensional models (Magar and Jothiprakash, 2011):

$$\text{AIC} = [n \times \ln(\text{RMSE})] + 2(p + q) \tag{10.31}$$

$$\text{BIC} = [n \times \ln(\text{RMSE})] + [(p + q) \times \ln(n)] \tag{10.32}$$

where RMSE = root mean square error and $n = $ number of parameters to be estimated.

10.4.6.2 Modified Nash-Sutcliffe Efficiency

The Nash-Sutcliffe efficiency (NSE) is a normalized statistic that determines the relative magnitude of the residual variance ("noise") compared to the measured data variance (Nash and Sutcliffe, 1970). A modified NSE is suggested that is less sensitive to high extreme values due to the squared differences (Legates and McCabe, 1999). The MNSE, indicating how well the plot of observed versus simulated data fits the 1:1 line, is computed as shown in following equation (Legates and McCabe, 1999):

$$\mathrm{MNSE} = 1 - \frac{\sum_{i=1}^{N} |O_i - P_i|}{\sum_{i=1}^{N} |O_i - \overline{O}|} \tag{10.33}$$

The MNSE ranging between 0 and 1.0 indicates an acceptable level of performance, whereas MNSE equal to 1 is considered as the best and $-\infty$ as the worst (Machiwal and Jha, 2015b).

10.4.6.3 Modified Index of Agreement

The index of agreement (IA) was developed by Willmott (1981) as a standardized measure of the degree of model prediction error and varies between 0 and 1. A computed value of 1 indicates a perfect agreement between the measured and predicted values, and 0 indicates no agreement at all. The IA can detect additive and proportional differences in the observed and simulated means and variances; however, it is oversensitive to extreme values due to the squared differences. The modified index of agreement (MIA) is less sensitive to high extreme values because errors and differences are given appropriate weighting by using the absolute value of the difference instead of using the squared differences (Legates and McCabe, 1999):

$$\mathrm{MIA} = 1 - \frac{\sum_{i=1}^{N} (O_i - P_i)}{\sum_{i=1}^{N} \left(|P_i - \overline{O}| + |O_i - \overline{O}| \right)} \tag{10.34}$$

where O_i = observed value; P_i = predicted value; $\overline{O}$ = mean of observed values; and N = number of observations.

The acceptable range of the MIA varies from 0.5 to 1 with higher values indicating a better fit of the model, whereas MIA equal to 1 is the best and 0 is the worst (Machiwal and Jha, 2015b).

10.4.7 Forecasting of the Subsurface Hydrologic Time Series

Once the best-fit stochastic model is identified, validated, and evaluated, the same can be used either for the generation of synthetic time series or forecasting future events for one or several time steps ahead. Box and Jenkins (1970) showed that the minimum mean square error forecast $x_t(l)$ is the conditional expectation of x_{t+l} at time t; when regarded as a function of l for fixed t, $x_t(l)$ may be termed the forecast function. An observation x_{t+l} at time $t+l$ can be expressed as an infinite weighted sum of previous random values. The forecasts are explicitly calculated for $l = 1, 2, \ldots,$ as given by following expression (Box et al., 2008):

$$x_t(l) = \sum_{j=0}^{p+d} \phi_j x_{t+l-j} + \sum_{j=l}^{q} \theta_j a_{t+l-j} \tag{10.35}$$

where $x_t(-j) = [x_{t-j}]$ denotes the observed value x_{t-j} for $j \geq 0$ and the moving average terms are not present for lead times $l > q$.

10.5 CURRENT STATUS OF STOCHASTIC TIME-SERIES MODELING APPLICATIONS IN SUBSURFACE HYDROLOGY

The application of stochastic time series modeling to hydrology was started about 6–7 decades ago, although it was restricted up to the problems of surface hydrology, such as floods and droughts, in the early days (McCuen, 2003).

The development of a large number of stochastic time series analysis methods and their application to studies considering surface hydrologic variables were witnessed during 1950s and 1960s (Sivakumar, 2017). However, stochastic time series modeling could not be adequately applied to subsurface hydrologic studies up until the end of 1990s (Shahin et al., 1993; Machiwal and Jha, 2006). Since 2000, the application of stochastic modeling to subsurface hydrology has been increasingly reported, and mainly groundwater level and groundwater quality time series are chosen by researchers for this purpose. Presently, the domain of stochastic time series modeling encompasses both surface and subsurface hydrology, and it is a proven and powerful tool for hydrologic forecasting.

It is revealed in the literature that several studies have tested hypotheses of the presence of normality (e.g., Mouser et al., 2005; Chou, 2006; Aguilar et al., 2007; Ayuso et al., 2009; Nas and Berktay, 2010; Narany et al., 2014; Noshadi and Ghafourian, 2016; Jovein and Hosseini, 2017) and trends (e.g., Taylor and Loftis, 1989; Loftis, 1996; Visser et al., 2009; Scanlon et al., 2010; Kaown et al., 2012; Machiwal and Jha, 2015a; Yazdanpanah, 2016; Koh et al., 2017) in groundwater quality time series as required prior to stochastic time series modeling. However, these studies did not extend their work to perform stochastic modeling. Also, it is observed that, excepting normality and trends, other hypotheses of the hydrologic time series, such as homogeneity, stationarity, periodicity, and persistence, have not been examined in studies dealing with subsurface hydrology.

An extensive literature search revealed that the first ever application of stochastic time series modeling to subsurface hydrologic variable was made by Law (1974), who estimated deterministic (trend and periodicity) and stochastic components in a monthly groundwater level time series for a total 84 wells in 22 states of the Western United States. First, deterministic components were removed if found significant and, second, the resulting stochastic series was tested for independence using Markov models (autoregressive models) up to three lags. Almost at the same time, Rao et al. (1975) successfully modeled a groundwater level series by using stochastic models. Their study utilized 30 years' data of two wells located in West Lafayette, Indiana, and of a third well located in Wool County in north central Wisconsin. The stochastic time series models explained 70%–80% of the groundwater table variation in Wisconsin and 30%–50% of the variation in Indiana, where the aquifers were affected by pumping and the influence of the Wabash River.

During the 1990s other studies dealing with stochastic time series modeling of groundwater hydrologic variables were reported. For example, Ahn and Salas (1997) used daily groundwater head time series (1985–1990) of Collier County of Florida, United States for stochastic time series modeling. An approach was presented to determine the uniform sampling time interval for monitoring serially correlated groundwater levels by developing the first-order difference ARIMA models for seven wells. Later on, using the same datasets that appeared in Ahn and Salas (1997), Ahn (2000) introduced an approach to build time series models of nonstationary data at different time intervals using second-order differencing. Vazquez-Amabile and Engel (2008) forecasted the groundwater depth of three observation wells located at the Muscatatuck Wildlife Refuge in the Storm Creek lower watershed, southeast Indiana. The groundwater depths were predicted for the period 1992–1996 using historical data as well as data generated by using the soil and water assessment tool (SWAT) model. The data simulated through SWAT and time series modeling were compared, and performance of time series modeling was found superior based on correlation coefficient and NSE criteria. Similarly, stochastic autoregressive (AR) models have been compared with back-propagation artificial neural network (BPANN) models in predicting groundwater levels in western Jilin province of China (Yang et al., 2009). Efficacy of all the models was assessed by using RMSE, mean absolute error, and NSE criteria. The results suggested that the groundwater level predictions made by the BPANN models were more precise in comparison to those predicted by AR models.

Since 2010, many studies have employed stochastic time series modeling to analyze subsurface hydrologic variables, mainly for groundwater levels. Panda and Kumar (2011) modeled the groundwater levels for five wells for time periods ranging from 13 years (1988–2000) to 31 years (1970–2000) in four blocks of the Balasore district of Orissa, India, using seasonal autoregressive integrated moving average (SARIMA) modeling in SPSS software. Results suggested that number and type of SARIMA model parameters do not differ much from each other for the five wells, and this further suggested that the aquifer system behaves consistently in the area. García-Díaz (2011) combined the stochastic ARIMA model with statistical process control to predict nitrate concentration in groundwater in the Valencian Community, East Spain. The capability of the best-fit ARIMA (0,1,1) model was verified for making forecasts using the t-ratio test, sample ACF and PACF for residuals, and the Ljung-Box test. Mirzavand et al. (2014) predicted groundwater levels in Kashan aquifer located in an arid region of the Esfahan province, Iran. Groundwater levels of 36 piezometers from 1999 to 2004 were used for model development and from 2005 to 2010 were used for model validation. The AIC and correlation coefficients adjudged AR (2) model as the best-fit, which was then used to predict the groundwater levels for the next 60 months. Patle et al. (2015) identified trends and forecasted pre- and postmonsoon groundwater levels for Karnal district of Haryana, India using 1974–2010 data. The best-fit ARIMA (0,1,2) model indicated a possible decline of about 13 and 12 m in the pre- and postmonsoon groundwater levels by 2050 over the groundwater levels in 2010. Groundwater depth in Kabudarahang aquifer located in Hamadan province of Iran was simulated and temporal change was predicted by Khorasani et al. (2016) using 2003–2017 data. The SARIMA models were developed using 12 years' data and verified using the remaining 3 years' data using the goodness-of-fit AIC criterion. Model results revealed a 5 m decline in groundwater depth over the next 3 years. Similarly, in coastal aquifer of South China, Yang et al. (2017) simulated groundwater levels using monthly data for the period 2000–2011 by developing SARIMA models in SPSS software. Periodic components were detected by plotting ACF and PACF of the time series and trend was removed by applying first-order differencing. The best-fit model was selected as ARIMA $(0,1,0) \times (0,1,1)^{12}$ based on three goodness-of-fit criteria, i.e., coefficient of determination, Nash-Sutcliffe model efficiency, and RMSE. Narany et al. (2017) predicted nitrate concentrations in groundwater for the period 2015–2030 using ARIMA models based on data collected for the period 1989–2014 in Northern Kelantan, Malaysia. The outcomes of the best-fit ARIMA (1,2,2) and (2,2,2) models suggested a linear increase in the nitrate concentrations from 2015 to 2030 at both sites. Groundwater levels in the White Volta River basin of Ghana were examined for the presence of trends and future forecasting using ARIMA modeling (Gibrilla et al., 2018). Trends in groundwater levels monitored during 2005–2014 at seven sites were identified using the M-K test and the best-fit ARIMA models were chosen based on the AIC and BIC criteria using R-software. Results indicated stable groundwater levels over most of the sites with increasing trends at a rate ranging from 0 to 0.006 m year^{-1}, although the groundwater level was found to be declining at one site by 1.008 m annually.

Recently, researchers have started comparing stochastic models with artificial intelligence techniques. For example, Adamowski and Chan (2011) compared ARIMA models with wavelet-transformed artificial neural network (ANN) models in predicting monthly groundwater levels at two sites in the Chateauguay watershed in Quebec, Canada. The study used data from November 2002 to October 2009 for 1 month lead time forecasting. The wavelet-transformed ANN models were found to provide more accurate forecasts compared to ARIMA models. On the other hand, Behnia and Rezaeian (2015) coupled stochastic time series models (ARMA, ARIMA, and SARIMA) with wavelet transformations to predict groundwater levels in two sub-basins of the Mashhad Plain, Iran. Results indicated that the hybrid models improved the groundwater level modeling, and wavelet-SARIMA hybrid model had the best performance due to the enhanced capability to cope with the nonlinearity and seasonality of data. Moreover, Choubin and Malekian (2017) compared nonlinear ANN and ARIMA models in forecasting groundwater levels up to 4 months ahead in the Shiraz Basin, southwest Iran. ARIMA (2,1,2) was the best-fit stochastic model selected based on the AIC and Schwarz Bayesian Criterion. The results indicated that the performance of the ARIMA model was appreciably better than the ANN.

10.6   SCOPE FOR FUTURE RESEARCH

It is evident that application of stochastic time series modeling in subsurface hydrology has received a great deal of attention from researchers since 2000. However, many subsurface hydrologic and hydrogeologic variables still have not been analyzed by the stochastic modeling. It is known that the groundwater level and groundwater quality are the only parameters that have been the main focus of studies dealing with stochastic time series modeling in subsurface hydrology. It also has been recognized from past studies that none of the hypotheses for stochastic modeling were tested before being applied for the prediction of groundwater levels. Mainly trends and periodicity were examined, and other necessary requirements, such as stationarity, homogeneity, and persistence, were neglected. Future studies will need to have tested all hypotheses in the groundwater level and water quality time series. Also, researchers should attempt to employ stochastic time series modeling to study the behavior of other subsurface hydrologic and hydrogeologic time series, such as groundwater recharge, draft, etc.

After 2000 many studies employed stochastic time series modeling to forecast/predict groundwater levels, although no significant advances have been made as yet. The application of stochastic time series modeling to subsurface hydrology is still in the embryonic stage as compared to their applications in surface hydrology, and there is a gap between their application in surface and subsurface hydrology. The main reason behind this gap is the priority given to surface-water resources in the earlier days due to their abundance and a lack of any water-quality issues. Stochastic modeling methods were frequently applied more than 50 years ago in the design of surface-water reservoirs and for the forecasting of river inflows. Researchers working in the field of subsurface hydrology should conquer the deficit in the studies by doing more and more research considering stochastic time series modeling for the prediction of subsurface hydrologic variables.

The primary requirement for applying stochastic time series models is a sufficient number of long data records, which are generally not available in many parts of the world, especially in developing nations like India where the monitoring of subsurface hydrologic data is limited due to inadequate funds. The cost of obtaining hydrologic and hydrogeologic data, e.g., groundwater level, water quality, etc., from the subsurface is usually very high because it involves drilling of wells and boreholes. Unfortunately, a large amount of data is necessary to understand the mechanisms of the groundwater stochastic processes and to validate or calibrate the stochastic models. Wherever data are available for groundwater levels and water quality, the period of availability is usually short, as the monitoring of subsurface data began very late in many countries and the sampling interval is quite large. Also, the subsurface hydrologic data at basin scale are difficult to acquire for the areas where groundwater basins have not been identified. A major priority for the future is to strengthen the monitoring of subsurface hydrologic data to ensure the availability of better-quality and long-term datasets.

Another reason for stochastic time series modeling being applied in solving problems of subsurface hydrology less frequently is the complex composition of micro- and macropores, which greatly influence the chemical, physical, and biological processes and are extremely difficult to analyze even with stochastic or statistical methods. In fractured aquifers the situation is further aggravated because the presence of heterogeneity and anisotropy makes it difficult to predict groundwater flow and compute flow velocity. To overcome this limitation, there is a need to improve understanding of groundwater flow systems especially in fractured aquifers by adopting modern and advanced tools and techniques.

Besides all the bottlenecks restricting a comprehensive application of stochastic time series modeling in subsurface hydrology, the number of studies on the application of stochastic models to subsurface variables has steadily increased over the last 2–3 decades. This clearly indicates that more and more researchers are engaged in this area and that stochastic time series modeling is an active subject of research for the management of the groundwater resources. In the future researchers will have to clarify three major points: (i) the relation between the deterministic and stochastic approaches while studying a natural subsurface system, (ii) the exact physical causes and mechanisms for the natural variability (heterogeneity and anisotropy) and stochasticity of the subsurface system, and (iii) the importance and effects of the scales in space and time while describing a natural subsurface system.

REFERENCES

Adamowski, J., Chan, H.F., 2011. A wavelet neural network conjunction model for groundwater level forecasting. J. Hydrol. 407, 28–40.

Adeloye, A.J., Montaseri, M., 2002. Preliminary streamflow data analyses prior to water resources planning study. Hydrol. Sci. J. 47 (5), 679–692.

Aguilar, J.B., Orban, P., Dassargues, A., Brouyère, S., 2007. Identification of groundwater quality trends in a chalk aquifer threatened by intensive agriculture in Belgium. Hydrogeol. J. 15, 1615–1627.

Ahn, H., 2000. Modeling of groundwater heads based on second-order difference time series models. J. Hydrol. 234, 82–94.

Ahn, H., Salas, J.D., 1997. Groundwater head sampling based on stochastic analysis. Water Resour. Res. 33 (12), 2769–2780.

Akaike, H., 1974. A new look at the statistical model identification. IEEE Trans. Autom. Control 19 (6), 716–723.

Ayuso, S.V., Acebes, P., López-Archilla, A.I., Montes, C., Guerrero, M.C., 2009. Environmental factors controlling the spatiotemporal distribution of microbial communities in a coastal, sandy aquifer system (Doñana, southwest Spain). Hydrogeol. J. 17, 767–780.

Behnia, N., Rezaeian, F., 2015. Coupling wavelet transform with time series models to estimate groundwater level. Arab. J. Geosci. 8 (10), 8441–8447.

Bethea, R.M., Rhinehart, R.R., 1991. Applied Engineering Statistics. Marcel Dekker, Inc., New York.

Bhanja, S.N., Mukherjee, A., Rodell, M., Wada, Y., Chattopadhyay, S., Velicogna, I., Pangaluru, K., Famiglietti, J.S., 2017. Groundwater rejuvenation in parts of India influenced by water-policy change implementation. Sci. Rep. 7. https://doi.org/10.1038/s41598-017-07058-2 (Article 7453).

Bhattacharya, P.K., Burman, P., 2016. Theory and Methods of Statistics. Academic Press, Amsterdam.

Box, G.E., Cox, D.R., 1964. An analysis of transformations. J. R. Stat. Soc. Ser. B: Methodol., 211–252.

Box, G.E.P., Jenkins, G.M., 1970. Time Series Analysis, Forecasting and Control. Holden-Day, San Francisco.

Box, G.E.P., Jenkins, G.M., Reinsel, G.C., 2008. Time Series Analysis: Forecasting and Control, fourth ed. Wiley Series in Probability and Statistics, Wiley, Hoboken, NJ.

Buishand, T.A., 1982. Some methods for testing the homogeneity of rainfall records. J. Hydrol. 58, 11–27.

Chou, C.J., 2006. Assessing spatial, temporal, and analytical variation of groundwater chemistry in a large nuclear complex, USA. Environ. Monit. Assess. 119, 571–598.

Choubin, B., Malekian, A., 2017. Combined gamma and M-test-based ANN and ARIMA models for groundwater fluctuation forecasting in semiarid regions. Environ. Earth Sci. 76, 538. https://doi.org/10.1007/s12665-017-6870-8.

Conover, W.J., 1980. Practical Nonparametric Statistics, second ed. John Wiley, New York.

D'Agostino, R.B., 1986. Tests for the Normal Distribution. In: D'Agostino, R.B., Stephens, M.A. (Eds.), Goodness-of-Fit Techniques. Marcel Dekker, New York, USA.

Dixon, W.J., Massey Jr., F.J., 1983. Introduction to Statistical Analysis, fourth ed. McGraw-Hill, New York, USA.

Douglas, E.M., Vogel, R.M., Kroll, C.N., 2000. Trends in floods and low flows in the United States: impact of spatial correlation. J. Hydrol. 240 (1–2), 90–105.

Filliben, J.J., 1975. The probability plot correlation coefficient test for normality. Technometrics 17, 111–117.

Fogg, G.E., LaBolle, E.M., 2006. Motivation of synthesis, with an example on groundwater quality sustainability. Water Resour. Res. 42, W03S05. https://doi.org/10.1029/2005WR004372.

García-Díaz, J.C., 2011. Monitoring and forecasting nitrate concentration in the groundwater using statistical process control and time series analysis: a case study. Stoch. Environ. Res. Risk Assess. 25, 331–339.

Gardner Jr., L.A., 1969. On detecting changes in the mean of normal variates. Ann. Math. Stat. 40, 116–126.

Gibrilla, A., Anornub, G., Adomako, D., 2018. Trend analysis and ARIMA modelling of recent groundwater levels in the White Volta River basin of Ghana. Groundwater Sustain. Dev. 6, 150–163.

Guganesharajah, K., Pavey, J.F., van Wonderen, J., Khasankhanova, G.M., Lyons, D.J., Lloyd, B.J., 2007. Simulation of processes involved in soil salinization to guide soil remediation. J. Irrig. Drain. Eng. ASCE 133, 131–139.

Haan, C.T., 2002. Statistical Methods in Hydrology, second ed. Iowa State University Press, Ames, IA, p. 496.

Hamed, K.H., Rao, A.R., 1998. A modified Mann-Kendall trend test for autocorrelated data. J. Hydrol. 204, 219–246.

Hameed, T., Marino, M.A., DeVries, J.J., Tracy, J.C., 1997. Method for trend detection in climatological variables. J. Hydrol. Eng. ASCE 2 (4), 157–160.

Han, D., Song, X., Currell, M.J., Cao, G., Zhang, Y., Kang, Y., 2011. A survey of groundwater levels and hydrogeochemistry in irrigated fields in the Karamay Agricultural Development Area, northwest China: implications for soil and groundwater salinity resulting from surface water transfer for irrigation. J. Hydrol. 405, 217–234.

Himmelblau, D.M., 1969. Process Analysis by Statistical Methods. John Wiley and Sons, New York.

Hipel, K.W., Mcleod, A.I., 1994. Time Series Modelling of Water Resources and Environmental Systems. Development in Water Science, vol. 45 Elsevier, New York.

Hipel, K.W., Mcleod, A.I., Lennox, W.C., 1977. Advances in Box Jenkins modelling: 1. Model construction. Water Resour. Res. 13, 567–575.

Hsu, K.L., Gupta, H.V., Sorooshian, S., 1995. Artificial neural network modelling of the rainfall-runoff process. Water Resour. Res. 31 (10), 2517–2530.

Jarque, C.M., Bera, A.K., 1987. A test for normality of observations and regression residuals. Int. Stat. Rev. 55 (2), 163–172.

Jovein, E.B., Hosseini, S.M., 2017. Predicting saltwater intrusion into aquifers in vicinity of deserts using spatio-temporal kriging. Environ. Monit. Assess. 189, 81. 16 pages, https://doi.org/10.1007/s10661-017-5795-8.

Kanji, G.K., 2001. 100 Statistical Tests. Sage Publication, New Delhi. 111 pp.

Kaown, D., Hyun, Y., Bae, G.-O., Oh, C.W., Lee, K.-K., 2012. Evaluation of spatio-temporal trends of groundwater quality in different land uses using Kendall test. Geosci. J. 16 (1), 65–75.

Kashyap, R.L., Rao, A.R., 1976. Dynamic stochastic models from empirical data. In: Mathematics in Science and Engineering. vol. 122. Academic Press, Inc., London, U.K. 334 p.

Kendall, M.G., 1973. Time Series. Charles Griffin and Co. Ltd., London, UK.

Khorasani, M., Ehteshami, M., Ghadimi, H., Salari, M., 2016. Simulation and analysis of temporal change of groundwater depth using time series modeling. Model. Earth Syst. Environ. 2, 90. https://doi.org/10.1007/s40808-016-0164-0.

Koh, E.-H., Lee, S.H., Kaown, D., Moon, H.S., Lee, E., Lee, K.-K., Kang, B.R., 2017. Impacts of land use change and groundwater management on long-term nitrate-nitrogen and chloride trends in groundwater of Jeju Island, Korea. Environ. Earth Sci. 76, 176. https://doi.org/10.1007/s12665-017-6466-3.

Konikow, L.F., Kendy, E., 2005. Groundwater depletion: a global problem. Hydrogeol. J. 13 (1), 317–320.

Konukcu, F., Gowing, J.W., Rose, D.A., 2006. Dry drainage: a sustainable solution to waterlogging and salinity problems in irrigation areas. Agric. Water Manage. 83, 1–12.

Kottegoda, N.T., 1980. Stochastic Water Resources Technology. McMillan & Co. Ltd., London, UK.

Law, A.G., 1974. Stochastic Analysis of Groundwater Level Time Series in the Western United States (Hydrology Paper No. 68). Colorado State University, Fort Collins, CO.

Legates, D.R., McCabe Jr., G.J., 1999. Evaluating the use of "goodness-of-fit" measures in hydrologic and hydroclimatic model validation. Water Resour. Res. 35, 233–241.

Li, F., Feng, P., Zhang, W., Zhang, T., 2013. An integrated groundwater management mode based on control indexes of groundwater quantity and level. Water Resour. Manage. 27, 3273–3292.

Lilliefors, H.W., 1967. On the Kolmogorov-Smirnov test for normality with mean and variance unknown. J. Am. Stat. Assoc. 62, 399–402.

Ljung, G.M., Box, G.E., 1978. On a measure of lack of fit in time series models. Biometrika 65 (2), 297–303.

Loftis, J.C., 1996. Trends in groundwater quality. Hydrol. Process. 10, 335–355.

Machiwal, D., Jha, M.K., 2006. Time series analysis of hydrologic data for water resources planning and management: a review. J. Hydrol. Hydromech. 54 (3), 237–257.

Machiwal, D., Jha, M.K., 2012. Hydrologic Time Series Analysis: Theory and Practice. Springer/Capital Publishing Company, the Netherlands/New Delhi, India. 303 pp.

Machiwal, D., Jha, M.K., 2015a. Identifying sources of groundwater contamination in a hard-rock aquifer system using multivariate statistical analyses and GIS-based geostatistical modeling techniques. J. Hydrol.: Reg. Stud. 4 (A), 80–110.

Machiwal, D., Jha, M.K., 2015b. GIS-based water balance modeling for estimating regional specific yield and distributed recharge in data-scarce hard-rock regions. J. Hydro Environ. Res. 9 (4), 554–568.

Magar, R.B., Jothiprakash, V., 2011. Intermittent reservoir daily-inflow prediction using lumped and distributed data multi-linear regression models. J. Earth Syst. Sci. 120, 1067–1084.

McCuen, R.H., 2003. Modeling Hydrologic Change: Statistical Methods. Lewis Publishers/CRC Press LLC, Florida. 433 pp.

McMohan, T.A., Mein, R.G., 1986. River and Reservoir Yield. Water Resources Publication, Littleton, CO.

Mirzavand, M., Sadatinejad, S.J., Ghasemieh, H., Imani, R., Motlagh, M.S., 2014. Prediction of ground water level in arid environment using a non-deterministic model. J. Water Resour. Prot. 6, 669–676.

Montgomery, D.C., Jennings, C.L., Kulahci, M., 2015. Introduction to Time Series Analysis and Forecasting. John Wiley & Sons Inc., Hoboken, NJ.

Moore, D.S., 1991. Statistics: Concepts and Controversies, third ed. W.H. Freeman, New York. 439 pp.

Mouser, P.J., Hession, W.C., Rizzo, D.M., Gotelli, N.J., 2005. Hydrology and geostatistics of a Vermont, USA kettlehole peatland. J. Hydrol. 301, 250–266.

Mujumdar, P.P., Kumar, D.N., 1990. Stochastic models of streamflow: some case studies. Hydrol. Sci. J. 35, 395–410.

Narany, T.S., Ramli, M.F., Aris, A.Z., Sulaiman, W.N.A., Fakharian, K., 2014. Spatiotemporal variation of groundwater quality using integrated multivariate statistical and geostatistical approaches in Amol-Babol Plain, Iran. Environ. Monit. Assess. 186, 5797–5815.

Narany, T.S., Aris, A.Z., Sefie, A., Keesstra, S., 2017. Detecting and predicting the impact of land use changes on groundwater quality: a case study in Northern Kelanta, Malaysia. Sci. Total Environ. 599–600, 844–853.

Nas, B., Berktay, A., 2010. Groundwater quality mapping in urban groundwater using GIS. Environ. Monit. Assess. 160, 215–227.

Nash, J.E., Sutcliffe, J.V., 1970. River flow forecasting through conceptual models, part I—a discussion of principles. J. Hydrol. 10 (3), 282–290.

Natrella, M.G., 1963. Experimental Statistics. National Bureau of Standards Handbook No. 91, US Government Printing Office, Washington, DC.

Noshadi, M., Ghafourian, A., 2016. Groundwater quality analysis using multivariate statistical techniques (case study: Fars province, Iran). Environ. Monit. Assess. 188, 419. 13 pages, https://doi.org/10.1007/s10661-016-5412-2.

Madansky, A., 1988. Prescriptions for Working Statisticians. Springer-Verlag, New York.

McGhee, J.W., 1985. Introductory Statistics. West Publishing Co., New York, USA.

Panda, D.K., Kumar, A., 2011. Evaluation of an over-used coastal aquifer (Orissa, India) using statistical approaches. Hydrol. Sci. J. 56 (3), 486–497.

Panda, D.K., Mishra, A., Jena, S.K., James, B.K., Kumar, A., 2007. The influence of drought and anthropogenic effects on groundwater levels in Orissa, India. J. Hydrol. 343, 140–153.

Patle, G.T., Singh, D.K., Sarangi, A., Rai, A., Khanna, M., Sahoo, R.N., 2015. Time series analysis of groundwater levels and projection of future trend. J. Geol. Soc. India 85, 232–242.

Phoon, K.-K., Quek, S.-T., An, P., 2003. Identification of statistically homogeneous soil layers using modified Bartlett statistics. J. Geotech. Geoenviron. Eng. ASCE 129 (7), 649–659.

Rao, A.R., Rao, R.G.S., Kashyap, R.L., 1975. Stochastic Models for Ground Water Levels. Water Resources Research Center, Purdue University, West Lafayette, IN.

Rissanen, J., 1978. Modeling by shortest data description. Automatica 14 (5), 465–471.

Ritzema, H.P., Satyanarayana, T.V., Raman, S., Boonstra, J., 2008. Subsurface drainage to combat waterlogging and salinity in irrigated lands in India: lessons learned in farmers' fields. Agric. Water Manage. 95, 179–189.

Rodell, M., Velicogna, I., Famiglietti, J.S., 2009. Satellite based estimates of groundwater depletion in India. Nature 460, 999–1002.

Sachs, L., 1972. Statistische Auswertungsmethoden, third ed. Springer-Verlag, Berlin.

Salas, J.D., 1993. Analysis and modeling of hydrologic time series (Editor-in-Chief). In: Maidment, D.R. (Ed.), Handbook of Hydrology. McGraw-Hill, Inc., USA. pp. 19.1–19.72.

Sang, Y.-F., Wang, Z., Liu, C., 2014. Comparison of the MK test and EMD method for trend identification in hydrological time series. J. Hydrol. 510, 293–298.

Scanlon, B.R., Reedy, R.C., Gates, J.B., 2010. Effects of irrigated agroecosystems: 1. Quantity of soil water and groundwater in the Southern High Plains, Texas. Water Resour. Res.. 46(9). https://doi.org/10.1029/2009WR008427.

Schwankl, L.J., Raghuwanshi, N.S., Walender, W.W., 2000. Time series modeling for predicting spatially variable infiltration. J. Irrig. Drain. Eng. ASCE 126 (5), 283–287.

Sen, P.K., 1968. Estimates of the regression coefficient based on Kendall's tau. J. Am. Stat. Assoc. 63 (324), 1379–1389.

Şen, Z., 2017. Innovative trend significance test and applications. Theor. Appl. Climatol. 127 (3–4), 939–947.

Shahin, M., Van Oorschot, H.J.L., De Lange, S.J., 1993. Statistical Analysis in Water Resources Engineering. A.A. Balkema, Rotterdam, the Netherlands. 394 pp.

Shapiro, S., 1980. How to test normality and other distributional assumptions. In: The ASQC Basic References in Quality Control: Statistical Techniques. vol. 3. American Society for Quality Control, Milwaukee, WI.

Shamsudduha, M., Chandler, R.E., Taylor, R.G., Ahmed, K.M., 2009. Recent trends in groundwater levels in a highly seasonal hydrological system: the Ganges-Brahmputra-Megha Delta. Hydrol. Earth Syst. Sci. 13, 2373–2385.

Sivakumar, B., 2017. Chaos in Hydrology—Bridging Determinism and Stochasticity. Springer, Netherlands.

Snedecor, G.W., Cochran, W.G., 1980. Statistical Methods. Iowa State University Press, Ames, IA.

Sophocleous, M., 2010. Groundwater management practices, challenges, and innovations in the High Plains aquifer, USA—lessons and recommended actions. Hydrogeol. J. 18 (3), 559–575 (Review).

Stephens, M.A., 1974. EDF statistics for goodness of fit and some comparisons. J. Am. Stat. Assoc. 69, 730–737.

Stephens, M.A., 1986. Tests based on EDF statistics. In: D'Agostino, R.B., Stephens, M.A. (Eds.), Goodness-of-Fit Techniques. Marcel Dekker, New York, USA.

Taylor, C.H., Loftis, J.C., 1989. Testing for trend in lake and ground water quality time series. J. Am. Water Resour. Assoc. 25 (4), 715–726.

USEPA, 1992. Guidance Document on the Statistical Analysis of Ground-Water Monitoring Data at RCRA Facilities (EPA/530/R-93/003). United States Environmental Protection Agency (USEPA), Office of Solid Waste, Washington, DC.

USEPA, 2006. Data Quality Assessment: Statistical Methods for Practitioners (Guidance Document EPA QA/G-9S). US Environmental Protection Agency (USEPA), Office of Environmental Information, Washington, DC.

Vazquez-Amabile, G.G., Engel, B.A., 2008. Fitting of time series models to forecast streamflow and groundwater using simulated data from SWAT. J. Hydrol. Eng. ASCE 13, 554–562.

Visser, A., Dubus, I., Broers, H.P., Brouyère, S., Korcz, M., Orban, P., Goderniaux, P., Batlle-Aguilar, J., Surdyk, N., Amraoui, N., Job, H., Pinault, J.L., Bierkens, M., 2009. Comparison of methods for the detection and extrapolation of trends in groundwater quality. J. Environ. Monit. 11, 2030–2043.

Walpole, R., Myers, R., 1985. Probability and Statistics for Engineers and Scientists, third ed. MacMillan, New York, USA.

Whittle, P., 1952. Tests of fit in time series. Biometrika 39 (3–4), 309–318.

Willmott, C.J., 1981. On the validation of models. Phys. Geogr. 2, 184–194.

Willmott, C.J., Ackleson, S.G., Davis, R.E., Feddema, J.J., Klink, K.M., Legates, D.R., O'Donnell, J., Rowe, C.M., 1985. Statistics for the evaluation and comparison of models. J. Geogr. Res. 90, 8995–9005.

WMO, 1966. Climatic Change (Technical Note 79). The World Meteorological Organization (WMO), Geneva, Switzerland.

Wu, J.C., Hu, B.X., Zhang, D.X., Shirley, C., 2003. A three-dimensional numerical method of moments for groundwater flow and solute transport in a nonstationary conductivity field. Adv. Water Resour. 26 (11), 1149–1169.

Yang, Z.P., Lu, W.X., Long, Y.Q., Li, P., 2009. Application and comparison of two prediction models for groundwater levels: a case study in Western Jilin Province, China. J. Arid Environ. 73, 487–492.

Yang, Q., Wang, Y., Zhang, J., Delgado, J., 2017. A comparative study of shallow groundwater level simulation with three time series models in a coastal aquifer of South China. Appl. Water Sci. 7, 689–698.

Yazdanpanah, N., 2016. Spatiotemporal mapping of groundwater quality for irrigation using geostatistical analysis combined with a linear regression method. Model. Earth Syst. Environ. 2. https://doi.org/10.1007/s40808-015-0071-9. 18 pages.

Yue, S., Wang, C.Y., 2002. Applicability of prewhitening to eliminate the influence of serial correlation on the Mann-Kendall test. Water Resour. Res.. 38 (6). https://doi.org/10.1029/2001WR000861 4-1–4-7.

Yue, S., Wang, C.Y., 2004. The Mann-Kendall test modified by effective sample size to detect trend in serially correlated hydrological series. Water Resour. Manage. 18, 201–218.

Yue, S., Pilon, P., Phinney, B., Cavadias, G., 2002. The influence of autocorrelation on the ability to detect trend in hydrological series. Hydrol. Process. 16, 1807–1829.

Zhang, G.P., Qi, M., 2005. Neural network forecasting for seasonal and trend time series. Eur. J. Oper. Res. 160 (2), 501–514.

Chapter 11

Intelligent Prediction Modeling of Water Quality Using Artificial Neural Networking: Nambiyar River Basin, Tamil Nadu, India

C. Gajendran*, K. Srinivasamoorthy[†] and P. Thamarai[‡]

*Department of Civil Engineering, Karunya Institute of Technology and Sciences, Coimbatore, India [†]Department of Earth Sciences, School of Physical, Chemical and Applied Sciences, Pondicherry University, Pondicherry, India [‡]Government College of Technology, Coimbatore, India

Chapter Outline

11.1 INTRODUCTION

Artificial neural network (ANN) is applied in the fields of hydrology, ecology, medicine, and other biological fields. The usage of ANNs in hydrology includes rainfall-runoff modeling, surface water-flow simulation and forecasting, groundwater-flow prediction, and water-quality issues. Many researchers have portrayed the significance of ANN in forecasting water quality in comparison with other models. The usage of neural networking has increased rapidly in the field of water-quality management and water-resource planning and management (Wen and Lee, 1998; Chakraborty et al., 1992; Lachtermacher and Fuller, 1994; Schizas et al., 1994). Reckhow (1999), using Bayesian probability network models, attempted to predict the water quality in the Neuse River in North Carolina. Zaheer and Bai (2003) carried out a study on the application of ANN for water-quality management in the Hanjiang River of China using a decision-making approach for water-quality management to control environmental pollution. Earlier research on water quality-management issues adopted traditional optimization techniques aided by an expert-system approach, which does not provide an appropriate solution in comparison with a decision-making approach, which relates the interpretation of data sets with certain rules and assumptions aided by a series of experiments. Diamantopoulou et al. (2005) predicted water-quality parameters in the Strymon River Basin near the Greek-Bulgarian border using ANN techniques and the prediction, which was attempted using 13 water-quality parameters, depicts the effectiveness of ANN in predicting river-water quality.

The Nambiyar Basin gains significance due to the presence of industrial activities with inadequate effluent treatment facilities and overextraction of groundwater resources, which results in the lowering of the water table along with quality degradation. Hence an attempt has been made in the present study to develop an ANN-based predictive model to predict important water-quality parameters like hardness, total dissolved solids (TDS), and chloride (Cl).

11.2 STUDY AREA

The Nambiyar River Basin is located along the southern most part of south India located between the latitudes of 8°08′ and 8°33′ north and longitudes of 77°28′ and 78°15′ east, with a total area of 2084 km² (Fig. 11.1). The Nambiyar River originates at an elevation of 1479 m above mean sea level (MSL) in Nalikkalmottai in Kallakadu Reserved Forest, traverses

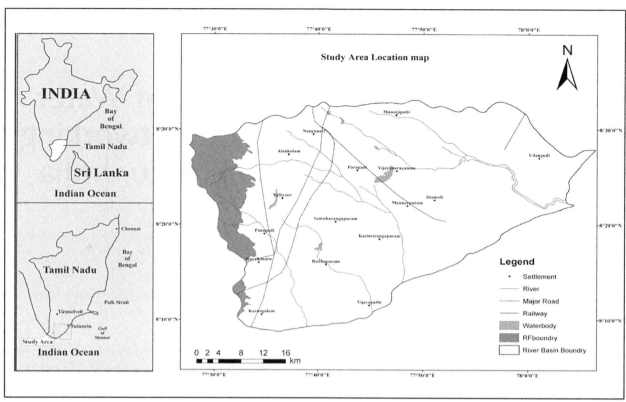

FIG. 11.1 Location of the study.

through Pudukulam and Pettaikulam, and, finally, confluences with the Gulf of Mannar at Thiruvambalampula. The geology of the area is comprised of rocks of khondalite and charnockite groups of the Archean age and the coastal region is comprised of litho units like sandstone and shell limestone confined to quaternary and recent formations. The weathered and fractured zone thickness in hard rock, irrespective of litho units, extends up to 25 and 30 m below ground level (BGL) and the yield of bore wells ranges between 45 and 295 L per minute (LPM). The transmissivity of aquifers ranges between 10 and 20 m² day⁻¹. In the sedimentary formation the groundwater level ranges between 33 and 90 m BGL and the transmissivity of the aquifer is 43 m² day⁻¹. The yield from the well ranges between 75 and 1045 LPM, irrespective of sedimentary formations.

11.3 INTELLIGENT PREDICTIVE MODEL

The present aim is to investigate the ability of the intelligent predictive model (IPM) to forecast the groundwater quality from the Nambiyar River Basin. IPM is used to forecast the water quality under historical and future scenarios. Due to various geochemical reactions and mechanisms that govern the availability of the water-quality parameters a domain-specific mechanism is required to correlate, predict, and forecast the water-quality variables. The isolation of such modeling tools will be of major use to water scientists to predict water-pollution levels, aided by suggested remedial aids in advance. Classical process-based modeling approaches can give good inference of water-quality parameters, but they lack in the data calibration process (Palani et al., 2008). The most important modeling approach used for nonlinear environmental relationships is the ANN (Zhang and Stanley, 1997). Hafizan et al. (2004) suggested the application of an ANN model insead of the auto-regressive integrated moving average (ARIMA) model in forecasting dissolved oxygen. Many authors have attempted water-quality forecasting using ANN (Huiqun and Ling, 2008; Elhatip and Kömür, 2008; Gorashi and Abdullah, 2012; Li et al., 2017; Hunter et al., 2018). However, few of the ANN models have tried to visually integrate a geographical information system (GIS) with the feed-forward back propagation network (BPN), a type of ANN, to create a GIS-BPN-based water-quality model (Gholami et al., 2016). An attempt has been made in this study to predict the groundwater quality in the Nambiyar River Basin using ANN modeling.

11.4 DATA COLLECTION AND ANALYSIS

Groundwater samples were collected from 32 open and bore wells (Fig. 11.2 and Table 11.1) during January and July indicating post- and premonsoon seasons, respectively, following the sampling procedure as suggested by Palmquist (1973). Field parameters such as EC, pH, and temperature were measured in the field using portable meters, and chemical constituents, like calcium, magnesium (Mg), sodium, potassium, bicarbonate, carbonate, and Cl, in the laboratory using standard procedures as suggested by APHA et al. (1984).

11.5 NEURAL NETWORK MODEL CONFIGURATION

The neural network system consists of three parts in which individual neurons are interconnected in several layers, of which the first layer is the input layer, which produces the network output (Fig. 11.3). The numerical data moves from the input to each unit and between the connections in a parallel fashion and the results are processed in the output layer. In this study the use of ANN modeling to predict TDS, Cl, and hardness in the Nambiyar River Basin is presented.

11.6 GIS INTEGRATION

The integration of GIS with ANN modeling can reduce the complicated process and increase the accuracy and efficiency of the model output (Liao and Tim, 1997; Basnyat et al., 2000; Jensen, 2000; He et al., 2001). When linked with GIS, the ANN can be useful for monitoring, evaluating, and decision making through the visual interpretation of the results (Solaimani et al., 2009). The GIS database consists of two data types, the first being the "spatial data" and the next the "attribute data." The spatial data include arc view shape files mainly representing the 32 measured points of the Nambiyar River Basin. The attributed data describes the input analytical elements of the sampling points, which include the concentration of TDS, electrical conductivity (EC), Cl, Mg, and hardness. The visual geographical distributions of the TDS, Cl, and Hardness models are presented in Figs. 11.4–11.6, respectively.

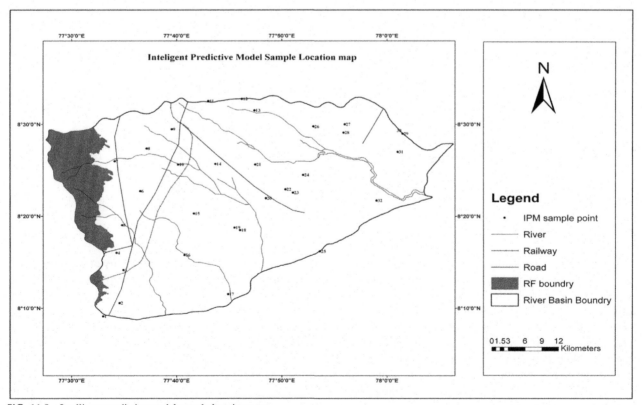

FIG. 11.2 Intelligent predictive model sample location map.

TABLE 11.1 Intelligent Predictive Model Sample Point Locations ID Details

Station Name	ID	Station Name	ID
Bagavathipuram	1	Vijayapathi	17
Karunkulam	2	Kausturirangapuram	18
Soundaralingapuram	3	Kasturirangapuram (a)	19
Kavalkinar	4	Mannarpuramvilakku	20
Panagudi	5	VadakkuVijayanarayanam	21
Valliyoor	6	Ittamozhipudur	22
Tirukkurungudi	7	Ittamozhi	23
Alangulam	8	Pudukulam	24
Nanguneri	9	Uvari	25
Tulukkarpatti	10	Karunkadal	26
Unnankulam	11	Anandapuram	27
Moolaikaraipatti	12	Ananda_puram	28
Munanjipatti	13	Tiruchendur	29
Parappadi	14	Sundarapuram	30
Samugarengapuram	15	Udankudi	31
Radhapuram	16	Padukkapathu	32

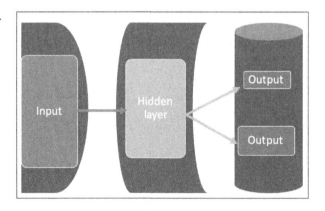

FIG. 11.3 Neural network flow diagram.

Understanding the quality of the groundwater is more significant than the quantity because the quality of water indicates its suitability for drinking, domestic, agricultural, and industrial purposes (Phiri et al., 2005; Chimwanza et al., 2006; Jain et al., 2006; Alam et al., 2007). The statistical analysis of the chemical constituents is represented in Table 11.2. The pH value ranges from 6.9 to 9.3, indicating the alkaline nature of the groundwater. The higher concentrations recorded in the meager wells might be due to the weathering of plagioclase feldspar by dissolved atmospheric carbon dioxide, which releases sodium and calcium and progressively increases the pH and alkalinity (Njitchoua et al., 1997; Chenini and Khemiri, 2009; Pejman et al., 2009; Ramkumar et al., 2010; Mohiuddin et al., 2010). The EC range is from 78.0 to 9800 μS/cm with higher ranges noted along the southeastern parts of the study area. Higher TDS is mainly due to dissolution and mixing whereas in coastal parts of the study area higher TDS indicates the presence of saline water (Aiuppa et al., 2000; Ramkumar et al., 2010).

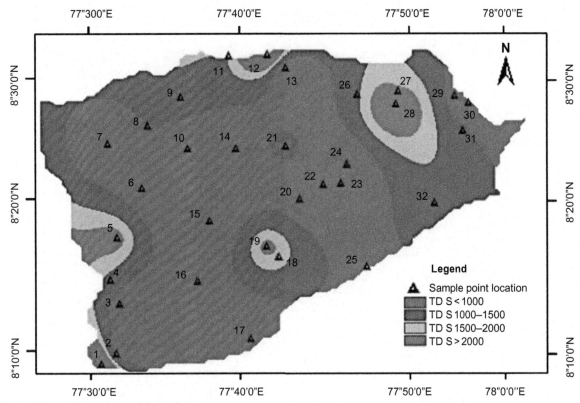

FIG. 11.4 TDS model output map of the study area.

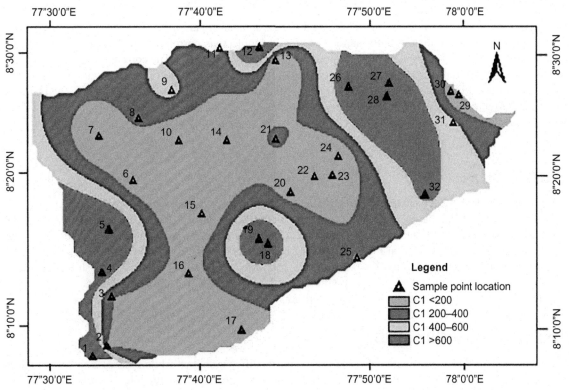

FIG. 11.5 Cl model output map of the study area.

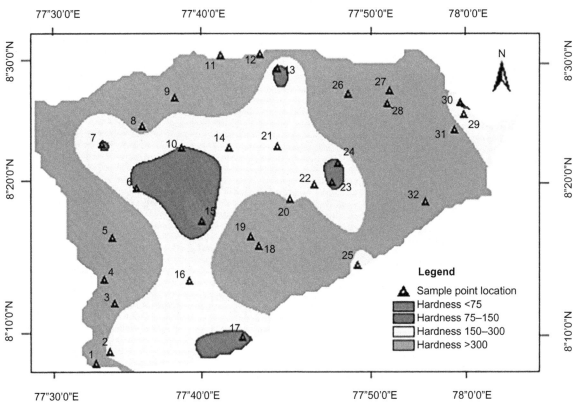

FIG. 11.6 Hardness model output map of the study area.

11.7 CORRELATION OF PHYSICOCHEMICAL PARAMETERS

One of the primal objectives of the present work is to reduce the number of parameters needed to carry out water-quality prediction without losing any important information. To meet this objective, correlation analysis was employed to investigate the relationship of each water quality parameter with the dependent variables. The correlation coefficient exhibits the relationship between the variables used for the water-quality analysis. Correlation matrices for 14 variables were attempted for the study area (Table 11.3) and the results signify a positive correlation between EC with TDS and Cl, and hardness with Mg. Significant correlations also exist between the pairs Na and EC, Mg and EC, hardness and EC, Na and TDS, Ca and TDS, Mg and TDS, Cl and Na, and Cl and Mg. The summary of the water-quality parameter in the study area is presented in Table 11.2.

11.8 REGRESSION

A linear regression model has been attempted to justify the relationship between the water-quality parameters, i.e., relating TDS with EC, Cl with EC, and hardness with Mg, using the training data set and also to compare the ANN model capabilities. The equations obtained were then tested with 642 test data to evaluate the predictability of the developed empirical relations. It was observed that the relation $Y = 0.6119X - 42.641$ with standard error of 0.00560, as shown in Fig. 11.7, can be used to estimate TDS, while the relation $Y = 0.2631X - 95.292$ with standard error of 0.01248, as shown in Fig. 11.8, can be used to estimate Cl, and $Y = 6.0098X + 141.84$ with standard error of 0.553901, as shown in Fig. 11.9. The regression model results are then compared with the neural network model results.

11.9 NEURAL NETWORK MODEL

Neural network models have been attempted for all the water-quality parameters while constructing single-hidden-layer ANN models (Bandyopadhyay and Chattopadhyay, 2007). But for the present study the four variables (TDS, Cl, Mg, and hardness) that met the requirement as good predictors for the IPM generation were considered for ANN modeling. To

TABLE 11.2 Summary of Water-Quality Parameters

Parameter		EC	pH	TDS	Hardness	Ca	Mg	Na	K	HCO$_3$	CO$_3$	Cl	SO$_4$	NO$_3$	F
Unit		µS/cm		ppm											
Primary (January 2009–July 2009)	Minimum	83.0	6.8	70.0	120.0	10.0	2.0	4.0	2.0	37.0	6.0	7.0	27.8	0.4	0.1
	Maximum	9140.0	9.5	1700.0	1060.0	923.0	578.0	976.0	97.0	568.0	85.0	2765.0	125.0	45.0	1.2
	Mean	1947.0	8.1	876.0	568.0	136.0	87.0	142.0	14.7	127.0	15.0	378.0	72.6	16.7	0.8
	Median	1343.0	8.1	912.0	673.0	68.0	94.0	117.0	11.3	198.0	17.0	198.0	71.3	17.0	0.7
	Standard deviation	1567.0	0.4	564.0	395.0	243.0	96.3	146.0	13.7	47.8	8.7	578.0	21.3	14.3	0.2
Secondary (1995–2008)	Minimum	78.0	7.0	150.0	70.0	6.0	3.0	4.0	1.0	32.0	5.0	7.9	22.0	0.0	0.1
	Maximum	9800.0	9.7	1500.0	978.0	780.0	542.0	780.0	248.0	432.0	73.0	2561.0	117.0	38.0	1.7
	Mean	1765.0	8.6	765.0	289.0	85.3	72.4	147.0	23.4	256.0	8.0	432.0	70.0	12.0	0.6
	Median	1456.0	8.6	780.0	321.0	63.0	47.6	112.0	10.0	273.0	5.0	275.0	65.0	9.0	0.7
	Standard deviation	1245.0	0.31	976.0	278.0	89.6	86.0	143.0	37.8	32.7	7.3	342.0	19.0	13.0	0.6
Total	Minimum	78.0	6.9	70.0	70.0	6.0	2.0	4.0	1.0	32.0	5.0	7.0	22.0	0.0	0.1
	Maximum	9800.0	9.3	1700.0	1060.0	923.0	578.0	976.0	248.0	568.0	85.0	2765.0	125.0	45.0	1.2
	Mean	1546.0	8.7	769.0	550.0	112.0	75.4	137.0	20.7	193.0	22.0	375.0	66.0	13.0	0.7
	Median	1445.0	8.7	765.0	376.0	89.0	63.0	112.0	10.0	124.0	34.0	264.0	60.0	12.0	0.6
	Standard deviation	1376.0	0.4	674.0	258.0	103.0	72.0	143.0	35.7	45.9	22.0	3.7	43.0	16.0	0.3

TABLE 11.3 Correlation Matrix of Physicochemical Parameters

	EC	pH	TDS	Na	K	Ca	Mg
EC	1.0000						
pH	0.14	1.00					
TDS	0.93	−0.11	1.00				
Na	0.83	−0.03	0.89	1.00			
K	0.39	0.00	0.45	0.35	1.00		
Ca	0.77	−0.17	0.82	0.57	0.27	1.00	
Mg	0.81	−0.12	0.86	0.64	0.40	0.69	1.00
Cl	0.95	−0.13	0.91	0.80	0.35	0.76	0.82
HCO_3	0.24	−0.02	0.35	0.39	0.24	0.22	0.23
CO_3	0.07	0.59	−0.03	0.02	0.09	−0.10	−0.04
SO_4	0.69	−0.09	0.76	0.69	0.35	0.56	0.64
NO_3	0.43	−0.16	0.38	0.25	0.30	0.34	0.35
F	0.03	−0.04	0.06	0.08	−0.00	0.04	0.06
Hardness	0.86	−0.16	0.91	0.66	0.37	0.90	0.93
	Cl	HCO_3	CO_3	SO_4	NO_3	F	Hardness
Cl	1.00						
HCO_3	0.14	1.00					
CO_3	−0.08	0.10	1.00				
SO_4	0.59	0.25	−0.03	1.00			
NO_3	0.37	0.05	−0.10	0.33	1.00		
F	0.02	0.11	−0.06	0.06	0.02	1.00	
Hardness	0.86	0.25	−0.07	0.66	0.38	0.05	1.00

FIG. 11.7 TDS regression model output.

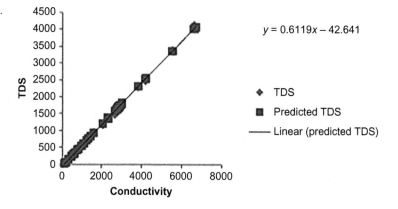

$y = 0.6119x − 42.641$

◆ TDS
■ Predicted TDS
— Linear (predicted TDS)

isolate the best predictor combination of ANN model, the data available from the Tamil Nadu Public Works Department (from 1995 to 2008) were used as the training and validation data set. To verify the data samples were collected and analyzed in the laboratory (2009) and used as the testing data set. In the present work a feed-forward BPN algorithm has been attempted using the Mat lab's neural network tool.

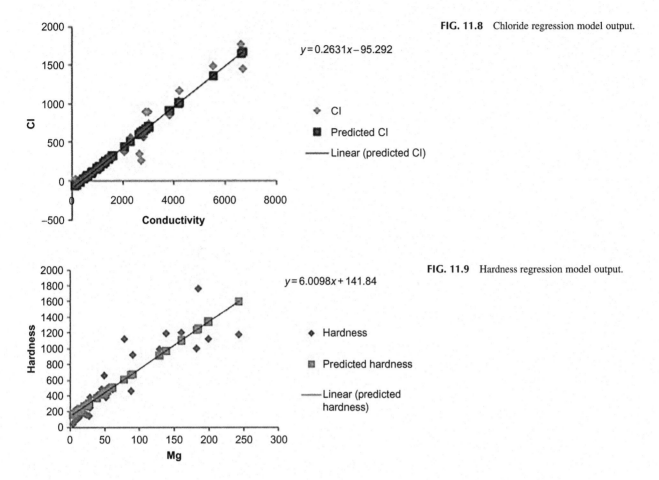

FIG. 11.8 Chloride regression model output.

$y = 0.2631x - 95.292$

◆ Cl

■ Predicted Cl

—— Linear (predicted Cl)

FIG. 11.9 Hardness regression model output.

$y = 6.0098x + 141.84$

◆ Hardness

■ Predicted hardness

—— Linear (predicted hardness)

11.9.1 Data Partition

The data in neural networks are grouped into three sets: learning, validation, and overfitting test sets. A total of 3210 data samples were grouped into training sets consisting of 1926 samples (60% of the total) (1995–2008) and 642 samples (20% of the total) (1995–2008) for validation and 642 samples (20% of the total samples) (2009) for testing. The results obtained showed that the proposed ANN-GIS based IPM had great ability to simulate and predict the TDS, Cl, and hardness with acceptable accuracies of mean square error (MSE): TDS MSE $= 1.5 \, \mathrm{E}^{-4}$; Cl MSE $= 3.2 \mathrm{E}^{-4}$; HARDNESS MSE $= 1.7 \, \mathrm{E}^{-4}$. The results are shown in Figs. 11.10–11.12. Based on the observation of the graphs obtained from these two models, it is evident that most errors for the regression model are little more than errors generated by the ANN model.

11.9.2 Model Development

An ANN-based IPM was developed to simulate and predict the TDS in the study area because of its effectiveness in determining the suitability of water for different utilities. The best results were generated after several trials using an ANN-architecture BPN algorithm with 750 hidden layer combinations. Cl concentrations were also simulated and predicted using the BPN algorithm, which generated 550 hidden layer combinations after several trials with a goodness of fit. Cl was considered for the present study due to its role as an effective pollution indicator. The best results for hardness were yielded after several trials generated 850 hidden layer combinations. Hardness was considered for the present study due to its importance in determining the quality of water for domestic, agricultural, and industrial utilities.

11.9.3 Model Performance Evaluation

To achieve the appropriate network architecture, several trials for each group were conducted until a suitable learning rate, number of hidden layers, and number of neurons per each hidden layer was reached. A suitable architecture is the one which produces minimal error terms in both the training and testing data of the model. The backpropagation algorithm minimizes

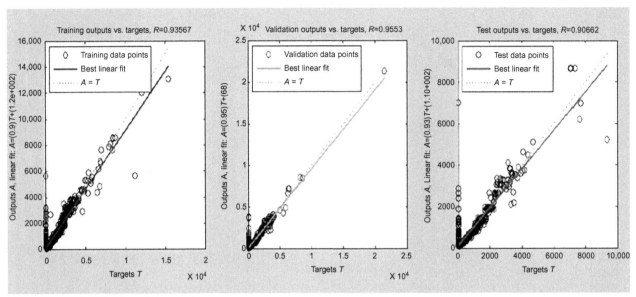

FIG. 11.10 ANN prediction model for TDS.

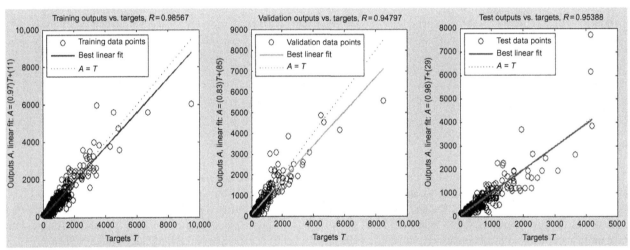

FIG. 11.11 ANN prediction model for Cl.

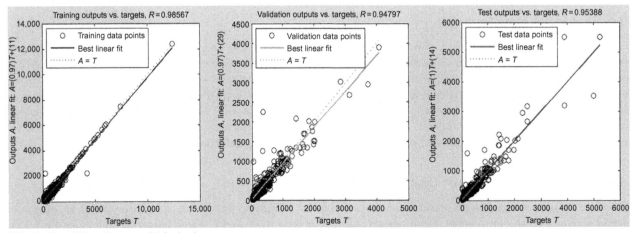

FIG. 11.12 ANN prediction model for hardness.

the MSE between the observed and the predicted output in the output layer. The performance of each network model is evaluated by computing the mean absolute percentage error (MAPE) and MSE. The structure that resulted in minimum errors was the one selected. Since the water-quality parameters were observed for 11 years (1997–2008), the implementation of the proposed ANN-based IPM can be an appropriate predictive tool. The outcomes of the models were also appraised using the laboratory real-time results (2009) from the same 32 monitoring points. In addition, visual analysis for the prediction data was carried out using ArcGIS. Neural networks are being used in a wide variety of applications as an important decision-making tool. The ANN-based GIS-coupled IPM exhibits robustness and reliable performance in predicting TDS, Cl, and hardness with EC and Mg as the input parameters. The prime source for these ions was identified as industrial effluents.

11.10   CONCLUSION

This study reveals the robustness and reliability of ANN-based GIS-coupled IPM in predicting TDS, Cl, and hardness with EC and Mg as the input parameters. Statistical modeling in comparison with ANN seconds the effective usage of ANN modeling. The model is capable of determining parameters like TDS, Cl, and hardness using EC and Magnesium. The BPN algorithm after several trials generated the best combinations of results depending on the correlation between the observed and predicted values. The results generated were spatially interpolated using GIS coupled with ANN and IPM suggests significant reliability for predicting TDS, Cl, and hardness with EC and magnesium as the input parameters. Industrial influence on ionic migration was isolated as the primary factor influencing the present scenario. It is also further suggested that the present methodology may be utilized for other river basin studies using ANN.

REFERENCES

Aiuppa, A., Allard, P.D., Alessandro, W., Michel, A., Parello, F., Treuil, M., Valenza, M., 2000. Mobility and fluxes of major, minor and trace metals during basalt weathering and groundwater transport at Mt. Etanvolcan (sieity). Geochem. Cosmochim. Acta 64, 1827–1841.

Alam, M.J.B., Muyen, Z., Islam, M.R., Islam, S., Mamun, M., 2007. Water quality parameters along rivers. Int. J. Environ. Sci. Technol. 4 (1), 159–167.

APHA, AWWA, WPCF, 1984. Standard Methods for the Examination of Water and Wastewater, 16th ed. APHA, Washington, DC, p. 1268.

Bandyopadhyay, G., Chattopadhyay, S., 2007. Single hidden layer artificial neural network models versus multiple linear regression model in forecasting the time series of total ozone. Int. J. Environ. Sci. Technol. 4 (1), 141–149.

Basnyat, P., Teeter, L.D., Lockaby, B.G., Flynn, K.M., 2000. The use of remote sensing and GIS in watershed level analyses of non-point source pollution problems. For. Ecol. Manage. 128, 65–73.

Chakraborty, K., Mehrotra, K., Mohan, C.K., Ranka, S., 1992. Forecasting the behaviour of multivariate time seires using neural network. Neural Netw. 5, 961–970.

Chenini, I., Khemiri, S., 2009. Evaluation of ground water quality using multiple linear regression and structural equation modeling. Int. J. Environ. Sci. Technol. 6 (3), 509–519.

Chimwanza, B., Mumba, P.P., Moyo, B.H.Z., Kadewa, W., 2006. The impact of farming on river banks on water quality of the rivers. Int. J. Environ. Sci. Technol. 2 (4), 353–358.

Diamantopoulou, M.J., Papamichail, D.M., Antonopoulos, V.Z., 2005. The use of a neural network technique for the prediction of water quality parameters. Oper. Res. Int. J. 5 (1), 115–125.

Elhatip, H., Kömür, M.A., 2008. Evaluation of water quality parameters for the Mamasin dam in Aksaray City in the central Anatolian part of Turkey by means of artificial neural networks. Environ. Geol. 53, 1157–1164. https://doi.org/10.1007/s00254-007-0705-y.

Gholami, V., Sebghati, M., Yousefi, Z., 2016. Integration of artificial neural network and geographic information system applications in simulating groundwater quality. Environ. Health Eng. Manage. J. 3 (4), 173–182.

Gorashi, F., Abdullah, A., 2012. Prediction of water quality index using back propagation network algorithm, case study: Gombak River. J. Eng. Sci. Technol. 7 (4), 447–461.

Hafizan, J., Sharifuddin, M.Z., Mohd Ekhwan, T., Mazlin, M., Hasfalina, C.M., 2004. Application of artificial neural network models for predicting water quality index. Jurnal Kejuruteraan Awam 16 (2), 42–55.

He, C., Shi, C., Yang, C., Agosti, B.P., 2001. A windows-based GIS-AGNPS interface. J. Am. Water Resour. Assoc. 37 (2), 395–406.

Huiqun, M., Ling, L., 2008. Water quality assessment using artificial neural network. In: International Conference on Computer Science and Software Engineering, Washington, DC.

Hunter, J.M., Maier, H.R., Gibbs, M.S., Foale, E.R., Grosvenor, N.A., Harders, N.P., Kikuchi-Miller, T.C., 2018. Framework for developing hybrid process-driven, artificial neural network and regression models for salinity prediction in river systems. Hydrol. Earth Syst. Sci. 22, 2987–3006. https://doi.org/10.5194/hess-22-2987-2018.

Jain, P., Sharma, J.D., Sohu, D., Sharma, P., 2006. Chemical analysis of drinking water of villages of Sanganer Tehsil, Jaipur District. Int. J. Environ. Sci. Technol. 2 (4), 373–379.

Jensen, J.R., 2000. Remote Sensing of the Environment: An Earth Resource Perspective, second ed. Prentice-Hall.

Lachtermacher, G., Fuller, J.D., 1994. Back Propagation in Hydrological Time Series Forecasting in Stocahstic and Statistical Methods in Hydrology and Environmental Engineering. vol. 3. Kluwer Academy, Norwell, MA, pp. 229–242.

Li, M.S., Wu, W., Chen, B.S., Guan, L.X., Wu, Y., 2017. Water quality evaluation using Back propagation artificial neural network based on self-adaptive particle swarm optimization algorithm and Chaos theory. Comput. Water Energy Environ. Eng. 6, 229–242. https://doi.org/10.4236/cweee.2017.63016.

Liao, H.H., Tim, U.S., 1997. An interactive modeling environment for nonpoint source pollution control. J. Am. Water Resour. Assoc. 33 (3), 1–13.

Mohiuddin, K.M., Zakir, H.M., Otomo, K., Sharmin, S., Shikazono, N., 2010. Geochemical distribution of trace metal pollutants in water and sediments of downstream of an urban river. Int. J. Environ. Sci. Technol. 7 (1), 17–28.

Njitchoua, R., Dever, L., Fontes, J.C., Naoh, E., 1997. Geochemistry origin and recharge mechanisms of groundwater from the Carona sandstone aquifer, Northern Cameroon. J. Hydrol. 190, 123–140.

Palani, S., Liong, S.-Y., Tkalich, P., 2008. An ANN application for water quality forecasting. Mar. Pollut. Bull. 56, 1586–1597.

Palmquist, W.N., 1973. Sampling for Groundwater Investigations. vol. 125. Dept. of Mines and Geology, Govt. of Karnataka, Groundwater Studies, p. 14.

Pejman, A.H., Nabi Bidhendi, G.R., Karbassi, A.R., Mehrdadi, N., Esmaeili Bidhendi, M., 2009. Evaluation of spatial and seasonal variations in surface water quality using multivariate statistical techniques. Int. J. Environ. Sci. Technol. 6 (3), 467–476.

Phiri, O., Mumba, P., Moyo, B.H.Z., Kadewa, W., 2005. Assessment of the impact of industrial effluents on water quality of receiving rivers in urban areas of Malawi. Int. J. Environ. Sci. Technol. 2 (3), 237–244.

Ramkumar, T., Venkatramanan, S., Anitha Mary, I., TamilSelvi, M., Ramesh, G., 2010. Hydro geochemical quality of groundwater in Vedaraniyam Town, Tamil Nadu, India. Res. J. Environ. Earth Sci. 2 (1), 44–48.

Reckhow, K.H., 1999. Water quality prediction and probability network models. Can. J. Fish. Aquat. Sci. 56, 1150–1158.

Schizas, C.N., Patticijis, C.S., Michaclides, S.C., 1994. Forecasting minimum temperature with short time length data using artificial neural network. Nerual Netw. World 4 (2), 209–219.

Solaimani, K., Modallaldoust, S., Lotfi, S., 2009. Investigation of land use changes on soil erosion process using geographical information system. Int. J. Environ. Sci. Technol. 6 (3), 415–424.

Wen, C.G., Lee, C.S., 1998. A neural network approach to multi objective optimization for water quality mangement in a river basin. Water Resour. Res. 34 (3), 427–436.

Zaheer, I., Bai, C.G., 2003. Application of artificial neural network for water quality management. Lowland Technol. Int. 5 (2), 10–15.

Zhang, Q., Stanley, S.J., 1997. Forecasting raw-water quality parameters for the North Saskatchewan River by neural network modelling. Water Res. 31 (9), 2340–2350.

Section C

Groundwater Quality Assessment Using GIS and Geostatistical Aspects

Chapter 12

Simulation of Seasonal Rainfall and Temperature Variation—A Case Study of Climate Change Projection in Ponnaiyar River Basin, Southern India

A. Jothibasu and S. Anbazhagan

Centre for Geoinformatics and Planetary Studies, Periyar University, Salem, India

12.1 INTRODUCTION

Extreme events have triggered massive consequences for human society and for the natural environment all over the world. Moreover, climate change has the potential to change the intensity and frequency of these events (Frias et al., 2012). Mahmood and Babel (2014) have explored future extreme temperature changes using the statistical downscaling model (SDSM) in the transboundary region of the Jhelum River Basin. Thoeun (2015) outlined observed and projected changes in temperature and rainfall in Cambodia. In the assessment of observed and projected changes in temperature and rainfall in Cambodia, it is noted that the PRECIS GCM output is on a monthly basis. In some cases, daily data are required to perform vulnerability and adaptation assessment.

The global climate models provide a laboratory for numerical experiments on climate transitions during the past, present, and future. General circulation models (GCMs) are a class of computer-driven models for weather forecasting, understanding of climate, and for projecting climate change (IPCC, 2007). Global climate models drive the regional climate models; time-dependent large-scale lateral boundary forcing is imposed from GCM simulations (Johns et al., 1997). The regional climate modeling (RCM) approach affords an increase of resolution over a region of the globe in comparison to the CGCMs, with regional grid-point spacing of a few tens of km in the horizontal, for operational use on climate timescales. The RCM approach could still be useful to reach a resolution of a few km for the same computational load (Laprise, 2008). A variation of this technique is to also force the large-scale component of the RCM solution throughout the entire domain (Evans et al., 2012). The main objective in the present research is to assess the simulated changes of temperature and rainfall for future predictions in the Ponnaiyar River Basin.

12.2 STUDY AREA

The Ponnaiyar River Basin is an interstate river and is one of the largest rivers of the state of Tamil Nadu, often reverently called "Little Ganga of the South." The river has supported many civilizations in peninsular India throughout history and continues to play a vital role in supplying precious water for drinking, irrigation, and industry to the people of the states of

Karnataka, Tamil Nadu, and Pondicherry. The study area extends over approximately of 11,595 km^2 and lies between latitudes 11°350 and 12°35'0'' north and longitudes 77°45'0'' and 79°55'0'' east (Fig. 12.1). The Ponnaiyar River originates on the south eastern slopes of the Chennakesava Hills, northwest of Nandidurg of the Kolar district in Karnataka State at an altitude of 1000 m above mean sea level (AMSL). The total length of the Ponnaiyar River is 432 km, of which 85 km lies in Karnataka State; 187 km in Dharmapuri, Krishnagiri, and Salem Districts; 54 km in Thiruvannamalai and Vellore Districts; and 106 km in Cuddalore and Villupuram Districts of Tamil Nadu. The Ponnaiyar Basin is predominantly built up with granite and gneisses rocks of the Archean period. The granite is of very good quality and extensive outcrops and masses of it are commonly found. The chief components of the rock are hornblende and feldspar. Foliation is seldom seen. In the plains of the reserve forest quartz is found commonly. Diamond granite is also found in scattered pockets in the area of the Chitteri hills in Dharmapuri and Krishnagiri sub-divisions. Charnockite rocks from the Archean period are also seen in some areas. Alluvium and sand-dunes of the Quaternary period are also seen at a few places. The 15 years' average annual rainfall for the period 2000–2014 in the basin was 969 mm. This catchment falls under the tropical belt; the climate in general is hot with April and May being the hottest months of the year when the temperature rises to 34°C.

12.3 METHODOLOGY

In the present study, the future climate conditions forced with the Special Report on Emission Scenarios (SRES) of AR4 of the IPCC at regional scale obtained throughout a dynamical downscaling from GCMs are available on the CCAFS website. The datasets contain a large range of RCMs developed by different countries and climate modeling communities with different spatial resolutions. The datasets are available in ARC GRID and ARC ASCII format, in decimal degrees and datum WGS84. This data format facilitates their integration into a geographical information system (GIS) environment for processing. The climate model downscaling data of the bccr_bcm2_0 was selected for this investigation due to its fine space scale (30 s) (Table 12.1). The fine space resolution is very important for this investigation due to the complex topography of the study area. The climatic baseline over the period 1961–1990 was used to compare and calculate projected changes for

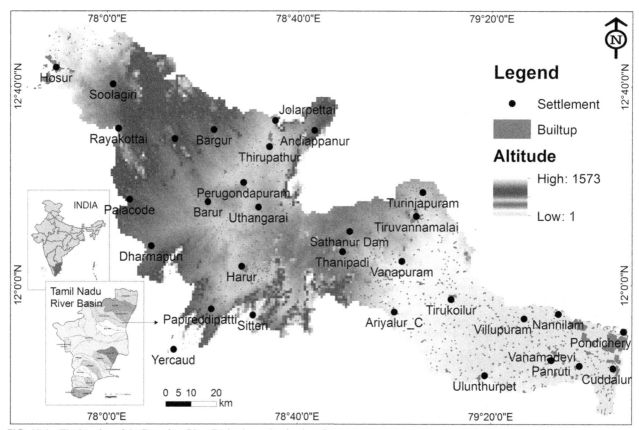

FIG. 12.1 The location of the Ponnaiyar River Basin shows the shuttle radar topography mission digital elevation model.

TABLE 12.1 Particulars of SRES A1B Global Circulation madel

File set	Delta method IPCC AR4	Spatial interpolation of anomalies (deltas) of original GCM outputs from IPCC CMIP2 applied to a Worldclim high-resolution baseline climate. Data were processed by CIAT
Scenario	SRES A1b	A very heterogeneous world with a continuously increasing global population and regionally oriented economic growth that is more fragmented and slower than in other storylines
Model	bccr_bcm2_0	sres_a1b_2040s_prec_30s_tile_a5_asc.zip
Extent	Global (A5 and B5)	
Format	ASCII grid format	ARC ASCII GRID refers to a specific interchange format developed for ARC/INFO rasters in ASCII format. The format consists of a header that specifies the geographic domain and resolution, followed by the actual grid cell values
Period	2020–2080	
Variables	Bioclimatic	Derived from the monthly temperature and rainfall values in order to generate more biologically meaningful variables. See http://www.worldclim.org/bioclim
	Maximum temperature	Monthly mean maximum temperature. Unit: °C * 10
	Mean temperature	Monthly mean average temperature. Unit: °C * 10
	Minimum temperature	Monthly mean minimum temperature. Unit: °C * 10
	Precipitation	Monthly accumulated rainfall. Unit: mm/month
Resolution		30 s

the average annual maximum temperature, average annual minimum temperature, and seasonal precipitation. The pattern of temperature and rainfall anomalies was calculated for each raster cell grid using ArcGIS software for short term (2011–2040) and medium term (2041–2080). Future climate simulation related to the IPCC's A1B greenhouse gas (GHG) emissions scenario was chosen to assess future temperature and rainfall projections. This scenario predicts an intermediate level of warming by the end of the century and a future where technology is shared between developed and developing nations in order to reduce regional economic disparities.

12.3.1   Baseline Climate

The hottest month of the year in the Ponnaiyar River Basin is May. During this month average maximum temperatures are between 28°C and 38°C. Coastal regions are hot and humid in summer with maximum temperatures around 38°C and humidity levels exceeding 90%. In the interior plains high temperatures in summer can exceed 35°C. The coldest month of the year in the study area is January. During this month average minimum temperatures are between 11°C and 21°C (Fig. 12.2). The coldest temperatures are encountered in highland and mountain areas in the northern and southern part of the study area.

Rainfall is caused by four principal mechanisms—convection, cold frontal troughs, monsoons, and tropical storms/cyclones—and their interactions with local topography and other meteorological conditions (Kwarteng et al., 2008). During the winter months of January and February, average rainfall is between 3 and 47 mm in the study area. Physiographic conditions significantly affect average winter rainfall (Charabi and Al-Hatrushi, 2010). In the summer months of March to May the average rainfall is between 61 and 242 mm for the overwhelming majority of the study area. During the summer months some parts of the study region are transformed into lush landscapes of green field and verdant vegetation (Charabi, 2009). During the northeast monsoon season the average baseline rainfall is between 278 and 774 mm. The hilly region and central part of the study area received the higher rainfall in this season. The baseline southwest monsoon average annual rainfall indicated that the eastern part of the coastal regions had higher rainfall (608–761 mm) (Fig. 12.3).

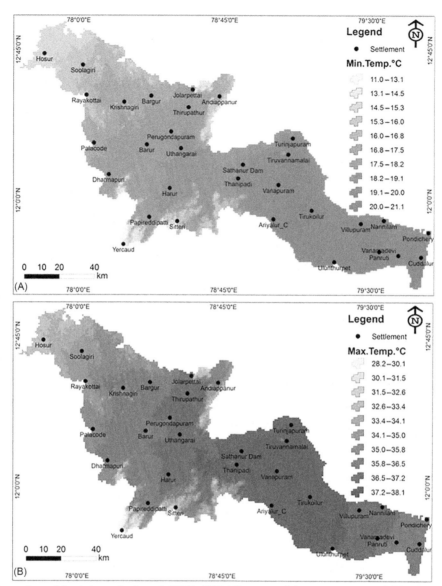

FIG. 12.2 Temperature patterns of the study area for the period 1961–1990: (A) average minimum temperature and (B) average maximum temperature.

12.3.2 Projection for 2011–2040

The simulated average maximum and minimum temperature changes during the period 2011–2040 are shown in Fig. 12.4. The A1B scenario clearly shows an increase in the future maximum temperature in the range of 1–2°C for the entire study area through 2040.

The projected average annual rainfall pattern for all four seasons is shown in Fig. 12.5. The winter season rainfall pattern changes from 8 to 49 mm in the study area. In the period of 2011–2040 the summer rainfall pattern deviated from 42 to 207 mm. The rainfall trends for northeast and southwest monsoons also increased in the projected period of 2011–2040.

12.3.3 Projection for 2041–2080

The average maximum and minimum temperatures are projected to increase in the range of 2–3°C, with the geographic distribution of these changes following similar patterns as in the earlier period by 2080. Fig. 12.6 shows the simulated maximum and minimum temperature changes during the period 2041–2080. The A1B scenario also clearly shows future

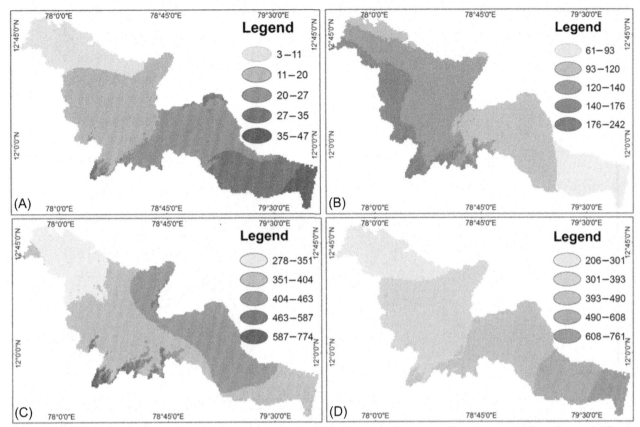

FIG. 12.3 Baseline rainfall patterns of study area for the period 1961–1990: (A) winter, (B) summer, (C) northeast, and (D) southwest rainfall.

minimum temperature increases that are similar to the results shown for maximum temperature changes. For both time periods there is a clear expansion compared to the results for maximum temperature change, suggesting that minimum temperatures will experience the greatest impact from climate change. Fig. 12.7 shows that the simulated average annual rainfall changes during the period 2041–2080. It indicates that the A1B scenario clearly shows that most of study area will become drier in the summer and winter seasons, with large portions of the study area receiving up to 40 mm less in annual rainfall throughout the projection period. On the other hand, the model results indicate that the southwest monsoons are likely to intensify, leading to increased rainfall in the southwestern parts of the country.

12.4 RESULTS AND DISCUSSION

GCMs lack the regional detail that impacts assessments on climate change. An RCM adds small-scale detailed information of future climate change to the large-scale projections of a GCM. Coarse resolution information from a GCM can be used to develop temporally and spatially fine scale information. The main advantage of an RCM is that it can provide high-resolution information on a large physically consistent set of climate variables and therefore a better representation of extreme events. Comparing the baseline (1961–1990) simulation output with the observed data for the period of 2011–2080, the RCM simulation was validated.

12.4.1 Simulation of Temperature 2011–2080

During the projected period the study area seems to show continually increasing temperatures from 1°C to 3.5°C. Significant annual average temperature increases of record high temperatures for around 2040 and 2080 indicate strong inter-decadel variability. Nevertheless, the significant increase recorded for 2080 is extremely severe compared to the rest. Highest (3.7°C) and lowest (1.3°C) temperature increases for the study area is obtained during the period of 2011–2080 (Fig. 12.8).

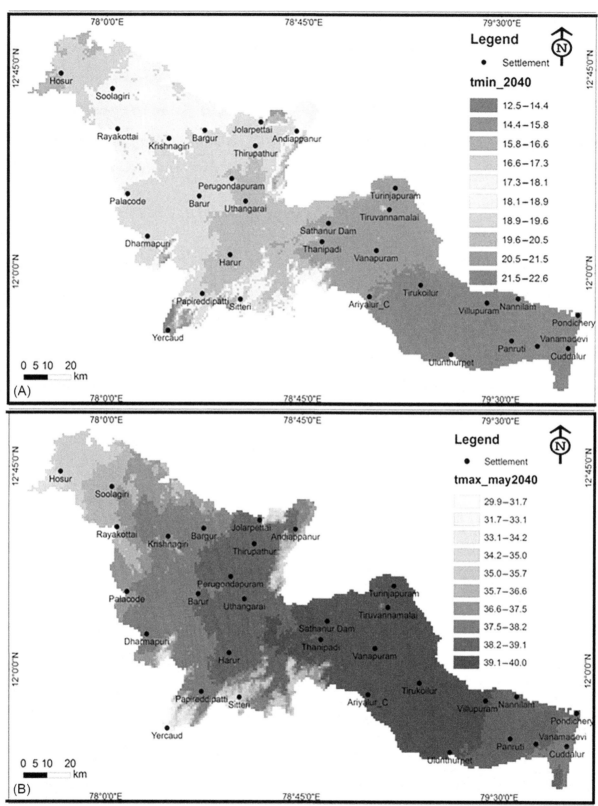

FIG. 12.4 Projected temperature pattern (2011–2040): (A) minimum temperature and (B) maximum temperature.

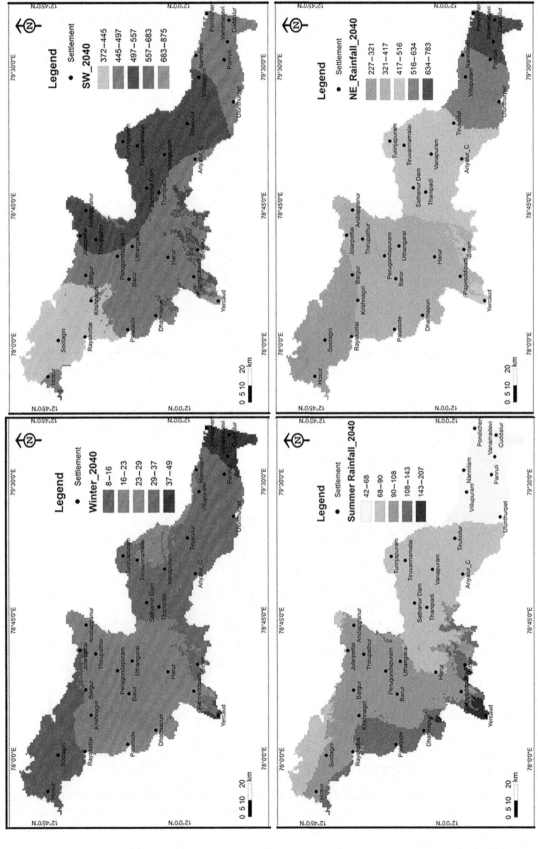

FIG. 12.5 Rainfall patterns of study area for the period 2011–2040.

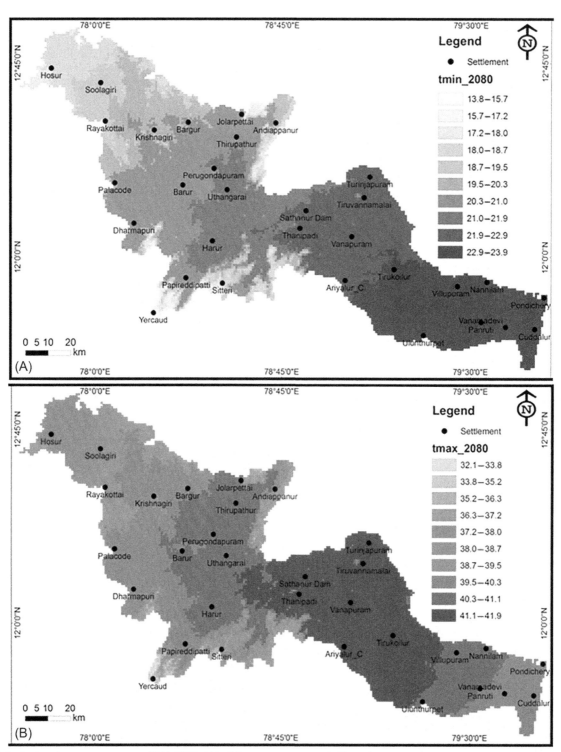

FIG. 12.6 Projected temperature pattern (2041–2080): (A) minimum temperature and (B) maximum temperature.

12.4.2 Rainfall Simulation (2011–2080)

In a region where the historic average annual rainfall levels are between 50 and 100 mm, climate change is expected to lead to between 20 and 40 mm less rainfall by 2080. This is equivalent to a reduction in the average annual rainfall of about 40%. With less future rainfall in northern areas, groundwater recharge and surface water flow are expected to also decrease.

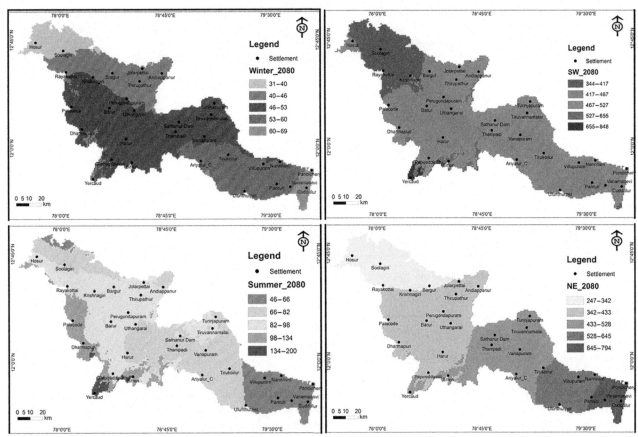

FIG. 12.7 Rainfall patterns of study area for the period 2041–2080.

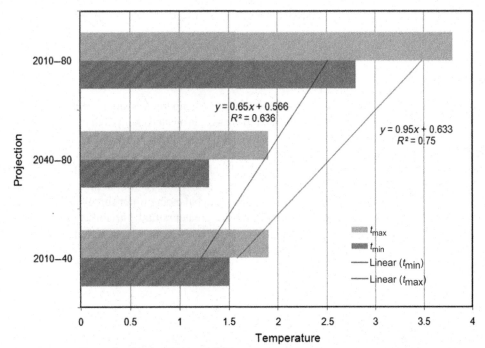

FIG. 12.8 Simulated temperature fluctuations between 2011 and 2080.

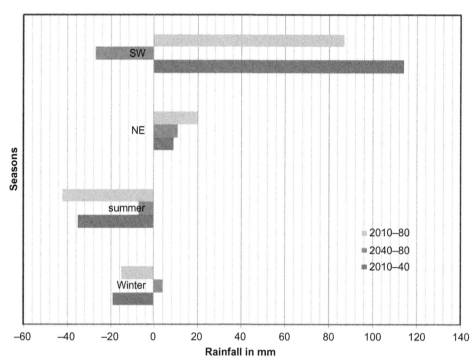

FIG. 12.9 Simulated seasonal average rainfall patterns between 2011 and 2080.

The SRES A1B scenario indicates that a slackening of the winter monsoon. The rainfall during the summer monsoon in 2080 is projected to be the worst over the study area. The negative annual rainfall trend is extremely evident in relation to climate change. Nevertheless, all of the regions seem to experience increased rainfall in the monsoons of the southwest toward the end of the century (2080). The significant increase in the annual temperatures simulated for 2080 corresponds to a significant reduction in the annual summer rainfall simulated for the same years (Fig. 12.9). Of these years, the highest rainfall reduction was simulated for 2080, which corresponds to the highest temperature increase simulated for annual temperature anomalies. This indication of significant reduction of rainfall together with significant increases in temperature is generally exhibited during El Nino events.

12.5 CONCLUSIONS

The Ponnaiyar River Basin baseline climate was evaluated relative to temperatures and rainfall patterns using proxy data. With climate change these baseline conditions are expected to change. Projected changes in temperatures and rainfall were assessed and the results of this assessment formed the basis by which the vulnerability of the study area to climate change could be understood. Water resources are already facing considerable threat in south India. With climate change it is expected that the prevention of groundwater degradation and balancing supply and demand will become even greater challenges. As discussed previously, northern Oman is expected to face decreasing rainfall in the coming decades. When combined with continued socioeconomic growth, the current challenges in balancing water supply and demand will increase, as will the difficulty in maintaining water-quality standards. Moreover, greater rainfall variability and longer drought episodes may adversely impact the already fragile and vulnerable mountain ecosystems of the region. Integrated water resource management (IWRM) is considered a fundamental organizing framework to identify and evaluate potential adaptation strategies for water resources.

REFERENCES

Charabi, Y., 2009. Arabian summer monsoon variability: teleconexion to ENSO and IOD. Atmos. Res. 91, 105–117.

Charabi, Y., Al-Hatrushi, S., 2010. Synoptic aspects of winter rainfall variability in Oman. Atmos. Res. 95, 470–486.

Evans, J., McGregor, J., McGuffie, K., 2012. Future regional climates. In: Henderson-Sellers, A., McGuffie, K. (Eds.), The Future of the World's Climate, pp. 223–250. https://doi.org/10.1016/B978-0-12-386917-3.00009-9.

Frias, M.D., Minguez, R., Gutierrez, J.M., Mendez, F.J., 2012. Future regional projections of extreme temperatures in Europe: a nonstationary seasonal approach. Clim. Change 113 (2), 371–392.

IPCC, 2007. Solomon, S., Qin, D., Manning, M., Chen, Z., Marquis, M., Averyt, K.B., Tignor, M., Miller, H.L. (Eds.), Contribution of Working Group I to the Fourth Assessment Report of the Intergovernmental Panel on Climate Change. Cambridge University Press, Cambridge, United Kingdom/New York, NY.

Johns, T.C., Carnell, R.E., Crossley, J.F., Gregory, J.M., Mitchell, J.F.B., Senior, C.A., Tett, S.F.B., Wood, R.A., 1997. The second Hadley centre coupled ocean-atmosphere GCM: model description, spinup and validation. Clim. Dyn. 13, 103–134.

Kwarteng, A.Y., Dorvlo, A.S., Ganiga, T.K., 2008. Analysis of a 27-year rainfall data (1977-2003) in the Sultanate of Oman. Int. J. Climatol. 29, 605–617.

Laprise, R., 2008. Regional climate modeling. J. Comput. Phys. 227, 3641–3666.

Mahmood, R., Babel, M.S., 2014. Future changes in extreme temperature events using the statistical downscaling model (SDSM) in the trans-boundary region of the Jhelum river basin. Weather Clim. Extrem. 5–6, 56–66.

Thoeun, H.C., 2015. Observed and projected changes in temperature and rainfall in Cambodia. Weather Clim. Extrem. 7, 61–71.

Chapter 13

Impact of Urbanization on Groundwater Quality

K. Brindha and Michael Schneider

Hydrogeology Group, Institute of Geological Sciences, Freie Universität Berlin, Berlin, Germany

Chapter Outline

13.1 INTRODUCTION

Urbanization is defined as the process where people living in rural areas move to urban areas. This is facilitated by the growing number of industries and job opportunities in the cities. To accommodate the increasing population and number of businesses the urban areas in many major cities extend beyond the administrative boundaries and into semiurban areas (e.g., Bangkok in Thailand, New Delhi in India, Berlin in Germany). Thus many rural areas are becoming "cities" and many cities are becoming "megacities." The increase in the population living in urban and rural areas from 1960 to 2016 is shown in Fig. 13.1. In 2016 about 54% of the world's population lived in urban areas, with a projected increase to 60% by 2030, and 66% by 2050 (United Nations, 2014). Tokyo tops the list with 38 million people followed by New Delhi, Shanghai, Mumbai, Sao Paulo, Beijing, Mexico City, Osaka, Cairo, and New York (United Nations, 2016). The pace of urbanization is faster in the developing nations of Asia and Africa.

Rapid increase of inhabitants in the urban areas requires proper infrastructure for water supply. At times the water supplies in countries like Bahrain, Denmark, Malta, Oman, Qatar, and Saudi Arabia are solely met by groundwater resources (UNESCO, 2004; Eslamian and Eslamian, 2017). Estimates indicate that half of the world's megacities are dependent on groundwater (Wolf et al., 2006). Many European cities are either completely dependent on groundwater (e.g., Budapest, Copenhagen, Hamburg, Munich, Rome, and Vienna) or receive more than half (e.g., Amsterdam and Brussels) of their water supplies from groundwater (UNESCO, 2004). In Asia, cities such as Lahore, Hanoi, and Kumamoto are solely dependent on groundwater (Takizawa, 2008; Basharat, 2016) and major cities like Dili, Beijing, New Delhi, Bangkok, Jakarta, Bandung, Hanoi, etc., are heavily dependent on groundwater.

The increased demand for water in urban areas is not often immediately met by the development of new or extended water supply systems. Many houses in the cities of developing nations are still deprived of piped water supplies or receive limited water supplies at fixed intervals (Brindha et al., 2014). Due to immediate needs and a deficiency of efficient water supply systems to provide for the growing population, people depend on alternate sources of water, which includes, (1) direct access to surface water or groundwater resources, (2) private water supplies in the form of bottled water, kiosks, water tankers, etc., and (3) public water supplies in the form of standpipes and public taps. A growing number of industries

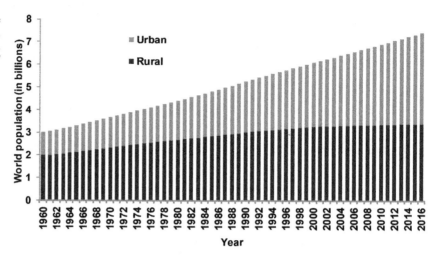

FIG. 13.1 Urban and rural population of the world, 1960–2016. (Source: *World Bank, 2018. World Population (Total, Urban and Rural). Available online at: https://data.worldbank.org/ (Accessed on 24 December 2018).*)

rely on groundwater as a cheap and quick option. In some cities groundwater extraction for industrial use exceeds that for drinking, such as in Bandung (80%) and Bangkok (60%) (Shrestha and Pandey, 2016).

Lately, an increased reliance on groundwater has led to groundwater depletion at very high rates, the adverse effects of which have been realized in recent decades. The rate of pumping of groundwater has overtaken the rate of replenishment. Groundwater is usually considered to be less vulnerable to pollution compared to surface water, but in recent times it has become increasingly contaminated due to human intervention. Due to the occurrence of groundwater in the subsurface it is vulnerable to pollution from sources located both on the surface and in the subsurface. More in urban areas than in rural areas, groundwater is prone to contamination from multiple sources (domestic, industrial and natural) and with more complex pollutants (fluoride, arsenic, nitrate, heavy metals, organic compounds, etc.).

In this chapter the various sources contaminating urban groundwater are discussed. Selected key contaminants from various sources are described in the subsections. Demarcation of groundwater pollution due to urbanization using geographical information system (GIS) tools is presented through a case study.

13.2   POTENTIAL SOURCES OF GROUNDWATER CONTAMINATION

Pollutants in groundwater are generally classified as (1) natural and anthropogenic or (2) physical, chemical, and biological. However, sources of groundwater contamination (Fig. 13.2) can be classified into various types such as (1) natural and anthropogenic, (2) surface and subsurface, (3) point and nonpoint, and (4) rural and urban. This chapter does not classify the sources based on these types, because a pollutant can reach the groundwater environment, especially in urban areas, by multiple pathways. Also, key contaminants from various sources are described within the subsection, but that does not imply that these contaminants are contributed solely from one particular source. For example, arsenic is a toxic pollutant in groundwater of many nations and is chiefly contributed by geochemical processes. Nevertheless, as a by-product of some human activities such as chemical industry, mining operations and agriculture, arsenic can be added to natural waters. The characteristics and pathways of arsenic in groundwater are explained under geogenic sources, as this is the largest contributing source of arsenic.

13.2.1   Natural Sources

Arsenic and fluoride are the most common and serious geogenic contaminants of groundwater causing pollution at a large-scale. The occurrence, origin, and chemical reactions of these ions in groundwater are influenced mostly by the local geology, hydrogeology, and geochemistry.

13.2.1.1   Arsenic

Arsenic in groundwater is largely contributed by natural weathering of minerals from rocks and soil. Urbanization and related human activities accelerate the release of arsenic from the natural environment. An estimated 150 million people from over 70 countries worldwide are exposed to arsenic poisoning by drinking groundwater containing high levels of arsenic (Ravenscroft et al., 2009). Of these, about 110 million people live in 10 Asian countries (South and South-East):

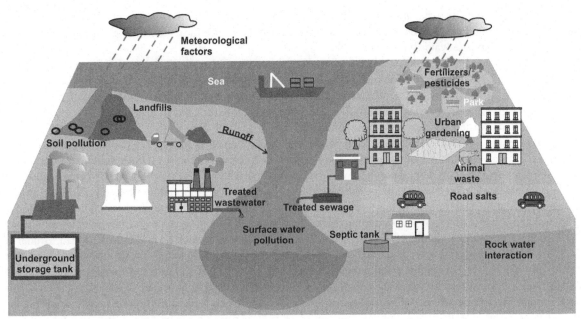

FIG. 13.2 Sources of groundwater contamination.

Bangladesh, Cambodia, China, India (West Bengal), Laos, Myanmar, Nepal, Pakistan, Taiwan, and Vietnam (Brammer and Ravenscroft, 2009). Smith et al. (2000) described the groundwater contamination by arsenic in Bangladesh as the largest and worst poisoning of a population in history. Cities such as Hanoi (Vietnam) located in one of these contaminated hotspots face serious consequences, as the Hanoi municipal water supply system depends on groundwater exploited from a polluted confined aquifer as the only water source for its domestic water supply (Nga, 2008). These highly contaminated aquifers contain arsenic at levels up to 112 µg/L (Nga, 2008), which exceeds the World Health Organization limit (10 µg/L) (WHO, 1993) by many times. An additional infrastructure is essential to treat the groundwater before supplying the local population.

Arsenic primarily occurs as arsenopyrite and is commonly associated with sulfide minerals. Orpiment, realgar, and arsenic-rich iron oxyhydroxide also release arsenic in the groundwater (Garelick et al., 2008). Four geochemical reactions namely, reductive dissolution, alkali desorption, sulfide oxidation, and geothermal activity, predominantly determine the occurrence and distribution of natural arsenic (Ravenscroft et al., 2009). Crops irrigated with arsenic contaminated groundwater lead to an accumulation of arsenic in the soil and in the end-products. In south and southeast Asia, the rice, which is the staple crop, is irrigated with groundwater, which has further enhanced the arsenic content of the rice (Williams et al., 2006; Pal et al., 2009). This may not only pose a risk to the population that consume these crops, but is also a threat to food production due to the toxic levels of arsenic accumulated in the irrigated soil.

Based on the pH and redox conditions, arsenic occurs in two oxidation states in the environment: arsenic (III) or arsenite and arsenic (V) or arsenate. Arsenite is more toxic than arsenate. Long-term exposure to low doses of arsenic (such as drinking arsenic contaminated groundwater) can lead to arsenicosis. This is a serious and irreversible health issue and the adverse health effects include skin lesions, circulatory disorders, neurological and respiratory complications, diabetes, and hepatic and renal dysfunction (Chen et al., 2009). Arsenic is a known carcinogen causing cancer of skin, lung, liver, bladder, and prostate. Acute toxicity may lead to death.

13.2.1.2 Fluoride

Fluoride, like arsenic, causes mass contamination of groundwater affecting around 200 million people from among 25 nations all over the world (Ayoob and Gupta, 2006). It is more pronounced in Africa (Kenya, Ghana, Tanzania, Malawi, Uganda, Sudan, South Africa, Ethiopia, and Algeria) and Asia (India, China, Pakistan, Thailand, Japan, and Sri Lanka). Fluoride naturally occurs in groundwater due to the weathering of fluoride-rich rocks. Long residence times of groundwater in aquifers with fluoride-rich rocks and slow movement of groundwater increases fluoride concentrations in groundwater (Brunt et al., 2004). Commonly occurring fluoride bearing minerals are fluorite, apatite, and mica. It is also present in biotite, amphibole (e.g., tremolite and hornblende), cryolite, epidote, fluorapatite, fluormica, topaz, sellaite, clays, villuanite, and phosphorite (Brindha and Elango, 2011). High fluoride concentrations in groundwater are associated with alkaline pH, high sodium, high bicarbonate, and low calcium contents. Hence most fluoride-rich groundwater has a characteristic sodium-bicarbonate water type.

Volcanic rocks enhance the concentration of fluoride and, thus, volcanic ash is also rich with fluoride. These volcanic ashes are highly soluble and cause fluoride contamination in groundwater. Though these geogenic sources are the largest contributor of fluoride, it also enters the groundwater through anthropogenic sources. Combustion of fossil fuels; production of cement, glass, ceramic, brick, plastics, tiles, and phosphate fertilizers; and industrial processes such as smelting, dyeing, and industrial wastewater, are additional sources.

Fluoride is essential for healthy bones and teeth, while in excess it causes fluorosis. As a public-health initiative, many countries have adopted fluoridation of the public water supply as an effective tool to reduce dental carries. Usually the fluoride level is adjusted to 1 mg/L in the public water supply. About 378 million people from 25 countries have access to artificially fluoridated water through the public water supply (The British Fluoridation Society, 2012). However, the percentage of population benefiting from this varies from <1% to 100%. Developed nations like Singapore and Hong Kong supply fluoridated water to 100% of the population. More than half of the population are consuming optimally fluoridated water (<1.5 mg/L) through the public water supply and from natural sources, such as in Australia (80%), Brunei (95%), Chile (70%), the Irish Republic (73%), Israel (70%), Malaysia (76%), New Zealand (61%), and the United States of America (USA) (66%) (The British Fluoridation Society, 2012). Though the positive effect of fluoride on teeth is accepted, the method used, i.e., mass medication of the population is argued against by certain groups as unethical. Dental products are also fluoridated to reduce dental carries and strengthen teeth. A moderate amount of fluoride intake causes dental fluorosis, i.e., discoloration of teeth from yellow to brown or black in the form of spots or streaks. But long-term exposure to fluoride-rich drinking water can cause crippling skeletal fluorosis, renal dysfunction, and kidney stones. In addition, high fluoride consumption disrupts the normal functioning of the respiratory, digestive, nervous, excretory, and reproductive systems (Meenakshi and Maheshwari, 2006).

13.2.1.3 Total Dissolved Solids

Salinization is an increase in the total dissolved solids (TDS) of the aquifer caused by natural or anthropogenic factors. The processes and sources of salinization vary for inland and coastal aquifers. In urban areas located inland, salinization may be due to geogenic or anthropogenic factors (Fig. 13.3). Saline water naturally underlies freshwater aquifers at greater depths in some regions (Martens and Wichmann, 2007). When the water from these saline aquifers is discharged onto the surface, the fresh water aquifers may also be contaminated. It is also possible that the salt water and fresh water mixes in the subsurface and the salinity of fresh water aquifer is increased. The salinity of aquifers depends on the distribution and rates of precipitation, evapotranspiration and recharge rates, type of aquifer material and its characteristics, residence time, flow velocities, and nature of the discharge areas (Richter and Kreitler, 1993). Aquifers in contact with salt deposits also turn saline due to natural rock-water interaction processes. Water pumped from these saline aquifers cannot be directly used for water supply or industrial purposes. Australia, being a dry continent, is highly affected by salinization, therefore groundwater in many parts of the country is naturally saline. Major anthropogenic sources of inland salinization in urban areas include irrigation of dry areas that lack proper drainage, increased evaporation and decreased precipitation facilitated by climate change, excessive groundwater pumping, wastewater with a high salt content being disposed of carelessly by industries onto the surface, etc (Foster and Chilton, 2003, Zimmermann-Timm, 2007). Irrigation and heavy groundwater pumping-induced salinization is commonly reported in India and Pakistan.

13.2.2 Heavy Groundwater Pumping

In coastal urban areas, the majority of salinization is due to seawater intrusion through human influence. Many megacities in the world are located in coastal areas. Nearly 44% of the world's population live within 150 km of the coast (Reed, 2010). The average population densities in these coastal regions are approximately three times higher than the average global population density (Small and Nicholls, 2003). It is estimated that in the year 2030, about 268 million to 286 million people will live in the low-elevation coastal zones with a large portion of the population housed in China, India, Bangladesh, Indonesia, and Vietnam (Neumann et al., 2015). Exponential population growth in these regions increases the demand for public water supply. In many urban coastal areas, groundwater forms the main freshwater source. Coastal aquifers are sensitive to changes in the environment, such as groundwater pumping and recharge due to their low topography. Thus overpumping of groundwater from the coastal aquifers results in seawater intrusion. This is caused by the variation in the density of fresh water and seawater. The high-density seawater moves into the coastal aquifer and forms a wedge shape (Fig. 13.4).

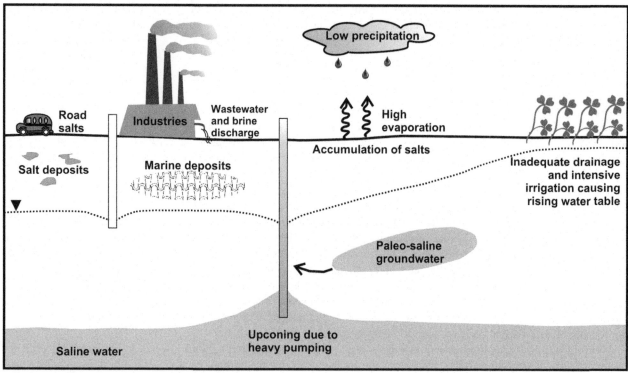

FIG. 13.3 Sources of inland salinization of groundwater.

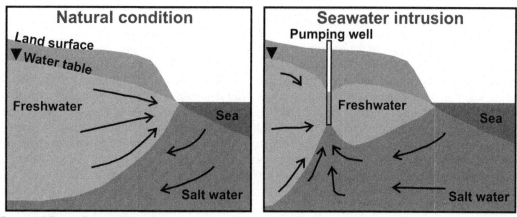

FIG. 13.4 Conceptual figure of seawater intrusion.

Depending on the intensity of pumping, this wedge can extend for several kilometers inland. These affected aquifers are characterized with high concentrations of chloride, bromide, and sodium.

In Chennai, a coastal city in southern India, the city's water requirement (about 10%) is met from pumping groundwater from the northern part of the city. Due to overpumping in the coastal areas, seawater intrusion has extended inland from 4 km in 1970–1975 (Subramanian, 1975) to 14 km in 2012 (Nair et al., 2015). In 2011 it was found that urban development and subsequent water supply withdrawals had resulted in seawater intrusion in approximately 1200 km^2 of the Biscayne aquifer in Florida, USA (Prinos et al., 2014). Seawater intrusion extends >1 km inland in a coastal area in eastern Greece and about 150 km^2 of the coastal aquifer has been intruded by saline water due to groundwater extraction through numerous borewells (Kazakis et al., 2016). Nearly 1520 km^2 of the coastal area near the Bohai Sea in northern China was affected by salinization due to seawater intrusion in the 1980s and this had increased to approximately 2457 km^2 in 2003 (Shi and Jiao, 2014).

Careful planning of groundwater extraction from the coastal aquifers is the preferred method to reduce seawater intrusion. Though alternative methods such as subsurface barriers and managed aquifer recharge are available to control seawater intrusion, they are cost-intensive and time consuming.

13.2.3 Industries

Several toxic heavy metals such as lead, cadmium, manganese, mercury, chromium, zinc, iron, aluminum, and copper are used in industrial processes. Industries withdraw a large portion of the groundwater in urban areas, but only a very small portion of this water is truly incorporated within the product. Most of the water used during fabricating is for cooling, washing, or transporting, and it returns to the environment as wastewater, only now it carries huge pollutant loads. Thermal power plants, engineering, paper, textile, and chemical-manufacturing industries are a few of the many industries that consume large amounts of water. Industries may extract water purely from the surface or from both the surface and groundwater, but the contamination from industrial effluents pollutes both these freshwater resources. The ideal practice to discard the effluents resulting from the manufacturing processes is by treating them on-site and disposing of them separately into receiving waters such as ponds, lakes, streams, rivers, and oceans. As a common practice the treated effluents are also disposed of into municipal sewers. The effluents from certain industries that are rich in nutrients (e.g., fertilizer manufacturing, food processing) are diluted and reused directly or by mixing with domestic wastewater for irrigation. This is practiced in countries like Australia, Canada, China, India, Mexico, USA, and the United Kingdom (UK). Regulatory authorities of many countries have stringent standards for the disposal of effluents. Nevertheless, illegal dumping of raw or partially treated effluents is a recurrent practice in developing nations. These effluents eventually find their way into the aquifers along with rainfall recharge and through interaction with surface water.

WHO (2017a) has listed four heavy metals (arsenic, cadmium, lead, and mercury) in its top-10 chemicals of major public concern, as they are carcinogenic. Exposure to arsenic is mainly through drinking groundwater containing high levels of arsenic or from food prepared using this water. Industrial processes contributing to arsenic in groundwater are smelting, metal extraction, processing and purification, manufacture of agro-chemicals (fertilizers, pesticides, herbicides, insecticides, and fungicides), wastewater from industries, etc.

13.2.3.1 Cadmium

Cadmium is a carcinogen and occurs in the environment as a resultant of industrial processes like smelting and electroplating during the manufacture of alloys, dyes, batteries, fertilizers, and fungicides. The main exposure route for cadmium is through the inhalation of air-borne cadmium particulates and chronic exposure causes lung cancer. Ingestion of cadmium through drinking water affects the renal (kidney disorders), skeletal (bone softening), and respiratory systems. One of the worst cases of large-scale cadmium poisoning was in the Toyama Prefecture in Japan in the early 1900s. The disease is prevalently called as the "itai-itai" (which means "it hurts, it hurts" in Japanese) disease. This caused osteomalacia (softening of bones) and renal tubular dysfunction in humans due to consumption of water with high concentrations of cadmium and of food crops grown with this water. This contamination was due to the disposal of tailings waste containing cadmium from mining and ore processing into the Jinzu River, which served as source for drinking and irrigation water. Groundwater in this area was also polluted due to the infiltration of effluents from unlined sewers and waste-storage ponds (Yoshida, 2002).

13.2.3.2 Lead

Lead is a toxic metal and its presence in drinking water is undesirable, as it is harmful even at low levels of exposure. Although lead occurs naturally in the Earth's crust, its high concentrations in groundwater are mainly due to anthropogenic activities. Lead is discharged in the wastewater as a byproduct from thermal power plants; some engineering processes like smelting and the recycling of batteries; and the manufacture of inks, pigments, cosmetics, utensils, pipes, paints, batteries, gasoline, ceramic, and a wide range of other products. Plumbing systems were in the past made of lead pipes and the corrosion of these pipes led to the occurrence of lead in the drinking water supply. Exposure to lead affects multiple organs in the cardiovascular, hematologic, gastrointestinal, neurologic, renal, and reproductive systems. It has adverse permanent consequences on children's health. Lead poisoning can affect the behavior and learning capabilities of children, reduce their intelligence quotient, delay growth, and cause hearing problems. Acute intake leads to permanent brain and nervous system damage, and even death. Mass lead poisoning is reported mainly in low- and middle-income countries, especially through the inhalation of lead particles from battery recycling and mining (WHO, 2017b).

13.2.3.3 Mercury

Mercury occurs naturally in the elemental form in the environment. When it is transformed into its organic form (methylmercury) it bioaccumulates, especially in aquatic species, i.e., fish and shellfish. Like lead, mercury is also a toxic heavy metal affecting children's neurological development. Fetuses are vulnerable to mercury poisoning through the ingestion of seafood by the mother. Ingestion of mercury affects digestive, immune, and nervous systems. It is used in the manufacture of electronic items (light emitting diode [LED] screens, switches, and batteries), thermometers, pharmaceutical items (dental amalgams, eye ointments, and drops), oil-based paints, and agro-chemicals (fertilizers, fungicides, and algaecides). The worst effects of mass mercury poisoning from industrial wastewater were seen in Minamata City, Japan. This disease, popularly called as the "Minamata disease" was first reported in 1956 (Harada, 1995). A chemical factory producing acetaldehyde was using mercury as a catalyst in its processes. Methylmercury was one of the components of the wastewater and this wastewater was released into the Minamata Bay, affecting the aquatic life. Humans and animals consuming the fish and shellfish were affected by this disease. Though the environmental hazard in Minamata was mostly to marine resources, the waterways carrying the wastewaters also contaminated the soil and groundwater (Ministry of the Environment, Japan, n.d.). Another incidence of mercury poisoning due to the illegal dumping of wastes from a mercury thermometer manufacturing factory was reported in Kodaikanal, India. Mercury contaminated glass scrap was disposed of for recycling and the scrap yards polluted the soil, air, surface water, and groundwater (IPEN, 2016).

13.2.3.4 Chromium

Textile, chrome pigments, paints and alloy production, and industrial processes such as electroplating and welding release chromium containing wastewater. The largest and most important sources of chromium pollution are tanneries. These industries process raw hides and skins to produce leather that is used for producing different finished products. The tanning process increases the durability of leather and this can be of two types: vegetable tanning and chemical tanning. Though using vegetable tannins is environmentally friendly, chemical tanning using chrome salts is preferred as it takes less time. The improper treatment of the industrial wastes leads to chromium contamination of the groundwater. Usually the effluents from many small-scale industries are treated in a common effluent treatment plant and are disposed of into receiving bodies. Nevertheless, the treated effluents do not comply with the standards for the disposal of wastewater and are rich in many ions (Brindha and Elango, 2012a). Many low- and middle-income countries from South Asia, Africa, and Central and South America are affected by chromium pollution from tanneries. India, Bangladesh, Nepal, Pakistan, China, Morocco, Turkey, Mexico, and Brazil are a few of the most affected countries in these regions. The health impacts due to the ingestion of chromium contaminated groundwater involve the gastrointestinal system, while occupational exposure to hexavalent chromium may lead to lung cancer.

13.2.3.5 Volatile Organic Compounds

Volatile organic compounds (VOCs) are carbon containing organic compounds that evaporate readily at room temperature. They are considered to be important environmental contaminants as they are mobile, persistent, and toxic (Squillace et al., 1999). With the multifold increase in the production of synthetic organic chemicals in the past few decades, the presence of VOCs in groundwater and soil has been increasingly reported. VOCs are an integral compound in a large variety of industrial and residential products including paints, inks, dyes, glue, fuel, gasoline, solvents, adhesives, deodorizers, refrigerants, and pesticides. They reach the groundwater mainly through human activities such as industrial wastewater, landfills, leakage and spills, pesticide use, etc. VOCs that are dense nonaqueous phase liquids (DNAPLs) with low solubility and high specific gravity are capable of penetrating deep below the water table and form persistent plumes. Due to their ability to occur in one or more of the four phases (gas, solid, water, and immiscible), remediation of aquifers contaminated with DNAPLs is often difficult, complex, expensive, and constrained by the extent of contamination, type of contaminant, and site-specific characteristics. The identification of VOCs in drinking water sources is of concern due to their carcinogenic, mutagenic, and teratogenic effect on humans. The United States Environmental Protection Agency (USEPA) has estimated that one-fifth of the nation's water supplies was contaminated with VOCs. In Switzerland, 14% of the monitoring sites in the urban areas had VOCs in the groundwater (FOEN, 2016).

13.2.3.6 Emerging Organic Contaminants

Emerging organic contaminants (EOCs) are defined as hazardous compounds that were previously not considered or known to be present in groundwater. Apart from industrial products, these include personal-care products, cosmetics, pharmaceuticals, veterinary products, food additives, engineered nanomaterials, and pesticides (Lapworth et al., 2012). Of these, pharmaceuticals are commonly reported in the groundwater of many countries including Italy, Canada, USA, Spain, Germany, and UK. Pharmaceutical wastes often reach the domestic wastewater disposal network and the wastewater treatment plants are not equipped with the infrastructure to remove these contaminants. Thus the treated wastewater along with the contaminants reach the surface water bodies and eventually pollute groundwater in urban areas. As these EOCs have been only recently discovered in groundwater, monitoring and guidelines for the presence of these compounds in drinking water has not been incorporated into the regulations of all nations. It is also possible that more contaminants have not yet been identified due to the limited availability of analytical facilities. The health implications for humans due to these EOCs are usually high, as they have low biodegradability and are therefore persistent (Kuroda and Fukushi, 2008).

13.2.4 Fertilizers

Fertilizers are used to enhance the plant growth and yield. Nitrogen, phosphorus, and potassium are the three major nutrients required by plants. Leaching of these nutrients from fertilizers is a common source of groundwater pollution. These nutrients are generally considered as an agricultural pollutant and are expected to be in high concentrations in non-urban agricultural areas rather than in cities. Nevertheless, use of fertilizers in gardens, parks, and lawns contribute to these pollutants in urban groundwater. Of these nutrients, nitrate contamination in groundwater used for drinking purpose is of highest concern due to the harmful effects on human health. Excess nitrate and phosphate applied to the soil are carried away by urban runoff and cause eutrophication in surface water bodies. This is a condition that causes algal blooms due to the availability of excess nutrients followed by reduced oxygen availability for aquatic flora and fauna, subsequently leading to the death of fish. Potassium-based fertilizers (e.g., potash) generally do not exert adverse effects on the groundwater environment and human health in comparison with other nutrients. Heavy metal and fluoride pollution from fertilizers have also been reported.

13.2.4.1 Nitrate

Nitrogen is ubiquitous in the environment and the nitrogen cycle, involving a series of biogeochemical processes, contributes to various nitrogen compounds in groundwater. Nitrate, a highly soluble nitrogen compound, occurs as a common surface and groundwater pollutant. An increase in the number of nonagricultural sources of nitrogen due to urban development has led to an increase in the nitrate concentrations in the groundwater of cities (Wakida and Lerner, 2005). Nitrogen bearing rocks containing ammonium-rich minerals and organic nitrogen compounds undergo nitrification under conducive conditions and contribute to nitrate in groundwater (Lowe and Wallace, 2001).

Anthropogenic sources of nitrogen compounds in groundwater are diverse ranging from point sources (e.g., sewage and wastewater treatment plants and landfills), multipoint sources (e.g., leaky sewers), and diffuse sources (e.g., fertilizers and atmospheric deposition) (Wakida and Lerner, 2005; Grimmeisen et al., 2017). Diffuse pollution from the application of fertilizers in agricultural areas is the predominant source of nitrate in groundwater. Agriculture intensification and the overall nature of farmers to utilize more than the required or prescribed level of fertilizers and pesticides expecting high yield is the principle cause of nitrate pollution. Asia (58%) is the largest consumer of nitrogen fertilizers and a large portion of the nitrogen demand comes from China (18%) and India (17%) (FAO, 2015). Nitrate from fertilizers is not just a problem in agricultural areas, but also in urban and peri-urban areas, where fertilizers are used in gardens and golf courses (Winter and Dillon, 2005). Animal waste also has a high nitrogen content. The improper disposal of these animal wastes as piles on unlined surfaces and using them as manure for agriculture can potentially pollute surface and groundwater through stormwater runoff and direct infiltration. Such instances have been reported in urbanized areas. Infiltration of leachate from landfills and from treated wastewater applied to agricultural lands also contaminate the soil and groundwater.

The WHO recommended limit for nitrate in drinking water is 50 mg/L (WHO, 1993). One of the most significant toxicological effects due to high nitrate consumption is methemoglobinemia, which particularly affects infants. This is a condition of the blood that results in cyanosis caused by the reduced ability of the red blood cells to carry oxygen (Fan and Steinberg, 1996). It also causes other health issues such as gastrointestinal infections, recurrent diarrhea, recurrent stomatitis, birth defects, deterioration of the immune system, hypertension, respiratory tract infection in children, and histopathological changes in cardiac muscles, alveoli of lungs, and adrenal glands (Gupta et al., 2008).

13.2.4.2 Phosphates

Phosphorus is one of the key elements necessary for plant growth and it is usually present along with other elements as phosphates, occurring naturally in rocks and mineral deposits. However, anthropogenic activities including sewage and wastewater runoff from agricultural and urban areas (especially from detergents) are the major source of phosphates in groundwater. Unlike nitrate, which primarily occurs in its organic form in soils, phosphorus occurs in organic and inorganic forms. Phosphates are comparatively less soluble than nitrates and are largely retained in soil by adsorption (Domagalski and Johnson, 2012). Therefore, in the past, phosphorus in soil was considered not highly mobile and hence not a major threat to groundwater quality. However, the wide occurrence and determination of phosphates in groundwater has led to the identification of leaching of phosphate fertilizers from the land surface to the groundwater. Africa is the largest exporter and South Asia is the largest consumer of phosphate fertilizers (FAO, 2011). Even at low concentrations, phosphates are likely to cause eutrophication of surface water bodies. Hence the monitoring of phosphate concentrations in the wastewater discharged to water bodies is essential. Permissible limits for phosphates in drinking water have not been proposed (WHO, 1993). They are toxic at very high levels and may cause digestive issues in humans.

13.2.5 Sewage Systems and Septic Tanks

Urban groundwater is artificially recharged to a larger percentage with water from leaky water supply systems and sewers. Nitrate and pathogens are the key contaminants from sewage. Grimmeisen et al. (2017) reported nitrate contamination in groundwater in Jordan due to infiltration of septic waste from leaky sewers. Leakage of human wastes from sewers was the major source of contamination from nitrogen compounds in the urban areas of Metro Manila, Philippines and Jakarta, Indonesia (Umezawa et al., 2009).

13.2.5.1 Microbial Contamination

Many water-borne diseases are caused due to ingestion of groundwater containing pathogenic microorganisms. These microscopic pathogens belong to four major groups, bacteria, viruses, protozoa and helminths. Drinking contaminated water is the major cause of large outbreaks of diseases such as cholera, typhoid, diarrhea, dysentery, etc. Nearly 2 billion people worldwide are exposed to drinking water with fecal contamination and this is estimated to cause about 502,000 diarrhea related deaths each year (WHO, 2017c). Children under the age of 5 years are frequent victims and nearly all deaths occur in developing nations.

The mixing of groundwater with sewage is the main source of fecal contamination. Thus microbial contamination of groundwater is closely associated with the local sanitation practices. Sewers in urban areas leak due to improper installation, development of cracks, damage due to disasters (e.g., earthquakes and floods) and land subsidence. The estimated wastewater leak from damaged sewer systems in Germany was several 100 million m^3/year (Eiswirth and Hötzl, 1997). Similarly, about 5–8 million m^3/year of sewage mixed with groundwater in Hannover, Germany (Mull et al., 1992). The estimated values were much higher in the USA, i.e., 950 million m^3/year of wastewater contaminated the groundwater (Pedley and Howard, 1997).

Septic tanks and cesspits are prevalent in developing countries and leaching of wastes from these tanks is another major source. Proper planning of the location, design, operation, and maintenance of these on-site disposal systems is crucial. Shallow groundwater is more susceptible to pollution from these structures compared to deep aquifers. The population density in an area, the density of the disposal systems, and the quality of the materials used for the construction of the on-site disposal systems determine the pollution risk. Discharge of raw or partially treated domestic wastewater into water bodies, its subsequent percolation through the unsaturated zone, and runoff from agricultural areas containing animal and human wastes also add to the contamination. Bacterial and viral contamination in shallow aquifers supplying the public water supply wells were reported in Finland and the USA. Certain microbes serve as indicators of pathogen contamination (e.g., *Escherichia coli*). However, these indicators vary for different pathogen groups. Fate and the transport of pathogens in the subsurface environment is poorly understood due to variation in the behavior of different microbial groups, complex biogeochemical processes, hydrogeological conditions, soil properties, soil heterogeneity, and multiple potential transport pathways.

Many countries disinfect the water before public supply. Chlorination is the simple, most reliable, and cost-effective method adopted. It is a prerequisite in some countries before public water supply (Japan, Israel, USA, UK, and Australia). Some European countries (e.g., Germany, the Netherlands, and Switzerland) have withdrawn this method after reports on the formation of potentially carcinogenic disinfection by-products during the chlorination process were published.

13.2.6 Landfills

Ideally the municipal waste generated in cities are recycled and the remnants are either incinerated or disposed of in landfills. The complexity in the management of these wastes is due to their miscellaneous composition including organic wastes, batteries, packaging materials, paper, metal containers, glass, plastics, clothes, electronics, furniture, etc. Often industrial and pharmaceutical wastes containing hazardous substances are disposed of along with municipal waste. The degree of municipal waste generation in a country is influenced by the economic development, urbanization rate, local habits, and climatic conditions (Hoornweg and Bhada-Tata, 2012). The levels of municipal-waste production in some countries are: USA produces 728 kg/person/year (USEPA, 2016a), Denmark produces 789 kg/person/year, Switzerland produces 725 kg/person/year, Germany produces 625 kg/person/year (Eurostat, 2017), Australia produces 565 kg/person/year (Pickin and Randell, 2017), India produces 124 kg/person/year, China produces 372 kg/person/year, Indonesia produces 190 kg/person/year, and Japan produces 624 kg/person/year (Hoornweg and Bhada-Tata, 2012).

Historically, landfills were places where municipal wastes were dumped without proper planning. During rainfall, the rainwater penetrates through the landfills and leachate is generated. The composition of leachate depends on the (1) climate (rainfall and snowmelt), (2) hydrogeological conditions (groundwater level), (3) waste composition and characteristics (particle size, density, permeability, initial moisture content, and biodegradability), (4) age of the landfill, (5) operation and maintenance (pretreatment, compaction, and vegetation cover), and (6) internal processes (organic matter decomposition) (Rapti-Caputo and Vaccaro, 2006). Typically, the leachate contains a potentially hazardous concentration of many ions, heavy metals, chemicals, toxic substances, organic compounds, and pathogens. Infiltration of the leachate through the vadose zone has resulted in surface and groundwater contamination in many countries, making the groundwater unsuitable. Such cases have been reported in both developed and developing countries, especially due to unlined landfills: Sant'Agostino in Italy (Rapti-Caputo and Vaccaro, 2006), Zhoukou in China (Han et al., 2014), Ranital in India (Samadder et al., 2017), Rishon Lezion in Israel (Aharoni et al., 2017), Londrina in Spain (Lopes et al., 2012), Guadalupe Victoria in Mexico (Reyes-López et al., 2008), and Augsburg, Munich and Gallenbach sites in Germany (Baumann et al., 2006). In recognition of the environmental hazard, landfills are now designed with liners and leachate collection systems that prevent the percolation of leachate and contamination of aquifers.

13.2.7 e-Waste

One of the current concerns in waste disposal is "e-waste," comprising of a vast variety of products, including computers, mobile phone, televisions, digital cameras, household appliances, etc. Due to the enhanced rate of production of electrical and electronic products, the short lifespan of these products, and their nonbiodegradable nature, it is a challenge to manage these wastes. Frequent innovations of upgraded models of mobile phones, laptops, and tablets every few months make the previous models obsolete within a short time of their introduction onto the consumer market. In 2014 global e-waste generation was about 42 million tons accounting for 5.9 kg of e-waste per person (Baldé et al., 2015). Developed countries in Europe (15.6 kg/person), Oceania (15.2 kg/person), and America (north, central, and south) (12.2 kg/person) generated the highest e-waste per person (Baldé et al., 2015). E-waste encompasses many toxic metals (e.g., mercury, lead, cadmium, etc.) and hazardous compounds (e.g., polyaromatic hydrocarbons [PAHs]). At present, only a few countries have enacted legislations for the proper disposal of e-wastes. Several less-developed countries have no strict regulations for e-waste management and, hence, many high-income countries get rid of their e-waste by shipping them illegally to low-income countries. These backyard processing centers are not professionally equipped to recycle the e-waste and the inefficient techniques endanger the local environment. Groundwater pollution with multiple pollutants from informal e-waste recycling areas have been widely reported in China, India, Ghana, and Nigeria. The remaining e-wastes from the processing centers also end up alongside municipal solid waste in landfills.

13.2.8 Storage Tanks

Storage tanks are tanks that are located either on the surface or subsurface. Underground storage tanks are preferred due to space-constraints on the ground and as they are less vulnerable to vehicular accidents and tampering (Hairston, 1995). These tanks normally hold liquids such as hydrocarbons (petroleum products, gasoline, crude oil, etc.), and other hazardous chemicals. Usually strict guidelines apply for the construction and maintenance of storage tanks. But accidental release of the toxic substances, especially from underground storage tanks, through leaks and cracks caused due to corrosion of the tanks, spillage from tank overfilling (Sacile, 2007), faulty installation, improper operation and maintenance, or negligence (Brindha and Elango, 2014) can lead to seepage through the soil and mixing with groundwater. Remediation of groundwater contaminated with petroleum products and chemicals is often difficult, costly, and time-consuming, taking up to several years. It also usually demands the clean-up of the soil that is soaked with the contaminants. In the USA >530,000 cases of the release of petroleum products from underground storage tanks have been reported, of which 71,000 areas are yet to be cleaned up (USEPA, 2016b). Leaks from an abandoned underground storage tank in southern

India holding petroleum products has led to contamination of groundwater that is used as a source of drinking water (Brindha and Elango, 2014).

13.2.9 Polluted Surface Water

Surface water is more vulnerable to anthropogenic pollution. Historically, surface water and groundwater were considered as separate resources and the management of these resources was addressed separately. However, with increased concerns over water resources and the environment, the importance of considering them as a single resource has become increasingly evident (Winter et al., 1998). Interaction between groundwater and surface water is common in all types of fresh surface water bodies such as rivers, streams, lakes, ponds, etc. Pollutants in surface water are prone to infiltration through the soil zone and cause groundwater contamination.

Surface runoff from urban areas carry chemicals from roads, leachate from landfills, nutrients and fertilizers from urban gardening, etc., resulting in surface water pollution. The release of treated or raw wastewater from industries into surface water bodies often results in pollution from toxic substances. Improper disposal of sewage and domestic wastewater into surface water bodies increases the pathogen load. Some of the worst polluted rivers that are of environmental concern include the Ganges and the Yamuna Rivers in India, Yellow River in China, Citarum River in Indonesia, the Marilao and the Pasig Rivers in Philippines, Buriganga River in Bangladesh, Jordan River in Israel, Sarno River in Italy, the Mississippi and the Cuyahoga Rivers in the USA, and the Matanza-Riachuelo River, Argentina.

13.2.10 Road Salts

Salt is added to roads and highways in winter to melt the snow and ice by lowering the freezing point of water. Rock salt (sodium chloride) is commonly used as it is effective, easily available, and inexpensive. Once the snow is melted, however, the saline water contaminates the soil, groundwater, and surface water (Fig. 13.3). This has also turned groundwater in many areas to be Na-Cl dominated (Godwin et al., 2003). Several surface water bodies in North America and Europe have reported increased concentrations of sodium and chloride after road salt application during the winter months. On average nearly 5 million tons of road salt is used in Canada every year for de-icing (Environment and Climate Change, Canada, 2017). In the USA, about 21 million tons of road salt is used every year by government and commercial bodies (Sander et al., 2007). Much of this is washed away and pollutes groundwater resources.

13.2.11 Impact of Climate Change

Climate influences all life forms and their activities on Earth. Variations in recent climatic conditions, such as global temperatures, extreme rainfall events, cyclones, storms, and heat waves, indicate that climate change is an undeniable reality. Furthermore, the impacts of climate change include extensive melting of snow and ice, sea level rises, and wide spatial and temporal changes in rainfall amounts, wind pattern, and ocean salinity (USGS, 2007). Changes in precipitation, recharge rates, and the availability of water resources will enhance groundwater pumping and the use of agro-chemicals. Sea level rise and its interaction with the groundwater is likely to increase the salinity in the aquifers. Urban areas are at high risk from these extreme weather events (Revi et al., 2014).

The Intergovernmental Panel on Climate Change predicts that increases in water temperature, rainfall intensity, and low-flow period in rivers will affect groundwater quality through an increase in the concentrations of nutrients, salts, pesticides, and pathogens (Bates et al., 2008). Rises in the temperature of surface water bodies will affect water quality through decreased oxygen levels, decreased biological processes, and increased chemical processes. Extreme rainfall events result in floods that may disrupt the sewer systems in cities and can critically pollute groundwater. Floods in southern India caused by high-intensity rainfall within a short timespan have led to the inundation of borewells with sewage contaminated surface water from a river. This increased the concentrations of nutrients and heavy metals, and introduced pathogens into the groundwater (Gowrisankar et al., 2017).

In contrast, extreme droughts increase the residence time of groundwater and enhance water-rock interaction. Decreased flow in rivers reduces the dilution of wastewater and treated effluents and increases toxic substances and fecal contaminants in groundwater. Reduced recharge, increased evapotranspiration, and increased groundwater pumping from shallow groundwater will enhance the salinity (Bates et al., 2008). These effects will be experienced more in the semiarid and arid regions. Short-term droughts, like that during the summer of 2000 in the USA, increased the nutrients and inorganic concentrations in groundwater (Kampbell et al., 2003). Both natural disasters, floods and droughts, caused by climate change increase the risk to human health. Floods increase the risk of cholera and droughts can lead to diarrhea.

13.3 CASE STUDY: GROUNDWATER CONTAMINATION IN AN URBAN AREA IN INDIA

Many megacities in India have faced rapid groundwater quality degradation in recent years. Groundwater quality is constantly under threat due to urbanization, improvements in living standards, and an exploding population. In New Delhi, the capital of the country, half of the water consumed is sourced from groundwater (World Bank, 2010). Nitrate, fluoride, and heavy metal pollution is reported consistently every year. Contamination caused by on-site sanitation systems in Indore and Kolkata (formerly Calcutta) (Pujari et al., 2012), tanneries in Chennai (Brindha and Elango, 2012a), and seawater and urban wastewater contamination in Visakapattnam (Rao et al., 2005) are a few examples indicating the diversity of pollutants in groundwater of urban areas in India. All these studies show that systematic monitoring of groundwater quality should be mandatory.

Of the many methods available for the assessment of pollution, remote sensing and GIS serve as useful tools for the interpretation of the data. Spatial maps help to identify the pollution extent and interpolation of the primary data assists in predicting pollutant concentrations even at locations that are inaccessible for sample collection. Additionally, the inherent characteristics of a study area, such as the geology, hydrogeology, and land use, can be overlapped with pollution maps and compared. Due to these advantages, GIS is widely used for groundwater quality mapping. The case study presented serves as an example for the application of GIS in groundwater quality mapping in an urban area in southern India.

13.3.1 Study Area

The Tiruchirappalli (also known as "Trichy") district is located centrally in the southern state of Tamil Nadu, India. The administrative headquarters is the city of Tiruchirappalli, which is the fourth most populated city in the state (Census of India, 2011) with a population density of about 5000 persons/km^2. With a subtropical climate, this area experiences temperature ranging from 15°C in winter up to 40°C in summer. Monsoon season is from June to September and the average rainfall is 818 mm/year (TWAD Board, 2015).

13.3.2 Water Demand and Supply

The water supply for the inhabitants of the city is provided by the Tiruchirappalli City Corporation. The Cauvery River, a perennial river flowing through the city, is the major source for the water supply and in addition to this, groundwater is pumped from borewells (Tiruchirappalli City Municipal Corporation, 2014). Together both sources account for the total water supply of 82.5 million L/day for domestic and commercial use. Irrigation is the predominant use in the areas surrounding the city. With many industries and agricultural activities, water demand is high, therefore there are many private borewells that are used to meet the local needs. Heavy pumping from these wells, especially for commercial use, has led to a decline in the groundwater levels. There are three canals drawing water from the Cauvery River and providing for the irrigation needs. This study focuses on the groundwater quality along one of the canals, the Uyyakondan. This canal flows through the city and it aggregates pollutants due to the dumping of wastes: raw sewage from households, biomedical wastes from hospitals, etc. (Brindha and Kavitha, 2015). Thus it is vital to assess the impact of surface water pollution from the canal on groundwater quality and the suitability of the groundwater for drinking purposes.

13.3.3 Sampling Methodology

Groundwater samples from household borewells located near the Uyyakondan canal were collected from 15 locations in January 2014. Samples were collected in precleaned high-density polyethylene bottles of 1 L capacity. Electrical conductivity (EC) of the samples were measured immediately after sampling in the field. Major cations (calcium, magnesium, sodium, and potassium) and anions (chloride and sulphate) were determined by standard procedures (APHA, 1998).

13.3.4 Groundwater Quality Mapping

Groundwater and surface water quality in this area has been studied by Brindha and Kavitha (2015). However, this study addressed the water quality at selected locations based on the water quality index (WQI) method. A clear demarcation of the areas contaminated due to poor surface water quality was not achieved. The present study uses GIS techniques to delineate areas with suitable and unsuitable drinking water quality.

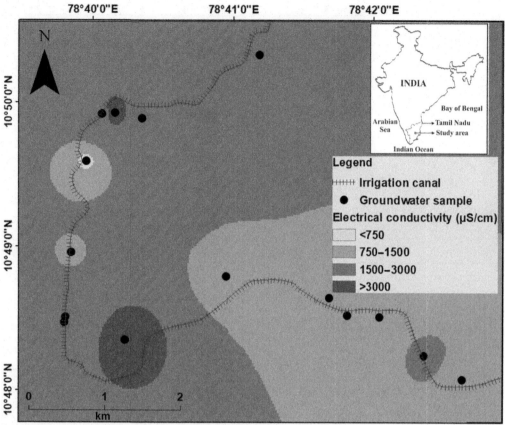

FIG. 13.5 Location of study area and spatial variation in electrical conductivity of groundwater.

The EC of groundwater varied from 630 to 4500 µS/cm. An EC of <750 µS/cm is suitable, from 750 to 1500 µS/cm is permissible, 1500 to 3000 µS/cm is not permissible, and >3000 µS/cm is hazardous for drinking. Two samples were hazardous, and the extent of pollution is given in Fig. 13.5. The order of dominance of cations was $Na^+ > Ca^{2+} > Mg^{2+} > K^+$ and anions was $Cl^- > SO_4^{2-}$. Sodium chloride was the dominant groundwater type. A comparison of cation and anion concentrations in groundwater with national and international drinking water standards (BIS, 2012; WHO, 1993) indicated a varied number of samples exceeding the permissible limit for each parameter: two locations for calcium, 14 locations for sodium, four locations for potassium, and one location for chloride. Magnesium and sulphate were within the permissible limits for drinking water. This type of classification of results for individual parameters makes it difficult to opt for a suitable management option in an area.

To overcome this, an overlay and index method was adopted for the preparation of a groundwater quality map based on all the parameters. Spatial maps for each parameter were prepared using the inverse distance weighted method, a widely used method for spatial interpolation. Suitable and unsuitable groundwater quality areas for drinking were classified based on the standards. A rank was assigned for the suitable and unsuitable range of measured ion concentrations. The groundwater quality map was arrived at by adding up the ranks of all of the parameters. A detailed description of the steps involved was reported by Brindha and Elango (2012b). Spatial variation in the concentration of various parameters (Fig. 13.6) shows common areas of groundwater contamination. The integrated groundwater quality map shows that groundwater is suitable for drinking purposes in <1 km² of the study area, poor in 32 km², and very poor in 5 km² (Fig. 13.7). Interpolation studies indicate 12 locations where groundwater is unsuitable whereas a previous study reported that groundwater was unsuitable in only six locations (Brindha and Kavitha, 2015). The underestimation was because WQI informs on the quality of the sampled location and does not take into consideration the interaction between the adjacent locations. Groundwater should be avoided for drinking use as demarcated in Fig. 13.7, which also highlights the areas where pollution sources may be introduced into the canal ("very poor" areas in Fig. 13.7) and subsequently into groundwater. However, an on-site field investigation is necessary to confirm this. The present study stresses the importance of cleaning the canal and the implementation of measures to prevent the dumping of waste and sewage into the canal.

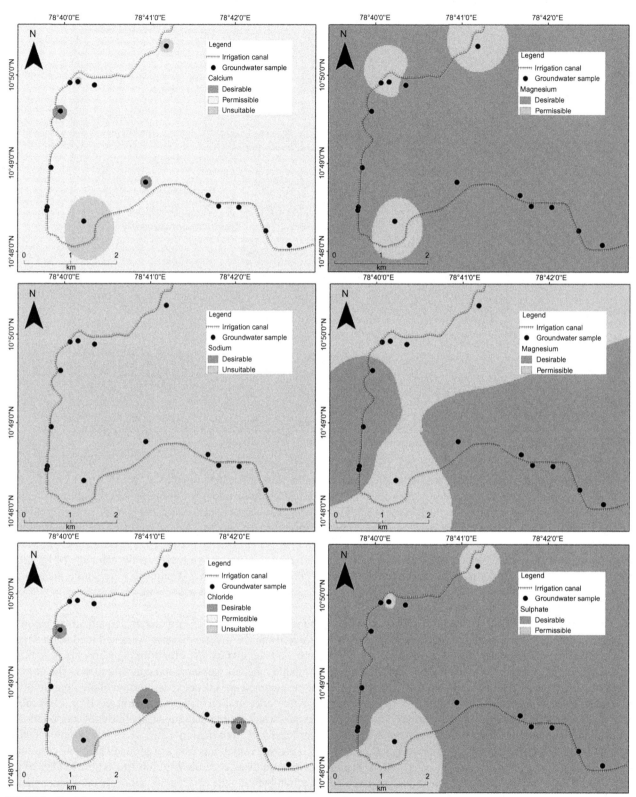

FIG. 13.6 Spatial variation in the concentration of various parameters in groundwater.

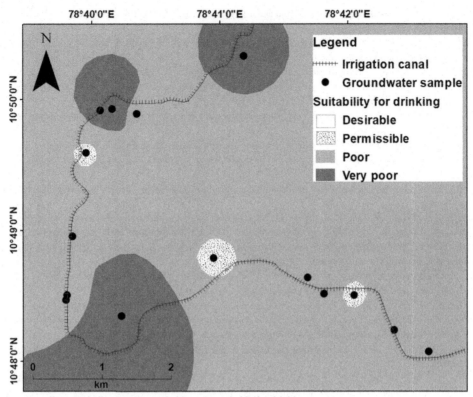

FIG. 13.7 Groundwater quality map indicating areas suitable and unsuitable for drinking use.

13.4 SUMMARY AND RECOMMENDATIONS

Groundwater contamination is a serious problem in many urban areas. Due to multiple pollutants and sources, the management of the resource is usually complex and challenging. With industrialization and urbanization a variety of new pollutants (emerging contaminants) have been discovered, and it is necessary for the municipal wastewater treatments systems to be consistently updated to remove the pollutants, before disposing of the treated wastewater into surface water bodies. Industries should adopt alternate technologies, where possible, by substituting highly toxic with less toxic chemicals and reducing the use of hazardous substances. Protection of the groundwater resource should not be considered as a stand-alone task and an integrated management of groundwater, surface water, and soil resources must be aimed for. All planning operations and polices developed should be coordinated with other policies on management of solid waste, land use, industrial wastewater disposal, etc. Regular monitoring is necessary to keep the pollution under control. Closing the urban water cycle by recycling and reusing wastewater can significantly reduce the excessive extraction of groundwater. Additionally, after appropriate treatment, recycled wastewater can be artificially recharged. Management of these resources must not be restricted to the administrative boundaries of the cities but should include the suburban areas, which are constantly being urbanized. Stringent policies and pollution penalties are required to motivate the rational use of water.

REFERENCES

Aharoni, I., Siebner, H., Dahan, O., 2017. Application of vadose-zone monitoring system for real-time characterization of leachate percolation in and under a municipal landfill. Waste Manage. 67, 203–213.

APHA (American Public Health Association), 1998. Standard Methods for the Examination of Water and Wastewater, 20th ed. American Public Health Association/American Water Works Association/Water Environment Federation, Washington, DC.

Ayoob, S., Gupta, A.K., 2006. Fluoride in drinking water: a review on the status and stress effects. Crit. Rev. Environ. Sci. Technol. 36, 433–487.

Baldé, C.P., Wang, F., Kuehr, R., Huisman, J., 2015. The Global e-Waste Monitor—2014. United Nations University, IAS—SCYCLE, Bonn, Germany.

Basharat, M., 2016. Groundwater environment in Lahore, Pakistan. In: Shrestha, S., Pandey, V.P., Shivakoti, B.R., Thatikonda, S. (Eds.), Groundwater Environment in Asian Cities- Concepts, Methods and Case Studies. Elsevier, USA, pp. 147–184.

Bates, B.C., Kundzewicz, Z.W., Wu, S., Palutikof, J.P. (Eds.), 2008. Climate Change and Water. IPCC Secretariat, Geneva (Technical Paper of the Intergovernmental Panel on Climate Change). 210 p.

Baumann, T., Fruhstorfer, P., Klein, T., Niessner, R., 2006. Colloid and heavy metal transport at landfill sites in direct contact with groundwater. Water Res. 40, 2776–2786.

BIS, 2012. Indian Standard Drinking Water Specification, Second Revision ISO: 10500:2012. Bureau of Indian Standards, Drinking Water Sectional Committee, FAD 25, New Delhi, India.

Brammer, H., Ravenscroft, P., 2009. Arsenic in groundwater: a threat to sustainable agriculture in South and South-East Asia. Environ. Int. 35, 647–654.

Brindha, K., Elango, L., 2011. Fluoride in groundwater: causes, implications and mitigation measures. In: Monroy, S.D. (Ed.), Fluoride Properties, Applications and Environmental Management. Nova Publishers, pp. 111–136.

Brindha, K., Elango, L., 2012a. Impact of tanning industries on groundwater quality near a metropolitan city in India. Water Resour. Manage. 26 (6), 1747–1761.

Brindha, K., Elango, L., 2012b. Groundwater quality zonation in a shallow weathered rock aquifer using GIS. Geo-Spat. Inf. Sci. 15 (2), 95–104.

Brindha, K., Elango, L., 2014. Polycyclic aromatic hydrocarbons in groundwater from a part of metropolitan city, India: a study based on sampling over a ten-year period. Environ. Earth Sci. 71, 5113–5120.

Brindha, K., Kavitha, R., 2015. Hydrochemical assessment of surface water and groundwater quality along Uyyakondan channel, South India. Environ. Earth Sci. 73, 5383–5393.

Brindha, K., Vaman, K.V.N., Srinivasan, K., Babu, M.S., Elango, L., 2014. Identification of surface water-groundwater interaction by hydrogeochemical indicators and assessing the suitability for drinking and irrigational purposes in Chennai, southern India. Appl. Water Sci. 4 (2), 159–174.

Brunt, R., Vasak, L., Griffioen, J., 2004. Fluoride in Groundwater: Probability of Occurrence of Excessive Concentration on Global Scale. International Groundwater Resources Assessment Centre (Report Nr. SP 2004-2).

Census of India, 2011. Provisional Population Totals. Available online at: www.censusindia.gov.in/2011-prov-results/paper2/data_files/India2/Table_3_PR_UA_Citiees_1Lakh_and_Above.pdf (Accessed 24 December 2018).

Chen, Y., Parvez, F., Gamble, M., Islam, T., Ahmed, A., Argos, M., Graziano, J.H., Ahsan, H., 2009. Arsenic exposure at low-to-moderate levels and skin lesions, arsenic metabolism, neurological functions, and biomarkers for respiratory and cardiovascular diseases: review of recent findings from the Health Effects of Arsenic Longitudinal Study (HEALS) in Bangladesh. Toxicol. Appl. Pharmacol. 239, 184–192.

Domagalski, J.L., Johnson, H., 2012. Phosphorus and Groundwater: Establishing Links Between Agricultural Use and Transport to Streams. U.S. Geological Survey Fact Sheet 2012-3004. 4 p.

Eiswirth, M., Hötzl, H., 1997. The impact of leaking sewers on urban groundwater. In: Chilton, J. et al., (Ed.), Groundwater in the Urban Environment. Problems, Processes and Management. vol. 1. Balkema, Rotterdam, pp. 399–404.

Environment and Climate Change, Canada, 2017. Five-Year Review of Progress: Code of Practice for the Environmental Management of Road Salts. Available online at: https://www.ec.gc.ca/sels-salts/default.asp?lang=En&n=45D464B1-1 (Accessed 24 December 2018).

Eslamian, S., Eslamian, F.A., 2017. Handbook of Drought and Water Scarcity: Environmental Impacts and Analysis of Drought and Water. CRC Press. 689 p.

Eurostat, 2017. Municipal Waste by Waste Operations. Available online at: http://appsso.eurostat.ec.europa.eu/nui/show.do?dataset=env_wasmun&lang=en (Accessed 24 December 2018).

Fan, A.M., Steinberg, V.E., 1996. Health implications of nitrate and nitrite in drinking water: an update on methemoglobinemia occurrence and reproductive and developmental toxicity. Regul. Toxicol. Pharmacol. 23, 35–43.

FAO, 2011. Current World Fertilizer Trends and Outlook to 2015. Available online at: http://www.fao.org/3/a-av252e.pdf (Accessed 24 December 2018).

FAO, 2015. World Fertilizer Trends and Outlook to 2018. Food and Agriculture Organization of the United Nations, Rome. Available online at: http://www.fao.org/3/a-i4324e.pdf (Accessed 24 December 2018).

FOEN, 2016. Volatile Organic Compounds in Groundwater. Federal Office for the Environment, Switzerland. Available online at: https://www.bafu.admin.ch/bafu/en/home/topics/water/info-specialists/state-of-waterbodies/state-of-groundwater/groundwater-quality/volatile-organic-compounds-in-groundwater.html (Accessed 24 December 2018).

Foster, S.S.D., Chilton, P., 2003. Groundwater: the processes and global significance of aquifer degradation. Philos. Trans. R Soc. Lond. B 358, 1957–1972.

Garelick, H., Jones, H., Dybowska, A., Valsami-Jones, E., 2008. Arsenic pollution sources. Rev. Environ. Contam. Toxicol. 197, 17–60.

Godwin, K.S., Hafner, S.D., Buff, M.F., 2003. Long-term trends in sodium and chloride in the Mohawk River, New York: The effect of fifty years of road-salt application. Environ. Pollut. 124, 273–281.

Gowrisankar, G., Chelliah, R., Ramakrishnan, S.R., Elumalai, V., Dhanamadhavan, S., Brindha, K., Antony, U., Elango, L., 2017. Chemical, microbial and antibiotic susceptibility analyses of groundwater after a major flood event in Chennai. Nat. Sci. Data 4, 170135.

Grimmeisen, F., Lehmann, M.F., Liesch, T., Goeppert, N., Klinger, J., Zopfi, J., Goldscheider, N., 2017. Isotopic constraints on water source mixing, network leakage and contamination in an urban groundwater system. Sci. Total Environ. 583, 202–213.

Gupta, S.K., Gupta, R.C., Chhabra, S.K., Eskiocak, S., Gupta, A.B., Gupta, R., 2008. Health issues related to N pollution in water and air. Curr. Sci. 94 (11), 1469–1477.

Hairston, J.E., 1995. Underground Storage Tanks (USTs) and NPS Pollution. USTs And How They Affect Water Quality, The Alabama Cooperative Extension System, U.S. Department of Agriculture. Available online at: http://www.aces.edu/pubs/docs/A/ANR-0790/WQ4.8.1.pdf (Accessed 24 December 2018).

Han, D., Tong, X., Currell, M.J., Cao, G., Jin, M., Tong, C., 2014. Evaluation of the impact of an uncontrolled landfill on surrounding groundwater quality, Zhoukou, China. J. Geochem. Explor. 136, 24–39.

Harada, M., 1995. Minamata disease: methylmercury poisoning in Japan caused by environmental pollution. Crit. Rev. Toxicol. 25 (1), 1–24.

Hoornweg, D., Bhada-Tata, P., 2012. What a Waste: A Global Review of Solid Waste Management. World Bank Urban Development Series Knowledge Papers. Available online at: http://siteresources.worldbank.org/INTURBANDEVELOPMENT/Resources/336387-1334852610766/What_a_Waste2012_Final.pdf (Accessed 24 December 2018).

IPEN, 2016. Guidance on the Identification, Management and Remediation of Mercury Contaminated Sites. International POPs Elimination Network. Available online at: https://ipen.org/sites/default/files/documents/IPEN%20Guidance%20on%20Mercury%20Contaminated%20Sites%20INC%207%202016.pdf (Accessed 24 December 2018).

Kampbell, D.H., An, Y.-J., Jewell, K.P., Masoner, J.R., 2003. Groundwater quality surrounding Lake Texoma during short-term drought conditions. Environ. Pollut. 125, 183–191.

Kazakis, N., Pavlou, A., Vargemezis, G., Voudouris, K.S., Soulios, G., Pliakas, F., Tsokas, G., 2016. Seawater intrusion mapping using electrical resistivity tomography and hydrochemical data. An application in the coastal area of eastern Thermaikos Gulf, Greece. Sci. Total Environ. 543, 373–387.

Kuroda, K., Fukushi, T., 2008. Groundwater contamination in urban areas. In: Takizawa, S. (Ed.), Groundwater Management in Asian Cities: Technology and Policy for Sustainability. Springer, Japan, pp. 125–149.

Lapworth, D.J., Baran, N., Stuart, M.E., Ward, R.S., 2012. Emerging organic contaminants in groundwater: a review of sources, fate and occurrence. Environ. Pollut. 163, 287–303.

Lopes, D.D., Silva, S.M.C.P., Fernandes, F., Teixeira, R.S., Celligoi, A., Dall'Antônia, L.H., 2012. Geophysical technique and groundwater monitoring to detect leachate contamination in the surrounding area of a landfill e Londrina (PR-Brazil). J. Environ. Manage. 113, 481–487.

Lowe, M., Wallace, J., 2001. Evaluation of Potential Geologic Sources of Nitrate Contamination in Ground Water, Cedar Valley, Iron County, Utah With Emphasis on the Enoch Area. Utah Department of Natural Resources, Utah Geological Survey. 50 p.

Martens, S., Wichmann, K., 2007. Groundwater salinisation. In: Lozán, J.L., Grassl, H., Hupfer, P., Menzel, L., Schönwiese, C.-D. (Eds.), Global Change: Enough Water for All. Wissenschaftliche Auswertungen, Hamburg, p. 384 S.

Meenakshi, Maheshwari, R.C., 2006. Fluoride in drinking water and its removal. J. Hazard. Mater. B137, 456–463.

Ministry of the Environment, Japan, n.d. Lessons From Minamata Disease and Mercury Management in Japan. Available online at: https://www.env.go.jp/chemi/tmms/pr-m/mat01/en_full.pdf (Accessed 24 December 2018).

Mull, R., Harig, F., Pielke, M., 1992. Groundwater management in the urban areas of Hanover, Germany. J. Inst. Water Environ. Manage. 6, 199–206.

Nair, I.S., Rajaveni, S.P., Schneider, M., Elango, L., 2015. Geochemical and isotopic signatures for the identification of seawater intrusion in an alluvial aquifer. J. Earth Syst. Sci. 124 (6), 1281–1291.

Neumann, B., Vafeidis, A.T., Zimmermann, J., Nicholls, R.J., 2015. Future coastal population growth and exposure to sea-level rise and coastal flooding— a global assessment. PLoS One 10(3).

Nga, T.T.V., 2008. Arsenic contamination in Hanoi City, Vietnam. In: Takizawa, S. (Ed.), Groundwater Management in Asian Cities: Technology and Policy for Sustainability. Springer, Japan, pp. 273–299.

Pal, A., Chowdhury, U.K., Mondal, D., Das, B., Nayak, B., Ghosh, A., Maity, S., Chakraborti, D., 2009. Arsenic burden from cooked rice in the populations of arsenic-affected and non-affected areas and Kolkata city in West Bengal, India. Environ. Sci. Technol. 43, 3349–3355.

Pedley, S., Howard, G., 1997. The public health implications of microbiological contamination of groundwater. Q. J. Eng. Geol. 30, 179–188.

Pickin, J., Randell, P., 2017. Australian National Waste Report 2016. Department of the Environment and Energy & Blue Environment Pty Ltd.

Prinos, S.T., Wacker, M.A., Cunningham, K.J., Fitterman, D.V., 2014. Origins and Delineation of Saltwater Intrusion in the Biscayne Aquifer and Changes in the Distribution of Saltwater in Miami-Dade County. U.S. Geological Survey Scientific Investigations Report 2014-5025, Florida. 101 p.

Pujari, P.R., Padmakar, C., Labhasetwar, P.K., Mahore, P., Ganguly, A.K., 2012. Assessment of the impact of on-site sanitation systems on groundwater pollution in two diverse geological settings—a case study from India. Environ. Monit. Assess. 184, 251–263.

Rao, N.S., Nirmala, I.S., Suryanarayana, K., 2005. Groundwater quality in a coastal area: a case study from Andhra Pradesh, India. Environ. Geol. 48, 543–550.

Rapti-Caputo, D., Vaccaro, C., 2006. Geochemical evidences of landfill leachate in groundwater. Eng. Geol. 85, 111–121.

Ravenscroft, P., Brammer, H., Richards, K.S., 2009. Arsenic Pollution: A Global Synthesis. Wiley-Blackwell, Hoboken.

Reed, D., 2010. Understanding the effects of sea-level rise on coastal wetlands: the human dimension. In: EGU General Assembly Conference Abstracts, vol. 12, p. 5480.

Revi, A., Satterthwaite, D.E., Aragón-Durand, F., Corfee-Morlot, J., Kiunsi, R.B.R., Pelling, M., Roberts, D.C., Solecki, W., 2014. Urban areas. In: Field, C.B., Barros, V.R., Dokken, D.J., Mach, K.J., Mastrandrea, M.D., Bilir, T.E., Chatterjee, M., Ebi, K.L., Estrada, Y.O., Genova, R.C., Girma, B., Kissel, E.S., Levy, A.N., MacCracken, S., Mastrandrea, P.R., White, L.L. (Eds.), Climate Change 2014: Impacts, Adaptation, and Vulnerability. Part A: Global and Sectoral Aspects. Contribution of Working Group II to the Fifth Assessment Report of the Intergovernmental Panel on Climate Change. Cambridge University Press, Cambridge, United Kingdom/New York, NY, pp. 535–612.

Reyes-López, J.A., Ramírez-Hernández, J., Lázaro-Mancilla, O., Carreón-Diazconti, C., Garrido, M.M.-L., 2008. Assessment of groundwater contamination by landfill leachate: a case in México. Waste Manage. 28, S33–S39.

Richter, B.C., Kreitler, C.W., 1993. Geochemical Techniques for Identifying Sources of Ground-Water Salinization. CRC Press. 272 p.

Sacile, R., 2007. Remote real-time monitoring and control of contamination in underground storage tank systems of petrol products. J. Clean. Prod. 15, 1295–1301.

Samadder, S.R., Prabhakar, R., Khan, D., Kishan, D., Chauhan, M.S., 2017. Analysis of the contaminants released from municipal solid waste landfill site: a case study. Sci. Total Environ. 580, 593–601.

Sander, A., Novotny, E., Mohseni, O., Stefan, H., 2007. Inventory of Road Salt Use in the Minneapolis/St. Paul Metropolitan Area (Project Report No. 503). Minnesota Department of Transportation, Minnesota. Available online at: https://conservancy.umn.edu/bitstream/handle/11299/115332/pr503.pdf?sequence (Accessed 24 December 2018).

Shi, L., Jiao, J.J., 2014. Seawater intrusion and coastal aquifer management in China: a review. Environ. Earth Sci. 72, 2811–2819.

Shrestha, S., Pandey, V.P., 2016. Groundwater as an environmental issue in Asian cities. In: Shrestha, S., Pandey, V.P., Shivakoti, B.R., Thatikonda, S. (Eds.), Groundwater Environment in Asian Cities: Concepts, Methods, and Case Studies. Elsevier, pp. 3–15.

Small, C., Nicholls, R.J., 2003. A global analysis of human settlement in coastal zones. J. Coast. Res. 19 (3), 584–599.

Smith, A., Lingas, E., Rahman, M., 2000. Contamination of drinking water by As in Bangladesh. Bull. World Health Organ. 78, 1093–1103.

Squillace, P.J., Moran, M.J., Lapham, W.W., Price, C.V., Clawges, R.M., Zogorski, J.S., 1999. Volatile organic compounds in untreated ambient groundwater of the United States, 1985–1995. Environ. Sci. Technol. 33 (23), 4176–4187.

Subramanian, S., 1975. Report on the Systemic Geological Mapping of Quaternary Formations in the Coastal Plains Between Minjur and Madras, Tamil Nadu (Unpublished Report). Geological Survey of India.

Takizawa, S., 2008. Groundwater problems in urban areas—introduction. In: Takizawa, S. (Ed.), Groundwater Management in Asian Cities Technology and Policy for Sustainability. Springer, Japan, pp. 3–11.

The British Fluoridation Society, 2012. One in a Million: The Facts About Water Fluoridation. Available online at: https://www.dentalwatch.org/fl/bfs.pdf (Accessed 24 December 2018).

Tiruchirappalli City Municipal Corporation, 2014. Water Supply. Available online at: http://cma.tn.gov.in/tiruchirappalli/en-in/Pages/Water-Supply.aspx (Accessed 24 December 2018).

TWAD Board, 2015. Rainfall. Tamil Nadu Water Supply and Drainage Board. Available online at: http://www.twadboard.gov.in/twad/tiruchirapalli_dist. aspx (Accessed 24 December 2018).

Umezawa, Y., Hosono, T., Onodera, S.-i., Siringan, F., Buapeng, S., Delinom, R., Yoshimizu, C., Tayasu, I., Nagata, T., Taniguchi, M., 2009. Sources of nitrate and ammonium contamination in groundwater under developing Asian megacities. Sci. Total Environ. 407, 3219–3231.

UNESCO, 2004. Groundwater resources of the world and their use. In: Zektser, I.S., Everett, L.G. (Eds.), IHP-VI, Series on Groundwater No. 6. Available online at: http://unesdoc.unesco.org/images/0013/001344/134433e.pdf (Accessed 24 December 2018).

United Nations, 2014. Department of Economic and Social Affairs, Population Division. World Urbanization Prospects: The 2014 Revision, Highlights (ST/ESA/SER.A/352). .

United Nations, 2016. Department of Economic and Social Affairs, Population Division. The World's Cities in 2016—Data Booklet (ST/ESA/SER.A/392). .

USEPA, 2016a. Advancing Sustainable Materials Management: 2014 Fact Sheet. Available online at: https://www.epa.gov/sites/production/files/2016-11/documents/2014_smmfactsheet_508.pdf (Accessed 24 December 2018).

USEPA, 2016b. Semiannual Report of UST Performance Measures End of Fiscal Year 2016. Available online at: https://www.epa.gov/sites/production/files/2016-11/documents/ca-16-34.pdf (Accessed 24 December 2018).

USGS, 2007. Climate Variability and Change. U.S. Geological Survey Fact Sheet 2007-3108. 2 p.

Wakida, F.T., Lerner, D.N., 2005. Non-agricultural sources of groundwater nitrate: a review and case study. Water Res. 39, 3–16.

WHO, 1993. Guidelines for Drinking Water Quality, second ed. vol. 1 Recommendations, WHO, Geneva, p. 130.

WHO, 2017a. Ten Chemicals of Major Public Health Concern. Available online at: http://www.who.int/ipcs/assessment/public_health/chemicals_phc/en/ (Accessed 24 December 2018).

WHO, 2017b. Lead Poisoning and Health. Available online at: http://www.who.int/mediacentre/factsheets/fs379/en/ (Accessed 24 December 2018).

WHO, 2017c. Drinking Water, Fact Sheet. Available online at: http://www.who.int/mediacentre/factsheets/fs391/en/ (Accessed 24 December 2018).

Williams, P.N., Islam, M.R., Adomako, E.E., Raab, A., Hossain, S.A., Zhu, Y.G., Feldmann, J., Meharg, A.A., 2006. Increase in rice grain arsenic for regions of Bangladesh irrigating paddies with elevated arsenic in groundwater. Environ. Sci. Technol. 40 (16), 4903–4908.

Winter, J.G., Dillon, P.J., 2005. Effects of golf course construction and operation on water chemistry of headwater streams on the Precambrian Shield. Environ. Pollut. 133, 243–253.

Winter, T.C., Harvey, J.W., Franke, O.L., Alley, W.M., 1998. Ground Water and Surface Water—A Single Resource. U.S. Geological Survey Circular 1139.

Wolf, L., Morris, B., Burn, S., 2006. AISUWRS: Urban Water Resources Toolbox. IWA Publishing, London, UK ISBN: 1843391384.

World Bank, 2010. Deep Wells and Prudence: Towards Pragmatic Action for Addressing Groundwater Overexploitation in India. The World Bank, Washington, DC, p. 97.

World Bank, 2018. World Population (Total, Urban and Rural). Available online at: https://data.worldbank.org/ (Accessed 24 December 2018).

Yoshida, F., 2002. Itai-Itai disease and countermeasures against cadmium pollution by the Kamioka Mine. In: The Economics of Waste and Pollution Management in Japan. Springer, Tokyo.

Zimmermann-Timm, H., 2007. Salinisation of inland waters. In: Lozán, J.L., Grassl, H., Hupfer, P., Menzel, L., Schönwiese, C.-D. (Eds.), Global Change: Enough Water for All?. Wissenschaftliche Auswertungen, Hamburg, p. 384 S.

Chapter 14

Groundwater and Surface Water Interaction

Haris Hasan Khan* and Arina Khan[†]

*Department of Geology, Aligarh Muslim University, Aligarh, India [†]Residential Coaching Academy, Aligarh Muslim University, Aligarh, India

Chapter Outline

14.1 INTRODUCTION

Surface water (including rivers, lakes, reservoirs, wetlands, estuaries, etc.) interacts with groundwater almost everywhere on Earth. This interaction takes place through the loss of surface water to groundwater, seepage of groundwater to surface water body, or a combination of both. The development or contamination of surface water or groundwater resources typically has an effect on each (Winter et al., 1998). Therefore a basic understanding of the interactions between surface water and groundwater is crucial for better management and sound policy making related to water-resource problems.

Knowledge of groundwater-surface water interactions is essential to address the following water-resource issues (Winter et al., 1998):

1. Conjunctive use of groundwater and surface water resources.
2. Water rights issues, especially accounting for the groundwater flows to and from surface water bodies, which can be difficult and controversial.
3. Assessment and minimization of losses and delays of water released from surface-water reservoirs.
4. Assessment and mitigation of floods along river valleys.
5. Assessment and control of contamination of surface water caused by groundwater and vice versa.
6. Integration of groundwater flows in watershed planning and management.
7. Evaluating the impacts of riparian zones on river-water quality.
8. Groundwater-surface water interactions control aquatic life and any changes in the magnitude and direction of these interactions may result in alterations.
9. Supporting dynamic habitats at the interface and sustenance of aquatic fauna that maintains diverse ecology and indicates the status of aquatic water quality.
10. Sustenance of wetlands and establishing new wetlands for conservation purposes.

Comprehensive literature reviews and knowledge synthesis on groundwater-surface water interactions are provided in Winter (1995, 1999), Woessner (2000), Sophocleous (2002), and Safeeq and Fares (2016). In addition, essential generalizations on the topic are presented in Winter et al. (1998) and Weight (2008).

14.2 SCALES OF GROUNDWATER-SURFACE WATER INTERACTION

Groundwater-surface water interaction takes place on a range of scales. Larkin and Sharp (1992), Brunke and Gonser (1997), and Woessner (2000) identified two scales of groundwater-surface water interactions: large-scale interactions, where the whole catchment or watershed influences the interaction process, and local-scale interactions within the hyporheic zone controlled mainly by stream-bed properties. An appropriate scale classification for studying groundwater-surface water interaction has emerged in recent years (Dahl et al., 2007) that divides the interaction into three scales: sediment scale (<1 m), reach scale (1–1000 m), and catchment scale (>1000 m). Although the scale boundaries are arbitrary, these scales incorporate the hierarchy of groundwater flow systems. The sediment scale is dominated by hyporheic processes, while the reach and catchment scales are dominated by local and regional groundwater flow systems respectively. These different levels of interactions are generally superimposed on each other and it is therefore necessary that the spatial scale be considered while investigating the interaction of groundwater and surface water in a region.

14.1.1 Large-Scale Interactions

On a regional and intermediate scale the recharge-discharge dynamics of groundwater are controlled by groundwater flow systems (Toth, 1963). Groundwater-flow systems are characterized by the boundary conditions imposed by their physiographic framework and by the distribution of recharge (Winter, 1999). The physiographic framework incorporates the topographic and geological conditions of a region, while recharge distribution is controlled by climate. Toth (1963) classified the flow systems into regional, intermediate, and local (Fig. 14.1). Regional-flow systems are recharged at regional water divides and discharge into regional (higher order) streams, while local-flow systems are recharged at local water divides and discharge into local (lower order) streams. Flow systems do not develop under extensive flat areas due to low hydraulic gradients. Increasing local topographic slopes increases the depth and intensity of local-flow systems, whereas increasing regional slopes increases the depth and intensity of regional-flow systems with a concomitant degeneration of local-flow systems. The hydrological response to recharge, and the water flux through the flow system diminish with increasing flow system scale, while the depth of penetration and residence time of groundwater increase with increasing flow system scale. Thus regional-flow systems tend to be deep, steady, slow (low flux), and more mineralized, while local-flow systems are shallow, unsteady (high variability), fast (greater flux), and less mineralized.

The interaction of groundwater and surface water on regional to local scales is dependent on the (Winter, 1976):

1. Position of the surface water body with respect to the groundwater flow systems.
2. Anisotropy and hydraulic conductivity contrasts of the groundwater system.

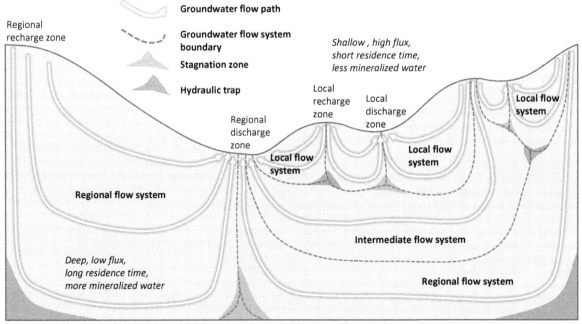

FIG. 14.1 Groundwater flow systems.

3. Configuration of the water table.

4. Depth of the surface-water body.

It is generally assumed that if the water table stands at a higher elevation than the elevation of the surface water bed, then the surface water body will receive groundwater. However, in many cases it is observed that deeper portions of the surface water body are actually losing water to the ground even when water table stands at a higher position. Winter (1976) demonstrated that the continuity of the local groundwater flow system boundary determines the nature of this interaction. If the boundary is continuous below the surface-water body, as shown in Fig. 14.2, there will be no seepage from the water body, while in case where the boundary is not continuous below the surface water body seepage to the groundwater will take place in the deeper portions of the water body. The continuity of the flow-system boundary is indicated by the presence of a stagnation point, which is the point of least head along a flow-system boundary (Winter, 1976).

The magnitude of seepage from a surface-water body is dependent on the difference between the head of the surface-water body and the head of the stagnation point. Several factors influence this head difference (Winter, 1976) and thus the degree of seepage:

1. Position of the water table (especially on the down-gradient side).

2. Aquifer anisotropy.

3. Hydraulic conductivity of high permeability zones.

4. Depth of high permeability zones beneath and down-gradient from the surface-water body.

5. Depth of the surface-water body.

Seepage to the groundwater from a surface-water body tends to increase or begin with a decrease in the water-table elevation; an increase in anisotropy, hydraulic conductivity, and surface-water depth; and the presence of high-permeability zones at shallow depths on the down-gradient side of surface bodies.

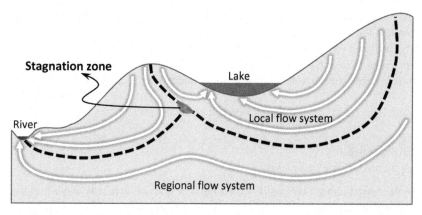

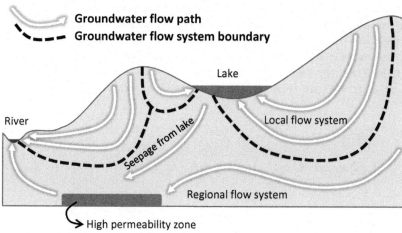

FIG. 14.2 Effect of continuity of groundwater flow system boundary on the seepage from a surface water body.

FIG. 14.3 Effect of groundwater table geometry on the occurrence of seepage zones.

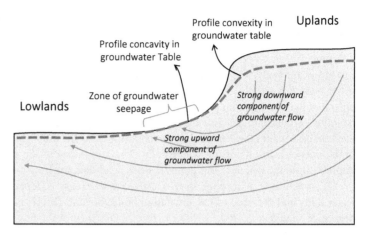

Interaction of groundwater with streams/lakes and wetlands is also regionally dictated by the configuration or shape of the water table (Fig. 14.3). In areas of water-table concavity (in a profile along the flow direction), the groundwater has a substantial upward component of flow, especially along the base of the water-table concavity. Conversely, in areas of water-table convexity, there is a substantial downward component of groundwater flux. Convexity or concavity of the water table is mostly a reflection of the convexity or concavity of the topographic surface. This results in the formation of wetlands or shallow lakes (in humid climates) and saline soils (in arid and semiarid climates) at the edges of river valleys and in flat landscapes adjacent to uplands. A water table sloping towards a surface-water body is relatively steep with respect to the flat surface of a water body leading to an overall concavity and thus there is seepage to the surface-water body.

14.1.2 Channel Scale (Hyporheic Zone) Interactions

Interaction of the surface waters of streams, lakes, and wetlands also takes place with the water present in the underlying sediments, i.e., the sediments immediately beneath the surface-water body. This interaction is independent of the large-scale interactions discussed above. This highly localized interaction between the groundwater and surface water is superimposed on the overall gaining and losing portions of a surface-water body (Woessner, 2000). Therefore an effluent (gaining) river reach may have localized zones where stream water infiltrates the underlying sediments (Brunke and Gonser, 1997).

The highly localized flow systems that develop within the sediments beneath surface-water bodies are mainly are controlled by (Woessner, 2000):

1. Irregularities or topography of the surface-water bed.
2. Hydraulic conductivity distribution within the underlying sediments.

Downwelling of stream water occurs where the longitudinal bed profile of streams is convex, whereas upwelling of hyporheic as well as deeper water takes place where the longitudinal bed profile is concave (Vaux, 1968). Convexity and concavity in stream beds may be present on a range of scales, e.g., it may be related to pool and riffle sequences in high gradient streams, and also to sediment bars, dunes, and ripples. Stream water enters channel sediments at the head of riffles and exits at the base of riffles into the pools (Harvey and Bencala, 1993) as shown in Fig. 14.4. Similarly, for sediment bars, dunes, etc., in alluvial rivers, stream-water enters and exits these bedforms on the stoss (upstream) side and exits on the lee (downstream) side, due to localized pressure-potential differences (Savant et al., 1987; Thibodeaux and Boyle, 1987). Upwelling and downwelling of stream water may also be related to obstacles present on the stream

FIG. 14.4 Effect of streambed heterogeneities on groundwater-stream water exchange.

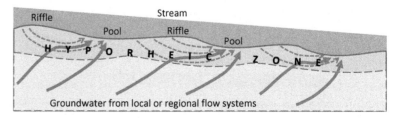

bed, e.g., boulders (Hutchinson and Webster, 1998). Channel sinuosity also influences the flow of water between the stream and the underlying sediments (Cardenas, 2009). stream water enters the sediments along the upstream portion of a meander and moves back to the stream along the downstream portion of the meander.

Heterogeneity of the sediments beneath a surface-water body leads to an enhanced and complex pattern of localized flow cells within the sediments. Woessner (2000), using numerical simulations, has shown that in homogenous sediments, the mixing of surface water and groundwater is limited to a depth of 0.7 m; however, with the introduction of horizontal high conductivity zones within the sediments, the zone of mixing extends to 1.5 m below the stream bed and the flow cells become more complex.

The mixing of groundwater and surface water within the underlying sediments creates a zone of intense biogeochemical activity known as the hyporheic zone (Stanford and Ward, 1988; Brunke and Gonser, 1997; Dahm et al., 1998; Lerner, 2009)), as shown in Fig. 14.4. The hyporheic zone is characterized by relatively high total organic carbon and microbial communities, thereby increasing the potential for biochemical reactions of pollutants derived from surface water or groundwater. The hyporheic zone is significant with respect to the surface and groundwater quality, and stream ecological functions. The passage of stream water into underlying sediments increases the residence time of water within the ground thereby enhancing the interaction of water with minerals and organic bio-films, leading to modifications in water quality and potential attenuation of pollutants. The downwelling and upwelling of stream water also dictates the faunal composition, distribution, and abundance within the hyporheic zone as well as surface-water body (Brunke and Gonser, 1997).

14.3 INTERACTION OF GROUNDWATER WITH RIVERS AND STREAMS

The interaction of rivers and streams with groundwater is controlled by the (Woessner, 2000):

1. Hydraulic conductivity and its distribution in channel and fluvial-plain sediments.
2. Position of the river stage with respect to the adjacent groundwater hydraulic heads.
3. Position and geometry of the stream channel within the fluvial plain.

River reaches interact with groundwater in a manner dictated by the abovementioned factors to produce the following conditions (Woessner, 1998; Winter et al., 1998; Hoehn, 1998), as shown in Fig. 14.5:

1. Effluent or fully gaining river reach.
2. Influent or fully losing river reach, hydraulically connected to the groundwater.
3. Influent or fully losing river reach, hydraulically disconnected from the groundwater.
4. Flow-through river reach.
5. Parallel-flow river reach.

Effluent or gaining river reaches are those in which the groundwater head at the stream channel interface is higher than the stream stage. Effluent reaches receive groundwater as baseflow. However, the flow of groundwater is primarily into the hyporheic zone below the stream and the hyporheic water, in turn, moves upward beneath pools and concavities in longitudinal profile of the stream bed. Influent river reaches lose water to the ground and are found where the stream stage is greater than the underlying and adjacent groundwater head. Influent river reaches may be hydraulically connected or disconnected to the groundwater beneath the river.

When the river stage is less than the groundwater head on one bank and greater than the groundwater head on the other bank, flow-through condition develops, where the groundwater seeps into the river through one bank and the river loses to groundwater through the opposite bank. This condition especially develops when the river reach is oriented perpendicular to the regional groundwater flow, which, in the case of fluvial plains, is along their axis (Woessner, 2000). In some cases, the groundwater head beneath and adjacent to a stream is almost equal to the stream stage, thereby prohibiting groundwater-stream water flows. Such reaches are called parallel-flow reaches or "zero exchange channels" (Woessner, 1998). Although large-scale interaction of groundwater and river water does not occur in parallel-flow reaches, hyporheic-scale flows do still take place.

Influent rivers and streams are further classified as losing connected and losing disconnected rivers. In losing connected streams there is hydraulic connection between the groundwater table and the stream, and the seepage loss (or induced infiltration) from stream bed increases as the water table is lowered. In losing disconnected streams, there is no hydraulic connection between the groundwater and stream, and lowering of the water table does not affect the seepage loss from the stream bed; however, lowering of the water table can extend the length of the stream that is disconnected (Brunner et al., 2011).

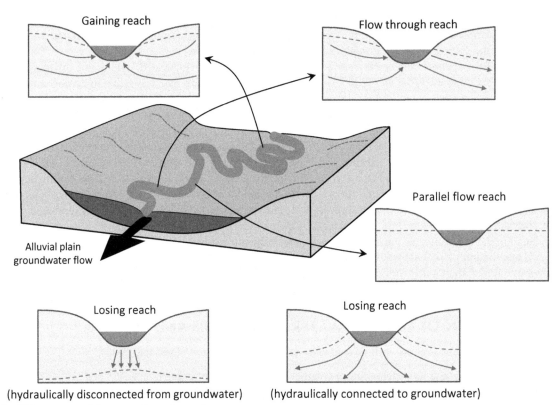

FIG. 14.5   Different types of river-groundwater interaction.

In all the possible conditions of groundwater-stream interaction discussed above, there is a hydraulic connection between groundwater and stream, except in the case of disconnected losing streams. However, the term "disconnected" does not imply that there is no exchange of water between the river and groundwater. In fact, a river in a losing disconnected state has the maximum seepage loss from its bed, referred to as the "maximum losing condition" by Parsons et al. (2008). Disconnection often results from clogging of the stream-bed sediments by biological (Treese et al., 2009) and sedimentary (Schalchli, 1992) processes, especially the formation of algal mats and the deposition of fine sediments. To verify whether a stream reach is disconnected or not, it must be shown that the seepage loss from the river bed does not change with a lowering of adjacent water table, and/or by the presence of an unsaturated zone beneath the stream (Brunner et al., 2011). The change from a losing connected to s losing disconnected stream is gradual and all transitional states are possible (Brunner et al., 2011).

It must also be mentioned that stream and river reaches may show temporal variations in the nature of their interactions with the groundwater. A reach may be losing in the dry season and gaining in the wet season, especially in the headwater zones of rivers and streams.

In additional to the large-scale groundwater-river interactions associated with flow systems, and also the small-scale hyporheic zone interactions, rivers also interact with groundwater on an intermediate scale (Winter et al., 1998), where the river water during flood conditions is stored in the banks and the floodplain sediments and later returned to the stream through baseflow (Fig. 14.6). This helps to moderate the floods and modify the water quality by passage through floodplain and bank sediments.

14.4   INTERACTION OF GROUNDWATER WITH LAKES AND RESERVOIRS

As far as groundwater-surface water interactions are concerned, lakes differ from rivers and streams in several ways (Winter, 1998):

1. Lake water level does not fluctuate as rapidly as in streams and rivers. Therefore bank storage is of lesser importance in lakes.
2. Evaporation commonly has a greater effect on lake levels than on stream levels. Because lakes have a larger surface area and are relatively less shaded than stream reaches, and also because lake water is not replenished as rapidly as streams.

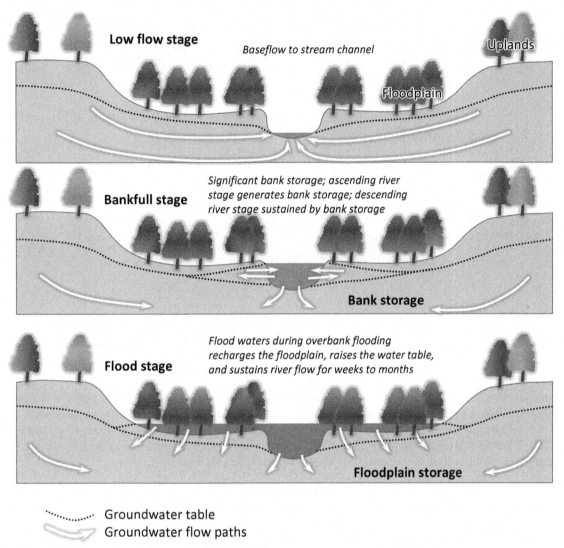

············· Groundwater table

Groundwater flow paths

FIG. 14.6 Interaction of rivers and groundwater at the floodplain scale.

3. Lakes can occur in many different landscape positions, whereas rivers are mainly found in lowlands or topographic lows.
4. Lake sediments typically contain more organic deposits than stream sediments, and these low permeability organic deposits affect the distribution and seepage of water and solutes more in lakes.

Reservoirs, which are artificial lakes, are designed primarily for flood control and distribution of surface water. Like lakes, reservoirs lose a significant amount of water by evaporation, and rapidly recycle biogeochemical materials within their waters and with the underlying organic rich hyporheic zone. Reservoirs are similar to streams as they have widely fluctuating water levels and thus significant bank storage, and they also have continuous flushing of water through them.

Like rivers and streams, lakes and reservoirs interact with groundwater either by losing or by gaining water or a combination of both in space and time, i.e., gaining on one bank and losing on the other bank, or gaining on the periphery and losing in the deeper part of the lake, and/or gaining in one season and losing in the other season. Lakes are dynamic bodies, and the movement of groundwater in their vicinity cannot be described in terms of static analysis (Sophocleous, 2002). Meyboom (1966, 1967) identified four commonly occurring conditions near permanent lakes as demonstrated in Fig. 14.7.

During spring the lakes receive discharge from local and intermediate groundwater systems (Fig. 14.7A). During summer the phreatophyte fringe surrounding the lakes receives shallow water from the local flow system as well as seepage from the lake periphery. This happens due to excessive withdrawal of groundwater by near-shore vegetation creating groundwater adjacent to the lake shore, as shown in Fig. 14.7B. As the groundwater levels decrease due to insufficient

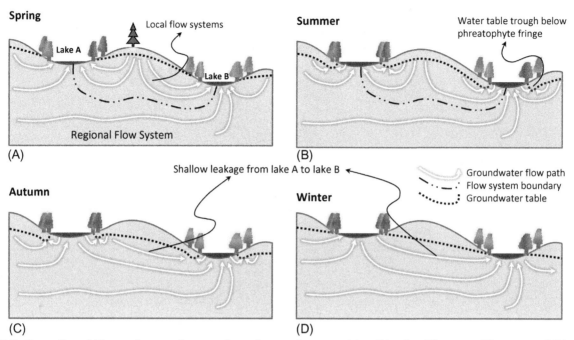

FIG. 14.7 Seasonally variable groundwater–surface water interaction around temperate lakes: (A) spring, (B) summer, (C) autumn, and (D) winter.

recharge in the dry season, local-flow systems begin to degenerate and shallow groundwater flow begins to occur from lake A to lake B (Fig. 14.7C). During autumn and winter recharge further diminishes and the local-flow system in the landscape between the two lakes degenerates completely and shallow groundwater flow from lake A to lake B is fully established (Fig. 14.7D). With the arrival of the following spring the snowmelt leads to an increase in the groundwater table and the consequent re-establishment of the local-flow system between the two lakes.

14.5 INTERACTION OF GROUNDWATER WITH WETLANDS

Wetlands occur in parts of the landscape where groundwater discharges to land surface and/or places that prevent rapid drainage of water from the land surface (Winter et al., 1998). Wetlands are characterized by a groundwater table near, at, or above the land surface and unique organic-rich soils indicating reduced conditions and ecological communities adapted to these conditions (Mitsch and Gosselink, 1986; Brinson et al., 2002). Based on the source of water, wetlands may be classified as:

1. Bogs, which are mainly precipitation dependant.
2. Fens, mainly groundwater dependant.
3. Swamps, marshes, and shallow open water, mainly surface-water dependant.

Fens can be present in lowlands or in landscapes with an appreciable slope, especially where the groundwater table intersects topographic slopes at the junction of uplands and lowlands. Some types of bogs can even occur on drainage divides (Winter et al., 1998). Marshes and swamps are found in lowlands.

In regions were the groundwater table has profile concavity, there is significant upward component of groundwater flow, leading to seepage of groundwater to the surface. Groundwater can also seep to the surface due to heterogeneities in the subsurface, especially the presence of high permeability zones within the aquifer (Winter et al., 1998). Wetlands also form adjacent to rivers and streams and along the coastal zone. These wetlands are subject to frequent water level changes due to riverine and coastal flooding and tides.

Groundwater-surface water interaction in wetlands differs from that in lakes in terms of the ease with which the water moves through their beds (Winter et al., 1998). The interaction between the groundwater and surface water is quite slow due to the high organic-matter content in wetland sediments; however, the exchange of surface water with the pore water of wetland sediments is quite strong due to the presence of fibrous root mats.

14.6 MEASUREMENT OF GROUNDWATER-SURFACE WATER INTERACTIONS

Extensive reviews regarding the measurement of interactions between groundwater and surface water at various scales are provided in Harvey and Wagner (2000), Kalbus et al. (2006), and Rosenberry and LaBaugh 2008. Weight (2008) discusses practical issues in the measurement of these interactions. The measurement methods have been classified by Kalbus et al. (2006) as follows:

1. Direct measurement.
2. Heat tracer methods.
3. Methods based on Darcy's law.
4. Mass balance approaches.

Direct measurement of groundwater-surface water interaction is possible by using seepage meters (Lee, 1977; Libelo and MacIntyre, 1994; Kelly and Murdoch, 2003). Seepage meters operate on the principle of isolating and covering a part of the groundwater-surface water interface with a chamber open at the bottom. Seepage meters are inexpensive to construct. Measurements at many locations along the stream bed are required to obtain a representative discharge; however, the discharge measured by seepage meters also includes the exchange flux of the hyporheic zone. Seepage-meter installation has the highest success in lakes or surface streams with low current velocities that do not have significant gravel or rocky armor fraction (Weight, 2008)

Groundwater discharge and recharge zones within surface water bodies can be identified using the temperature differences between the groundwater and surface water. Groundwater temperatures are relatively stable throughout the year, while surface-water temperatures fluctuate strongly on daily and seasonal time scales. Where surface water is seeping into the groundwater, bottom sediment temperatures mimic the surface water temperatures, whereas bottom sediment temperatures resemble more closely the groundwater temperatures in zones were groundwater seepage is occurring.

Temperature differences can be utilized by plotting a time series of temperatures at a particular location and identifying the direction of the flux based on the penetration of cyclic temperature changes (Constantz, 1998; Constantz and Stonestrom, 2003). Temperature differences can also be employed for the identification and quantification of fluxes by mapping the spatial variability of bottom sediment temperatures and then relating them to the fluxes obtained from other methods, e.g., minipiezometer data (Conant, 2004) or by heat-transport modeling (Schmidt et al., 2006). Important reviews regarding the measurement of fluxes using temperature differences are provided in Anderson (2005), Constantz (2008), and Rau et al. (2014).

Methods based on Darcy's law involve the estimation of hydraulic gradients to assess the direction of groundwater-surface water interaction and the estimation of hydraulic conductivity of bottom sediments and aquifers to assess the magnitude of the fluxes. Hydraulic gradients are measured using wells, piezometers, or minipiezometers. Wells and piezometers are relatively larger, permanent installations, and are generally employed for assessment of large-scale groundwater-surface water interactions, while minipiezometers are smaller, temporary installations employed mainly for the estimation of hyporheic exchange fluxes near streams and lakes (Lee and Cherry, 1978). Piezometers and minipiezometers, when installed in a group having their screened interval at successively greater depths, give an indication of the vertical movements of groundwater or hyporheic zone waters. A lower hydraulic head in piezometers screened at shallow depths and a greater hydraulic head in piezometers screened at greater depths indicates groundwater upwelling or discharge, and vice versa.

Mass balance approaches require the careful measurement of the total inflow and outflow of surface water in a surface-water body or a part of it. The deficit or excess between the inflows and outflows are attributed to groundwater-surface water fluxes. In stream reaches, usually, the stream discharge is measured at the beginning and end of a stream reach, by velocity gauging, using current meters (Carter and Davidian, 1968) or gauging flumes (Kilpatrick and Schneider, 1983), or by dilution gauging methods (Kilpatrick and Cobb, 1985). It is desirable that the selected stream reach should not have any tributary or canal confluence or divergence from the main stream, if it does, then all large and small inflows and outflows of surface water must be accounted for. Discharge measurements should be performed under low-flow conditions, so that any increase in discharge is not attributed to overland flow and shallow subsurface flows. Other methods employing a mass balance approach are hydrograph separation, and environmental tracers.

The different methods used for identifying and quantifying groundwater-surface water fluxes differ in the resolution, sampled volume, and time scales represented by them. There can be substantial differences in the fluxes obtained from assimilating point measurements and fluxes obtained by using large-scale techniques. Hence, it is desirable that measurements may be made at a range of scales by using multiple techniques. Specifically, the distinction between large-scale interactions and channel-scale (hyporheic zone) interactions must be considered while measuring these fluxes.

REFERENCES

Anderson, M.P., 2005. Heat as a ground water tracer. Ground Water 43, 951–968.

Brinson, M.M., MacDonnell, L.J., Austen, D.J., Beschta, R.L., Dillaha, T.A., Danahue, D.L., Gregory, S.V., Harvey, J.W., Molles, M.C., Rogers, E.I., Stanford, J.A., 2002. Riparian Areas Functions and Strategies for Management. National Academy Press, Washington, DC.

Brunke, M., Gonser, T., 1997. The ecological significance of exchange processes between rivers and groundwater. Freshw. Biol. 37 (1), 1–33.

Brunner, P., Cook, P.G., Simmons, C.T., 2011. Disconnected surface water and groundwater: from theory to practice. Ground Water 49 (4), 460–467.

Cardenas, M.B., 2009. A model for lateral hyporheic flow based on valley slope andchannel sinuosity. Water Resources Research. 45.

Carter, R.W., Davidian, J., 1968. General procedures for gauging streams. In: U.S. Geol. Surv. Techniques of Water Resources Investigations. Book 3 (Chapter A-6).

Conant, B., 2004. Delineating and quantifying ground water discharge zones using streambed temperatures. Ground Water 42 (2), 243–257.

Constantz, J., 1998. Interaction between stream temperature, streamflow, and groundwater exchanges in Alpine streams. Water Resour. Res. 34 (7), 1609–1615.

Constantz, J., 2008. Heat as a tracer to determine streambed water exchanges. Water Resour. Res. 44, W00D10.

Constantz, J. and Stonestrom, D. (2003) Heat as a tracer of water movement near streams in: Heat as a Tool for Studying the Movement of Ground Water Near Streams, edited by: Stonestrom, D. and Constantz, J., U.S. Geol. Surv. Circular 1260.

Dahl, M., Nilsson, B., Langhoff, J.H., Refsgaard, J.C., 2007. Review of classification systems and new multi-scale typology of groundwater–surface water interaction. J. Hydrol. 344, 1–16.

Dahm, C.N., Grimm, N.B., Mamonier, P., Valett, H.M., Vervier, P., 1998. Nutrient dynamics at the interface between surface waters and groundwaters. Freshw. Biol. 40, 427–451.

Harvey, J.W., Bencala, K.E., 1993. The effect of streambed topography on surface-subsurface water exchange in mountain catchments. Water Resour. Res. 29 (1), 1608–1620.

Harvey, J.W., Wagner, B.J., 2000. Quantifying hydrologic interactions between streams and their subsurface hyporheic zones. In: Jones, J.A., Mulholland, P.J. (Eds.), Streams and Groundwaters. Academic Press, New York, pp. 3–44.

Hoehn, E., 1998. Solute exchange between river water and groundwater in headwater environments. In: Hydrology, Water Resources and Ecology in Headwaters.Proceedings of the HeadWater '98 Conference Held at Meran/Merano, Italy, April 1998, pp. 165–171. IAHS Publ. No. 248.

Hutchinson, P.A., Webster, I.T., 1998. Solute uptake in aquatic sediments due to current-obstacle interactions. J. Environ. Eng. 124 (5), 419–426.

Kalbus, E., Reinstorf, F., Schirmer, M., 2006. Measuring methods for groundwater—surface water interactions: a review. Hydrol. Earth Syst. Sci. 10, 873–887.

Kelly, S.E., Murdoch, L.C., 2003. Measuring the hydraulic conductivity of shallow submerged sediments. Groundwater 41 (4), 431–439.

Kilpatrick, F.A., Cobb, E.D., 1985. Measurement of discharge using tracers. In: Techniques of Water-Resources Investigations. U.S. Geological Survey. Book 3 (Chapter A-16).

Kilpatrick, F.A., Schneider, V.R., 1983. Use of flumes in measuring discharge. In: Techniques of Water-Resources Investigations. U.S. Geological Survey. Book 3 (Chapter A-14).

Larkin, R.G., Sharp, J.M., 1992. On the relationship between river basin geomorphology, aquifer hydraulics, and groundwater flow direction in alluvial aquifers. Geol. Soc. Am. Bull. 104, 1608–1620.

Lee, D.R., 1977. Device for measuring seepage flux in lakes and estuaries. Limnol. Oceanogr. 22 (1), 140–147.

Lee, D.R., Cherry, J.A., 1978. A field exercise on groundwater flow using seepage meters and mini-piezometers. J. Geol. Educ. 27, 6–10.

Lerner, D.N., 2009. The Hyporheic Handbook. A handbook on the groundwater–surface water interface and hyporheic zone for environment managers. Integrated catchment science programme. Environment Agency Science report.264, pp.

Libelo, E.L., MacIntyre, W.G., 1994. Effects of surface-water movement on seepage-meter measurements of flow through the sediment-water interface. Hydrogeol. J. 2, 49–54.

Meyboom, P., 1966. Unsteady groundwater flow near a willow ring in a hummocky moraine. J. Hydrol. 4, 38–62.

Meyboom, P., 1967. Mass transfer studies to determine the groundwater regime of permanent lakes in hummocky moraine of western Canada. J. Hydrol. 5 (2), 117–142.

Mitsch, W.J., Gosselink, J.G., 1986. Wetlands. Van Nostrand Reinhold Company, New York.

Parsons, S., Evans, R., Hoban, M., 2008. Surface–groundwater connectivity assessment. A report to the Australian Government from the CSIRO Murray-Darling Basin Sustainable Yields Project. 35, CSIRO, Australia.

Rau, G.C., Anderson, M.S., McCallum, A.M., Roshan, H., Acworth, R.I., 2014. Heat as a tracer to quantify water flow in near-surface sediments. Earth Sci. Rev. 129, 40–58.

Rosenberry, D.O., LaBaugh, J.W., 2008. Field techniques for estimating water fluxes between surface water and ground water: U.S. Geological Survey Techniques and Methods 4–D2. p. 128.

Safeeq, M., Fares, A., 2016. Groundwater and surface water interactions in relation to natural and anthropogenic environmental changes. In: Fares, A. (Ed.), Emerging Issues in Groundwater Resources, Advances in Water Security. Springer, Prairie View, TX.

Savant, S.A., Reible, D.D., Thibodeaux, L.J., 1987. Convective transport within stable river sediments. Water Resour. Res. 23 (9), 1763–1768.

Schalchli, U., 1992. The clogging of coarse gravel river beds by fine sediment. Hydrobiologia 235, 189–197.

Schmidt, C., Bayer-Raich, M., Schirmer, M., 2006. Characterization of spatial heterogeneity of groundwater-stream water interactions using multiple depth streambed temperature measurements at the reach scale. Hydrol. Earth Syst. Sci. 10, 849–859.

Sophocleous, M., 2002. Interactions between groundwater and surface water: the state of the science. Hydrogeol. J. 10, 52–67.

Stanford, J.A., Ward, J.V., 1988. The hyporheic habitat of river ecosystems. Nature 335, 64–66.

Thibodeaux, L.J., Boyle, J.D., 1987. Bedform-generated convective transport in bottom sediments. Nature 325, 341–343.

Toth, J., 1963. A theoretical analysis of groundwater flow in small drainage basins. J. Geophys. Res. 68, 4785–4812.

Treese, S., Meixner, T., Hogan, J.F., 2009. Clogging of an effluent dominated semiarid river: a conceptual model of stream–aquifer interactions. J. Am. Water Resour. Assoc. 45 (4), 1047–1062.

Vaux, W.G., 1968. Intra-gravel flow and interchange of water in a streambed. Fish. Bull. 66, 479–489.

Weight, W.D., 2008. Groundwater surface water interactions. In: Weight, W.D. (Ed.), Hydrogeology Field Manual. McGraw Hill, Butte, Montana, pp. 231–282.

Winter, T.C., 1976. Numerical Simulation Analysis of the Interaction of Lakes & Groundwater. USGS Professional Paper 1001.

Winter, T.C., 1995. Recent advances in understanding the interaction of groundwater and surface water. Rev. Geophys. (Suppl.) 985–994.

Winter, T.C., 1999. Relation of streams, lakes, and wetlands to groundwater flow systems. Hydrogeol. J. 7, 28–45.

Winter, T.C., Harvey, J.W., Franke, O.L., Alley, W.M., 1998. Groundwater and Surface Water. A Single Resource. US Geological Survey, Denver. CO Circular 1139.

Woessner, W.W., 1998. Changing views of stream-groundwater interaction. In: Van Brahana, J., Eckstein, Y., Ongley, L.W., Schneider, R., Moore, J.E. (Eds.), Proceedings of the Joint Meeting of the XXVIII Congress of the International Association of Hydrogeoloists and the Annual Meeting of the American Institute of Hydrology. American Institute of Hydrology, St. Paul, MN, pp. 1–6.

Woessner, W.W., 2000. Stream and fluvial plain groundwater interactions: rescaling hydrogeological thought. Ground Water 38 (3), 423–429.

FURTHER READING

Toth, J., 1999. Groundwater as a geologic agent: an overview of the causes, processes, and manifestations. Hydrogeol. J. 7, 1–14.

Winter, T.C., 2001. The concept of hydrological landscapes. J. Am. Water Resour. Assoc. 37 (2), 335–349.

Chapter 15

Evaluation of Vulnerability Zone of a Coastal Aquifer Through GALDIT GIS Index Techniques

G. Gnanachandrasamy*,†, T. Ramkumar‡, J.Y. Chen*,†, S. Venkatramanan§,¶,‖,#, S. Vasudevan‡ and S. Selvam**

*School of Geography and Planning, Sun Yat-sen University, Guangzhou, China †School of Earth Sciences and Engineering, Sun Yat-sen University, Guangzhou, China ‡Department of Earth Sciences, Annamalai University, Chidambaram, India §Department of Earth and Environmental Sciences, Pukyong National University, Busan, South Korea ¶Department for Management of Science and Technology Development, Ton Duc Thang University, Ho Chi Minh City, Vietnam ‖Faculty of Applied Sciences, Ton Duc Thang University, Ho Chi Minh City, Vietnam #Department of Geology, Alagappa University, Karaikudi, India **Department of Geology, V.O. Chidambaram College, Thoothukudi, India

Chapter Outline

15.1 INTRODUCTION

In coastal regions groundwater resources are very important where drinking water is scare. The present condition of the world's ground water is widely monitored and reported on by the world water-assessment program. Seawater incursion (SWI) into coastal aquifer is a common phenomenon occurring in several coastal areas throughout the world (Todd and Mays, 2005). The problems of groundwater contamination include: ground vertical movements, climate change, both anthropogenic and natural influences, the fresh-salt water interface, and SWI. Of these, SWI is a key problem in most of the coastal areas around the world (Santucci et al., 2016). Hence, the identification of vulnerable areas in the coastal aquifers is necessary to allow them to be studied. Various methods, such as isotope techniques, geophysical methods, water quality, SEAWAT, SINTACS, EPIK, PESTICIDE, DRASTIC, and GALDIT, are used to find the vulnerable areas (Mahlknecht et al., 2017; Santucci et al., 2016; Surinaidu et al., 2016; Moghaddam et al., 2017; Gnanachandrasamy et al., 2016). For each method different data were used. Of all the methods GALDIT was the most extensively used. Globally the GALDIT method has been used in various coastal areas, including the United States (Tasnim and Tahsin, 2016), Tunisia (Gontara et al., 2016), Iran (Moghaddam et al., 2017), Greece (Lappas et al., 2016; Recinos et al., 2015), and in India on the Goa coast (Chachadi and Lobo-Ferreira, 2005; Lenin Kalyana Sundaram et al., 2008), Ramanathapuram coast (Santha Sophiya and Syed, 2013), and Mangalur coast (Mahesha et al., 2012).

The deltaic aquifer is the key source of water for human consumption and agricultural activities for the development of economic activities in the coastal part of the India. The problems of SWI are most notable in the coastal aquifers in India. Earlier research on these coastal aquifer systems has dealt with various hydrological characteristics in east coast of India (Venkatramanan et al., 2014, Gnanachandrasamy et al., 2012; Anithamary et al., 2012; Ramanathan et al., 1993; Thiyagarajan and Baskaran, 2011; Chidambaram et al., 2014; Selvam et al., 2014), assessment of SWI using geophysical methods (Surinaidu et al., 2012), and groundwater-quality mapping (Seshadri et al., 2013).

GIS and Geostatistical Techniques for Groundwater Science. https://doi.org/10.1016/B978-0-12-815413-7.00015-8

The current study aims to identify and demarcate the vulnerable areas for SWI using the GALDIT modeling index. This index relates to the analysis of SWI and the identification of the vulnerable areas within the study site. The results of the current study can be used for coastal areas development and management, as well as for environmental prediction and monitoring.

15.2 STUDY SITE

The study site lies between longitude 79° 35′ 0″ to 79° 50′ 0″ east and latitude 10° 35′ 0″ to 11° 25′ 0″ north on the east coast of India, which is the coastal part of the Cauvery deltaic region. The area is bounded by the Bay of Bengal coast on the east, which acts as a major source of SWI in the area. The total areal extent of the region is 2715.83 km². The major river, the Coleroon, is in the northern part of the study area. Minor rivers such as the Tirumalairajanar, Vennar, Vettar, and Arasalar rivers drain the southern part of the study site and these are the branches and tributaries of the Cauvery river. The major aquifer type is discontinuous and fairly thick, with confined fresh water overlaid by saline water towards the coast and the level of water ranging from 1 to 8 m. In total 30 locations are selected in the area and their locations are numbered in Fig. 15.1. The dominant soil types of the area are either sandy or clayey with low alkalinity/salinity. They are normally high in sodium and poorly drained and the nutrient content of the soil is low (Sivanappan, 2007). Geologically the area comprises recent formations with alluvium and cuddalore. These are of clayey, sandy clay, silty clay, and sandy characteristic. These textures are predominant in the Cauvery delta. Geomorphologically the coastal stretch is covered with beach ridges, beaches, swamps, mud flats, and back waters. The general drainage type is dendritic and trellis pattern and stream orders are categorized on the basis of origin. The groundwater level varies from 1 to 7 m. During the summer season the water level goes down to 6–7 m below the ground; when water level drops below 7–10 m during the summer, the quality of water also changes and becomes more brackish and salty, resulting in a severe shortage of potable water in the study area. The temperatures ranged from 40.6°C to 19.3°C, and the humidity varies between 70% and 77%.

15.3 METHODOLOGY

Groundwater resources are being polluted as a result of SWI and this situation has been amplified immensely in recent years. The levels of groundwater have decreased due to the fact that the study area is now classified as having high salinity. Therefore it was imperative to monitor the interface between the freshwater and saltwater in this current study. The GALDIT method of vulnerability assessment of SWI, projected by Chachadi and Lobo-Ferreira (2001), was successfully used to assess the areas vulnerable to SWI. In this method six parameters were used. The parameters were: occurrence of groundwater (G), aquifer conductivity (A), level of groundwater above sea (L), distance from the sea (D), impact of existing sea water (I), and aquifer thickness (T). All the individual (GALDIT) parameters were evaluated and calculated with respect to seawater incurison. The numerical ranking system used to assess the potential for SWI in the hydrologeological setting comprised GALDIT factors (Lobo Ferreira et al., 2005). The weightage and ranking of the parameters, as decided by expert consideration of the hydrogeology of the study area, are presented in Table 15.1. The GALDIT modeling index values were obtained using the following equation:

$$\text{GALDIT} \quad \text{index} = \sum_{i=1}^{6} \{(W_i)R_i\} / \sum_{i=1}^{6} W_i$$

where
  W_i = the ith parameter weight
  R_i = the rating of the ith parameter.

The minimum ranking and maximum ranking of the GALDIT index were 2.5 and 10, respectively. The maximum values of the GALDIT index denoted higher vulnerability to SWI. The SWI vulnerability classification index is presented in Table 15.2. Based on this classification the vulnerable areas were classified into three classes, namely high vulnerability, moderate vulnerability, and low vulnearability (Chachadi and Lobo-Ferreira, 2005).

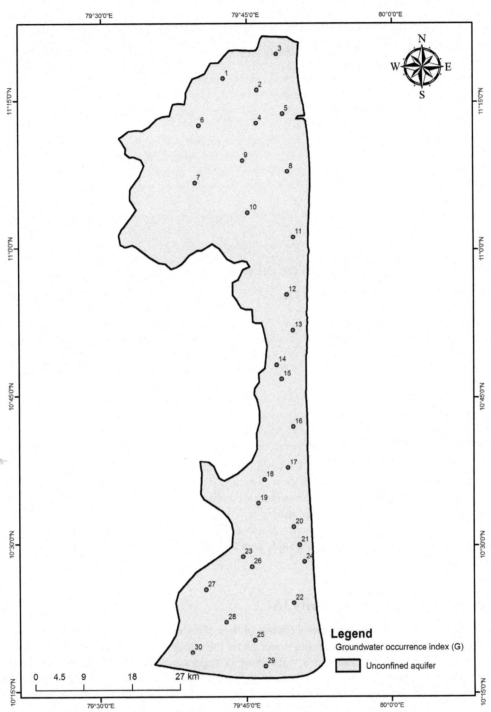

FIG 15.1 Groundwater occurrence (G).

TABLE 15.1 Weightage and Ranking of GALDIT Parameters Index (Chachadi and Lobo-Ferreira, 2005)

Parameters	Weight (W_i)	Ranking (R_i)			
		2.5	5	7.5	10
Groundwater occurrence (G)	1		Leaky	Unconfined	Confined
Aquifer hydraulic conductivity (A)	3	<5	5–10	10–40	>40
Height of groundwater above sea level (L)	4	>2	1.5–2	1–1.5	<1
Distance from the shore (D)	4	>30	20–30	15–20	<15
Impact of existing status of seawater intrusion (I)	3	<1	1–1.5	1.5–2	>2
Thickness of aquifer (T)	2	<5	5–7.5	7.5–10	>10

TABLE 15.2 Vulnerability Classification Based on GALDIT Index (Chachadi and Lobo-Ferreira, 2005)

S. No.	GALDIT Index Range	Vulnerability Classes
1	≥ 7.5	High vulnerability
2	5–7.5	Moderate vulnerability
3	< 5	Low vulnerability

15.4 RESULT AND DISCUSSIONS

15.4.1 Groundwater Occurrence (G)

Groundwater occurrence (parameter G) represents the types of aquifer in the study site. In this case the aquifer in the study site, Nagapattinam, is marked as unconfined and shallow in nature. It contains sandy and clayey soils. Groundwater occurrence affects the extent of SWI into the groundwater; in natural conditions a confined aquifer is less affected by SWI than an unconfined aquifer. Unconfined aquifers are under atmospheric pressure while confined aquifers are situated under aquitards and the pressure is higher than the atmospheric pressure (Savariya and Bhatt, 2014). So, ranking of groundwater occurrence corresponds to the value 7.5 in the entire study area according to the GALDIT modeling index. Fig. 15.1 shows the representation of parameter (G) (Table 15.3).

15.4.2 Aquifer Hydraulic Conductivity (A)

Aquifer hydraulic conductivity (A) is the measure of the rate of flow of water in the aquifer. By definition, aquifer hydraulic conductivity is the ability of the aquifer to transmit water under the effect of a hydraulic gradient (Lobo Ferreira et al., 2005). Higher hydraulic conductivity results in a wider cone of depression and a larger extent of SWI (Savariya and Bhatt, 2014). In the present study the hydraulic conductivity rating varied between 5 and 7.5, which indicates that the hydraulic conductivity of the study area varies between 5 and 40 m/day. From the known values we interpolated the values for the aquifer area using a kriging method, which is used in geographical information system (GIS) technology (Joao, 2005), and near the adjacent areas smaller ranking values were assigned. Spatial maps (Fig. 15.2) showed a representation of the hydraulic conductivity (A) of the study area.

15.4.3 Height of Groundwater Level Above Sea Level (L)

The groundwater level adopted in this parameter was provided by the public groundwater department (PWD) of Tamil Nadu; this value was considered as the parameter of height of the groundwater level above sea level (L). This is a very important factor in the evaluation of the SWI in the coastal area, as the height of the groundwater determines the hydraulic pressure. Hydraulic pressure has the ability to push seawater back. If the groundwater level is below the mean sea level then

TABLE 15.3 Vunerability Class

S. No	G	A	L	D	I	T	GALDIT Index	Vulnerability Classes
1	7.5	15	10	40	30	5	6.3	Moderate vulnerability
2	7.5	15	10	40	30	20	7.2	Moderate vulnerability
3	7.5	15	20	40	30	20	7.8	High vulnerability
4	7.5	15	10	40	30	20	7.2	Moderate vulnerability
5	7.5	15	20	40	30	20	7.8	High vulnerability
6	7.5	15	10	40	30	20	7.2	Moderate vulnerability
7	7.5	15	10	40	30	20	7.2	Moderate vulnerability
8	7.5	22.5	20	40	30	20	8.2	High vulnerability
9	7.5	15	10	40	30	20	7.2	Moderate vulnerability
10	7.5	15	10	40	22.5	5	5.9	Moderate vulnerability
11	7.5	15	30	40	30	10	7.8	High vulnerability
12	7.5	22.5	20	40	30	10	7.6	High vulnerability
13	7.5	22.5	20	40	30	20	8.2	High vulnerability
14	7.5	22.5	10	40	30	20	7.6	High vulnerability
15	7.5	22.5	10	40	30	20	7.6	High vulnerability
16	7.5	22.5	20	40	30	10	7.6	High vulnerability
17	7.5	22.5	20	40	30	10	7.6	High vulnerability
18	7.5	22.5	20	30	30	20	7.6	High vulnerability
19	7.5	22.5	20	40	30	15	7.9	High vulnerability
20	7.5	22.5	30	40	30	15	8.5	High vulnerability
21	7.5	15	20	40	30	20	7.8	High vulnerability
22	7.5	15	20	40	30	20	7.8	High vulnerability
23	7.5	15	20	40	30	20	7.8	High vulnerability
24	7.5	22.5	30	40	30	10	8.2	High vulnerability
25	7.5	22.5	20	40	30	10	7.6	High vulnerability
26	7.5	22.5	20	40	30	20	8.2	High vulnerability
27	7.5	22.5	20	40	30	10	7.6	High vulnerability
28	7.5	22.5	20	40	30	10	7.6	High vulnerability
29	7.5	22.5	20	40	30	10	7.6	High vulnerability
30	7.5	22.5	30	40	30	10	8.2	High vulnerability

this leads to the strongest possible vulnerability to SWI (Joao, 2005). The height of groundwater above mean sea level was calculated for the summer season in the year of 2012. The maximum height of groundwater above mean sea level was recorded at Nagapattinam (4.5 m) and the ranking assigned in the groundwater level parameter was between 2.5 and 7.5. Fig. 15.3 shows a representation of the GALDIT parameter (L) for the study area. Along the coast the water level varied from 1 to 1.5 m. Whereas away from the coastal areas the water level was shown to be >2 m. The central part of the study area was shown to have a water level between 1.5 and 2 m.

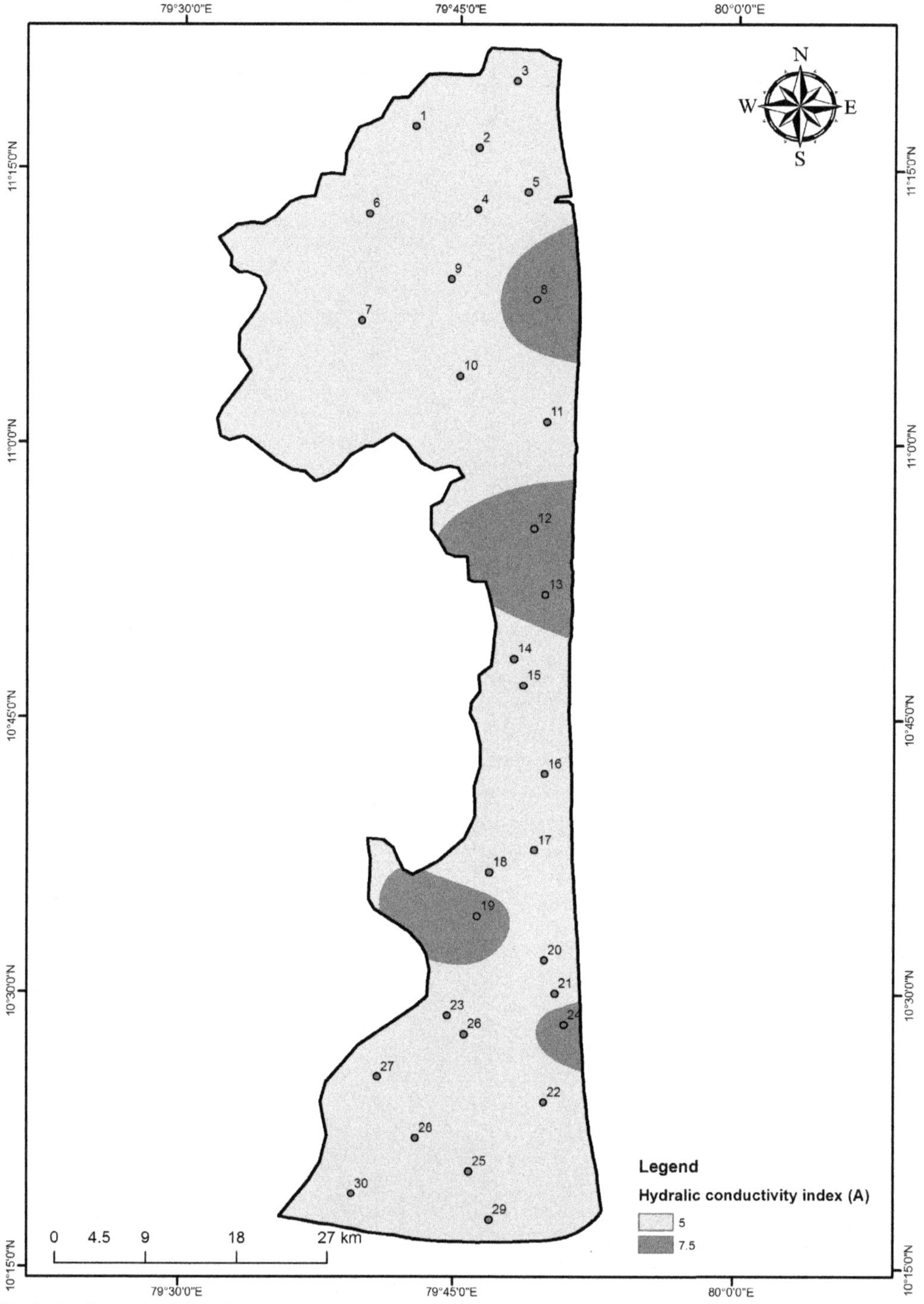

FIG 15.2 Aquifer hydraulic conductivity (A).

15.4.4 Distance From the Shore (D)

This is an important parameter within the GALDIT index. The impact of SWI decreases when moving perpendicularly from shore towards the land. The magnitude of SWI is directly related to the perpendicular distance from the coast (Honnanagoudar et al., 2014). The maximum distance from the coast is 19 km at Enankudi village (location 18) and the ranking assigned was 7.5. Most of the locations showed a minimum distance (<15 km) and the ranking assigned was 10. The spatial distribution map of GALDIT D parameter is shown in Fig. 15.4.

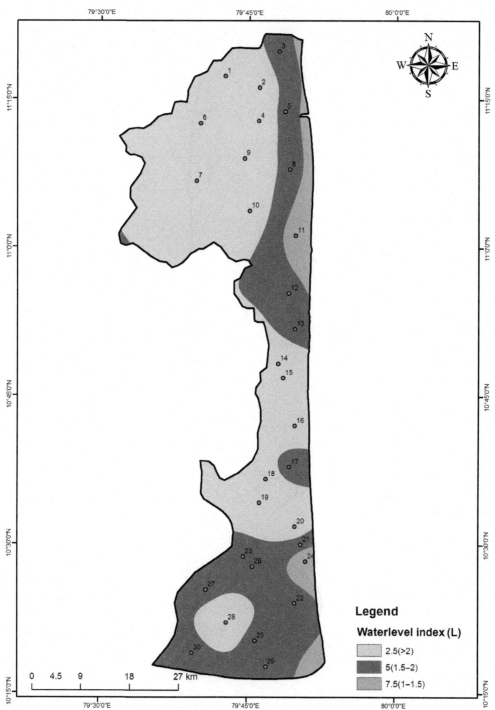

FIG 15.3 Groundwater level (L).

15.4.5   Impact of the Existing Status of Saltwater Intrusion (I)

The ratio of Cl/HCO_3^- was used to determine SWI into the coastal aquifer if the area under mapping is invariably under stress and this stress has already modified the hydraulic balance between seawater and fresh water. The ranking 7.5 was assigned at Cenkodai (location 10) and Maruthur therkku (location 27), the Cl/HCO_3^- ratio for which was below 2 mg/L. The ranking assigned to the remaining locations was 10. The areas near the coast had a Cl/HCO_3^- ratio of more than 2 mg/L. The impact of the existing status of SWI is shown in Fig. 15.5.

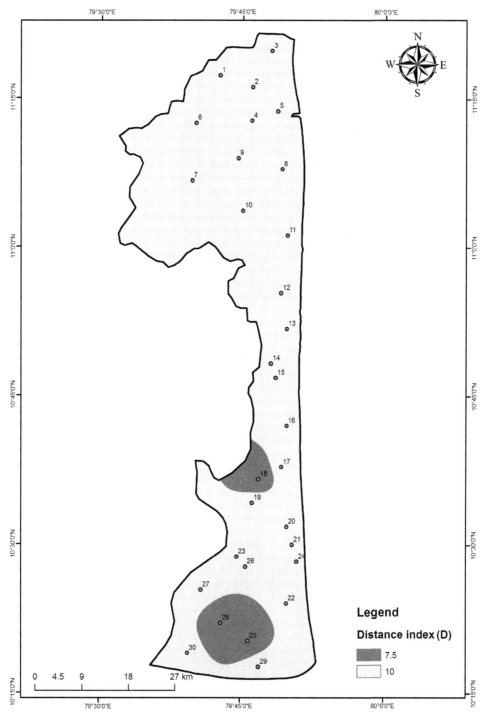

FIG 15.4 Distance from shore (D).

15.4.6 Aquifer Thickness (T)

Aquifer thickness (T) is one of the most significant parameters in determining SWI for the coastal aquifer. The higher the aquifer thickness, the greater the extent of the SWI. Minimum and maximum thickness of ranking varied from 2.5 to 10 and 5 to 7.5, respectively. A ranking of 2.5 was assigned to an aquifer thickness of <5, a ranking of 5 to a thickness of 5–7.5, a ranking of 7.5 to a thickness of between 7.5 and 10, and a ranking 10 to a thickness greater than 10. Fig. 15.6 shows the GALDIT index for aquifer thickness (T).

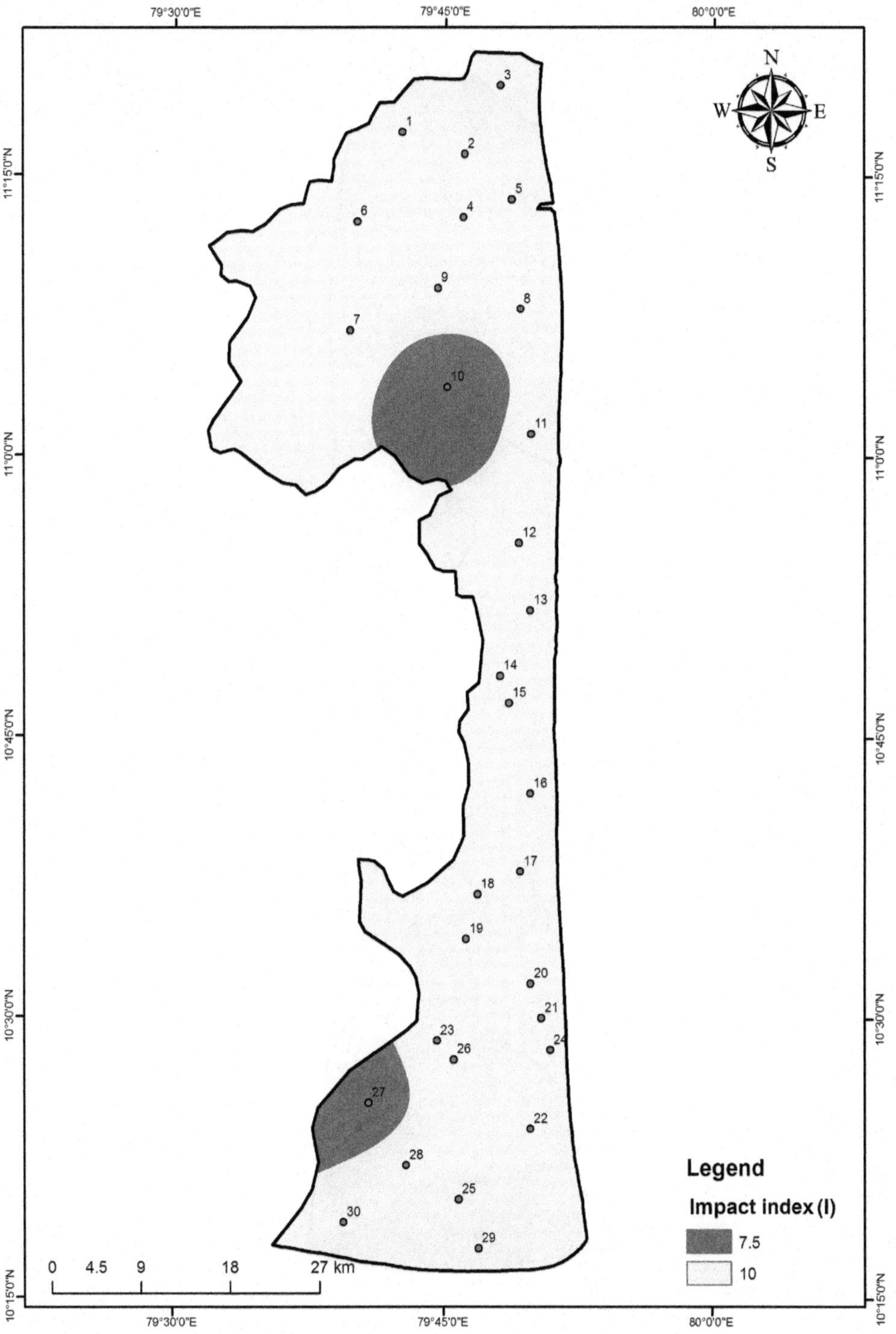

FIG 15.5   Impact of existing status of seawater intrusion (I).

15.5   CONCLUSIONS

The assessment of SWI in the coastal aquifers of the Cauvery deltaic region was successfully carried out through the application of the GALDIT index method. Using the GALDIT index method the results indicated that the aquifer is within a high-to-moderate category of vulnerability for SWI. The GALDIT index map of the proposed area clearly showed that the area near to the coast had high vulnerability and areas further away from the coast had moderate vulnerability. The GALDIT

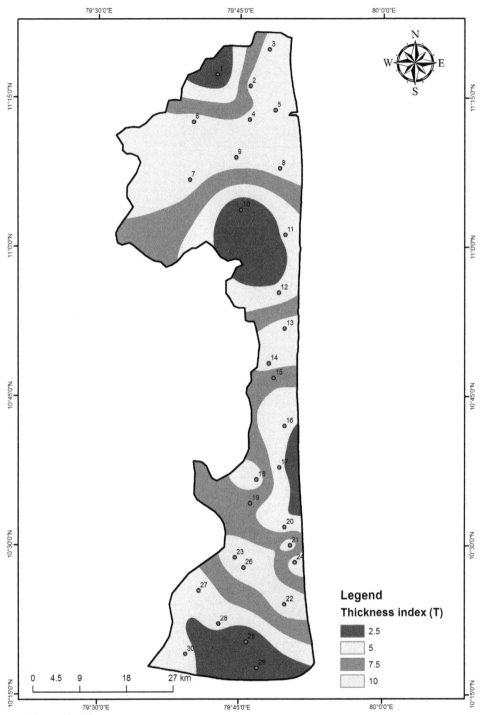

FIG 15.6 Thickness of the aquifer (T).

index map is shown in Fig. 15.7. From the results, the total study area was classified as 92.8% being highly vulnerable and 7.2% being moderately vulnerables. The modern technique of GALDIT index aquifer vulnerability mapping is used for the identification of SWI in coastal areas. Finally, the vulnerability maps can be used as a tool for the management of the coastal groundwater resources and to predict SWI.

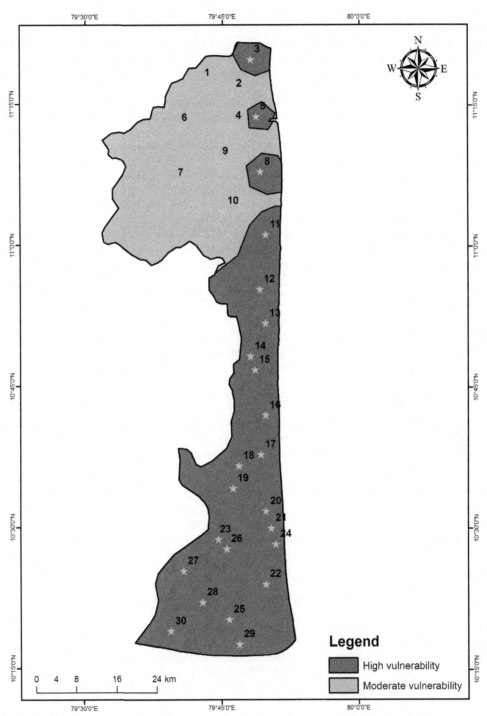

FIG 15.7 GALDIT index map.

REFERENCES

Anithamary, I., Ramkumar, T., Venkatramanan, S., 2012. Application of statistical analysis for the hydrogeochemistry for saline groundwater in Kodia-karai, Tamilnadu, India. J. Coast. Res. 281, A89–A98. https://doi.org/10.2112/JCOASTRES-D-09-00156.

Chachadi, A.G., Lobo-Ferreira, J.P., 2001. Sea water intrusion vulnerability mapping of aquifers using GALDIT method. In: Proc. Workshop Modeling in Hydrogeology. Anna University, Chennai, India, pp. 143–156.

Chachadi, A.G., Lobo-Ferreira, J.P., 2005. Assessing aquifer vulnerability to sea-water intrusion using GALDIT method: part 2 – GALDIT indicator descriptions. In: IAHS and LNEC, Proc. The Fourth Inter Celtic Colloquium on Hydrology and Management of Water Resources. Universidade do Minho, Guimarães, Portugal.

Chidambaram, S., Karmegam, U., Prasanna, M.V., Sasidhar, P., 2014. A study on evaluation of probable sources of heavy metal pollution in groundwater of Kalpakkam region, South India. Environmentalist 32, 371–382.

Gnanachandrasamy, G., Ramkumar, T., Venkatramanan, S., Anitha Mary, I., Vasudevan, S., 2012. GIS based hydrogeochemical characteristics of ground-water quality in Nagapattinam district, Tamilnadu, India. Carpath. J. Earth Environ. Sci. 7 (3), 205–210. ISSN-1842-4090.

Gnanachandrasamy, G., Ramkumar, T., Venkatramanan, S., Chung, S.Y., Vasudevan, S., 2016. Identification of saline water intrusion in part of Cauvery Deltaic region Tamil Nadu, Southern India-Using GIS and VES methods. Mar. Geophys. Res. 37, 113–126.

Gontara, M., Allouche, N., Jmal, I., Bouri, S., 2016. Sensitivity analysis for the GALDIT method based on the assessment of vulnerability to pollution in the northern Sfax coastal aquifer, Tunisia. Arab. J. Geosci. 9, 416. https://doi.org/10.1007/s12517-016-2437-3.

Honnanagoudar, S.S., Venkat Reddy, D., Mahesha, A., 2014. Analysis of vulnerability assessment in the Coastal Dakshina Kannada District, Mulki to Talapady Area, Karnataka. Int. J. Comput. Eng. Res. 04 (5), 2250–3005.

Lappas, I., Kallioras, A., Pliakas, F., Rondogianni, T., 2016. Groundwater vulnerability assessment to seawater intrusion through gis—based galdit method. Case study: atalantic coastal aquifer, central Greece. Bull. Geol. Soc. Greece 50, 798–807.

Lenin Kalyana Sundaram, V., Dinesh, G., Ravikumar, G., Govindarajalu, D., 2008. Vulnerability assessment of seawater intrusion and effect of artificial recharge in Pondicherry coastal region using GIS. Indian J. Sci. Technol. 1 (7), 1–7.

Lobo Ferreira, J.P., Chachadi, A.G., Diamantino, C., Henriques, M.J., 2005. Assessing Aquifer vulnerability to seawater intrusion using GALDIT method: part I application to the Portuguese Aquifer of Monte Gordo.The Fourth Inter Celtic Colloquium on Hydrology and Management of Water Resources, Guimaraes, Portugal, July 11–14, 2005, pp. 1–12.

Mahesha, A., ASCE, A.M., Vyshali, Lathashri, U.A., Ramesh, H., 2012. Parameter estimation and vulnerability assessment of coastal unconfined aquifer to saltwater intrusion. J. Hydrol. Eng. 17, 933–943.

Mahlknecht, J., Merchán, D., Rosner, M., Meixner, A., Ledesma-Ruiz, R., 2017. Assessing seawater intrusion in an arid coastal aquifer under high anthro-pogenic influence using major constituents, Sr and B isotopes in groundwater. Sci. Total Environ. 587–588, 282–295.

Moghaddam, H.K., Jafari, F., Javadi, S., 2017. Vulnerability evaluation of a coastal aquifer via GALDIT model and comparison with DRASTIC index using quality parameters. Hydrol. Sci. J. 62 (1), 137–146.

Ramanathan, A.L., Vaithiyanatham, P., Subramanian, V., Das, B.K., 1993. Geochemistry of the Cauvery East Coast of India Estuary. Estuaries 16 (3A), 459–474.

Recinos, N., Kallioras, A., Pliakas, F., Schuth, C., 2015. Application of GALDIT index to assess the intrinsic vulnerability to seawater intrusion of coastal granular aquifers. Environ. Earth Sci. 73, 1017–1032. https://doi.org/10.1007/s12665-014-3452-x.

Santha Sophiya, M., Syed, T.H., 2013. Assessment of vulnerability to seawater intrusion and potential remediation measures for coastal aquifers: a case study from eastern India. Environ. Earth Sci. 70, 1197–1209. https://doi.org/10.1007/s12665-012-2206-x.

Santucci, L., Carol, E., Kruse, E., 2016. Identification of palaeo-seawater intrusion in groundwater using minor ions in a semi-confined aquifer of the Río de la Plata littoral (Argentina). Sci. Total Environ. 566–567, 1640–1648.

Savariya, P., Bhatt, N., 2014. Assessing groundwater vulnerability to seawater intrusion in Morbi-Maliya using GALDIT method. Int. J. Comput. Eng. Res. 04 (5), 2250–3005.

Selvam, S., Manimaran, G., Sivasubramanian, P., Balasubramanian, N., Seshunarayana, T., 2014. GIS-based evaluation of water quality index of ground-water resources around Tuticorin coastal city, south India. Environ. Earth Sci. 71, 2847–2867.

Seshadri, H., Kaviyarasan, R., Sasidhar, P., Balasubramaniyan, V., 2013. Effect of saline water bodies on the hydrogeochemical evaluation of groundwater at Kalpakkam coastal site, Tamil Nadu. J. Geol. Soc. India 82 (5), 535–544.

Sivanappan, R.K., 2007. Mapping and Study of Coastal Water Bodies in Nagapattinam District. NGO Co-Ordination and Resource Centre (NCRC), Nagapattinam.

Surinaidu, L., Gurunadha Rao, V.V.S., Tamma Rao, G., et al., 2012. An integrated approach to investigate saline water intrusion and to identify the salinity sources in the Central Godavari delta Andhra Pradesh, India. Arab. J. Geosci. https://doi.org/10.1007/s12517-012-0634-2.

Surinaidu, L., Muthuwattab, L., Amarasinghe, U.A., Jain, S.K., Ghosh, N.C., Kumar, S., Singh, S., 2016. Reviving the Ganges water machine: accelerating surface water and groundwater interactions in the Ramganga sub-basin. J. Hydrol. 540, 207–219.

Tasnim, Z., Tahsin, S., 2016. Application of the method of Galdit for groundwater vulnerability assessment: a case of South Florida. Asian J. Appl. Sci. Eng. 5, 27–40.

Thiyagarajan, M., Baskaran, R., 2011. Groundwater quality in the coastal stretch between Sirkazhi and Manampandal, Tamil Nadu, India using Arc GIS Software. Arab. J. Geosci. https://doi.org/10.1007/s12517-011-0500-7.

Todd, D.K., Mays, L.W., 2005. Groundwater Hydrology. third ed.Wiley, Hoboken.

Venkatramanan, S., Chung, S.Y., Ramkumar, T., Gnanachandrasamy, V.S., Lee, S.Y., 2014. Application of GIS and hydrogeochemistry of groundwater pollution status of Nagapattinam district of Tamil Nadu, India. Environ. Earth Sci. https://doi.org/10.1007/s12665-014-3728-1.

FURTHER READING

Chachadi, A.G., Lobo Ferreira, J.P., Noronha, L., Choudri, B.S., 2002. Assessing the impact of the sea level rise on saltwater intrusion in coastal aquifers using GALDIT model, NIH, Roorkee. Coast. Policy Res. Newsl. 7, 27–31.

Ferreira, L., Chachadi, A.G., 2005. Assessing aquifer vulnerability to saltwater intrusion using GALDIT method: part 1—application to the portuguese aquifer to Monte Gordo. In: Proceedings of this 4th Interseltic Colloquim on Hydrol. Mangt. Water Resor., Portugal, pp. 1–12.

Kumar, C.P., Chachadi, A.G., Purandara, P.K., Kumar, S., Juyal, R., 2007. Modelling of Seawater Intrusion in Coastal Area of North Goa. Goa University, Goa.

Chapter 16

A Statistical Approach to Identify the Temporal and Spatial Variations in the Geochemical Process of a Coastal Aquifer, South East Coast of India

M.V. Prasanna*, S. Chidambaram[†], R. Thilagavathi[‡], C. Thivya[§], S. Venkatramanan[¶,‖,#,**] and N. Murali Krishnan[††]

*Department of Applied Geology, Faculty of Engineering and Science, Curtin University Malaysia, Miri, Malaysia [†]Water Research Center, Kuwait Institute for Scientific Research, Safat, Kuwait [‡]Department of Earth Sciences, Annamalai University, Chidambaram, India [§]Department of Geology, University of Madras, Chennai, India [¶]Department of Earth and Environmental Sciences, Pukyong National University, Busan, South Korea [‖]Department for Management of Science and Technology Development, Ton Duc Thang University, Ho Chi Minh City, Vietnam [#]Faculty of Applied Sciences, Ton Duc Thang University, Ho Chi Minh City, Vietnam [**]Department of Geology, Alagappa University, Karaikudi, India [††]Media and Communication Department, Faculty of Humanities, Curtin University Malaysia, Miri, Malaysia

16.1 INTRODUCTION

The global upsurge in populations and rapid urbanization has resulted in an increased dependence on groundwater as a major source of water supply. Therefore it is imperative to understand the different hydrogeochemical forms that occur in the aquifer system. When water reaches the saturated aquifer it undergoes various geochemical reactions (Coetsiers and Walravens, 2006; Chidambaram et al., 2008, 2009, 2015; Prasanna et al., 2010; Manivannan et al., 2011; Vasanthavigar et al., 2012; Thialgavathi et al., 2012; Singaraja et al., 2014; Adithya et al., 2016; Banajarani et al., 2018). Later these geochemical changes can be identified using: hydrogeochemical methods (Srinivasamoorthy et al., 2008; Subramani et al., 2010; Subba Rao et al., 2012), statistical methods (Prasanna et al., 2010; Vasanthavigar et al., 2013; Thivya et al., 2013, 2016; Singaraja et al., 2013), isotopical methods (Prasanna et al., 2010; Chidambaram et al., 2014; Thivya et al., 2016), the water saturation index (Chidambaram et al., 2011a,b,c), hydrogeochemical modeling (Karmegam et al., 2011; Chidambaram et al., 2012), and flow modeling (Gnanasundar and Elango, 2000; Senthilkumar and Elango, 2004; Elango and Senthilkumar, 2006). But statistical methods in particular have become an important technique to identify the hidden geochemical processes (Raghunath et al., 2002; Nosrati and Van Den Eeckhaut, 2012; Venkatramanan et al., 2018; Thivya et al., 2016; Thilagavathi et al., 2017). An attempt was made to evaluate the water quality of the present coastal aquifer around the Kalpakkam region by Karmegam et al. (2010). Later a few samples were studied in an attempt to assess the mixing proportions of the groundwater in this region (Karmegam et al., 2011). The importance of the saturation index of clay minerals for groundwater samples (Chidambaram et al., 2011a,b,c) and the heavy metal concentration along with the pollution and evaluation indices were studied (Chidambaram et al., 2012). But a detailed hydrogeochemical study with seasonal variations in this region using a statistical method was still lacking. Hence this study attempts to evaluate the seasonal and temporal variations of different water types and hydrogeochemical processes which substantiates the statistical method of interpretation.

16.2 STUDY AREA

The proposed study area, Kalpakkam, falls in the Kanchepuram district located in the northern part of Tamilnadu between latitude $12°37'$ to $12°25'$ north and longitude $80°00'$ and $80°12'$ east (Fig. 16.1). It comprises of both hard rock and sedimentary formation. The study area is in the coastal region and receives more rain compared with the interior. In general, the northeast monsoon brings more rainfall during the months of October, November, and December, which is the major donor, with a 64% contribution, to the total annual rainfall.

The Sadras backwater of the Bay of Bengal is situated in the east and landfill sites were noted along the northeast side. Sites of waste disposal were also identified in a few locations in the study area. The study area comprises two major lithologies, i.e., charnockite (hard rock) and alluvium (fluvial or marine). River Palar flows from the northwest with southeastern sides and the banks are covered with quaternary alluvium. The famous Indira Gandhi Atomic Research (IGCAR) station is located in this Kalpakkam region.

The depth of the hard rock aquifer varies from 12.9 to 46.0 m beneath the ground surface. It is shallow on the northern and western sides, but deeper in the central part of the study area. The weathered/fractured charnockite and alluvium are considered as significant aquifer systems. The greater thickness of alluvium is on the southern and eastern sides, and lenses for clays were also encountered in this formation (Karmegam et al., 2010).

16.3 METHODOLOGY

Twenty-nine groundwater samples were collected in each season, namely, February—post monsoon (POM), October—northeast monsoon (NEM), May—summer (SUM), and July—southwest monsoon (SWM), from available bore wells in the study area. Six groundwater samples were collected from alluvium, five groundwater samples from fluvial, and eighteen groundwater samples from charnockite. Sampling and chemical analysis were carried out using standard procedures (APHA, 1995). pH, electrical conductivity (EC), and total dissolved solids were measured in the field using a field analysis kit (Eutech Handheld Instruments). Then the collected samples were sealed and brought to the laboratory and stored at 4°C. The samples were filtered with 0.45-micron filter paper for further elemental analysis.

Ca, Mg, HCO_3, and Cl were analyzed using the titrimetric method. Na and K were obtained through flame photometry (ELICO CL 378) and the range for flame photometry is 1–100 ppm for a precision of about ±1 digit. H_4SiO_4, SO_4 and PO_4 were measured by UV spectrophotometry (ELICO SL171 minispec) and the range of spectrophotometry was 340–1000 nm with a precision of ±2.5 nm. Fluoride (F^-) concentration was measured using the Orion fluoride ion electrode model (94–09, 96–09). The error percentage of the analytical results were resolved by the ionic balance of the groundwater samples and was found to be 5%–10%. Then the analytical results were processed by the Statistical Package for the Social Sciences (SPSS) 16.0. The principal component analysis was performed using Kaiser's Varimax Rotation.

16.4 RESULTS AND DISCUSSION

The groundwater in Kalpakam region is generally alkaline with pH ranging from 6.10 to 8.10. The pH is lower in SWM, but higher in SUM, NEM, and POM seasons (Table 16.1). A few abnormal values were noted in certain locations. EC values ranged from 200.90 to 6780 µs/cm. The highest EC value was noted in SWM when compared with other seasons. Total dissolved solids (TDS) ranged from 111.3 to 5240 mg/L and higher in SWM followed by SUM, NEM, and POM. The order of dominance of cations and anions in the study area during the four different seasons are given in (Table 16.2).

16.4.1 Correlation Matrix

Correlation analysis was performed utilizing the Pearson's correlation coefficient for all the samples from the four seasons. The Pearson's relationship coefficient values ranged from +1 to −1. A value close to +1 indicated a perfect correlation, ±0.75 to ±1 is a high degree of correlation, ±0.25 to ±0.75 is a moderate degree of correlation, and ±0 to ±0.25 shows a low degree of correlation.

In POM, a positive correlation existed between Cl-Ca, Cl-Mg, Cl-HCO_3, HCO_3-Mg, HCO_3 pH, HCO_3-Ca and Ca-Mg (Table 16.3). A poor correlation existed between K, SO_4, PO_4, NO_3, and H_4SiO_4 with other ions. Cl revealed a useful relationship with Ca, and Mg and HCO_3 demonstrated leaching of secondary salts. Also, a critical relationship was found between HCO_3 and Ca, and Mg and pH, demonstrating chemical weathering (Chidambaram et al., 2008). No solid negative

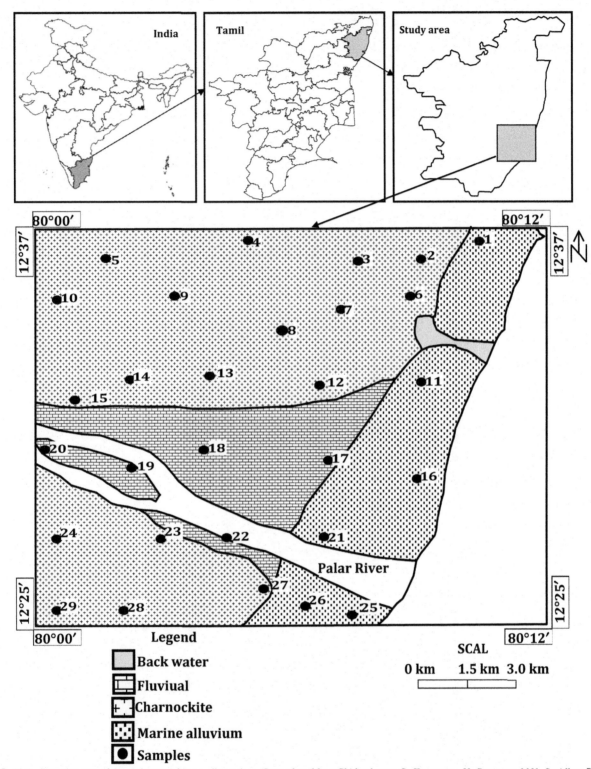

FIG. 16.1 Location map of the study area with sampling points. *(Reproduced from Chidambaram, S., Karmegam, U., Prasanna, M.V., Sasidhar, P. 2012. A study on evaluation of probable sources of heavy metal pollution in groundwater of Kalpakkam region, South India. Environmentalist 32, 371–382.)*

TABLE 16.1 Maximum, Minimum and Average of the Chemical Constituents in Groundwater Representing All Four Sampling Seasons (All Values in mg/L Except EC in µS/cm and pH)

Seasons		Temp (°C)	pH	EC	TDS	Ca^{2+}	Mg^{2+}	Na^+	K^+	Cl^-	HCO_3^-	SO_4^{2-}	H_4SiO_4	PO_4^-	NO_3^-	F^-
POM	Min	27.60	6.85	255.70	129.30	4.00	2.40	32.10	3.20	35.45	109.80	0.20	27.00	0.10	3.36	0.33
	Max	32.60	8.10	3273.00	1643.00	120.00	96.00	212.00	38.90	455.80	489.00	29.80	95.00	3.40	166.00	2.10
	Ave	30.22	7.55	1218.02	610.81	39.69	41.16	123.67	8.08	179.50	308.45	8.02	61.12	0.73	31.19	0.74
NEM	Min	29.00	6.85	200.90	111.30	0.11	1.39	41.67	2.17	42.01	61.00	7.66	14.00	0.00	0.05	0.13
	Max	33.10	8.06	3039.00	1668.00	73.53	129.98	164.03	33.83	521.52	268.40	231.46	102.00	6.25	76.24	3.86
	Ave	31.21	7.31	1169.73	646.00	11.67	43.06	90.86	8.18	169.48	139.28	50.38	51.81	0.73	15.69	0.92
SWM	Min	30.20	6.50	223.00	192.0	12.00	7.20	28.10	0.40	17.73	97.60	8.24	6.00	0.10	0.03	0.37
	Max	34.5	7.4	6780.00	5240.00	116.00	103.20	1150.00	65.10	2127.00	597.80	103.89	99.00	2.68	2.40	0.96
	Ave	31.8	7.56	1146.83	884.76	46.25	24.08	170.72	11.24	268.70	296.61	30.90	53.00	0.59	0.76	0.60
SUM	Min	29.80	6.10	256.50	200.00	12.00	2.40	21.20	0.80	53.18	97.60	2.80	14.00	0.08	0.04	0.33
	Max	38.90	8.00	5190.00	3800.00	176.00	72.00	615.50	122.80	1120.65	597.80	127.70	101.00	1.20	8.88	1.92
	Ave	31.74	6.90	1183.38	846.13	49.63	25.28	128.44	19.76	227.03	267.14	36.36	51.31	0.40	1.40	0.70

TABLE 16.2 The Order of Dominance of Major Ions in Different Seasons

Season	Dominant Anions
POM	$HCO_3^- > Cl^- > NO_3^- > SO_4^{2-} > F^- > PO_4^{3-}$
NEM	$Cl^- > HCO_3^- > SO_4^{2-} > NO_3^- > F^- > PO_4^{3-}$
SWM	$HCO_3^- > Cl^- > SO_4^{2-} > NO_3^- > F^- > PO_4^{3-}$
SUM	$HCO_3^- > Cl^- > SO_4^{2-} > NO_3^- > F^- > PO_4^{3-}$

Season	Dominant Cations
PRM	$Na^+ > Mg^{2+} > Ca^{2+} > K^+$
SWM	$Na^+ > Mg^{2+} > Ca^{2+} > K^+$
NEM	$Na^+ > Ca^{2+} > Mg^{2+} > K^+$
POM	$Na^+ > Ca^{2+} > Mg^{2+} > K^+$

TABLE 16.3 Correlation Analysis of Groundwater Samples Collected During POM

	Ca^{2+}	Mg^{2+}	Na^+	K^+	Cl^-	HCO_3^-	SO_4^{2-}	H_4SiO_4	PO_4^{3-}	NO_3^-	F^-	pH
Ca^{2+}	1.00											
Mg^{2+}	0.88	1.00										
Na^+	0.06	0.21	1.00									
K^+	−0.07	0.06	0.15	1.00								
Cl^-	0.73	0.76	0.46	−0.09	1.00							
HCO_3^-	0.59	0.75	0.46	0.33	0.51	1.00						
SO_4^{2-}	0.41	0.42	−0.01	0.04	0.25	0.21	1.00					
H_4SiO_4	0.05	0.08	0.01	−0.25	−0.10	0.01	0.11	1.00				
PO_4^{3-}	−0.04	−0.05	0.19	−0.09	−0.05	0.09	−0.05	0.31	1.00			
NO_3^-	0.02	0.18	0.12	0.43	0.16	0.20	0.02	−0.25	−0.25	1.00		
F^-	0.366	0.328	−0.353	−0.165	0.117	0.234	−0.031	−0.002	0.060	−0.121	1.000	
pH	0.18	0.39	0.29	0.27	0.18	0.64	0.04	−0.02	0.01	0.16	0.288	1.00

correlations were established, although a poor correlation of SO_4, PO_4, NO_3, and H_4SiO_4 with other ions indicated the impact of agricultural practices. Cl and HCO_3 demonstrated a critical correlation with Ca and Mg, which leads to the leaching of secondary salts during chemical weathering (Prasanna et al., 2008).

During NEM, a positive correlation was observed between Cl-Na, Cl-Mg, Cl-F, HCO_3-Mg, HCO_3-K, H_4SiO_4-K, Na-Cl, Na-K, Na-F, and Ca-K (Table 16.4), demonstrating the predominance of the weathering and leaching processes. Cl reveals a good correlation with Na and Mg demonstrated the leaching of secondary salts. A good correlation of HCO_3 with Mg and K indicated chemical weathering (Srinivasamoorthy et al., 2009). A poor positive correlation of SO_4, PO_4, and NO_3 with other ions might be due to the impact of dilution.

In general, the weathering and leaching of secondary salts with minor anthropogenic inputs are the predominant processes throughout this season. A solid negative correlation is noted between K and H_4SiO_4, demonstrating the decrease of hydrogen and silicate in the aquifer along with the decrease of K (Manivannan, 2010).

In SUM, a good correlation existed between Cl-Ca, Cl-Mg, Cl-Na, Cl-HCO_3, and Cl-NO_3, HCO_3-Mg, HCO_3-NO_3, Ca-Mg, Mg-Na, Mg-NO_3, Mg-K, K-NO_3, and Na-NO_3 (Table 16.5). A poor correlation existed between SO_4, PO_4, and H_4SiO_4 with all other ions. Cl showed good correlation with Ca, Mg, Na, and HCO_3, which suggests the leaching of secondary salts,

TABLE 16.4 Correlation Analysis of Groundwater Samples Collected During NEM

	Ca^{2+}	Mg^{2+}	Na^+	K^+	Cl^-	HCO_3^-	SO_4^{2-}	H_4SiO_4	PO_4^{3-}	NO_3^-	F^-	pH
Ca^{2+}	1.00											
Mg^{2+}	0.20	1.00										
Na^+	0.38	0.38	1.00									
K^+	0.56	0.43	0.52	1.00								
Cl^-	0.43	0.64	0.75	0.46	1.00							
HCO_3^-	0.41	0.55	0.45	0.65	0.34	1.00						
SO_4^{2-}	−0.04	0.36	0.20	−0.01	0.21	−0.03	1.00					
H_4SiO_4	−0.12	−0.49	−0.34	−0.55	−0.27	−0.33	−0.36	1.00				
PO_4^{3-}	0.23	−0.10	−0.01	0.23	−0.18	0.20	−0.21	−0.22	1.00			
NO_3^-	0.13	0.40	0.10	−0.07	0.22	0.12	0.13	−0.18	0.15	1.00		
F^-	0.34	0.22	0.57	0.34	0.60	0.30	0.02	0.01	−0.03	0.09	1.00	
pH	0.35	0.19	0.20	0.46	0.30	0.23	0.00	−0.22	0.13	0.01	0.42	1.00

TABLE 16.5 Correlation Analysis of Groundwater Samples Collected During SUM

	Ca^{2+}	Mg^{2+}	Na^+	K^+	Cl^-	HCO_3^-	SO_4^{2-}	H_4SiO_4	PO_4^{3-}	NO_3^-	F^-	pH
Ca^{2+}	1.00											
Mg^{2+}	0.65	1.00										
Na^+	0.42	0.90	1.00									
K^+	0.19	0.59	0.71	1.00								
Cl^-	0.50	0.94	0.97	0.64	1.00							
HCO_3^-	0.59	0.64	0.64	0.59	0.55	1.00						
SO_4^{2-}	0.11	0.13	0.08	0.20	0.08	0.03	1.00					
H_4SiO_4	−0.03	−0.26	−0.31	−0.22	−0.33	−0.11	0.12	1.00				
PO_4^{3-}	−0.20	−0.10	−0.02	0.10	−0.07	0.01	−0.05	0.11	1.00			
NO_3^-	0.47	0.52	0.50	0.65	0.50	0.57	0.14	−0.03	0.38	1.00		
F^-	−0.19	0.06	0.12	−0.02	0.02	0.21	−0.27	0.18	0.00	−0.20	1.00	
pH	−0.24	−0.16	−0.11	−0.06	−0.08	−0.20	−0.19	−0.10	−0.09	−0.19	0.12	1.00

and there was a significant correlation of HCO_3 with Ca, Mg, Na, and K, indicating chemical weathering. A major correlation of NO_3 with Cl, Mg, HCO_3, Na, and K indicated the anthropogenic influence on the aquifer system, such as domestic sewage and/or agricultural impact (Vengosh and Keren, 1996).

In SWM, a good correlation existed between Cl-Ca, Cl-Mg, Cl-Na, Cl-HCO$_3$, Cl-K, Cl-F and Cl-NO$_3$, HCO$_3$-Mg, HCO$_3$-NO$_3$, HCO$_3$-Ca, HCO$_3$-Na, HCO$_3$-K, Ca-Mg, Ca-Na, Ca-NO$_3$, Ca-F, Mg-Na, Mg-F, K-NO$_3$, K-SO$_4$, K-Na and Na-NO$_3$, Na-F, NO$_3$-SO$_4$ Cl showed good correlation with Ca, Mg, Na, K, F, and HCO$_3$ (Table 16.6), which indicates leaching of secondary salts. A significant correlation of HCO_3 with Ca, Mg, Na, and K indicates chemical weathering. A good correlation was observed in Ca, Cl, HCO$_3$, Na, K, and SO$_4$, which indicates an anthropogenic influence on the

TABLE 16.6 Correlation Analysis of Groundwater Samples Collected During SWM

	Ca^{2+}	Mg^{2+}	Na^+	K^+	Cl^-	HCO_3^-	SO_4^{2-}	H_4SiO_4	PO_4^{3-}	NO_3^-	F^-	pH
Ca^{2+}	1.00											
Mg^{2+}	0.62	1.00										
Na^+	0.50	0.71	1.00									
K^+	0.21	0.35	0.71	1.00								
Cl^-	0.62	0.79	0.97	0.66	1.00							
HCO_3^-	0.55	0.75	0.78	0.64	0.77	1.00						
SO_4^{2-}	0.43	0.26	0.43	0.62	0.43	0.43	1.00					
H_4SiO_4	0.17	0.22	−0.10	−0.03	0.02	0.10	0.14	1.00				
PO_4^{3-}	0.00	−0.06	0.06	0.48	0.02	0.14	0.48	0.08	1.00			
NO_3^-	0.68	0.47	0.58	0.52	0.60	0.55	0.71	0.06	0.38	1.00		
F^-	0.56	0.60	0.64	0.37	0.68	0.45	0.11	−0.10	−0.04	0.53	1.00	
pH	−0.07	−0.02	0.10	0.24	0.04	0.27	−0.04	−0.08	−0.03	−0.14	−0.16	1.00

aquifer. NO_3 also had a good correlation with other ions, so that a higher concentration of nitrogen indicates the dominance of anthropogenic impacts (Jacks, 1973).

The major ions that exhibited a good-to-moderate correlation with other ions in almost all seasons were HCO_3, Cl^-, Na^+, Ca^{2+}, and Mg^{2+}. This is mainly due to the impact of the weathering and leaching of the secondary salts into the permeable zones of the formations (Chidambaram et al., 2009) or to the impact of the industrial effluents along the river alluvium brought from the Palar River.

16.4.2 Factor Analysis

Factor analysis (FA) was used for the hydrochemical datasets of the four seasons. The sorted FA investigation results included variable loadings, eigen values, and variance toward each component for the four seasons. The factor loadings were sorted as stated by the criteria of Liu et al. (2003), i.e., strong, moderate, and weak, compared to >0.75, 0.75–0.50 and 0.50–0.30, respectively.

16.4.3 Post Monsoon

FA presented four significant factors explaining 60.9% of the total variance of the post monsoon dataset. The association of the ions in Factor I, representing Ca^{2+}, Mg^{2+}, Cl^- and HCO_3^-, indicated the leaching of secondary salts or backwater recharge. Factor II shows (Table 16.7) the influence of pH in the hydrogeochemical environment. It is associated with a positive loading of K^+ and HCO_3, which indicates the weathering of potash feldspar. Factor III indicates the enrichment of Na^+ (Vengosh and Keren, 1996; Wayland et al., 2003) may be due to the ion exchange process. The negative loadings of K^+, Cl^-, and NO_3 indicate a low anthropogenic effect due to agriculture impact. Factor IV indicates that the enrichment of NO_3 represent an anthropogenic influence over the study area.

16.4.4 Northeast Monsoon

FA rendered four significant factors explaining 71% of the total variance of the northeast monsoon dataset. The ions in Factor I indicates the influence of high TDS water with a dominance of Ca^{2+}, Mg^{2+}, Na^+, K^+, Cl^-, HCO_3^-, and F^-, which indicates the leaching of secondary salts due to backwater recharge (Table 16.8). Factor II shows the influence of PO_4^{3-} due to the anthropogenic inputs from domestic sewage or agricultural activities (Vengosh and Keren, 1996). In addition, it should be noted that PO_4 has a negative loading with Mg^{2+}, Na^+, Cl^-, SO_4^{2-}, and NO_3^- (Table 16.8). The third factor associated with H_4SiO_4 and F indicates the weathering of silicate and fluoride minerals with less anthropogenic effect

TABLE 16.7 Factor Analysis for the POM Samples (Varimax Rotated)

	1	2	3	4
Ca^{2+}	0.909	0.094	−0.099	−0.010
Mg^{2+}	0.884	0.294	0.022	0.072
Na^+	0.096	0.311	0.936	−0.102
K^+	−0.124	0.546	0.025	0.274
Cl^-	0.871	−0.083	0.450	0.179
HCO_3^-	0.597	0.780	0.162	−0.092
SO_4^{2-}	0.345	0.034	−0.088	−0.001
H_4SiO_4	−0.013	−0.006	−0.049	−0.383
PO_4^{3-}	−0.052	0.088	0.146	−0.344
NO_3^-	−0.020	0.361	0.101	0.919
F^-	0.389	0.080	−0.444	−0.097
pH	0.229	0.613	0.092	−0.064

TABLE 16.8 Factor Analysis for NEM Samples (Varimax Rotated)

	1	2	3	4
Ca^{2+}	0.610	0.403	0.071	0.166
Mg^{2+}	0.709	−0.454	−0.214	0.085
Na^+	0.767	−0.096	0.275	−0.040
K^+	0.794	0.330	−0.171	−0.321
Cl^-	0.802	−0.258	0.367	0.097
HCO_3^-	0.699	0.205	−0.215	−0.018
SO_4^{2-}	0.224	−0.715	−0.139	−0.238
H_4SiO_4	−0.555	0.209	0.602	0.309
PO_4^{3-}	0.123	0.570	−0.565	0.251
NO_3^-	0.262	−0.310	−0.295	0.813
F^-	0.607	0.117	0.576	0.171
pH	0.504	0.332	0.101	−0.130

due to the negative loading of SO_4, PO_4, and NO_3. Factor IV enriched with NO_3 represent the anthropogenic sources, but Factor IV still shows the negative behavior of Ca^{2+}, K^+, SO_4^-, and H_4SiO_4.

16.4.5 Summer

The summer season showed four significant factors explaining 74.5% of the total variance of the dataset. Factor I is represented by Ca^{2+}, Mg^{2+}, Na^+, K^+, Cl^-, HCO_3^-, and NO_3 with the total data variability (TDV) of 39.7% (Table 16.9). This factor indicates the leaching of secondary salts due to backwater recharge. The association of nitrate in this factor indicates the mixing of the domestic waste in the backwaters during this season. Factor II indicates the enrichment of F^- with

TABLE 16.9 Factor Analysis for SUM Samples (Varimax Rotated)

	1	2	3	4
Ca^{2+}	0.634	−0.259	−0.316	0.362
Mg^{2+}	0.943	−0.045	−0.129	0.000
Na^+	0.937	0.049	0.023	−0.144
K^+	0.741	−0.110	0.359	−0.142
Cl^-	0.928	−0.036	-0.053	−0.175
HCO_3^-	0.796	0.139	0.056	0.268
SO_4^{2-}	0.081	−0.627	−0.024	0.257
H_4SiO_4	−0.291	0.173	0.113	0.758
PO_4^{3-}	−0.051	0.036	0.905	0.080
NO_3^-	0.646	−0.261	0.542	0.171
F^-	0.095	0.890	−0.041	0.174
pH	−0.164	0.307	−0.028	−0.585

TABLE 16.10 Factor Analysis for SWM Samples (Varimax Rotated)

	1	2	3	4
Ca^{2+}	0.677	0.158	−0.288	0.352
Mg^{2+}	0.848	−0.028	0.044	0.308
Na^+	0.899	0.248	0.201	−0.197
K^+	0.482	0.705	0.407	−0.246
Cl^-	0.939	0.196	0.105	−0.047
HCO_3^-	0.771	0.263	0.378	0.225
SO_4^{2-}	0.274	0.793	−0.039	0.227
H_4SiO_4	0.013	0.053	−0.014	0.457
PO_4^{3-}	−0.092	0.673	0.021	0.015
NO_3^-	0.574	0.637	−0.343	0.137
F^-	0.772	0.006	−0.303	−0.231
pH	0.018	−0.001	0.513	−0.020

negative loadings of Ca^{2+}, Mg^{2+}, K^+, Cl^-, SO_4^{2-}, and NO_3^- may be due to the dissolution and enrichment of F bearing minerals, which can be seen in the negative association of Ca and F (Chidambaram, 2000). Factor III is represented by PO_4 and NO_3^- indicating the anthropogenic impacts from the leaching of landfills (Ghabayen et al., 2006). Factor IV indicates the predominance of H_4SiO_4, representing silicate dissolution.

16.4.6 Southwest Monsoon

In SWM, four factors were extracted with 70.92% of (TDV). Factor 1 is represented by Ca^{2+}, Mg^{2+}, Na^+, Cl^-, HCO_3^-, F^-, and NO_3^- (Table 16.10) with a TDV of 39.2%. This factor exhibits the dominant of the dissolution of secondary salts during

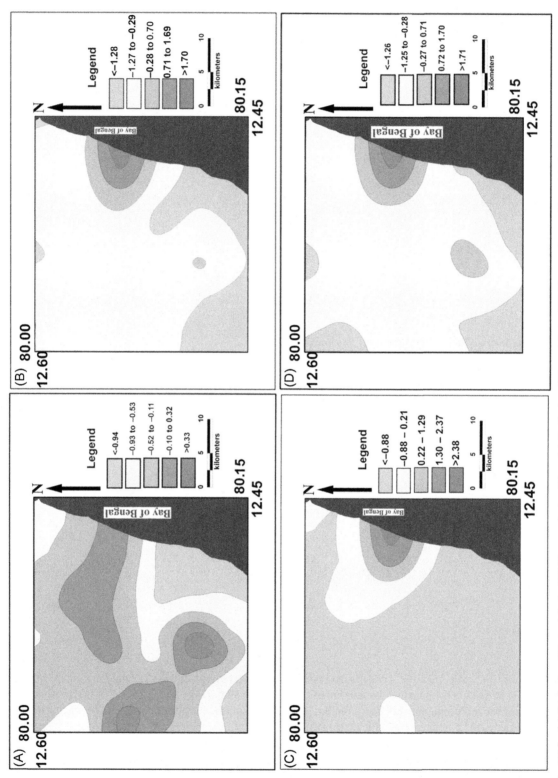

FIG. 16.2 (A) Spatial distribution of Factor 1 during POM in groundwater; (B) spatial distribution of Factor 1 during NEM in groundwater; (C) spatial distribution of Factor 1 during SUM in ground-water; and (D) spatial distribution of Factor 1 during SWM in groundwater.

the monsoon into the aquifers due to backwater recharge. NO_3^- contamination of groundwater suggests that the mixing of domestic sewage in the coastal aquifers. Factor II is represented by K^+, SO_4^{2-}, PO_4^3, and NO_3, which indicates the anthropogenic impacts from the agricultural practices like fertilizers (Vengosh et al., 1996), Factor III is represented by pH and indicates the dominance of base ion exchange and factor IV is H_4SiO_4, which indicates the predominance of silicate dissolution (Chidambaram et al., 2008).

16.4.7 Factor Scores

The factor scores are projections of information by comparing eigen vectors. High positive factor scores ($>+1$) reflect regions that are the most influenced and negative (<-1) reflect those unaffected by the geochemical process by that factor. Close to zero scores reflect regions that are influenced to a normal level of the process or factor. The factor score was additionally evaluated to discover the spatial variety of the factor and to recognize the zone of its representation. They were usually achieved using two methodologies, like weighted least square technique and the regression method.

The positive zones indicated the hydrogeochemical regime of that factor. The hydrological system is too complex and it is not possible to list out all the factors responsible for the hydrogeochemistry of this region. Hence the first factor for all the four seasons was plotted to obtain the active zone representing the hydrogeochemical regime. The first factor consisted of Ca^{2+}, Mg^{2+}, Cl^-, HCO_3^-, Na^+, and K^+. This factor exhibits the dominance of dissolution of secondary salts during the monsoon into the aquifers due to the influence of Sadras backwater. The highest values were observed in the northeastern part of study area during SUM, SWM, and NEM. In POM, the western region with hard rock showed the higher values (Fig 16.2A–D).

16.5 CONCLUSION

The statistical analysis of the hydrochemical data demonstrated that higher scores were noted in the northeastern portion of the study region, indicating the leaching of secondary salts due to the Sadras backwater. Saline water incursion into the coastal aquifers and local and rural anthropogenic activities were found to be the major controlling elements in the groundwater geochemistry. More areas with positive spatial scores were delineated in the northeastern, western, and southern parts of the study area. The western and the southern parts of the region fall along the flood plain alluvium. Overall, leaching of secondary salts, weathering, saline water intrusion into the coastal aquifers, and domestic, industrial, and agricultural anthropogenic activities were found to be the major controlling factors in the study area. Based on these findings, a proper management system must be determined before the geochemistry of groundwater reaches hazardous levels. Hence the groundwater of this region needs to be managed by considering the geochemical nature and its seasonal fluctuations in order to save this fragile coastal environment.

REFERENCES

Adithya, V.S., Chidambaram, S., Thivya, C., Thilagavathi, R., 2016. A study on the impact of weathering in groundwater chemistry of a hard rock aquifer. Arab. J. Geosci. 2, 158.

APHA, (Ed.), 1995. Standard Methods for the Examination of Water and Waste Water. 19th ed. APHA, USASS, Washington, DC.

Banaja Rani, Panda, Chidambaram, S., Ganesh, N., Adithya, V.S., Prasanna, M.V., Pradeep, K., Vasudevan, U., 2018. A hydrochemical approach to estimate mountain front recharge in an aquifer system in Tamil Nadu, India. Acta Geochim. 37 (3), 465–488.

Chidambaram, S., 2000. Hydrogeochemical Studies of Groundwater in Periyar District, Tamil Nadu, India. (Unpublished PhD thesis). Department of Geology, Annamalai University.

Chidambaram, S., Ramanathan, A.L., Prasanna, M.V., Loganatan, D., Badri narayanan, T.S., Srinivasamoorthy, K., Anandhan, P., et al., 2008. Study on the impact of tsunami on shallow groundwater from Portnova to Pumpuhar, using geoelectrical technique—south east coast of India. Indian J. Mar. Sci. 37 (2), 121–131.

Chidambaram, S., Prasanna, M.V., Ramanathan, A.L., Vasu, K., Hameed, S., Warrier, U.K., Manivannan, R., Srinivasamoorthy, K., Ramesh, R., 2009. Stable isotopic signatures in precipitation of 2006 southwest monsoon of Tamilnadu. Curr. Sci. 96 (9), 10.

Chidambaram, S., Karmegam, U., Prasanna, M.V., Sasidhar, P., Vasanthavigar, M., 2011a. A study on hydrochemical elucidation of coastal groundwater in and around Kalpakkam region, Southern India. Environ. Earth Sci. 64 (5), 1419–1431.

Chidambaram, S., Karmegam, U., Sasidhar, P., Prasanna, M.V., Manivannan, R., Arunachalam, S., Manikandan, S., Anandhan, P., 2011b. Significance of saturation index of certain clay minerals in shallow coastal groundwater, in and around Kalpakkam, Tamil Nadu, India. J. Earth Syst. Sci. 120 (5), 897–909.

Chidambaram, S., Ramanathan, A.L., Prasanna, M.V., Manivannan, R., 2011c. Application of hydro-geochemical and geo-electrical techniques to identify the impact of tsunami in the coastal groundwater. In: The Tsunami Threat-Research and Technology. https://doi.org/10.5772/14616.

Chidambaram, S., Anandhan, P., Prasanna, M.V., Ramanathan, A.L., Srinivasamoorthy, K., Senthil Kumar, G., 2012. Hydrogeochemical modelling for groundwater in Neyveli aquifer, Tamil Nadu, India, using PHREEQC: a case study. Nat. Resour. Res. (3), 311–324.

Chidambaram, S., Paramaguru, P., Prasanna, M.V., Karmegam, U., Manikandan, S., 2014. Chemical characteristics of coastal rainwater from Puducherry to Neithavasal, Southeastern coast of India. Environ. Earth Sci. 72 (2), 557–567.

Chidambaram, S., Prasad, M.B.K., Prasanna, M.V., 2015. Evaluation of metal pollution in groundwater in the industrialized environs in and around Dindigul, Tamil Nadu, India. Water Qual. Expo. Health 7 (3), 307–317.

Coetsiers, M., Walravens, K., 2006. Chemical characterization of the Noegene Aquifer, Belgium. Hydrogéologie 14, 1556–1568.

Elango, L., Senthilkumar, M., 2006. Modelling the effect of subsurface barrier on groundwater flow regime. In: MODFLOW and More 2006: Managing Ground-Water Systems – Conference Proceedings. Poeter, Hill, & Zheng. www.mines.edu/igwmc/.

Ghabayen, M.S., Mac, M.K., Kemblowski, M., 2006. Ionic and isotopic ratios for identification of salinity sources and missing data in the Gaza aquifer. J. Hydrol. 318, 360–373.

Gnanasundar, D., Elango, L., 2000. Groundwater flow modeling of a coastal aquifer near Chennai City, India. J. Indian Water Res. Soc. 20, 162–164.

Jacks, G., 1973. Chemistry of ground water in a district in Southern India. J. Hydrol. 18, 185–200.

Karmegam, U., Chidamabram, S., Sasidhar, P., Manivannan, R., Manikandan, S., Anandhan, P., 2010. Geochemical Characterization of Groundwater. Res. J. Environ. Earth Sci. 2 (4), 170–177.

Karmegam, U., Chidambaram, S., Prasanna, M.V., Sasidhar, P., Manikandan, S., Johnsonbabu, G., Dheivanayaki, V., Paramaguru, P., Manivannan, R., Srinivasamoorthy, K., Anandhan, P., 2011. A study on the mixing proportion in groundwater samples by using Piper diagram and Phreeqc model. Chin. J. Geochem. 30 (4), 490–495.

Liu, C.W., Lin, K.H., Kuo, Y.M., 2003. Application of factor analysis in the assessment of groundwater quality in a blackfoot disease area in Taiwan. Sci. Total Environ. 313, 77–89.

Manivannan, R., 2010. Hydrogeochemistry of groundwater in Dindigul district, Tamil Nadu, India. Unpublished Ph.D thesis, Department of Earth Sciences, Annamalai University, India.

Manivannan, R., Chidambaram, S., Anandhan, P., Karmegam, U., Sinagaraja, C., Johnsonbabu, G., Prasanna, M.V., 2011. Study on the significance of temporal ion chemistry in groundwater of Dindigul District, Tamilnadu, India. E-J. Chem. 0973-49458 (2), 938–944.

Nosrati, K., Van Den Eeckhaut, M., 2012. Assessment of groundwater quality using multivariate statistical techniques in Hashtgerd Plain, Iran. Environ. Earth Sci. 65, 331–344.

Prasanna, M.V., Chidambaram, S., Pethaperumal, S., Srinivasamoorthy, K.J., Peter, A., Anandhan, P., Vasanthavigar, M., 2008. Integrated geophysical and chemical study in the lower subbasin of Gadilam River, Tamilnadu, India. Environ. Geosci. 15 (4), 145–152.

Prasanna, M.V., Chidambaram, S., Shahul Hameed, A., Srinivasamoorthy, K., 2010. Study of evaluation of groundwater in Gadilam basin using hydrogeochemical and isotope data. Environ. Monit. Assess. 168, 63–90.

Raghunath, R., Murthy, T.R.S., Raghavan, B.R., 2002. The utility of multivariate statistical techniques in hydrogeochemical studies: an example from Karnataka, India. Water Res. 36, 2437–2442.

Senthilkumar, M., Elango, L., 2004. Three-dimensional mathematical model to simulate groundwater flow in the lower Palar River basin, southern India. Hydrogeol. J. 12 (4), 197–208.

Singaraja, C., Chidambaram, S., Anandhan, P., Prasanna, M.V., Thivya, C., Thilagavathi, R., 2013. A study on the status of fluoride ion in groundwater of coastal hard rock aquifers of south India. Arab. J. Geosci. 6 (11), 4167–4177.

Singaraja, C., Chidambaram, S., Anandhan, P., Prasanna, M.V., Thivya, C., Thilagavathi, R., Sarathidasan, J., 2014. Geochemical evaluation of fluoride contamination of groundwater in the Thoothukudi District of Tamilnadu, India. Appl Water Sci 4, 241–250. https://doi.org/10.1007/s13201-014-0157-y.

Srinivasamoorthy, K., Chidambaram, M., Prasanna, M.V., Vasanthavigar, M., John Peter, A., Anandhan, P., 2008. Identification of major sources controlling groundwater chemistry from a hard rock terrain—a case study from Mettur taluk, Salem district, Tamilnadu, India. J. Earth Syst. Sci. 117 (1), 49–58.

Srinivasamoorthy, K., Chidambaram, S., Sarma, V.S., Vasanthavigar, M., Vijayaraghavan, K., Rajivgandhi, R., Anandhan, P., Manivannan, R., 2009. Hydrogeochemical characterisation of groundwater in Salem District of Tamilnadu, India. Res. J. Environ. Earth Sci. 1 (2), 22–33.

Subba Rao, N., Surya Rao, P., Venktram Reddy, G., Nagamani, M., Vidyasagar, G., Satyanarayana, N.L.V.V., 2012. Chemical characteristics of groundwater and assessment of groundwater quality in Varaha River Basin, Visakhapatnam District, Andhra Pradesh, India. Environ. Monit. Assess. 184, 5189–5214. https://doi.org/10.1007/s10661-011-2333-y.

Subramani, T., Rajmohan, N., Elango, L., 2010. Groundwater geochemistry and identification of hydrogeochemical processes in a hard rock region, Southern India. Environ. Monit. Assess. 162, 123–137.

Thilagavathi, R., Chidambaram, S., Prasanna, M.V. Thivya, Singaraja, C., 2012. A study on groundwater geochemistry and water quality in layered aquifers system of Pondicherry region, southeast India. Appl. Water. Sci. 2, 253. https://doi.org/10.1007/s13201-012-0045-2.

Thilagavathi, R., Chidambaram, S., Panda, B.R., Tirumalesh, K., Devaraj, N., Kurmana, A., 2017. Understanding the decadal variation of the groundwater resources along the coastal aquifers of Pondicherry—a climate change perspective. J. Clim. Chang. 3 (2), 25–42.

Thivya, C., Chidambaram, S., Singaraja, C., Thilagavathi, R., Mohan Viswanathan, P., Anandhan, P., Jainab, I., 2013. A study on the significance of lithology in groundwater quality of Madurai district, Tamil Nadu (India). Environ. Dev. Sustain. 15 (5), 1365–1387.

Thivya, C., Chidambaram, S., Rao, M.S., Gopalakrishnan, M., Thilagavathi, R., Prasanna, M.V., Nepolian, M., 2016. Identification of recharge processes in groundwater in hard rock aquifers of Madurai District using stable isotopes. Environ. Process. (2), 463–477.

Vasanthavigar, M., Srinivasamoorthy, K., Rajiv Ganthi, R., et al., 2012. Characterisation and quality assessment of groundwater with a special emphasis on irrigation utility: Thirumanimuttar sub-basin, Tamil Nadu, India. Arab. J. Geosci. 5 (2), 245–258.

Vasanthavigar, M., Srinivasamoorthy, K., Prasanna, M.V., 2013. Identification of groundwater contamination zones and its sources by using multivariate statistical approach in Thirumanimuthar sub-basin. Tamil Nadu, India. Environ. Earth Sci. 68, 1783–1795.

Vengosh, A., Keren, R., 1996. Chemical modifications of groundwater contaminated by recharge of treated sewage effluent. Contam. Hydrol. 23, 347–360.

Vengosh, A., Artzi, Y., Zirlin, I., Pankratov, I., Harpaz, H., Rosenthal, E., Ayalon, A., 1996. Monitoring water quality in the Coastal Plain aquifer of Israel (Givat Brener, Gedera, Yavne) (in Hebrew), Hydrol. Rep. 4/96, 34 pp. Hydrol. Serv., Jerusalem, Israel.

Venkatramanan, S., Chung, S.Y., Kim, T.H., Prasanna, M.V., Hamm, S.Y., 2018. Assessment and distribution of metals contamination in groundwater: a case study of Busan City, Korea. Water Qual. Expo. Health 7, 219–225. https://doi.org/10.1007/s12403-014-0142-6.

Wayland, K., Long, D., Hyndman, D., Pijanowski, B., Woodhams, S., Haack, K., 2003. Identifying relationships between base flow geochemistry and land use with synoptic sampling and R-mode factor analysis. J. Environ. Qual. 32, 180–190.

FURTHER READING

Elango, L., Sivakumar, C., 2008. Regional simulation of a groundwater flow in coastal aquifer, Tamil Nadu, India. In: Ahmed, S., Jayakumar, R., Salih, A. (Eds.), Groundwater Dynamics in Hard Rock Aquifers. Springer, Dordrecht.

Panda, B., Chidambaram, S., Thilagavathi, R., Thivya, C., 2017. Geochemical signatures of groundwater along mountain front and riparian zone of Courtallam, Tamil Nadu. Groundw. Sustain. Dev. 1–17.

Sivakumar, C., Elango, L., 2006. Groundwater pollution due to tsunami in Kalpakkam, Tamil Nadu. In: International Conference in Environmental Geosciences, Chennai.

Chapter 17

Hydrogeochemistry of Groundwater From Tamil Nadu and Pondicherry Coastal Aquifers, South India: Implication for Chemical Characteristics and Sea Water Intrusion

S. Gopinath*, K. Srinivasamoorthy*, R. Prakash*, K. Saravanan* and D. Karunanidhi[†]

*Department of Earth Sciences, School of Physical, Chemical and Applied Sciences, Pondicherry University, Pondicherry, India [†]Department of Civil Engineering, Sri Shakthi Institute of Engineering and Technology, Coimbatore, India

Chapter Outline

17.1 INTRODUCTION

The measurement of groundwater quality in coastal regions is complex due to the various input sources including precipitation, seawater, and domestic, industrial, and agricultural influences. Among the abovementioned sources of influence seawater intrusion is the most general and widespread problem to make the groundwater resources unfit for domestic utilities. To understand seawater intrusion and other related issues it is vital to reveal the basis of salinity aided by hydrogeochemical detailing (Mahesha and Nagaraja, 1996). Hydrochemistry comprises the chemistry and quality of water in a complex environment. The major and minor ions in groundwater will aid in determining the alteration in water quality both spatially and temporally (Wang and Jiao, 2012). The changes in freshwater composition through mixing and induced chemical reactions like the ion exchange process are generally aided by hydrochemical activity (Elango et al., 2003). Seawater intrusion into coastal aquifers can be identified using hydrogeochemical signatures like electrical conductivity (EC), ionic ratios, isotopic techniques, and resistivity surveys (Gnanasundar and Elango, 1999; Nair et al., 2013). Groundwater quality is influenced not only by seawater intrusion, but also by from natural and anthropogenic factors like evaporation, dissolution, leaching of fertilizer and pesticides from irrigated water, domestic waste water, and salt pan activities.

The hydrochemical facies evolution diagram is more useful in demarcating saline water and fresh groundwater resources (Forcada, 2010). The seawater intrusion into the aquifers can be aided by the ion-exchange process, which is activated by an alteration in the groundwater chemistry (Saxena et al., 2003; Singh et al., 2009; Gopinath et al., 2015).

GIS and Geostatistical Techniques for Groundwater Science. https://doi.org/10.1016/B978-0-12-815413-7.00017-1

The demarcation of saline groundwater is also attempted by using the "seawater mixing index" (SMI), which is calculated using ions like Na, Mg, Cl, and SO_4 (Parka et al., 2005; Omonona et al., 2014). Analyses of saline intrusion and water-quality issues have been attempted along global coastal stretches (Batayneh et al., 2012; Fengshan et al., 2014; Mogren et al., 2011; Papazotos and Koumantakis Vasileiou, 2016; Gopinath et al., 2016, 2018) The primary intention of the current study was to differentiate the causes of groundwater salinization along the coastal aquifers of the Tamilnadu and Pondicherry regions.

17.2 STUDY AREA

The demarcated study area is the coastal Nagapattinam region of Tamilnadu and Karaikkal, the UT of Pondicherry, India. The area falls within the latitudes $10°85''$ and $11°40''$ north and longitudes $79°01''$ and $80°01''$ east with a total extent of $1000\,km^2$. The temperatures in the study region range between $19.3°C$ and $40.6°C$. Precipitation to the study area is mainly from southwest and northeast monsoons during the months June to September and October to December. The average annual precipitation in the study region is $1230\,mm$ (CGWB, 2008). The maximum ground elevation observed was $21\,m$ above mean sea level (AMSL) in the western regions, then descending along the coast. The major occupation of people is agriculture, including aquaculture and fish catching. The main industries isolated in the study area are petrochemical, fertilizer, and salt production.

17.2.1 Geology

The study area encompasses lithostratigraphic units confined to sedimentary formations from the Miocene to Quaternary periods. The Quaternary formation covers the major portions of the study area encompassing alluvium, sandstones, clay marls, silt sand, and clay. Alluvial deposits are widespread along western portions of the study region, fluviomarine sediments along the central parts and marine sands are found along the eastern portions of the study regime (Fig. 17.1).

FIG. 17.1 Location, geology and groundwater sampling location points of the study area.

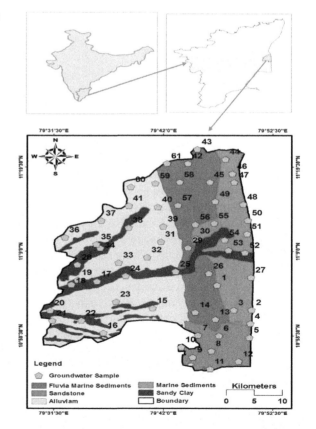

17.2.2 Drainage

The river Cauvery crisscrosses the study area with numerous main streams and minor irrigation channels. The main rivers are the Varasalare, Uppanar, Thirumullairajan, Arasalar, and Vettar, which converge upon the Bay of Bengal within the study domain.

17.3 METHODOLOGY

Groundwater samples were collected from bore and dug wells for two diverse times: pre monsoon (PRM) season during the month of August and post monsoon season (POM), collected during the month of February. The samples were collected using polyethylene bottles that were treated prior to sampling with an acid wash, washed with respective samples, and shifted to the hydrogeology laboratory, Department of Earth Sciences, Pondicherry University. The samples were filtered using 0.45-µm filter paper before analysis for physical and chemical constituents. Physical parameters like total dissolved solids (TDS), EC, and pH were analyzed using a Systronics-371 water-quality analyzer and major ions like bicarbonate, chloride, Ca, and Mg were analyzed by the titrimetric method. Sodium (Na^+) and potassium were analyzed using a Flame photometer (Thermo scientific). Nitrate, phosphate, sulphate, and fluoride were investigated using a UV-Spectrophotometer (Merck). Bromide (Br^-) was determined using a Hanna HI 4010 electro chemical analyzer. The ionic charge balance error for the anions and cations were generally within $\pm 10\%$. The analysis was attempted adopting standard procedures (APHA, 1995).

17.4 RESULT AND DISCUSSION

The pH ranged 6.00 to 8.40 with an average of 7.91 indicating the alkaline nature of groundwater. Elevated pH was observed during POM (8.40) and a lower pH (6.00) was observed during PRM, which was found to fluctuate in certain locations with few abnormalities. EC in the groundwater ranged from 176.00 to $12,870.00\,\mu S\,cm^{-1}$ and 133.00 to $14,500.00\,\mu S\,cm^{-1}$ with averages of 2554.98 and $2331.30\,\mu S\,cm^{-1}$ during PRM and POM, respectively (Table 17.1). Higher EC ($14,500.00\,\mu S\,cm^{-1}$) observed in Tarangambadi, located near to Bay of Bengal, during POM indicating the influence of seawater incursion (Srinivasamoorthy et al., 2011; Gopinath et al., 2018).

17.4.1 Cations and Anions

Bicarbonate in groundwater samples ranged from 54.00 to $657.00\,mg\,L^{-1}$ and 101.00 to $751.00\,mg\,L^{-1}$, with average values of 226.39 and $333.50\,mg\,L^{-1}$ during the PRM and POM seasons, respectively. Higher concentrations ($751.00\,mg\,L^{-1}$) were noted during POM, which might be due to silicate weathering, carbonate dissolution from atmospheric, and soil CO_2 (Gopinath et al., 2015). Chloride in groundwater ranged from 106.35 to $3870.90\,mg\,L^{-1}$ and 70.89 to $2942.30\,mg\,L^{-1}$ with averages of 638.66 and $487.00\,mg\,L^{-1}$ during the PRM and POM seasons, respectively. Higher Cl^- was observed during PRM along northeastern parts of the study area indicating seawater influence due to groundwater overextraction. Other possible sources might also be due to agricultural return flows, clay layers trapped with saline water, base exchanges and a longer migration of groundwater along the flow path (Freez and Cherry, 1979; Bower and Heaton, 1978). SO_4^{2-} in the groundwater ranged from 10.00 to $149.00\,mg\,L^{-1}$ and 11.0 to $234.00\,mg\,L^{-1}$ with averages of 69.24 and $57.18\,mg\,L^{-1}$ during the PRM and POM seasons, respectively. Higher values ($234.00\,mg\,L^{-1}$) were recorded during POM, which might be associated with enrichment from seawater intrusion, weathering, and dissolution of gypsum and anhydride (Miller, 1979). Nitrate in groundwater ranged from 12.00 to $134.00\,mg\,L^{-1}$ and 6.00 to $124.00\,mg\,L^{-1}$ with averages of 65.25 and $50.22\,mg\,L^{-1}$ during PRM and POM seasons, respectively. Higher nitrates during PRM indicates the agricultural application of nitrogen fertilizers as a source (Jalali, 2007; Andrade and Stigter, 2009; Gopinath et al., 2018). Silicate in groundwater ranged from 21.00 to $162.00\,mg\,L^{-1}$ and 6.00 to $124.00\,mg\,L^{-1}$ with averages of 63.89 and $40.83\,mg\,L^{-1}$ during the PRM and POM seasons, respectively. Higher H_4SiO_4 was noted during PRM, which signifies incongruent silicate dissolution (Guglielmi et al., 2000) from the lithostratigraphic units of the study area. Phosphate in groundwater ranged from 0.12 to 15.0, 0.58 to $47.00\,mg\,L^{-1}$ with averages of 1.90 and $6.10\,mg\,L^{-1}$ during the PRM and POM seasons, respectively, which were found to be fluctuating irrespective of space and time and without a definite trend. Higher concentrations were observed during NEM, which indicates leaching from the phosphate-rich fertilizers that are applied as part of agricultural practices (Krishna Kumar et al., 2009). Br^- in groundwater ranged from 0.14 to $8.14\,mg\,L^{-1}$ and 0.11 to $9.47\,mg\,L^{-1}$ with averages of 1.43 and $1.44\,mg\,L^{-1}$ during the PRM and POM seasons, respectively. Higher Br^- levels were observed during POM, indicating the influence of seawater intrusion (Davis et al., 2004) along with pesticide application (Taylor, 1994).

TABLE 17.1 Statistical Parameters of Chemical Constituents (All Values in mgL^{-1} Except EC in $\mu S\,cm^{-1}$ and pH in Standard Unit) for Groundwater During PRM and POM Seasons

Parameters	PRE				POM			
	Max	Min	Avg	St Dev	Max	Min	Avg	St Dev
Ca	173.00	35.00	90.02	32.76	324.00	35.00	117.49	67.69
Mg	152.00	1.20	53.78	26.79	336.00	12.00	69.35	53.47
Na	1235.00	37.00	261.24	261.89	1068.39	58.81	236.02	190.03
K	100.00	6.00	30.72	23.56	142.00	8.00	56.85	33.06
HCO$_3$	657.00	54.00	226.39	148.74	751.00	101.00	333.50	155.22
PO$_4$	15.00	0.12	1.90	2.50	47.00	0.58	6.10	9.13
H$_4$SiO$_4$	162.00	21.00	63.89	33.96	124.00	6.00	40.83	27.90
Cl	3870.90	106.35	638.66	781.81	2942.30	70.89	487.00	437.37
SO$_4$	149.00	10.00	69.24	28.42	234.00	11.00	57.18	45.24
NO$_3$	134.00	12.00	65.25	29.46	124.00	6.00	50.22	29.56
Br	8.14	0.14	1.4	1.27	9.47	0.02	1.54	1.45
pH	7.69	6.08	7.01	0.34	7.83	6.10	7.08	0.36
EC	12870.00	176.00	2554.98	3069.85	14500.00	133.00	2331.30	3620.09
TDS	11590.00	107.00	1560.08	2176.77	13800.00	266.00	3730.80	3433.64

17.4.2 Hydrochemical Facies Evolution Diagram

The plot proposed by Piper (1953) has certain limits in demarcating the evolutionary sequences of water from fresh to saline (Ghiglieri et al., 2012). The hydrochemical facies evolution diagram is more effective in demarcating fresh and seawater intrusion facies evolution (Forcada, 2010). During seawater intrusion, the aquifer is prejudiced by two major processes namely the rise in salinity (line I) to inverse exchange reactions (line II) resulting in the Ca-Cl facies and the groundwater chemistry due to movement towards seawater line (line III), which signifies the Na-Cl facies. During the freshening process, recently recharged water mixes with the existing water indicating Na-HCO$_3$ (lines I and II) and evolves along line III' towards the recently recharged water (Ahmad et al., 2003). About 58% of the groundwater samples during PRM (Fig. 17.2) and 45% during POM suggested seawater intrusion and the rest of the samples suggested a freshening process, signifying the nonsaline nature of the water during PRM due to infiltered precipitated water during the POM season (Fig. 17.2). Groundwater samples of a freshwater nature were noted along the northwestern parts of the study area and saline category samples were abundant along the eastern parts of the study area near to the coast. A higher percentage of samples suggesting a freshening process was noted during POM and salinization was noted to be higher during PRM indicating the influence of the monsoons.

17.5 IONIC RATIO PLOTS

17.5.1 Sodium Versus Chloride

The Na$^+$-Cl$^-$ ratios have been effectively utilized to isolate the mechanisms controlling water salinity in coastal regions globally. The Na$^+$-Cl$^-$ ratio remained unaltered (Jankowski and Acworth, 1997) with horizontal lines indicating the effective contribution of evaporation and evapotranspiration. If dissolution of halite is the chief process responsible for Na$^+$, then the ratio would be equal to 1, and ratios greater than 1 indicates silicate weathering and/or seawater as the sources of Na (Mercado, 1985; Meyback, 1987). The Na$^+$-Cl$^-$ ratio plot (Fig. 17.3) suggests samples representing the northeastern parts of the study area, irrespective of season, are influenced by seawater intrusion and also possibly the reverse exchange process (Vengosh et al., 1999; Martos et al., 1999; Gopinath et al., 2015, 2018). These samples from the northwestern side of the study region, irrespective of season, seem to be influenced by rock-water interaction (Abdalla, 2016; Omonona et al., 2014).

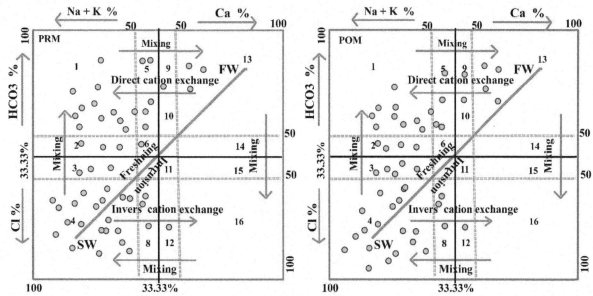

FIG. 17.2 Hydrochemical facies diagrams for groundwater sample from the study area.

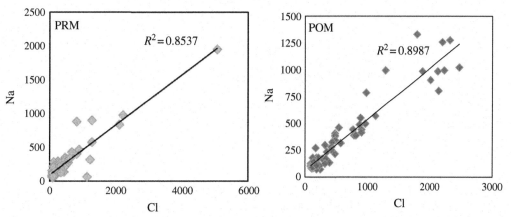

FIG. 17.3 The ratio plot for sodium vs chloride during groundwater samples from the study area.

17.5.2 Chloride vs Chloride/Bromide

The Cl^- vs Cl^-/Br^- ratio plot is considered to be a good indicator to use in the isolation of groundwater samples, as Cl^- and Br^- have chemically conservative natures in the aqueous environment, they do not form insoluble precipitates, they are nonabsorbed onto mineral or organic surfaces, they do not participate in redox reactions, and they are also effective in isolating the origin of salinity in coastal aquifers (Hsissou et al. 1999; Fetter, 1993). The most abundant ion in seawater is the Cl^- ions (19,000 $mg L^{-1}$) along with Br^- (65 $mg L^{-1}$) and when the mixing of seawater with freshwater takes place, the Br^--Cl^- ratio remains unchanged but the concentration of Cl^- and Br^- ions inversely correlates to the distance from the coast (Fig. 17.4). Therefore if salinization of freshwater is mainly due to the mixing of seawater without any anthropogenic influences, then the Cl^--Br^- ratio should be constant (Kim et al., 2003). The Cl vs Cl-Br ratio ranged between 78.13 to 2092 and 143.2 to 2532.14 during PRM and POM seasons. The ratio plot suggests the samples representing northeastern parts of the study area, irrespective of season, are influenced by seawater intrusion and that the southwestern samples taken during PRM and POM seasons might also be affected by anthropogenic sources like sewage discharge and agricultural processes (Flury and Prapitz, 1993; Nair et al., 2012).

17.5.3 Ion Exchange and Seawater Mixing Ratio

To discriminate the sources of salinity and other hydrochemical processes in aquifers during fresh/saltwater mixing the ionic delta (D) values were calculated by considering the theoretical mixing of fresh/saltwater and comparing the calculated

FIG. 17.4 The ratio plot for chloride vs chloride/bromide in during pre monsoon and post monsoon.

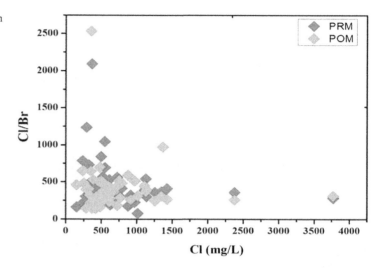

concentrations with actual chloride concentrations of the samples (Pennisi et al., 2006). The concentration of each ion (i) in the theoretical mixing of fresh/saltwater (m_i, mix) is calculated using the equation as suggested by (Fidelibus, 2003):

$$f_{sea} = \frac{m_{Cl^-,sample} - m_{Cl^-,fresh}}{m_{Cl^-,sea} - m_{Cl^-,fresh}} \qquad (17.1)$$

Here m_{cl} − Sea and m_{cl} −, fresh (in $meqL^{-1}$) are the species, (i) concentrations in seawater and freshwater, and f_{sea} is the fraction of seawater in the mixed fresh/seawater. The seawater fraction is calculated on the basis of the Cl^- ion represented in $meqL^{-1}$, due to its conservative behavior (Tellam, 1995; Appelo and Postma, 2005), as given below:

$$m_{i.mix} = {}_{sea} X mi._{sea} + (1 - {}_{sea}) X m_{i.fresh} \qquad (17.2)$$

For every individual ion m_i, negative values represent depletion and positive values indicate enrichment when compared with simple nonreactive mixing, as calculated by:

$$m_{i.react} = m_{i.sample} - m_{i.mix} \qquad (17.3)$$

For the present study, the chemical composition of rainwater collected from the study area has been used for the calculation of the freshwater end member. The percentage of seawater fraction (sw %) and ionic delta (D $_{mi}$) of Na^+, Ca^{2+}, Mg^{2+}, K^+, Cl^-, HCO_3^-, and SO_4^{2-} were calculated for the study area's groundwater samples. Fig. 17.5 illustrates the complex hydrochemical changes that might have been activated in the groundwater of the study area. The results show a small range of seawater contribution. During PRM the groundwater samples contained seawater fractions ranging between 0.003 to 0.252 with an average of 0.024. Higher seawater fractions were identical near to the coast and southern parts of the study area. Lower seawater fractions were observed in the other areas indicating the lower significance of saline water intrusion. During POM the seawater fraction ranged from 0.004 to 0.124 with an average of 0.033 and a higher seawater fraction followed the same trend as that of PRM season. When the percentage of seawater increases, the groundwater becomes enriched with Ca^{2+} and Mg^{2+} and Na^+ ions are depleted. There are two different trends along which the cation evolution takes place. A first group of samples from the coastal tracts of the study area displayed Na^+ react and K^+ deficiency with an enrichment of Ca^{2+} and Mg^{2+} content. These patterns signify the direct contact between Na^+-enriched seawater with Ca^{2+}-enriched freshwater. Regarding anions, samples that were enriched with HCO_3^- showed depletion in SO_4^{2-} indicating that the water samples were influenced by seawater intrusion (Barker et al., 1998; Barbecot et al., 2000; Andersen et al., 2005). The group of samples taken during the PRM season that displayed enrichment of Na and the most negative values of Ca^{2+} and Mg^{2+} indicate the significance of the Na^+-Ca^- exchange/mineral dissolution process (Grassi et al., 2007). These samples were from the northwestern parts of the study area. Na^+ and K^+, Ca^{2+}, and Mg^{2+} were found to be higher in the groundwater samples taken from the central parts of the study area, suggesting nonsaline sources irrespective of seasons (Appelo and Postma, 2005). These samples were recorded to have elevated HCO_3^- and SO_4^{2-} concentrations. Finally, the ratio of saline water intrusion was found to be dominant during PRM compared with other seasons and freshening was prominent during the POM seasons.

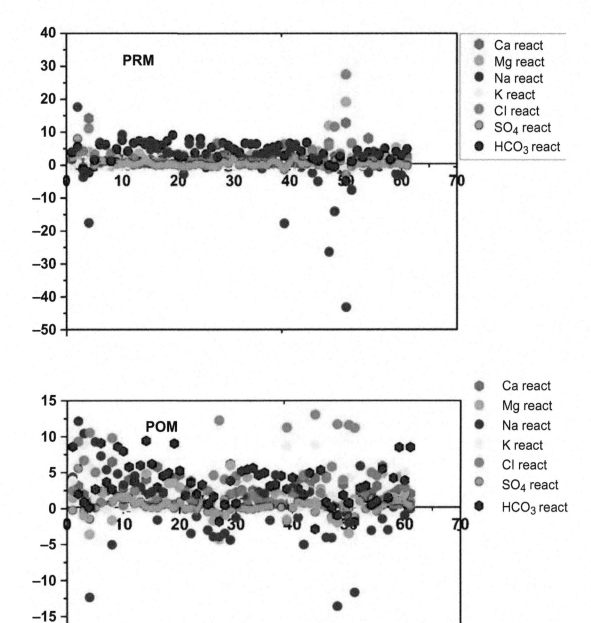

FIG. 17.5 Seawater mixing ratios for groundwater samples.

17.5.4 Seawater Mixing Index

In the present study SMI (Parka et al., 2005) was attempted for the effective isolation of seawater mixing in the groundwater samples . This parameter is mainly based on the contributions of major ions like Na^+, Mg^{2+}, Cl^-, and SO_4^{2-} found in the seawater. The SMI is calculated as:

$$SMI = a \times \frac{C_{Na}}{T_{Na}} + b \times \frac{C_{Mg}}{T_{Mg}} + c \times \frac{C_{Cl}}{T_{Cl}} + d \times \frac{C_{SO_4}}{T_{SO_4}} \tag{17.4}$$

The constants a, b, c, and d denote the relative concentration proportion of Na^+, Mg^{2+}, Cl^-, and SO_4^{2-} assumed as ($a = 0.31$, $b = 0.04$, $c = 0.57$, and $d = 0.08$) in seawater (Parka et al., 2005), respectively; C represents the respective ions measured (mg/L^{-1}) in the groundwater samples from the study area and T represents the calculated regional threshold values of selected ions in the groundwater interpreted from the cumulative probability curves. If the SMI calculated was greater than 1, the water was found to be influenced by seawater mixing and a value less than 1 indicates the effect of freshwater (Mondal et al., 2011). Spatial representation of SMI is represented in Fig. 17.6. During PRM and POM about 38% and 32% of the samples were found to be influenced by seawater intrusion and this was specifically noted along the eastern parts of the study area, indicating the influence of seawater intrusion.

17.6 WATER QUALITY INDEX

Demarcating groundwater quality is essential since it controls the suitability of water for drinking (Yidana et al., 2010; Oinam et al., 2012). The water quality index (WQI) is a tool that is used to predict the combined impact of groundwater quality and its suitability for drinking (Mitra and ASABE Member, 1998). The WQI calculated was equated with WHO (2004) and BIS (1991) standards to assess the water's suitability for drinking purposes. For computed WQI four steps are monitored.

17.6.1 Step I

The chemical parameters (TDS, HCO_3^-, Cl^-, SO_4^{2-}, PO_4^{3-}, NO^-_3, F^-, Ca^{2+}, Mg^{2+}, Na^+, K^+, and Si) were assigned weightages (w_i) based on its influence in determining the quality of water for drinking use (Table 17.2). A maximum weightage of 5 was assigned to nitrate, TDS, Cl^-, Na^+, and SO_4^{2-} due to their key significance in water-quality assessment (Srinivasamoorthy et al., 2008). Parameters like HCO_3^- and PO_4^{3-} were assigned a minimum weightage of 1 due to their less-important role in water-quality valuation. Ions like Ca^{2+}, Mg^{2+}, Na^+, and K^+ were assigned weightages ranging between 1 and 5 depending on their implications on water quality (Vasanthavigar et al., 2010).

17.6.2 Step II

In the subsequent step, the relative weight (W_i) is calculated using the formula:

$$W_i = w_i / \sum_{i=1}^{n} w_i \qquad (17.5)$$

where W_i indicates the relative weight, w_i is the weight of every factor, and n signifies the parameters number. The relative weight (W_i) calculated for each parameter is represented in Table 17.2.

17.6.3 Step III

A quality ranking scale (q_i) for every single parameter is given by isolating its attention in the individual groundwater sample by its relevant standard conferring to the guiding principle as per WHO (2004) and the result is increased by 100:

$$q_i = (C_i - S_i) \times 100 \qquad (17.6)$$

where q_i represents the quality rating, C_i is the absorption of each chemical parameter in each water sample expressed in $mg\,L^{-1}$, and S_i is the Indian drinking water standard for individual chemical parameters expressed in $mg\,L^{-1}$ as per the classifications by (BIS, 1991; WHO, 2004).

17.6.4 Step IV

For calculating the WQI, the SI is calculated for each chemical parameter, which is used to regulate the WQI as per the following equation:

$$SI_i = W_i \times q_i \qquad (17.7)$$

$$WQI = \sum SI_i \qquad (17.8)$$

where SI_i indicates the subindex of Ith parameter, q_i represents ratings based on the absorption of the Ith parameter, and n represents the number of parameters.

The water-quality types, which were calculated from the WQI during PRM and POM seasons are represented in Table 17.3. During PRM, the excellent category was represented by 26% of the groundwater samples, good by 37% of

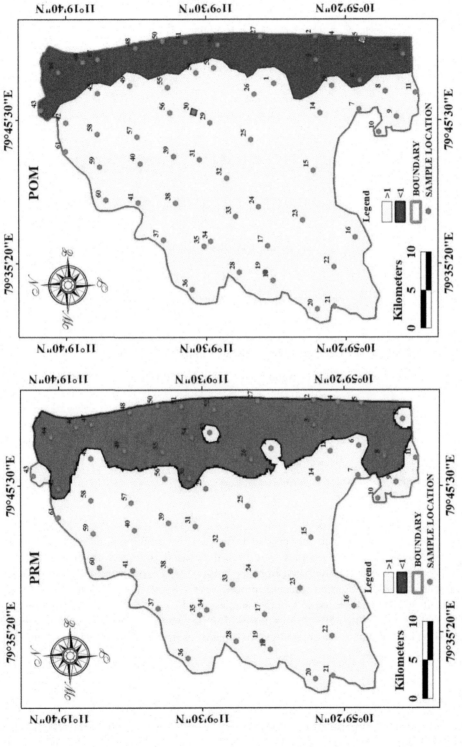

FIG. 17.6 Spatial plots for seawater mixing index.

TABLE 17.2 Weightages and Relative Weightages Assigned for Hydrochemical Parameters

Chemical Parameters (mg L^{-1})	BIS 10500 (1991) Indian Standard	WHO (2004)	Weight (w_i)	Relative Weight $W_i \frac{w_i}{\sum_{i=1}^{n} w_i}$
Total dissolved solids	500	1000	5	0.121
Bicarbonate	–	125	1	0.024
Chloride	250	250	5	0.121
Sulphate	200	200	5	0.121
Phosphate	–	–	1	0.024
Nitrate	45	45	5	0.121
Fluoride	1	1.5	4	0.097
Calcium	75	75	3	0.073
Magnesium	30	30	3	0.073
Sodium	–	200	5	0.121
Potassium	–	20	2	0.048
Silicate	–	–	2	0.048
			$\sum w_i = 41$	$\sum w_i = 0.992$

TABLE 17.3 The Categorization of Water Samples Based on WQI

Range	Quality of Water	PRM (%)	POM (%)
<50	Excellent water	26	45
50–100	Good water	37	20
100–200	Poor water	25	26
200–300	Very poor water	5	4
>300	Water unsuitable for drinking	7	5

the samples, poor by 25% of the samples, and very poor and unsuitable for drinking purposes was represented by 5% and 7% of the samples, respectively, and these samples were from the coastal zones already demarcated as saline intruded zones. Good and excellent water samples originated from the northwestern parts of the study area that had already been isolated as zones influenced by rock-water/ion-exchange processes with domestic/agricultural influences. Of the samples taken during POM, excellent water represented 45% of the samples, good 20% of the samples, poor 26%, very poor 4%, and water unsuitable for drinking represented 5% of the samples. The percentage of samples categorized as excellent water increased during POM and the samples categorized as poor-quality water decreased, indicating the dilution influence of rainfall.

17.7 CONCLUSION

The main aim of the current study was to isolate the sources of groundwater salinization for the Nagapattinam and Karaikal coastal aquifers. A higher pH was observed due to the influence of atmospheric CO_2. Variations in EC are due to saline water, ion exchange, and anthropogenic sources. The facies plot suggests the dominance of saline intrusion during PRM and freshening of water during POM seasons. The Na vs Cl plot suggests Na from silicate weathering and Cl from seawater. The Cl vs Cl/Br plot indicates seawater intrusion and anthropogenic influences. Higher concentrations of Na$^+$, K$^+$, Ca^{2+}, and

Mg^{2+} were found in groundwater samples confined to the central and western parts of the study area, suggesting nonsaline sources irrespective of the season. The saline water intrusion ratio was found to be dominant during PRM rather than the POM season. SMI suggests a higher percentage of samples were influenced by seawater intrusion during PRM in comparison with POM indicating the influence of seawater intrusion. The WQI calculated for both the seasons signified unsuitable to poor water quality was noted in samples taken from the area proximal to the sea and good to excellent water samples were from the northwestern part of the study area indicating the influence of rock-water/ion-exchange processes and domestic/agricultural influences.

ACKNOWLEDGMENTS

The authors thank University Grants Commission (UGC), India for supporting this work by major research project (Grant No. 41-1036/2012 (SR) dated 01/07/2012). The First author acknowledges UGC for granting Project fellow position.

REFERENCES

Abdalla, F., 2016. Ionic ratios as tracers to assess seawater intrusion and to identify salinity sources in Jazan coastal aquifer, Saudi Arabia. Arab. J. Geosci. 9, 40. https://doi.org/10.1007/s12517-015-2065-3.

Ahmad, N.Z., Sen, Z., Ahmad, M., 2003. Ground water quality assessment using multirectangular diagrams. Ground Water 41 (6), 828–832.

Andersen OB, Seneviratne SI, Hinderer J and Viterbo P (2005) GRACE-derived terrestrial water storage depletion associated with the 2003 European heat wave. Geophys. Res. Lett. 32: L18405 https://doi.org/10.1029/2005GL023574.

Andrade, A., Stigter, T.Y., 2009. Multi-method assessment of nitrate and pesticide contamination in shallow alluvial groundwater as a function of hydrogeological setting and land use. Agric. Water Manag. 96 (12), 1751–1765.

APHA, 1995. Standard Methods for the Examination of Water and Wastewater, 19th ed. American Public Health Association, Washington, p. 1467.

Appelo, C.A.J., Postma, D., 2005. Geochemistry, Groundwater and Pollution, second ed. Taylor and Francis, Great Britain.

Barbecot, F., Marlin, C., Gibert, E., Dever, L., 2000. Hydrochemical and isotopic characterisation of the Bathonian and Bajocian coastal aquifer of the Caen area (northern France). Appl. Geochem. 15, 791–805.

Barker, A.P., Newton, R.J., Bottrell, S.H., Tellam, J.H., 1998. Processes affecting groundwater chemistry in a zone of saline intrusion into an urban sandstone aquifer. Appl. Geochem. 13, 735–749.

Batayneh, A., Elawadi, E., Mogren, S., Ibrahim, E., Qaisy, S., 2012. Groundwater quality of the shallow alluvial aquifer of Wadi Jazan (Southwest Saudi Arabia) and its suitability for domestic and irrigation purpose. Sci. Res. Essays 7 (3), 352–364.

BIS, 1991. Indian Standard Drinking Water Specification, first rev. pp. 1–8.

Bower, D.R., Heaton, K.C., 1978. Response of an aquifer near Ottawa to tidal forcing and the Alaskan earthquake of 1964. Can. J. Earth Sci. 15, 331340.

CGWB, 2008. District groundwater brochure Nagapattinam district, Tamil Nadu. Technical Report Seriespp. 1–21.

Davis, S.N., Moysey, S., Cecil, L.D., Zreda, M., 2004. Chlorine-36 in groundwater of the United States. Empirical data. Hydrogeol. J. 11, 217–227.

Elango, L., Kannan, R., Senthil Kumar, M., 2003. Major ion chemistry and identification of hydrogeochemical processes of groundwater in a part of Kancheepuram district, Tamil Nadu. Environ. Geosci. 1 (4), 157–166.

Fengshan, M., Wei, A., Qinghai, D., Haijun, Z., 2014. Hydrochemical characteristics and the suitability of groundwater in the coastal region of Tangshan, China. J. Earth Sci. 25 (6), 1067–1107.

Fetter, C.W., 1993. Contaminant Hydrogeology. Macmillan Publishing Co., New York.

Fidelibus, M.D., 2003. Environmental tracing in coastal aquifers: old problems and new solutions. In: Coastal aquifers intrusion technology: Mediterranean countries. vol. II. IGME, Madrid, pp. 79–111.

Flury, M., Prapitz, A., 1993. Bromide in the natural environment: occurrence and toxicity. J. Environ. Qual. 22, 747–758.

Forcada, E., 2010. Dynamic of sea water interface using hydrochemical facies evolution diagram. Ground Water 48 (2), 212–216.

Freez, R.A., Cherry, J.A., 1979. Groundwater. Prentice Hall, New Jersey, p. 604.

Ghiglieri, G., Carletti, A., Pittalis, D., 2012. Analysis of salinizationprocesses in the coastal carbonate aquifer of Porto Torres (NWSardinia, Italy). J. Hydrol. 432–433, 43–51.

Gnanasundar, D., Elango, L., 1999. Groundwater quality assessment of a coastal aquifer using geoelectrical techniques. J. Environ. Hydrol. 7 (2), 2133.

Gopinath, S., Srinivasamoorthy, K., Saravana, K., Prakash, R., Suma, C.S., Faizal Khan, A., Senthilnatha, D., Sarma, V.S., Devi, P., 2015. Hydrogeochemical characteristics of coastal groundwater in Nagapattinam and Karaikal aquifers implications for saline intrusion and agricultural suitability. J. Coast. Sci. 2 (2), 1–11 ISSN: 2348–6740.

Gopinath, S., Srinivasamoorthy, K., Vasanthavigar, M., Saravanan, K., Prakash, R., Suma, C.S., Senthilnathan, D., 2016. Hydrochemical characteristics and salinity of groundwater in parts of Nagapattinam district of Tamil Nadu and the Union Territory of Puducherry, India. Carbonates Evaporites. https://doi.org/10.1007/s13146-016-0300-y.

Gopinath, S., Srinivasamoorthy, K., Saravanan, K., Prakash, R., 2018. Discriminating groundwater salinization processes in coastal aquifers of southeastern India: geophysical, hydrochemical and numerical modeling approach. Environ. Dev. Sustain. https://doi.org/10.1007/s10668-018-0143-x.

Grassi, S., Cortecci, G., Squarci, P., 2007. Groundwater resource degradation in coastal plains: the example of the Cecina area (Tuscany—Central Italy). Appl. Geochem. 2, 2273–2289.

Guglielmi, Y., Bertrand, C., Compagnon, F., Follacci, J.P., Mudry, J., 2000. Acquisition of water chemistry in mobile fissured basement massif: its role in the hydrogeological knowledge of the La Clapiere landslide (Mercantour massif, southern Alps, France). J. Hydrol. 229, 138–148.

Hsissou, Y., Mudry, J., Mania, J., Bouchaou, L., Chauve, P., 1999. Utilisation du rapport Br/Cl pour determiner l'origine de la salinite´ des eaux souterraines: exemple de la plaine du Souss (Maroc). C.R. Acad. Sci. 328, 381–386.

Jalali, M., 2007. Hydrochemical identification of groundwater resources and their changes under the impacts of human activity in the Chah Basin in western Iran. Environ. Monit. Assess. 33, 347–367.

Jankowski, J., Acworth, R.I., 1997. Impact of debris-flow deposits on hydrogeochemical processes and the development of dryland salinity in the Yass River catchment, New South Wales, Australia. Hydrogeol. J. 5 (4), 71–88.

Kim, Y., Lee, K.S., Koh, D.C., Lee, D.H., Lee, S.G., Park, W.B., Koh, G.W., Woo, N.C., 2003. Hydrogeochemical and isotopic evidence of groundwater salinization in a coastal aquifer—a case study in Jeju volcanic island, Korea. J. Hydrol. 270, 282–294.

Krishna Kumar, S., Ram Mohan, V., Dajkumar Sahayam, J., Jeevanandam, M., 2009. Assessment of groundwater quality and hydrogeochemistry of Manimuktha River basin, Tamil Nadu, India. Environ. Monit. Assess. 159 (1–4), 341–351.

Mahesha, A., Nagaraja, S.H., 1996. Effect of natural recharge on seawater intrusion in coastal aquifers. J. Hydrol. 174, 211–220.

Martos, F.S., Bosch, A.P., Calaforra, J.M., 1999. Hydrogeochemical processes in an arid region of Europe (Almeria, SE Spain). Appl. Geochem. 14 (6), 1735–1745.

Mercado, A., 1985. The use of hydrogeochemical patterns in carbonate sand and sandstone aquifers to identify intrusion and flushing of saline waters. Ground Water 23 (5), 635–664.

Meyback, M., 1987. Global chemical weathering of surficial rocks estimated from river dissolved loads. Am. J. Sci. 287, 401–428.

Miller, G.T., 1979. Living in the Environment. Wadsworth, Belmond, CA, p. 470.

Mitra, B.K., ASABE Member, 1998. Spatial and temporal variation of ground water quality in sand dune area of aomori prefecture in Japan. Paper number 062023, 2006 ASAE Annual Meeting.

Mogren, S., Batayneh, A., Elawadi, E., Al-Bassam, A., Ibrahim, E., Qaisy, S., 2011. Aquifer boundaries explored by geoelectrical measurements in the Red Sea coastal plain of Jazan area, southwest Saudi Arabia. Int. J. Phys. Sci. 15, 3768–3776.

Mondal, N.C., Singh, V.P., Singh, S., Singh, V.S., 2011. Hydrochemical characteristic ofcoastal aquifer from Tuticorin, Tamil Nadu, India. Environ. Monit. Assess. 175, 531–550.

Nair, I.S., Parimala Renganayki, S., Elango, L., 2012. Identification of seawater intrusion by geochemical signatures in North Chennai Coastal aquifer and mitigation measures through Managed aquifer Recharge. In: Proceedings of fifth international groundwater conference (IGWC-2012), Aurangabad.

Nair, I.S., Parimala Renganayaki, S., Elango, L., 2013. Identification of seawater intrusion by Cl/Br ratio and mitigation through managed aquifer recharge in aquifers North of Chennai, India. JGWR. 2321-47832(1).

Oinam, J.D., Ramanthan, A.L., Singh, G., 2012. Geochemical and statistical evaluation of groundwater in Imphal and Thoubal district of Manipur, India. J. Asian Earth Sci. 48, 136–149.

Omonona, V.O., Okogbue, C.O., Isreal, G.O., Ayuba, R., Akpah, F.A., 2014. Hydrochemical characteristics of groundwater of a coastal aquifer in Southsouth Nigeria. Adv. Appl. Sci. Res. 5 (6), 77–90.

Papazotos, P., Koumantakis Vasileiou, E., 2016. lSeawater intrusion and nitrate pollution in coastal aquifer of marathon basin. Bull. Geol. Soc. Greece 50, 927–937.

Parka, S.-C., Yuna, S.-T., Chaea, G.-T., Yooa, I.-S., Shina, K.-S., Heoa, C.-H., Leeb, S.-K., 2005. Regional hydrochemical study on salinization of coastal aquifers, western coastal area of South Korea. J. Hydrol. 313, 182–194.

Piper, A.M., 1953. A graphic procedure in the geochemical interpretation of water analysis, US Geological Survey, Groundwater,1953, Note No.12.

Pennisi, M., Bianchini, G., Muti, A., Kloppmann, W., Gonfiantini, R., 2006. Behaviour of boron and strontium isotopes in groundwater-aquifer interactions in the Cornia Plain(Tuscany, Italy). Appl. Geochem. 21, 1169–1183. https://doi.org/10.1013/j. apgeochem.2006.03.001.

Saxena, V.K., Singh, V.S., Mondal, N.C., Jain, S.C., 2003. Use of chemical parameters to delineation fresh groundwater resources in Potharlanka Island, India. Environ. Geol. 44 (5), 516–521.

Singh VS, Sarwade DV, Mondal NC, Nanadakumar MV, Singh B (2009) Evaluation of groundwater resources in a tiny Andrott Island, Union Territory of Lakshadweep, India, Environ. Monit. Assess. 158 (1–4), 145–154.

Srinivasamoorthy, K., Chidambaram, S., Prasanna, M.V., 2008. Identification of major sources controlling Groundwater Chemistry from a hard rock terrain—a case study from Mettur taluk, Salem district, Tamilnadu, India. J. Earth Syst. Sci. 117 (1), 49–58.

Srinivasamoorthy, K., Vasanthavigar, M., Vijayaraghavan, K., Sarathidasan, R., Gopinath, S., 2011. Hydrochemistry of groundwater in a coastal region of Cuddalore district, Tamilnadu, India: implication for quality assessment. Arab. J. Geosci. https://doi.org/10.1007/s12517011-0351-2.

Taylor, R.W.D., 1994. Methyl bromide—is there any future for this noteworthy fumigant? J. Stored Prod. Res. 30 (4), 253–260.

Tellam, J.H., 1995. Hydrochemistry of the saline ground-waters of the lower Mersey Basin Permo-Triassic sandstone aquifer, UK. J. Hydrol. 165, 45–84. https://doi.org/10.1073/pnas.1101242108.

Vasanthavigar, M., Srinivasamoorthy, K., Vijayaragavan, K., Rajiv Ganthi, R., Chidambaram, S., Anandhan, P., Manivannan, R., Vasudevan, S., 2010. Application of water quality index for groundwater quality assessment: Thirumanimuttar sub-basin, Tamilnadu, India. Environ. Monit. Assess. https://doi.org/10.1007/s10661-009-1302-1.

Vengosh, A., Spivack, A.J., Artzi, Y., Ayalon, A., 1999. Geochemical and Boron, Strontium, and Oxygen isotopic constraints on the origin of the salinity in groundwater from the Mediterranean coast of Israel. Water Resour. Res. 35 (6), 1877–1894.

Wang, Y., Jiao, J.J., 2012. Origin of groundwater salinity and hydrogeochemical processes in the confined Quaternary aquifer of the Pearl River Delta, China. J. Hydrol. 438–439, 112–124. https://doi.org/10.1016/j. J Hydrology.2012.03.008.

WHO, 2004. World Health Organization, Guidelines for Drinking Water Quality Recommendations. vol. 1. WHO, Geneva, p. 515.

Yidana, S.M., Banoeng-Yakubo, B., Akabzaa, T.M., 2010. Analysis of groundwater quality using multivariate and spatial analyses in the Keta basin, Ghana. J. Afr. Earth Sci. 58, 220–234.

FURTHER READING

Gopinath, S., Srinivasamoorthy, K., 2015. Application of geophysical and hydrogeochemical tracers to investigate salinisation sources in Nagapatinam and Karaikal Coastal Aquifers, South India. Aquat. Proc. 4, 65–71.

Hussein, M., Loni, O., 2011. Major ionic composition of Jizan thermal springs, Saudi Arabia. J. Emerg. Trends Eng. Appl. Sci. 2 (1), 190–196.

Lagudu, S., Rao, V.G., Prasad, P.R., Sarma, V., 2013. Use of geophysical and hydrochemical tools to investigate seawater intrusion in coastal alluvial aquifer, Andhra Pradesh, India. In: Wetzelhuetter, C. (Ed.), Groundwater in Coastal Zones of Asia-Pacific. Springer, Dordrecht, New York, pp. 49–65.

Singh, A.K., Mondal, G.C., Kumar, S., Singh, T.B., Tewary, B.K., Sinha, A., 2008. Major ion chemistry, weathering processes and water quality assessment in upper catchments of Damodar River basin, India. Environ. Geol. 54, 745–758.

Chapter 18

Organic (Hydrocarbon) Contamination: Nonaqueous Phase Liquids

Cüneyt Güler

Department of Geological Engineering, Mersin University, Mersin, Turkey

Chapter Outline

18.1 INTRODUCTION

It would not be wrong to say that, today, petroleum (a.k.a., crude oil) has much more importance than any other traded commodity, and is strategically used by major players (e.g., corporations, investors, commodity traders, governments, etc.) to shape the global economy, technology, and politics on a daily basis. As attested by the archeological findings, liquid petroleum and its semisolid form called bitumen were also used in antiquity for many purposes, including illumination, mummification, waterproofing, as medicine, and as a chemical-warfare agent (Speight, 2002). Although, organic chemicals produced from liquid petroleum or other sources, as we know them today, have been around for less than 200 years (i.e., since the Industrial Revolution), they have made an unprecedented impact on modern civilizations. In the modern age hydrocarbon compounds have become an indispensable part of our daily lives and form the basis of a wide variety of industrial products and raw materials, including pharmaceuticals (drugs) and countless petrochemical/chemical products intended to be used for many specific purposes and applications (e.g., fuels, solvents, glues, lubricants, dyes, paints, plastics, rubbers, synthetic fibers, propellants, refrigerants, cleaning and degreasing agents, polishing and coating formulations, wood preservatives, pesticides, adhesives, explosives, etc.). These commercial products have so deeply and extensively penetrated our daily lives that most of us today could not even imagine a world without them.

We all know that each technological advancement comes with a price, and that there is not only the risk of creating environmental problems, but also the consequences of exposing a large number of people to that technology, which may not become obvious even after several decades or generations. As sadly illustrated by the well-publicized "Love Canal" incident (New York, United States) in the late 1970s (Hironaka, 2014; Newman, 2016; US EPA, 2017a), hydrocarbon compounds can have detrimental health effects on not only the individuals directly exposed to them via various routes, but also on generations to come of an entire community. The Love Canal site had been used from 1940 to 1952 as a chemical dumpsite, where more than 200 different types of toxic hydrocarbon compounds (including halogenated hydrocarbons) were buried in barrels, the total amount reaching 21,800 tons (Hironaka, 2014; Newman, 2016). After the closure of the dumpsite, the Love Canal area became a residential neighborhood (starting from 1955) comprised of about 1000 families (Newman, 2016), for several decades no one suspected what lay beneath the surface and its inherent dangers for their health. Eventually the effects became visible (through cancers, birth defects, body anomalies, miscarriages, etc.) and this incident not only mobilized a group of protesters (so-called Nimbys) under the slogan "Not In My Back Yard" (Hironaka, 2014), but also resulted in the creation of a federal government program called "Superfund"

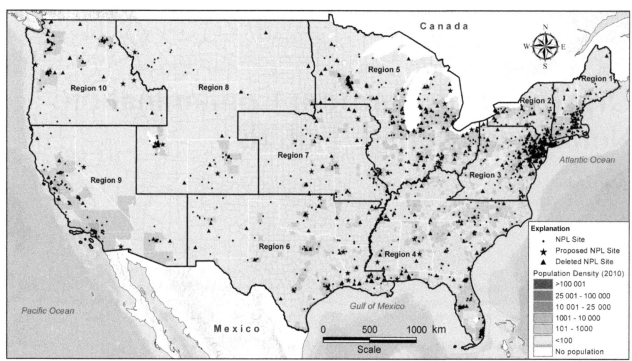

FIG. 18.1 Spatial distribution of hazardous-waste sites registered on the Superfund National Priorities List (NPL) overlaid on the US population-density map (2010 census) where only the part covering conterminous US was presented for visualization purposes. Data related to NPL sites and the US population retrieved from US EPA (2017c) and US Census Bureau (2010), respectively. Polygon boundaries (*in black*) represent the US EPA Regions 1 to 10.

(officially the Comprehensive Environmental Response, Compensation, and Liability Act of 1980 or CERCLA) (CERCLA, 1980; US EPA, 2017b). CERCLA was designed to fund the cleanup of contaminated sites (a.k.a., Superfund sites) placed on the National Priorities List (NPL) by the US Environmental Protection Agency (US EPA, 2017b). Later, CERCLA was amended and reauthorized by the Superfund Amendments and Reauthorization Act (SARA) on 17 October 1986 (SARA, 1986). As of September 2017, 1342 hazardous-chemical sites are on the Superfund NPL, with an additional 49 new sites having been proposed for the list and 394 sites being de-listed after all of the appropriate response actions under CERCLA were completed (US EPA, 2017c). The distribution of Superfund NPL sites found in different US EPA regions (Regions 1 to 10) is depicted in Fig. 18.1. It is clear from the map in Fig. 18.1 that the majority of the NPL sites are located in the eastern states or US EPA Regions 1 to 5, where the population density, and hence the potential for anthropogenic contamination, is much higher than the majority of western states.

In this chapter, organic contaminants, particularly petroleum hydrocarbons and hydrocarbon derivatives, collectively known as nonaqueous phase liquids (NAPLs), will be discussed as regards their sources and production, common causes of contamination, physical/chemical properties, as well as their behavior and fate in the subsurface environment (e.g., in soil and groundwater).

18.2 CLASSIFICATION OF HYDROCARBONS

Hydrocarbon chemistry has a special place in organic (i.e., carbon) chemistry. Hydrocarbons can be described as organic chemical compounds of both natural and manmade (i.e., synthetic) origin that are mainly composed of carbon (C) and hydrogen (H) atoms arranged in many different ways to form a practically unlimited number of chain- or ring-like structures (Olah and Molnár, 2003). Petroleum (e.g., crude oil) in its natural state may contain thousands of different hydrocarbon constituents and varying amounts of some other elemental impurities (e.g., N, O, S, Cu, Fe, Ni, V, etc.) (Speight, 2002). The composition of crude oils from different fields or areas generally show characteristic differences due to the nature of the initial organic material (e.g., phytoplankton, zooplankton, higher plants, and bacteria) buried in sediments and their burial depth (which delimits pressure and temperature conditions). General classification of hydrocarbons can be seen in Fig. 18.2, where they generally fall into two major classes: aliphatics and aromatics.

Aliphatic hydrocarbons can be divided into two types based on the nature of the chain structure, such as open (straight) chain (a.k.a., acyclic) and closed chain (a.k.a., alicyclic). There are three types of open-chain aliphatic hydrocarbons,

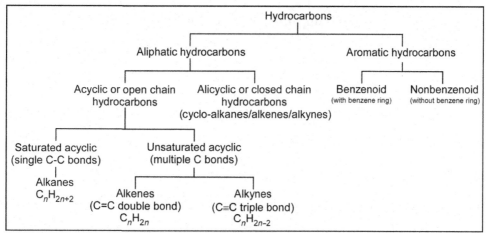

FIG. 18.2 General classification of hydrocarbons (where "*n*" is the number of carbon (C) atoms in the general molecular formula).

including alkanes (with only C—C single bonds), alkenes (with one or more C=C double bonds), and alkynes (with one or more C≡C triple bonds) (Fig. 18.2). Alkanes (a.k.a., paraffins), alkenes (a.k.a., olefins), and alkynes (a.k.a., acetylenes) can be easily distinguished by their suffix (i.e., "-ane," "-ene," and "-yne," respectively), each representing their class and nature of carbon-carbon bonding. Based on the nature of the carbon-carbon bonds (i.e., single, double, or triple) hydrocarbons are either classified as saturated (e.g., alkanes) or unsaturated (e.g., alkenes and alkynes). Open-chain aliphatic hydrocarbons, such as alkanes, alkenes, and alkynes, can also have branched chains, with one or more C atoms branching off the otherwise straight chain. Due to branching, the chain structures of the hydrocarbons become shorter, generally resulting in changes in their physical properties (e.g., lower boiling point and viscosity, greater volatility, etc.). Therefore knowing the structure of a hydrocarbon compound is important, since it provides an insight into its physical properties. Some hydrocarbon compounds, called isomers, can have exactly the same elemental composition (i.e., molecular formula) but differ in the structural or spatial (in 3D space) arrangements of the atoms or atom groups within the molecule. These compounds will display different chemical and physical properties (e.g., isomers of xylene: *o-*, *m-*, and *p-*xylenes). There are many different types of isomerism, which can be classified into two main categories: structural isomers (i.e., constitutional isomers) and stereoisomers. For example, hydrocarbon compounds *cis*-1,2-dichloroethene and *trans*-1,2-dichloroethene are classified as stereoisomers or more specifically "cis-trans isomers." They are chlorinated organic substances (containing Cl), both with a molecular formula $C_2H_2Cl_2$ but with different arrangements of atoms in the 3D space. The isomerism subject will not be discussed further here; however, interested readers can find detailed information on the subject in most organic chemistry books. Closed-chain aliphatic compounds have the prefix "cyclo" (e.g., cyclopentene) and the end carbon atoms of a carbon chain join together to form a ring. Hydrocarbons in this category can be either saturated or unsaturated. Benzenoid hydrocarbons (a.k.a., arenes) have alternating double and single bonds between their carbon atoms forming a cyclic (i.e., ring-like) molecular structure, of which benzene (C_6H_6) is the simplest (see Fig. 18.3) and the most stable one. The more complex ones, such as polycyclic aromatic hydrocarbons (PAHs), contain at least two or more benzene rings fused together (e.g., naphthalene). Aromatic hydrocarbons not containing any benzene ring are called nonbenzenoid compounds. All aromatic hydrocarbons are characterized by their distinct (e.g., aromatic) scents. Examples for some of the aliphatic and aromatic hydrocarbons are shown in Fig. 18.4.

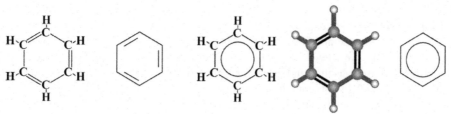

FIG. 18.3 Different molecular representations used to show benzene (C_6H_6) ring structure. Note that three of the carbon atoms have double bonds and three of them have single bonds.

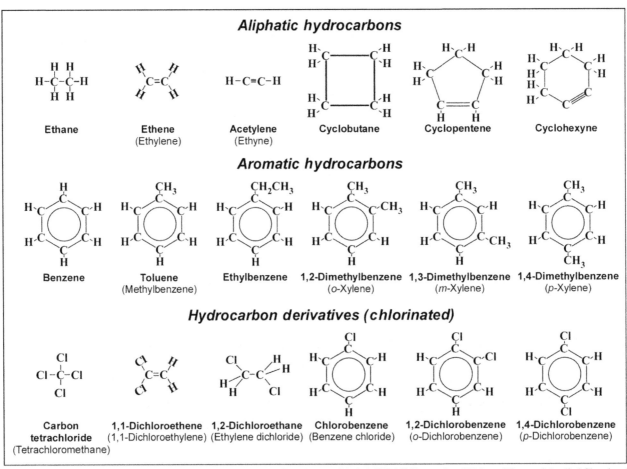

FIG. 18.4 Examples for molecular structures of different hydrocarbon classes with their IUPAC (International Union of Pure and Applied Chemistry) names in bold and commonly used other names in parenthesis.

18.3 PRODUCTION OF HYDROCARBON COMPOUNDS

Starting from the mid-1840s, nonrenewable natural organic materials (mainly fossil fuels), such as coal, natural gas, oil shale, tar sand, and crude oil, were exploited to produce a variety of hydrocarbon compounds in gas, liquid, and semisolid (e.g., wax) forms for many uses. Among these, liquid kerosene has a special place, it replaced whale oil after 1850s and is still used today as a lighting fuel in kerosene lamps and for many other purposes. Advances in the petroleum-extraction technologies (e.g., drilling and pumping), the invention of the internal combustion engine, the mass production of automobiles, and later developments in the associated industries have made a great impact on the organic chemistry field, where the number and variety of hydrocarbon compounds identified and produced have shown a dramatic increase. Although, the term "organic" implies living things (e.g., plants and animals), hydrocarbons can also be synthesized from purely inorganic elements and substances via methanation (e.g., CO or $CO_2 + H_2$) and oxidative condensation of methyl alcohol or methane (Olah and Molnár, 2003). Recently developed technologies have also made the production of renewable biofuels (e.g., bioethanol and biodiesel) more feasible using a variety of methods. For example, bioethanol is produced by yeast fermentation of sugar-containing plants or their wastes (e.g., corn, soybean, sugar beet, sugarcane, wood, sawdust, etc.) and biodiesel production requires vegetable oil (e.g., palm, peanut, soybean, sunflower, etc.) or animal-fat sources. Use of biofuels in modern motor vehicles is on the rise and US is the world leader in production (see Table 18.1). However, the use of biofuels is currently restricted to "low-level blending" with petroleum-based fuels. For instance, currently the most common blends for petroleum-based gasoline and diesel fuels are E10 (e.g., 10% bioethanol and 90% petroleum gasoline) and B20 (e.g., 20% biodiesel and 80% petroleum diesel), respectively (US DOE, 2017). For higher-level blends, generally specific modifications are required to prevent damage to certain parts of the vehicles' engines (e.g., hoses and gaskets).

Crude oil in its natural state has not much use; therefore, it is subjected to a multistep refining process to produce a wide spectrum of products that are marketed today. Oil-refining technology exploits the dissimilar physical and chemical

TABLE 18.1 Total Biofuel (Bioethanol and Biodiesel) Production (×1000 Barrels Per Day) of Top 10 Countries Between the Years 2005 and 2014

Country	2005	2006	2007	2008	2009	2010	2011	2012	2013	2014
United States	261.0	335.0	457.0	650.0	747.0	881.5	967.1	921.3	953.6	1018.8
Brazil	276.0	307.0	396.0	486.0	478.0	467.2	404.1	423.4	494.3	489.5
Germany	36.0	59.0	64.0	65.0	58.0	49.8	66.7	65.3	68.5	71.9
Argentina	0.2	0.7	3.9	14.0	24.0	21.3	31.7	53.4	47.3	61.4
Indonesia	0.2	0.5	1.2	2.2	5.9	7.3	28.6	34.7	39.8	59.7
China	22.0	32.0	31.0	39.0	48.0	43.0	50.1	49.4	54.8	58.6
France	11.0	17.0	28.0	50.0	58.0	45.6	58.9	58.3	57.6	54.5
Thailand	1.6	2.6	4.2	13.0	18.0	13.0	22.4	25.5	34.7	39.2
Canada	4.6	5.2	15.0	17.0	22.0	24.8	31.3	31.6	33.2	36.2
Spain	8.2	8.2	11.0	10.0	21.0	17.2	19.5	16.0	19.6	22.0

(Data from US EIA (Energy Information Administration), 2017a. International Energy Statistics. https://www.eia.gov/beta/international/data/browser (accessed 29 November 2017).)

properties of different petroleum hydrocarbon constituents to separate them into products that have many uses (e.g., fuels for vehicles, heating, illumination, petrochemical feedstocks, lubricants, waterproofing materials, waxes for candles and packaging, solvents, etc.) (US EIA, 1999). The majority of hydrocarbon compounds are used as "fuels" and can be produced using a number of processes, for example, fractional distillation, removing/adding H atoms to the molecules (e.g., dehydrogenation and hydrogenation), and breaking off the carbon-carbon bonds of the long-chain hydrocarbons in the presence of catalysts/solvents under high temperatures and pressures (e.g., catalytic and thermal cracking) (Speight, 2002). In addition, fuel hydrocarbons are also used as feedstock to produce a myriad of synthetic organic chemicals called "hydrocarbon derivatives." These organic chemicals are produced from aliphatic and aromatic hydrocarbon series by substituting their molecular H atom(s) with other atoms or atom groups (e.g., halogens including Cl, Br, F, and I) (see Fig. 18.4). From an organic-contamination point of view, both fuel (e.g., gasoline and diesel range) hydrocarbons and hydrocarbon derivatives (e.g., chlorinated hydrocarbons) are of a major public health concern due to their toxicity, persistent nature, distinctly different physical and chemical properties, and extensive use throughout the world. These chemical products form the basis of or raw materials for many chemical and industrial processes or products that we daily come across in modern life. Hence their presence and concentrations in the atmosphere (air), biosphere (plants and animals), lithosphere (soil and rock), and hydrosphere (water) is strongly affected by anthropogenic activities involving their production, transportation, storage, usage, and disposal.

18.4 RELEVANT INFORMATION SOURCES ON HYDROCARBON COMPOUNDS

Organic chemistry is an important and dynamic field, where the number of identified organic compounds are immense and increase each day. Naming of these organic chemical compounds are made according to the rules defined in IUPAC (International Union of Pure and Applied Chemistry) Nomenclature of Organic Chemistry, a.k.a., Blue book (Favre and Powell, 2013; IUPAC, 2017). To prevent ambiguities, each organic chemical is also assigned a unique code called InChI, the IUPAC International Chemical Identifier (Heller et al., 2015). Nevertheless, Chemical Abstracts Service registry number or CAS RN (CAS, 2017) assigned by the American Chemical Society is probably the most commonly used chemical-substance identifier among the others available (see Apodaca, 2011; Heller et al., 2015). CAS registry database of the American Chemical Society currently contains more than 134 million unique organic and inorganic substances, of which in excess of 348,000 are subject to regulations by various international, national, and state agencies (CAS, 2017). A great majority of these chemical substances do not occur naturally and have been added to the long list of chemicals only during the last two centuries (i.e., since the Industrial Revolution). Although the number of organic chemical compounds that are currently manufactured and commercially available is rather vague, the total number (including both organic and inorganic compounds) is probably around 84,000 (IOM, 2014). CAS registry number provides an important starting point, especially when searching for the physical and chemical properties of the regulated organic compounds, which can have multiple (sometimes several thousand!) synonyms and/or commercial names. Specific information on compound structures, formulae, chemical and physical properties, exposure limits, health effects, current regulations and standards, cleanup technologies, production and consumption data, etc., can be obtained from the online websites or databases of various governmental agencies listed below:

- *Royal Society of Chemistry (RSC)*: ChemSpider database lists compound structure, formula, systematic name, synonyms, trade names, literature references, physical properties, spectra (RSC, 2015).
- *The National Institute of Standards and Technology (NIST)*: NIST Chemistry WebBook lists compound structure, formula, name, IUPAC identifier, CAS registry number, reaction, molecular weight, ion energetics properties, vibrational and electronic energies (NIST, 2017).
- *Centers for Disease Control and Prevention (CDC)*: CDC database lists medical signs and symptoms, protection, treatment, toxicological profile and adverse health effects, medical management guidelines (MMG), prevention and personal protective equipment, sampling and analytical methods, decontamination, spillage disposal, packaging, and labeling information (CDC, 2016).
- *Superfund Chemical Data Matrix (SCDM)*: SCDM Query database lists name, synonyms, hazard ranking system (HRS) factor values and benchmarks, toxicity, persistence, degradation, mobility, bioaccumulation, physical characteristics, and class information (US EPA, 2017d).
- *US National Library of Medicine Toxicology Data Network (TOXNET)*: TOXNET database allows simple and advanced searches for a variety of information from 15 different databases, either simultaneously or individually (US NLM, 2017a).
- *Agency for Toxic Substances and Disease Registry (ATSDR)*: ATSDR's Substance Priority List database lists CAS registry number, affected organ systems, cancer classification, chemical classification, and summary data (ATSDR, 2017).
- *US EPA National Primary Drinking Water Regulations (NPDWR)*: NPDWR site lists legally enforceable primary standards and treatment techniques that apply to public water systems, as well as maximum contaminant level goals (MCLGs), maximum contaminant levels (MCLs), potential health effects from long-term exposure, and sources of contaminants in drinking water (US EPA, 2009).
- *US EPA Integrated Risk Information System (IRIS)*: IRIS database lists compound structure, name, synonyms, CAS registry number, noncancer and cancer assessments, affected organs, etc. (US EPA, 2015).
- *US EPA Contaminated Site Clean-Up Information (CLU-IN)*: CLU-IN website provides detailed information about currently used innovative contaminant treatment/remediation and site-characterization technologies (US EPA, 2017e).
- *US EIA Official Energy Statistics, Data, Analysis and Forecasting (EIA)*: EIA website provides a wide variety of free and open data (related to all energy sources) in the form of downloadable tables, graphs, maps, as well as raw data. Data made available through an Application Programming Interface (API) and open data tools. The data in the API is also available in bulk file, in spreadsheet programs via an add-in, and via widgets that embed interactive data visualizations of EIA data on any website (US EIA, 2017a,b).

In addition to these online databases (e.g., simple or relational), geographic information system (GIS) technology, especially online mapping platforms (e.g., web GIS technology) are being used increasingly for data query, dissemination, analysis, and visualization purposes. GIS can allow even nonexpert GIS users to organize, analyze, and visualize vast amounts of complex spatial data and can help one to see spatial distribution (i.e., arrangement across the earth surface) of events, features, changes, and trends with an exceptional precision and clarity. As earlier illustrated in Fig. 18.1, GIS-based maps from different online platforms (e.g., governmental, nongovernmental, private, educational, commercial, etc.) can be combined to produce "new information" that can be effectively used for decision-making purposes, especially in emergency situations, such as accidental releases and spills of organic contaminants. For example, Division of Specialized Information Services of the US National Library of Medicine provides an online GIS database called TOXMAP (US NLM, 2017b), which enable internet users to visually explore data retrieved from the US EPA Superfund (US EPA, 2017b) and Toxics Release Inventory (TRI) programs (US EPA, 2017f). In the TOXMAP site, web GIS-based queries can be made at a variety of levels (i.e., state, metropolitan area, watershed, and tribal) (US NLM, 2017b). Data and maps on the US EPA Superfund National Priorities List (NPL) website can also be viewed and downloaded separately through the web GIS application designed for that purpose (US EPA, 2017c). US EIA website called "U.S. Energy Mapping System" provides very detailed information related to all energy infrastructures and sources available in the United States (US EIA, 2017b), as well as raw data and statistics related to other countries. This information can be visualized through a web GIS application, where internet users can overlay data of their choice on different basemap layers, add/remove layers, and change legends to customize their maps for a variety of uses. Of course the list of institutions mentioned above is not exhaustive and they should be viewed as mere examples of the type (e.g., simple tables, relational databases, web GIS applications, etc.) and level of information that might be available and useful for many different purposes. Online data sources are also available in other languages and from other counties (e.g., Canada, China, European Union, etc.) that might also be worth exploring.

18.5 CAUSES OF NONAQUEOUS PHASE LIQUID CONTAMINATION

Liquid-phase contaminants, such as petroleum-based fuels (e.g., gasoline, kerosene, fuel oil, jet fuel, and diesel) and hydrocarbon derivatives (e.g., halogenated hydrocarbons), are typically characterized by their immiscible (i.e., hydrophobic) nature in the presence water (e.g., they separate into two phases). These hydrocarbon compounds are collectively known as NAPLs and their use by industrial, agricultural, urban (e.g., transportation), commercial, and military activities has shown a steady increase since the World War II. It is not surprising to see that 141 out of 1342 hazardous chemical sites listed on the Superfund NPL (see US EPA, 2017c) are located in places where various past military activities have been documented. Most of these sites are still actively used military areas. The gasoline constituents, especially BTEX compounds (i.e., benzene, toluene, ethylbenzene, and isomers of xylene; see Fig. 18.4), are the ones that most frequently occur together at contaminated sites, due to their significant volatility, water solubility, and mobility (Coates et al., 2002; Güler, 2009; Odermatt, 1994). The widespread occurrence of gasoline contamination is not so surprising; for example, fuel hydrocarbons comprise 90% of the petroleum products marketed in the United States (US EIA, 1999). On the other hand, hydrocarbon derivatives, especially the chlorinated ones (e.g., 1,2-dichloroethane, 1,1,1-trichloroethane, dichloromethane, carbon tetrachloride, 1,2-dichloropropane, etc.), pose an even greater challenge in terms of contaminant mitigation owing to their uniquely different physical properties (e.g., higher volatility, water solubility, and density) compared to that of most fuel hydrocarbons.

NAPL release into the subsurface can be continuous, episodic, and impulsive in character and the causes may be unintentional/intentional. NAPL contamination incidents are generally associated with historic and/or existing anthropogenic activities or land use in the contaminated sites and they present a great challenge in terms of achieving the cleanup goals enforced by law or remedial action objectives set for the site, both of which profoundly depend on the currently available contaminant treatment and site-characterization technologies (see US EPA, 2017e). Most NAPL compounds are classified as "human carcinogens" or "possible human carcinogens" by governmental agencies (ATSDR, 2017; CDC, 2016; OSHA, 2017; US EPA, 2009, 2015; US NLM, 2017a) due to their mutagenic effects, even at low concentrations (Dean, 1985; Molson et al., 2002). Much work is needed to be done here, since out of over 1000 petroleum components, specific toxicity data exist for only 95, most of which is inadequate for the development of exposure criteria (Miller et al., 2000). Humans can be exposed to these toxic chemicals or chemical mixtures through a wide variety of pathways, including direct ingestion, dermal absorption through the skin and other organs in direct contact with the contaminants, ingestion of contaminated groundwater, inhalation of volatile organic compound (VOC)-contaminated air and fugitive dusts, and consumption of foodstuffs obtained from a contaminated source.

NAPL contaminants are ubiquitous in the environment and there are many potential point and nonpoint (e.g., diffuse) sources at and below the ground surface. Some of the most commonly occurring NAPL contaminant sources are depicted in Fig. 18.5, the majority of which can be classified as concentrated point sources. Since, NAPLs are used in virtually every sector, the contamination is almost inevitable, the majority of which occurs underground and often can go unnoticed for extended periods of time until their discovery. According to a recent survey (US EPA, 2017g) over 535,000 underground storage tank (UST) releases have been confirmed in the United States (as of March 2017), of which 465,000 sites have been cleaned up within a decadal time frame. Some 70,000 UST sites are still on the list and waiting for remedial action to take place (US EPA, 2017g). According to an estimation, 95% of the USTs are used to store petroleum fuels (Lund, 1995), where gasoline is the predominant product.

Hydrocarbon contamination incidents start with the oil exploration stage and may also take place during extraction, production, refining, processing, blending, transportation, distribution, storage, consumption, and disposal operations involving liquid-phase petroleum hydrocarbons and hydrocarbon derivatives. The most common causes of NAPL contamination can be generalized as follows (Domenico and Schwartz, 1990; Güler, 2009; Weiner, 2008):

- Oil exploration- and production-related activities, accidents (e.g., spills and well-blow outs), and discharges (e.g., drilling fluids, formation water, fluids used for enhanced oil recovery, etc.).
- Large-scale oil-refinery and chemical-industry operations.
- Small-scale commercial activities (e.g., gasoline stations, dry cleaners, mechanic workshops).
- Leaking aboveground and underground storage tanks.

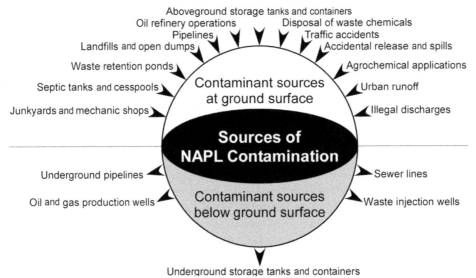

FIG. 18.5 Major sources of nonaqueous-phase liquid contamination at and below the ground surface.

- Damaged utility lines (e.g., hazardous product pipelines, sewer lines, reclaimed water lines, etc.).
- Traffic accidents.
- Accidental release and spills.
- Military operations.
- Waste storage and disposal operations (e.g., slop tanks, ponds, landfills, dumpsites, septic tanks, cesspools, waste injection wells, etc.).
- Agricultural activities (e.g., agrochemical application).
- Urban runoff.
- Illegal discharge and disposal of waste petroleum products.

These contamination sources have had a substantial impact on air, soil, and water quality throughout the world, limiting the beneficial uses of these valuable natural resources.

18.6 PHYSICAL PROPERTIES OF NONAQUEOUS PHASE LIQUID CONTAMINANTS

Compositionally, NAPLs can either consist of a single chemical compound (e.g., ethene, benzene, chlorobenzene, carbon tetrachloride, etc.) or exist as complex (i.e., multicomponent) mixtures (e.g., gasoline and diesel fuels) containing several hundred different organic chemicals. For example, fuels such as gasoline or diesel may contain between 100 to 500 different types of hydrocarbon compounds, depending on their formulation. Fuel formulations generally show changes depending on the manufacturer, geographic location, season/climate, regulations, restrictions, etc. Gasoline-range hydrocarbons are those that contain 4–12 carbon atoms (C_4-C_{12}) in their molecular structure (e.g., alkanes, alkenes, cycloalkanes, and aromatics) (Speight, 2002). Typically, gasoline-range hydrocarbons are highly flammable and volatile compounds (e.g., BTEX compounds) with boiling points between $-1°C$ and $216°C$ (Speight, 2002). Gasoline also contains chemical additives, which are used to improve its specific properties. For example, organometallic compounds such as tetraethyl lead ($C_8H_{20}Pb$) and tetramethyl lead ($C_4H_{12}Pb$), and oxygenates such as ethers (e.g., methyl *tert*-butyl ether (MTBE), ethyl *tert*-butyl ether (ETBE), *tert*-amyl methyl ether (TAME)), and alcohols (e.g., ethanol and *tert*-butyl alcohol (TBA)), are frequently used as antiknock agents, as octane enhancers, or to reduce exhaust emissions. For instance, tetraethyl lead (TEL) was added to gasoline mainly between early 1920s and mid-1990s to increase its octane rating. However, due to their toxic nature and growing concerns about the contamination of groundwater resources, the use of TEL and MTBE as "fuel additives" was banned in most countries (Weiner, 2008). Diesel-range hydrocarbons, however, generally contain 8–18 carbon atoms (C_8-C_{18}) in their molecular structures, therefore they are less flammable and volatile, and have higher boiling points (between $126°C$ and $258°C$) compared to gasoline range hydrocarbons (Speight, 2002). Most fuel hydrocarbons (e.g., BTEX) have densities lower than that of water (see Table 18.2); hence they are called "light NAPLs" or LNAPLs (Bedient et al., 1999) and tend to float on the surface once they reach the water table (Fig. 18.6).

TABLE 18.2 Important Physical and Chemical Properties of Selected Light and Dense Nonaqueous Phase Liquid Compounds

CAS RN	Compound Name	Molecular Formula	Type	Molecular Weight (g/mol)	Density at 20°C (ρ, g/cm^3)^a	Vapor Pressure at 25°C (P_v, mm Hg)^b	Melting Point (T_m, °C)^b	Boing Point (T_b, °C)^b	Water Solubility (S_w, mg/L)^b	Henry's Law Constant at 25°C (K_H, atm m^3/mol)^b	US EPA (MCL) (mg/L)^c
Aromatic hydrocarbons											
71-43-2	Benzene	C$_6$H$_6$	Aromatic	78.112	0.876	94.8	5.5	80.0	1790	5.55E−03	0.005
108-88-3	Toluene	C$_7$H$_8$	Aromatic	92.138	0.862	28.4	−94.9	110.6	526	6.64E−03	1
100-41-4	Ethylbenzene	C$_8$H$_{10}$	Aromatic	106.165	0.867	9.6	−94.9	136.1	169	7.88E−03	0.7
95-47-6	o-Xylene (2-Xylene)	C$_8$H$_{10}$	Aromatic	106.165	0.880	6.61	−25.2	144.5	178	5.18E−03	*Total 10*
108-38-3	m-Xylene (3-Xylene)	C$_8$H$_{10}$	Aromatic	106.165	0.864	8.29	−47.8	139.1	161	7.18E−03	*for all*
106-42-3	p-Xylene (4-Xylene)	C$_8$H$_{10}$	Aromatic	106.165	0.861	8.84	13.2	138.3	162	6.90E−03	*Xylenes*
100-42-5	Styrene	C$_8$H$_8$	Aromatic	104.149	0.906	6.4	−31.0	145.0	310	2.75E−03	0.1
Polinuclear aromatic hydrocarbons (PAHs)											
50-32-8	Benzo[a]pyrene	C$_{20}$H$_{12}$	Aromatic	252.309	1.351	5.49E−09^a	176.5	496.0^a	0.00162	4.57E−07	0.0002
91-20-3	Naphthalene	C$_{10}$H$_8$	Aromatic	128.171	1.162	0.085	80.2	217.9	31	4.40E−04	–
Halogenated (chlorinated) hydrocarbon derivatives											
108-90-7	Chlorobenzene	C$_6$H$_5$Cl	Haloaromatic	112.557	1.106	12	−45.2	131.7	498	3.11E−03	0.1
95-50-1	1,2-Dichlorobenzene	C$_6$H$_4$Cl$_2$	Haloaromatic	147.001	1.306	1.36	−16.7	180.0	156	1.92E−03	0.6
106-46-7	1,4-Dichlorobenzene	C$_6$H$_4$Cl$_2$	Haloaromatic	147.001	1.248	1.74	52.7	174.0	81.3	2.41E−03	0.075
120-82-1	1,2,4-Trichlorobenzene	C$_6$H$_3$Cl$_3$	Haloaromatic	181.446	1.459	0.46	17.0	213.5	49	1.42E−03	0.07

Continued

TABLE 18.2 Important Physical and Chemical Properties of Selected Light and Dense Nonaqueous Phase Liquid Compounds—cont'd

CAS RN	Compound Name	Molecular Formula	Type	Molecular Weight (g/mol)	Density at 20°C (ρ, g/cm³)[a]	Vapor Pressure at 25°C (P_v, mm Hg)[b]	Melting Point (T_m, °C)[b]	Boiling Point (T_b, °C)[b]	Water Solubility (S_w, mg/L)[b]	Henry's Law Constant at 25°C (K_H, atm m³/mol)[b]	US EPA (MCL) (mg/L)[c]
107-06-2	1,2-Dichloroethane	$C_2H_4Cl_2$	Haloalkane	98.959	1.235	78.9	−35.5	83.5	8600	1.18E−03	0.005
71-55-6	1,1,1-Trichloroethane	$C_2H_3Cl_3$	Haloalkane	133.403	1.338	124	−30.4	74.0	1290	1.72E−02	0.2
79-00-5	1,1,2-Trichloroethane	$C_2H_3Cl_3$	Haloalkane	133.403	1.442	23	−36.6	113.8	4590	8.24E−04	0.005
75-09-2	Dichloromethane	CH_2Cl_2	Haloalkane	84.932	1.326	435	−95.1	40.0	13000	3.25E−03	0.005
56-23-5	Carbon tetrachloride	CCl_4	Haloalkane	153.822	1.594	115	−23.0	76.8	793	2.76E−02	0.005
78-87-5	1,2-Dichloropropane	$C_3H_6Cl_2$	Haloalkane	112.985	1.158	53.3	−100.0	95.5	2800	2.82E−03	0.005
75-01-4	Vinyl chloride	C_2H_3Cl	Haloalkene	62.498	0.911	2980[a]	−154.0	−13.3	8800	2.78E−02	0.002
75-35-4	1,1-Dichloroethene	$C_2H_2Cl_2$	Haloalkene	96.943	1.213	600	−123.0	31.6	2420	2.61E−02	0.007
156-59-2	cis-1,2-Dichloroethene	$C_2H_2Cl_2$	Haloalkene	96.943	1.284	200	−80.0	60.1	6410	4.08E−03	0.07
156-60-5	trans-1,2-Dichloroethene	$C_2H_2Cl_2$	Haloalkene	96.943	1.257	331[a]	−49.8	48.7	4520	9.38E−03	0.1
79-01-6	Trichloroethene	C_2HCl_3	Haloalkene	131.387	1.464	69	−84.7	87.2	1280	9.85E−03	0.005
127-18-4	Tetrachloroethene	C_2Cl_4	Haloalkene	165.832	1.623	18.5	−22.3	121.3	206	1.77E−02	0.005

CAS RN, Chemical Abstracts Service registry number (CAS, 2017). Nonaqueous-phase liquid molecular weight calculated from the atomic weights (i.e., C = 12.0107, H = 1.00794, and Cl = 35.4527). Boiling point (T_b) at 760 mm Hg. Henry's Law constant (K_H) is also known as air-water partition coefficient. *US EPA (MCL)*, maximum contaminant level allowed in drinking water by US EPA National Primary Drinking Water Regulations.
[a]*US NLM (2017c).*
[b]*US NLM (2017a).*
[c]*US EPA (2009).*

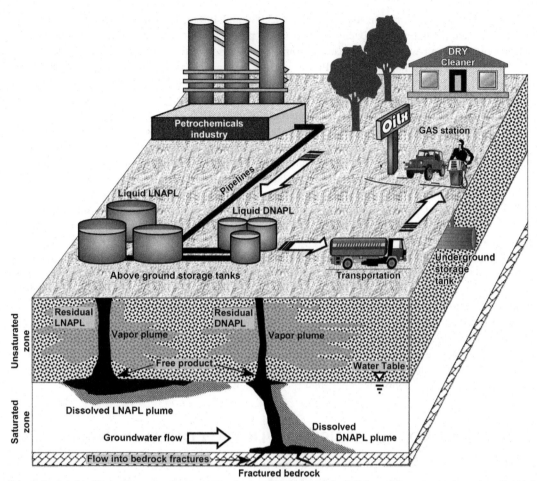

FIG. 18.6 Schematic drawing illustrating a complex contamination scenario involving both light and dense nonaqueous phase liquid (LNAPL and DNAPL) compound leakages from aboveground storage tanks at an industrial site and their behavior in the subsurface environment after release. NAPLs in the unsaturated zone mainly occur in four phases: (1) immiscible free-phase (pure NAPL product), (2) residual phase (adsorbed on solid particles), (3) gaseous phase (VOCs in soil gas), and (4) aqueous phase (dissolved in water). The composition of the free-phase NAPL contaminant continuously changes due to preferential loss of its components through volatilization, solubilization, and adsorption processes. *(Modified from Güler, C., Alpaslan, M., 2006. In-situ remediation technologies applied for petroleum hydrocarbon contaminated aquifers. Jeol. Mühendis. Derg. 30-2 (65), 33–50.)*

Halogenated hydrocarbons, on the other hand, form one of the most diverse and significant groups of contaminants with respect to groundwater and air contamination due to their relatively high volatility and water solubility (see Table 18.2). Halogenated hydrocarbons, especially chlorinated ones, are mostly used as solvents in dry-cleaning and industrial-degreasing sectors. In the past, particularly the beginning of the late 1920s, halogenated hydrocarbons containing chloride (Cl) and bromide (Br), e.g., 1,2-dichloroethane and 1,2-dibromoethane, were added to leaded gasoline (gasoline having TEL) to serve as lead (Pb) scavengers. The presence of Pb in petroleum hydrocarbon-contaminated groundwater therefore may indicate an old spill, especially if it occurs together with aforementioned halogenated hydrocarbons, and this can be used to age-date (in a relative sense) the contamination. However, these halogenated hydrocarbons are also present in some agrochemicals, and therefore need to be approached with caution. The majority of halogenated hydrocarbons (e.g., carbon tetrachloride, chlorobenzene, dichloromethane, 1,2-dichloropropane) have densities higher than that of water (Table 18.2); hence, they are called "dense NAPLs" or DNAPLs (Bedient et al., 1999), and tend to sink bottom of the water column in the aquifer (Fig. 18.6). In consequence, when they are released into the ground in large amounts, the distinctly different physical and chemical properties of LNAPLs and DNAPLs force them to behave differently in the subsurface environment, especially in the saturated zone of the aquifer (Fig. 18.6).

Although both LNAPLs and DNAPLs are immiscible (because of their hydrophobic and nonionic nature) and sparingly soluble in water (Table 18.2), their concentrations can easily reach above levels that are considered toxic for most organisms. The water solubility of LNAPL and DNAPL compounds is generally inversely related to their molecule size or the number of carbon atoms in the molecules (see Table 18.2). For example, the water solubility of DNAPLs dichloromethane (CH_2Cl_2) and chlorobenzene (C_6H_5Cl) is 13,000 and 498 mg/L, respectively, where the former contains only one carbon atom and the latter have six carbon atoms. The processes causing partitioning of NAPL contaminants (in the subsurface) between the air-water-solid phases are mostly governed by volatilization, solubilization, and adsorption/desorption mechanisms. A great majority of NAPLs is volatile under normal temperature and pressure conditions and therefore they are often called VOCs (Larrañaga et al., 2016). The volatility of NAPLs is directly related to their vapor pressures (see Table 18.2). Hydrocarbon compounds with small molecules (i.e., lighter hydrocarbon fractions) can volatilize more readily in ambient air. After their release into the environment, NAPLs or NAPL mixtures will tend to slowly dissolve in water and volatilize into the soil gas found within the pore spaces or open fractures existing below ground surface. The ratio of the partial pressure of a NAPL in air to its concentration in water is called the Henry's Law constant (K_H, atm m^3/mol). The soil-screening level (SSL) equations used for the evaluation and cleanup of NAPL-contaminated sites require the dimensionless form of Henry's Law constant, which is calculated by dividing K_H (atm m^3/mol) with RT (R = universal gas constant, atm m^3/mol K; and T = temperature in degrees Kelvin, K) (US EPA, 1991). Henry's Law constant (K_H) is temperature dependent and directly related to NAPL's vapor pressure (P_v, atm) and molecular weight (MW, g/mol), but inversely related to its water solubility (S_w, mg/L or g/m^3). Generally, highly volatile or water-soluble NAPLs are also the most toxic ones (see Table 18.1). NAPLs, such as BTEX compounds (i.e., benzene, toluene, ethylbenzene, and isomers of xylene), have Henry's Law constants between 5.18×10^{-3} and 7.88×10^{-3} atm m^3/mol, which indicate they can be transported along the groundwater flow path, farther than most organic compounds. All halogenated (i.e., chlorinated) alkanes (haloalkanes) and alkenes (haloalkenes) have moderate to relatively high Henry's Law constant values (Table 18.2), therefore, they can be expected to form vapor (gaseous) phase NAPL plumes that emanate from the source area (residual phase) and the aqueous (dissolved) phase plume. Important physical and chemical properties of selected NAPL (i.e., LNAPL and DNAPL) compounds are presented in Table 18.2. The physical and chemical properties of NAPLs are known to play an important role in their behavior and fate in the subsurface environment. Since composition of NAPL constantly changes after a spill event due to its partitioning behavior, measurement of the site-specific physical and chemical properties is very important, rather than relying solely on literature data.

18.7 BEHAVIOR OF NONAQUEOUS PHASE LIQUID CONTAMINANTS IN THE SUBSURFACE

Factors affecting the behavior and fate of NAPLs (i.e., LNAPL and DNAPL) in an aquifer are mainly related to properties of and interactions between the contaminant(s), air (e.g., in atmosphere and soil gas), water (in pore spaces), and the aquifer media itself. After a near-surface release of a large volume of NAPL contaminant(s) into an unsaturated zone, NAPLs will tend to follow an ever-widening (e.g., cone shaped) downward path through the entire thickness of the unsaturated zone (Fig. 18.6) under gravitational and capillary forces. However, if the volume of release is insignificant or small, NAPL contaminants will be held in place due to attractive forces (e.g., capillarity), where they will be partitioned between the air, water, and solid phases found within the unsaturated zone. Therefore NAPLs in the unsaturated zone mainly occur in four phases: immiscible free-phase product (i.e., pure NAPL), residual phase (adsorbed on solid particles or other aquifer materials), gaseous phase (in soil gas), and aqueous phase (dissolved in water) (DiGiulio and Cho, 1990). The NAPL distribution among these four phases can be described empirically by partition coefficients, which depend on properties and types of soil and contaminant(s). Except for the residual phase, the other three NAPL phases move separately, each forming their own contaminant plumes (Fig. 18.6). As downward migration of NAPL into the unsaturated zone progresses, it gradually becomes entrapped in pores (as residual blobs and pools) but continues to be a source of contamination (Grathwohl, 2000). A great portion of the NAPL contaminants found in the unsaturated zone (as residual phase) are adsorbed onto soil particles. Since most NAPLs have high vapor pressures (i.e., volatile) (see Table 18.2), residual NAPLs in the unsaturated zone are gradually lost through the vaporization process in the soil pore spaces (Fig. 18.6), forming gaseous (vapor)-phase VOCs. Especially in NAPL-contaminated sites, VOC concentrations in the air, soil gas (filling the pore spaces in the unsaturated zone), and underground openings (e.g., sewer lines, storm drains, basements of buildings, etc.) can reach harmful levels causing an additional health concern. Toxicological assessments on various VOCs, including BTEX compounds, can be found in Calabrese and Kenyon (1991).

Although the downward migration of LNAPLs and DNAPLs into the unsaturated zone is very similar in character, they display highly contrasting behaviors near the saturated zone due to differences in their relative densities compared to that of water. For instance, when LNAPLs reach the water table they tend to float on top of it as an immiscible free-phase product (Fig. 18.6). However, when DNAPLs reach the water table they will tend to penetrate through the entire thickness of the saturated zone of the aquifer in a downward direction, and only stop when they come across impervious strata (Fig. 18.6).

The DNAPL migration through fractured rocks (in saturated zone) largely depends on the existence of primary and secondary high-permeability preferential pathways (e.g., bedding planes, fractures, faults, etc.). As LNAPLs and DNAPLs slowly dissolve in groundwater, they form aqueous-phase contaminant plumes (Fig. 18.6), which slowly propagate and migrate along the flow path, defined by the hydraulic gradient and advection-dispersion-diffusion mechanisms. If the residual saturation is close to the water table, more NAPL contaminants may enter the aqueous (dissolved) phase due to the periodic rise and fall of the water table. If they are present in the NAPL mixture, water-miscible organic solvents, such as ethanol, methanol, and acetone, tend to increase the water solubility of otherwise immiscible NAPL contaminants (Nkedi-Kizza et al., 1985). In the saturated zone, soil pore spaces do not contain any air and therefore there only exist three NAPL phases (i.e., immiscible free-phase, residual phase, and aqueous [dissolved] phase).

18.8 ABIOTIC AND BIOTIC PROCESSES AFFECTING NONAQUEOUS PHASE LIQUID CONTAMINANTS

At NAPL contaminated sites, each individual NAPL component is subjected to a variety of processes, which can be collected under two broad categories: distribution processes and transformation processes. Distribution processes (e.g., dilution, dispersion, adsorption-desorption, dissolution, immiscible phase separation, volatilization, etc.) affect only the form and state of the NAPL contaminants and do not change significantly their chemical properties or toxicity (Boulding, 1995). Transformation processes (e.g., hydrolysis, oxidation-reduction, aerobic/anaerobic biodegradation, etc.), on the other hand, in most cases irreversibly change the chemical structure of the NAPL contaminants (Boulding, 1995), under naturally occurring chemical and biological conditions. Among others, microbially mediated aerobic/anaerobic biodegradation processes and oxidation-reduction reactions are the most important ones. A great majority of the NAPL contaminants is more or less susceptible to transformation processes (see Table 18.3) and, except for a few cases, newly formed byproducts will be all harmless substances. All these abiotic and biotic processes play an important part in the control of the downgradient migration of the NAPL contaminants and assist in decrease the mass, volume, concentration, and toxicity of contaminants through time in the affected aquifer(s) (Bockelmann et al., 2003; Chen et al., 2005; Cho et al., 1997; Güler, 2009). For example, Fig. 18.7 depicts a LNAPL contamination case from the Karaduvar area (Mersin, SE Turkey) where the plume size decreased by 87% (in area) between March 2006 and May 2010 due to the aforementioned abiotic and biotic natural attenuation processes (for details see Güler, 2009). This NAPL plume had an unusual length (i.e., 1500 m), since according to the literature, in 90% of the cases the plume boundaries (defined by a concentration of 10 ppb) do not extend further than 76 m downgradient of the source area (Weiner, 2008). This indicated that a large volume (unknown) of NAPL had been released into the environment from a continuous source over a long period of time.

In relatively shallow aquifer systems, autochthonous microorganisms play a critical role in the degradation of fuel hydrocarbons and hydrocarbon derivatives. As it was demonstrated by previous studies, most fuel hydrocarbons (NAPLs)

TABLE 18.3 Relative Susceptibilities of Petroleum Hydrocarbons and Hydrocarbon Derivatives to Biodegradation

Type of Organic Contaminant	Biodegradation Susceptibility	Condition
Open-chain (short) aliphatics	High	Aerobic
Open-chain (branched) aliphatics	High	Aerobic
Open-chain (long) aliphatics	Low	Aerobic
Closed-chain aliphatics (cycloalkanes)	Low	Aerobic
Aromatics (single benzene ring) (e.g., toluene)	High	Aerobic
PAHs (2–3 benzene rings) (e.g., naphthalene)	High	Aerobic
PAHs (4–6 benzene rings) (e.g., benzo[a]pyrene)	Low	Aerobic
Halogenated aliphatics (chlorinated alkanes)	High	Anaerobic/aerobic
Halogenated aromatics (low-Cl chlorobenzenes)	High	Anaerobic/aerobic
Halogenated aromatics (high-Cl chlorobenzenes)	Low	Anaerobic

(Modified from ICSS (International Centre for Soil and Contaminated Sites), 2006. Manual for Biological Remediation Techniques, ICSS at the German Federal Environmental Agency, Dessau. Available at: https://www.umweltbundesamt.de/sites/default/files/medien/publikation/long/3065.pdf.)

FIG. 18.7 Change in the extent of the aqueous (dissolved)-phase LNAPL plume through time in the Karaduvar coastal aquifer (Mersin, Turkey). Initial contaminant plume was more than 1500 m in length and nearly 320 m in width and contained both gasoline and diesel range hydrocarbons. *(Data from Güler, C., 2009. Site characterization and monitoring of natural attenuation indicator parameters in a fuel contaminated coastal aquifer: Karaduvar (Mersin, SE Turkey). Environ. Earth Sci. 59 (3), 631–643; Güler, C., Alpaslan, M., Kurt, M.A., Temel, A., 2010. Deciphering factors controlling trace element distribution in the soils of Karaduvar industrial-agricultural area (Mersin, SE Turkey). Environ. Earth Sci. 60 (1), 203–218. Güler, C., Kaplan, V., Akbulut, C., 2013. Spatial distribution patterns and temporal trends of heavy-metal concentrations in a petroleum hydrocarbon-contaminated site: Karaduvar coastal aquifer (Mersin, SE Turkey). Environ. Earth Sci. 70 (2), 943–962.)*

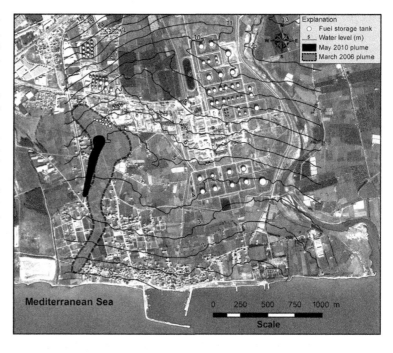

and hydrocarbon derivatives (DNAPLs) can be biodegraded, to some extent, by subsurface microorganisms under both aerobic and anaerobic conditions (Atlas, 1981; Borden, 1994; Chapelle, 1999; Lee et al., 1988; Rifai et al., 1995). The susceptibility of NAPL contaminants to biodegradation can be generally ordered as follows: open-chain alkanes > branched-chain alkanes > aromatics with single benzene ring > closed-chain (cyclic) alkanes > chlorinated aliphatics > chlorinated aromatics (see Table 18.3). High-molecular-weight NAPLs, such as open-chain (long) aliphatics, highly chlorinated aromatics, and complex PAHs (e.g., benzo[*a*]pyrene) are more resistant to biodegradation and may not be degraded noticeably by microorganisms.

Autochthonous microorganisms to a certain degree can degrade organic contaminants, such as LNAPLs and DNAPLs, by facilitating oxidation-reduction (a.k.a. redox) reactions. The biodegradation process typically involves a consortium of bacteria (e.g., *Acinetobacter*, *Anthrobacter*, *Corynebacterium*, *Flavobacterium*, *Microcossus*, *Nocardia*, *Pseudomonads*, etc.) rather than a single one (Chapelle, 1993; Riser-Roberts, 1992). They metabolize NAPLs by breaking carbon-carbon and carbon-hydrogen covalent bonds, converting them into energy, and using the carbon for cell growth. During this process, NAPLs act as electron donors. Electrons from the NAPL molecules are transferred to certain compound(s) (called terminal electron acceptors or TEAs) found in the soil and water media. In the aerobic biodegradation of NAPLs, dissolved oxygen (O_2) in groundwater is preferentially used as a TEA. As a result, in NAPL-contaminated parts of the aquifer, dissolved oxygen concentrations decrease very rapidly (Chapelle, 1999). When dissolved oxygen levels decrease further (e.g., below 0.5 mg/L), anoxic conditions will start to develop within the dissolved NAPL plume, where anaerobic bacterial populations start to dominate the contaminated zone and anaerobic biodegradation prevails. The anaerobic bacterial populations begin to flourish and use other TEAs in a well-defined sequence including nitrate (NO_3^-), Mn(IV) and Fe(III) in Mn/Fe oxyhydroxides, and sulfate (SO_4^{2-}) (Chapelle, 1999; Ponnamperuma, 1972). When dissolved oxygen is completely exhausted within the contaminant plume, methanogenesis commences, wherein carbon dioxide (CO_2) gas is used as an electron donor and methane (CH_4) gas is produced (Chapelle, 1999). However, it should be noted that the anaerobic biodegradation progresses much more slowly than the aerobic (e.g., O_2 reduction) (Borden, 1994; Rifai et al., 1995). In aquifers contaminated by NAPLs, microbially mediated aerobic/anaerobic biodegradation processes utilize TEAs in a sequential manner and convert organic contaminants into mostly harmless substances, such as carbon dioxide (CO_2) and water (H_2O) (Chapelle, 1999). As an example of LNAPL biodegradation, the aerobic/anaerobic reactions for one of the most commonly occurring BTEX component, toluene (C_7H_8) is given below:

$$\text{Oxygen reduction}: \quad C_7H_8 + 9O_2 \rightarrow 7CO_2 + 4H_2O \ (\text{Aerobic}) \tag{18.1}$$

$$\text{Denitrification}: \quad 5C_7H_8 + 36NO_3^- + H^+ \rightarrow 35HCO_3^- + 3H_2O + 18N_2 (\text{Anaerobic}) \tag{18.2}$$

$$\text{Mn(IV) reduction}: \quad C_7H_8 + 18MnO_2 + 29H^+ \rightarrow 7HCO_3^- + 8Mn^{2+} + 15H_2O \ (\text{Anaerobic}) \tag{18.3}$$

$$\text{Fe(III)reduction}: \quad C_7H_8 + 36FeOOH + 65H^+ \rightarrow 7HCO_3^- + 36Fe^{2+} + 51H_2O \text{ (Anaerobic)} \tag{18.4}$$

$$\text{Sulfate reduction}: \quad 2C_7H_8 + 9SO_4^{2-} + 6H_2O \rightarrow 14HCO_3^- + 5H_2S + 4HS^- \text{ (Anaerobic)} \tag{18.5}$$

$$\text{Methanogenesis}: \quad 2C_7H_8 + 10H_2O \rightarrow 5CO_2 + 9CH_4 \text{ (Anaerobic)} \tag{18.6}$$

As shown by the set of stoichiometric reactions given for toluene (Eqs. 18.1–18.6), in LNAPL-contaminated aquifers, the concentrations of dissolved oxygen (O_2), nitrate (NO_3^-), and sulfate (SO_4^{2-}) will tend to decrease, whereas manganese (Mn^{2+}), iron (Fe^{2+}), bicarbonate (HCO_3^-), and methane (CH_4) concentrations will tend to increase substantially within the contaminant plume. Meanwhile, due to sequential reduction of electron acceptors, LNAPL concentrations decrease and the redox potential (Eh) of the groundwater becomes gradually negative. Monitoring of these physicochemical indicators in groundwater can give clues about the occurrences, as well as the types and locations of active biodegradation processes within a LNAPL plume in the saturated zone (Cheon et al., 2004; Cozzarelli et al., 1995; Güler, 2009). If naturally occurring biodegradation reaction rates are too slow, oxygen and inorganic nutrients (e.g., containing nitrogen and phosphorus) can be added to NAPL-contaminated part(s) of the aquifer to increase the reaction rates.

Although the final products are mostly innocuous substances, some of the intermediate reaction steps may yield compounds that are more toxic than the original NAPL contaminant (Weiner, 2008). For example, the DNAPL contaminant, tetrachloroethene (C_2Cl_4), can be anaerobically biodegraded to vinyl chloride (C_2H_3Cl) (Eqs. 18.7–18.10), of which solubility, mobility, and toxicity exceeds the former compound (see Table 18.2). In this set of reactions, reductive dechlorination of chlorinated ethenes occur in the following order: PCE (tetrachloroethene, a.k.a. perchloroethene), TCE (trichloroethene), DCE (dichloroethene) isomers, vinyl chloride, and ethene, where chloride (Cl) atoms sequentially replaced one by one with hydrogen (H) atoms (Wiedemeier et al., 1999). Among the three DCE isomers, *cis*-1,2-DCE is typically the most abundant chlorinated ethene compared to the other two given in Eq. 18.9 (Wiedemeier et al., 1999). In PCE-contaminated aquifers the rate of reductive dechlorination decreases with the decreasing number of Cl atoms in the byproduct DNAPL molecules (Wiedemeier et al., 1999):

$$\text{PCE}\,(CCl_2 = CCl_2) \rightarrow \text{TCE}\,(CCl_2 = CHCl) \tag{18.7}$$

$$\text{TCE}\,(CCl_2 = CHCl) \rightarrow cis - \text{or}\,trans - 1,2 - \text{DCE}\,(CHCl = CHCl)\,\text{or}\,1,1 - \text{DCE}\,(CCl_2 = CH_2) \tag{18.8}$$

$$cis - \text{or}\,trans - 1,2 - \text{DCE}\,(CHCl = CHCl)\,\text{or}\,1,1 - \text{DCE}\,(CCl_2 = CH_2) \rightarrow \text{Vinyl chloride}\,(CHCl = CH_2) \tag{18.9}$$

$$\text{Vinyl chloride}\,(CHCl = CH_2) \rightarrow \text{Ethene}\,(CH_2 = CH_2) \tag{18.10}$$

Naturally occurring aerobic/anaerobic biodegradation processes occurring in NAPL-contaminated sites were accepted as a passive remediation method known as "monitored natural attenuation" (ASTM, 2004; US EPA, 1999), and employed in a number of NAPL-contaminated sites with suitable site conditions (Chen et al., 2005; Cho et al., 1997; Cozzarelli et al., 2001; Suarez and Rifai, 2002). Monitored natural attenuation method is generally more economical, operative, and easier to implement than most engineered cleanup approaches (e.g., soil-vapor extraction and permeable reactive barriers) (Güler and Alpaslan, 2006). However, at some sites, the abovementioned natural processes may not be effective and engineered remedial actions are needed to reduce risks to human and environmental receptors (see Güler and Alpaslan, 2006; US EPA, 2017e). In general, multiple technologies are combined into what is known as the "treatment train," to effectively remediate the contaminated site. These remediation technologies use three principal strategies to reduce the harmful effects of contaminants, including the:

- extraction or separation of the contaminant from the environmental media (e.g., soil and water)
- destruction or alteration of the contaminant's chemical structure (e.g., degradation)
- immobilization of the contaminant in place to prevent its further migration (e.g., containment)

Monitoring of the prevailing groundwater physicochemical regime in NAPL-contaminated aquifers can give important clues about the underlying natural attenuation processes actively occurring at the site. GIS technology provides the tools needed to create, manage, analyze, and visualize the data associated with the site and can be used to develop an integrated conceptual model (in 2D and 3D) for the understanding of contaminant behavior in the subsurface. GIS-based map layers depicting aquifer characteristics (e.g., well locations, water levels, physicochemical properties of groundwater, aquifer lithology, mineral composition, amount of organic matter, porosity, soil hydraulic conductivity and granulometry, etc.), contaminant distribution and concentrations, and land use/cover information is of the utmost importance when predicting the source of contamination and the fate of individual contaminant plumes, as well as identifying possible receptors located downgradient of the contamination source(s). However, the information related to water levels, water chemistry, contaminant distribution and concentrations should be updated periodically to monitor the physical and chemical changes occurring in the subsurface water and contaminant(s).

18.9 CONCLUDING REMARKS

GIS technology can be used to integrate a variety of data types that are much needed for the effective planning and implementation of remedial solutions. When the capabilities of GIS are combined with other online/offline data and information sources, it can play a central role in rendering the subsurface distribution of aquifer, groundwater, and contaminant (e.g., NAPL) characteristics, properties, and features, and can be used to develop an integrated conceptual model (in 2D and 3D) of the contaminated site. NAPLs present a great challenge in terms of cleanup efforts due to their diverse chemical/physical and toxicity impacts to the natural environment and living organisms. Even though some of these contaminants are fairly well characterized with respect to their behavior and fate in the surface and subsurface environments, such information on a great majority of them is lacking and the gains cannot keep pace with the developments in the field of organic chemistry. Every day tens of new organic substances are added to the existing list and we know practically nothing about their potential effects on living and nonliving things. Much work is needed to close the information gaps that exist among the scientific community, practitioners, and decision makers who currently engaged in NAPL-contamination research and remediation. Contamination caused by LNAPLs can be more easily remediated when compared with DNAPLs; however, a successful cleanup can be achieved only after several decades and spending large sums of money. The most effective way to protect natural resources (e.g., biota, air, soil, and water) is certainly to not contaminate the environment that we all depend on for our existence in first place. Science is advancing every day and scientists are continually seeking, by all the means in their power, innovative and sustainable technologies (e.g., green chemistry) to save the mother Earth.

REFERENCES

Apodaca, R.L., 2011. Sixty-Four Free Chemistry Databases. http://depth-first.com/articles/2011/10/12/sixty-four-free-chemistry-databases. (Accessed 18 October 2017).

ASTM, 2004. Standard E1943-98(2004), Standard Guide for Remediation of Ground Water by Natural Attenuation at Petroleum Release Sites. American Society for Testing and Materials, West Conshohocken, PA.

Atlas, R.M., 1981. Microbial degradation of petroleum hydrocarbons: an environmental perspective. Microbiol. Rev. 45 (1), 180–209.

ATSDR (Agency for Toxic Substances and Disease Registry), 2017. The ATSDR 2017 Substance Priority List. https://www.atsdr.cdc.gov/spl. (Accessed 19 October 2017).

Bedient, P.B., Rifai, H.S., Newell, C.J., 1999. Ground Water Contamination: Transport and Remediation, second ed. PTR Publications, Upper Saddle River, NJ.

Bockelmann, A., Zamfirescu, D., Ptak, T., Grathwohl, P., Teutsch, G., 2003. Quantification of mass fluxes and natural attenuation rates at an industrial site with a limited monitoring network: a case study. J. Contam. Hydrol. 60 (1–2), 97–121.

Borden, R.C., 1994. Natural bioremediation of hydrocarbon-contaminated ground water. In: Norris, R.D., Hinchee, R.E., Brown, R., McCarty, P.L., Semprini, L., Wilson, J.T. et al. (Eds.), Handbook of Bioremediation. Lewis Publishers, Boca Raton, FL, pp. 177–199.

Boulding, J.R., 1995. Practical Handbook of Soil, Vadose Zone, and Ground-Water Contamination: Assessment, Prevention, and Remediation. Lewis Publishers, CRC Press, Boca Raton, FL.

Calabrese, E.J., Kenyon, E.M., 1991. Air Toxics and Risk Assessment. Lewis Publishers, Chelsea, MI.

CAS (Chemical Abstracts Service), 2017. Regulated Chemicals Information. http://www.cas.org/content/regulated-chemicals/substance. (Accessed 15 October 2017).

CDC (Centers for Disease Control and Prevention), 2016. Emergency Preparedness and Response. https://emergency.cdc.gov/chemical. (Accessed 19 October 2017).

CERCLA, 1980. Comprehensive Environmental Response, Compensation, and Liability Act of 1980, Public Law No. 96-510, 94 Stat. 2767.

Chapelle, F.H., 1993. Ground-Water Microbiology and Geochemistry. John Wiley, New York.

Chapelle, F.H., 1999. Bioremediation of petroleum hydrocarbon-contaminated ground water: the perspectives of history and hydrology. Ground Water 37 (1), 122–132.

Chen, K.F., Kao, C.M., Wang, J.Y., Chen, T.Y., Chien, C.C., 2005. Natural attenuation of MTBE at two petroleum-hydrocarbon spill sites. J. Hazard. Mater. 125 (1–3), 10–16.

Cheon, J.Y., Lee, J.Y., Lee, K.K., 2004. Characterization of the hydrogeologic environment at a petroleum hydrocarbon contaminated site in Korea. Environ. Geol. 45 (6), 869–883.

Cho, J.S., Wilson, J.T., DiGiulio, D.C., Vardy, J.A., Choi, W., 1997. Implementation of natural attenuation at a JP-4 jet fuel release after active remediation. Biodegradation 8 (4), 265–273.

Coates, J.D., Chakraborty, R., McInerney, M.J., 2002. Anaerobic benzene biodegradation—a new era. Res. Microbiol. 153 (10), 621–628.

Cozzarelli, I.M., Herman, J.S., Baedecker, M.J., 1995. Fate of microbial metabolites of hydrocarbons in a coastal plain aquifer: the role of electron acceptors. Environ. Sci. Technol. 29 (2), 458–469.

Cozzarelli, I.M., Bekins, B.A., Baedecker, M.J., Aiken, G.R., Eganhouse, R.P., Tuccillo, M.E., 2001. Progression of natural attenuation processes at a crude-oil spill site: I. Geochemical evolution of the plume. J. Contam. Hydrol. 53 (3–4), 369–385.

Dean, B.J., 1985. Recent findings on the genetic toxicology of benzene, toluene, xylenes and phenols. Mutat. Res. 154 (3), 153–181.

DiGiulio, D.C., Cho, J.S., 1990. Conducting field tests for evaluation of soil vacuum extraction application. In: Proceedings of Fourth National Outdoor Action Conference on Aquifer Restoration, Ground Water Monitoring, and Geophysical Methods. National Ground Water Association, Dublin, OH, pp. 587–601.

Domenico, P.A., Schwartz, F.W., 1990. Physical and Chemical Hydrogeology. John Wiley & Sons, New York.

Favre, H.A., Powell, W.H., 2013. Nomenclature of Organic Chemistry. IUPAC Recommendations and Preferred Name 2013, The Royal Society of Chemistry, Cambridge.

Grathwohl, P., 2000. Time scales of remediation of complex source zones. In: Johnston, C.D. (Ed.), Contaminated Site Remediation: From Source Zones to Ecosystems. CSIRO Land and Water Center for Groundwater Studies, Melbourne, pp. 635–650.

Güler, C., 2009. Site characterization and monitoring of natural attenuation indicator parameters in a fuel contaminated coastal aquifer: Karaduvar (Mersin, SE Turkey). Environ. Earth Sci. 59 (3), 631–643.

Güler, C., Alpaslan, M., 2006. In-situ remediation technologies applied for petroleum hydrocarbon contaminated aquifers. Jeol. Mühendis. Derg. 30-2 (65), 33–50.

Heller, S.R., McNaught, A., Pletnev, I., Stein, S., Tchekhovskoi, D., 2015. InChI, the IUPAC international chemical identifier. J. Cheminform. 7 (23), 1–34.

Hironaka, A., 2014. Greening the Globe, World Society and Environmental Change. Cambridge University Press, New York.

IOM (Institute of Medicine), 2014. Identifying and Reducing Environmental Health Risks of Chemicals in Our Society: Workshop Summary. The National Academies Press, Washington, DC. https://www.ncbi.nlm.nih.gov/books/NBK268889. (Accessed 21 October 2017).

IUPAC (International Union of Pure and Applied Chemistry), 2017. IUPAC Nomenclature. https://iupac.org/what-we-do/nomenclature. (Accessed 22 October 2017).

Larrañaga, M.D., Lewis Sr., R.J., Lewis, R.A., 2016. Hawley's Condensed Chemical Dictionary. 16th ed John Wiley & Sons, Hoboken, NJ.

Lee, M.D., Thomas, J.M., Borden, R.C., Bedient, P.B., Ward, C.H., Wilson, J.T., Conway, R.A., 1988. Biorestoration of aquifers contaminated with organic compounds. Crit. Rev. Environ. Control. 18 (1), 29–89.

Lund, L., 1995. Changes in UST and LUST: the federal perspective. Tank Talk 10 (2–3), 7.

Miller, D., Rivera, R.G., Travis, C.C., Solis, R., Calva, L., 2000. An approach to soil restoration of hydrocarbon contamination using a carbon fractionation approach. In: 7th International Petroleum Environmental Conference, Albuquerque, New Mexico, November 2000. National Energy Technology Laboratory, pp. 1264–1286.

Molson, J.W., Barker, J.F., Frind, E.O., Schirmer, M., 2002. Modeling the impact of ethanol on the persistence of benzene in gasoline-contaminated groundwater. Water Resour. Res. 38 (1), 1–12.

Newman, R.S., 2016. Love Canal: A Toxic History From Colonial Times to the Present. Oxford University Press, New York.

NIST (The National Institute of Standards and Technology), 2017. NIST Chemistry WebBook. http://webbook.nist.gov/chemistry. (Accessed 15 October 2017).

Nkedi-Kizza, P., Rao, P.S.C., Hornsby, A.G., 1985. Influence of organic cosolvents on sorption of hydrophobic organic chemicals by soils. Environ. Sci. Technol. 19 (10), 975–979.

Odermatt, J.R., 1994. Natural chromatographic separation of benzene, toluene, ethylbenzene and xylenes (BTEX compounds) in a gasoline contaminated ground water aquifer. Org. Geochem. 21 (10–11), 1141–1150.

Olah, G.A., Molnár, Á., 2003. Hydrocarbon Chemistry, second ed. John Wiley & Sons, Hoboken, NJ.

OSHA (Occupational Safety and Health Administration), 2017. Carcinogens—OSHA Standards. https://www.osha.gov/SLTC/carcinogens/standards.html. (Accessed 11 December 2017).

Ponnamperuma, F.N., 1972. The chemistry of submerged soils. Adv. Agron. 24, 29–96.

Rifai, H.S., Borden, R.C., Wilson, J.T., Ward, C.H., 1995. Intrinsic bioattenuation for subsurface restoration. In: Wilson, J.T., Hinchee, R.E., Downey, D.C. (Eds.), Intrinsic Bioremediation. Battelle Press, Columbus, OH, pp. 1–29.

Riser-Roberts, E., 1992. Bioremediation of Petroleum Contaminated Sites. CRC Press, Boca Raton, FL.

RSC (Royal Society of Chemistry), 2015. RSC ChemSpider Database. http://www.chemspider.com/Default.aspx. (Accessed 15 October 2017).

SARA, 1986. Superfund Amendments and Reauthorization Act of 1986, Public Law No. 99-499, 100 Stat. 1613.

Speight, J.G., 2002. Handbook of Petroleum Product Analysis. John Wiley & Sons, Hoboken, NJ.

Suarez, M.P., Rifai, H.S., 2002. Evaluation of BTEX remediation by natural attenuation at a coastal facility. Ground Water Monit. Rem. 22 (1), 62–77.

US Census Bureau, 2010. US Census Bureau 2010 Census. https://www.census.gov/2010census. (Accessed 5 November 2016).

US DOE (Department of Energy), 2017. Alternative Fuels Data Center: Alternative Fuels and Advanced Vehicles. https://www.afdc.energy.gov/fuels. (Accessed 16 October 2017).

US EIA (Energy Information Administration), 1999. Petroleum: An Energy Profile. Department of Energy, Washington, DC DOE/EIA-0545.

US EIA (Energy Information Administration), 2017a. International Energy Statistics. https://www.eia.gov/beta/international/data/browser. (Accessed 29 November 2017).

US EIA (Energy Information Administration), 2017b. U.S. Energy Mapping System. https://www.eia.gov/state/maps.php. (Accessed 29 November 2017).

US EPA (Environmental Protection Agency), 1991. In: Risk Assessment Guidance for Superfund (Ed.), Volume I–Human Health Evaluation Manual (Part B, Development of Risk-Based Preliminary Remediation Goals). EPA/540/R-92/003. United States Environmental Protection Agency, Office of Emergency and Remedial Response, Washington, DC.

US EPA (Environmental Protection Agency), 1999. Use of Monitored Natural Attenuation at Superfund, RCRA Corrective Action, and Underground Storage Tank Sites. OSWER Directive 9200. Office of Solid Waste and Emergency Response, Washington, DC, pp. 4–17P.

US EPA (Environmental Protection Agency), 2009. National Primary Drinking Water Regulations (NPDWRs), EPA 816-F-09-004. https://www.epa.gov/ground-water-and-drinking-water/national-primary-drinking-water-regulations. (Accessed 23 October 2017).

US EPA (Environmental Protection Agency), 2015. Integrated Risk Information System (IRIS). https://www.epa.gov/iris. (Accessed 24 October 2017).

US EPA (Environmental Protection Agency), 2017a. EPA History. Press Releases and Articles, Love Canal.https://www.epa.gov/history/love-canal. (Accessed 20 October 2017).

US EPA (Environmental Protection Agency), 2017b. Superfund. https://www.epa.gov/superfund. (Accessed 20 October 2017).

US EPA (Environmental Protection Agency), 2017c. Superfund: National Priorities List (NPL). https://www.epa.gov/superfund/superfund-national-priorities-list-npl. (Accessed 20 October 2017).

US EPA (Environmental Protection Agency), 2017d. Superfund Chemical Data Matrix (SCDM Query). https://www.epa.gov/superfund/superfund-chemical-data-matrix-scdm-query. (Accessed 19 October 2017).

US EPA (Environmental Protection Agency), 2017e. Contaminated Site Clean-Up Information (CLU-IN). https://clu-in.org. (Accessed 24 October 2017).

US EPA (Environmental Protection Agency), 2017f. Toxics Release Inventory (TRI) Program. https://www.epa.gov/toxics-release-inventory-tri-program. (Accessed 24 October 2017).

US EPA (Environmental Protection Agency), 2017g. Cleaning Up Underground Storage Tank (UST) Releases. https://www.epa.gov/ust/cleaning-underground-storage-tank-ust-releases. (Accessed 11 December 2017).

US NLM (National Library of Medicine), 2017a. Toxicology Data Network (TOXNET). https://toxnet.nlm.nih.gov. (Accessed 19 October 2017).

US NLM (National Library of Medicine), 2017b. TOXMAP: Environmental Health Maps. https://toxmap.nlm.nih.gov. (Accessed 5 November 2017).

US NLM (National Library of Medicine), 2017c. PubChem Open Chemistry Database. https://pubchem.ncbi.nlm.nih.gov. (Accessed 5 November 2017).

Weiner, E.R., 2008. Applications of Environmental Aquatic Chemistry: A Practical Guide, second ed. CRC Press, Taylor & Francis Group, Boca Raton, FL.

Wiedemeier, T.H., Rifai, H.S., Newell, C.J., Wilson, J.T., 1999. Natural Attenuation of Fuels and Chlorinated Solvents in the Subsurface. John Wiley & Sons, New York.

FURTHER READING

Güler, C., Alpaslan, M., Kurt, M.A., Temel, A., 2010. Deciphering factors controlling trace element distribution in the soils of Karaduvar industrial-agricultural area (Mersin, SE Turkey). Environ. Earth Sci. 60 (1), 203–218.

Güler, C., Kaplan, V., Akbulut, C., 2013. Spatial distribution patterns and temporal trends of heavy-metal concentrations in a petroleum hydrocarbon-contaminated site: Karaduvar coastal aquifer (Mersin, SE Turkey). Environ. Earth Sci. 70 (2), 943–962.

ICSS (International Centre for Soil and Contaminated Sites), 2006. Manual for Biological Remediation Techniques. ICSS at the German Federal Environmental Agency, Dessau. Available at https://www.umweltbundesamt.de/sites/default/files/medien/publikation/long/3065.pdf.

Chapter 19

Prediction of Spatial Fluoride Concentrations by a Hybrid Artificial Neural Network in Complex Aquifers

Ata Allah Nadiri*, Rahman Khatibi† and Frank T.-C. Tsai‡

*Department of Earth Sciences, Faculty of Natural Sciences, University of Tabriz, Tabriz, Iran †GTEV-ReX Limited, Swindon, United Kingdom
‡Department of Civil and Environmental Engineering, Louisiana State University, Baton Rouge, LA, United States

Chapter Outline

19.1 INTRODUCTION

This research is focused on the spatial distribution of fluoride in aquifers by using artificial neural networks (ANNs) to develop a predictive model for complex aquifers, where there are some measurements available but with known regional variations. The application of ANNs to hydrological and environmental problems is diverse and includes prediction, estimation, classification, and pattern recognition; they are equally applicable to generating spatial distributions of contaminants. The chapter presents 10 variations of the ANN model used to predict spatial fluoride concentrations in complex aquifers including basaltic, nonbasaltic, and mixing zones. The novelty of the chapter is on formulating a hybrid ANN (HANN) model to predict spatial variations in the fluoride concentrations in aquifers, as well as to take account of the inherent complexity in the data.

The spatial generation of hydrochemical contaminants distributed in aquifers can be analyzed using flow and transportation equations, but their use is precluded in this chapter as they require a large amount of measured data. Also, construction and calibration of these models are time-consuming and often costly (Nadiri et al., 2014; Nadiri and Hassan, 2015). A lack of adequate data is often a critical problem so that the resolution of the underlying problems requires considerable time. However, artificial intelligence (AI) provides the tools to deal with a variety of these problems, including the prediction of the spatial distribution of contaminants in aquifers, by training their models and enabling the learning of parameters from locally measured data.

Although trace amounts of fluoride (less than 1.5 mg/L) are necessary in food and drinking water to avoid weak bones and teeth, concentrations greater than this allowable amount can have harmful impacts on human health, such as osteoporosis and endemic fluorosis (Kharb and Susheela, 1994; Jacks et al., 2005). Different causes of fluorosis are discussed in the literature, including the impact of long-term ingestion of fluoride-bearing water and food (Dissanayake, 1991; Rao Nagendra, 2003). A permissible concentration is set to less than 1.5 mg/L for drinking water by the USEPA (Environmental Protection Agency, 2009) and by WHO (World Health Organization, 2006). Notably, the mean concentration of the fluoride ions in the study area was 2.8 mg/L and its maximum value was approximately 8.1 mg/L. This was the driver for the investigation presented in this chapter.

GIS and Geostatistical Techniques for Groundwater Science. https://doi.org/10.1016/B978-0-12-815413-7.00019-5

High fluoride concentrations in groundwater occur in different countries around the world, e.g., east Africa, southeastern Korea, northern China, Turkey, India, and Iran (Gizaw, 1996; Oruc, 2003; Jacks et al., 2005; Kim and Jeong, 2005; Guo et al., 2007; Nadiri et al., 2014; Chitsazan et al., 2015). The origins of fluoride ions are often geogenic due to fluoride-bearing minerals (e.g., fluorite, fluorapatite, topaz, mica, cryolite and amphiboles). Water-rock interactions are often regarded as the main factor in fluoride occurrence in the aquifers within various geological settings (Saxena and Ahmed, 2003; Nadiri, 2015). Fluoride contamination in groundwater may vary significantly due to the contribution from: (i) dry and rainy seasons (Saxena and Ahmed, 2001), (ii) physicochemical parameters (Ozsvath, 2006), (iii) types of geological settings (Gupta et al., 2005), (iv) tectonic factor (Handa, 1975), and (v) regional climate (Davraz et al., 2008). The presence of water in aquifers triggers dissolution and release of fluoride ions when an alkaline medium (pH = 7.6 to 8.6) and moderate electric conductivity (750 to 1750 µS/cm) prevail (Saxena and Ahmed, 2001).

The spatial distribution of fluoride in aquifers is a key parameter for the risk-based management of water resources. Techniques used for estimating the hydrological and hydrochemical characteristics of fluoride include statistics (Shrestha and Kazama, 2007), geostatistics (Diodato et al., 2010), geographic information systems (GIS) (Hudak and Sanmanee, 2003), and a combination of the aforementioned techniques (Xie et al., 2008). The spatial variations in fluorosis and the fluoride content in the water in Bophuthatswana (South Africa) were studied by Zietsman (1991) using statistical methods, which indicated that the differences in the occurrence of fluorosis did not adequately depend on the fluoride concentration in the water. Hudak and Sanmanee (2003) deduced that volcanic ash deposits in north-central Texas influenced fluoride concentrations in the Woodbine aquifer. Xie et al. (2008) used the geostatistical method to depict the spatial variations in the fluoride concentrations in the soil of the Hang-Jia-Hu Plain, China.

ANNs are empirical and black-box models with the capability to recognize a nonlinear relationship between the input and output for given datasets. Their successful applications to model diverse nonlinear systems in water sciences are wide-ranging (ASCE, 2000), including spatiotemporal groundwater level predictions (Coppola et al., 2005; Nadiri, 2007; Nadiri et al., 2007; Nourani et al., 2008a,b), parameter estimation (Merdun et al., 2006; Garcia and Shigidi, 2006; Nadiri et al., 2014, 2017a,b,c), and water-quality estimation (Sharma et al., 2003; Almasri and Kaluarachchi, 2005; Nadiri et al., 2014; Chitsazan et al., 2015). More than 90% of research has studied water-quantity variables and less than 10% of research has focused on water-quality variables (Maier et al., 2010). This chapter is focused on water quality.

Integrating surface water- and groundwater-quality modeling in complex aquifer systems is challenging. This study developed a HANN model to estimate the spatial distribution of fluoride concentrations in a complex aquifer system in the Poldasht and Bazargan Plains in the Maku area of West Azerbaijan, Iran, and its efficiency was compared with ANNs, a geostatistical method, and a multiple linear regression model. The HANN was applied to estimate fluoride concentrations in both surface water and groundwater. The aquifer under study has complex regional variations, which can be demarcated as basaltic and nonbasaltic zones. Previous studies on high fluoride concentrations in the aquifers of the study area include Asghari Moghaddam and Fijani (2009), Nadiri et al. (2014), and Chitsazan et al. (2015). This chapter identifies correlated variables that affect fluoride concentration distributions. The study uses 10 ANN variations to develop predictive models and investigate their performance.

19.2 VARIATIONS OF ARTIFICIAL NEURAL NETWORKS—MODEL SPECIFICATIONS

Khatibi et al. (2017) reviewed the developmental context of ANNs and multilayer perceptron (MLP), which emerged as working tools in the late 1980s and early 1990s. Their topology comprised an input layer, hidden layers, and an output layer, and their processing units in each layer were known as neurons or perceptrons. The number of neurons in the input and output layers were set by the number of input and output variables; however, the numbers of neurons in the hidden layers were identified through a trial-and-error procedure (ASCE, 2000).

ANN models establish a relationship between input and output data and therefore they are operated in the prediction mode, which is referred to as the feedforward ANN mode. The incoming signal, X_{ij} for neuron j from n inputs is transformed into the output signal through a polynomial function in the form of a weighted summation expressed by: $S_j = \sum_{i=1}^{n} W_{ij} X_{ij} + b_j$, where X_{ij} is a n inputs ($X_{1j}, X_{2j}, X_{3j}, ..., X_{nj}$), W_{ij} are weights and b_j is bias; and S_j is the (linear or nonlinear) transfer function accounting for the outgoing signal of the neuron at j.

Feedforward formulations require the values of their inherent parameters (weights and bias values) to be identified and this is an inverse problem of feedforward models, which is referred to as the backpropagation problem and modelers refer to the process as model training activities (Nadiri et al., 2014; Nadiri, 2015; Khatibi et al., 2017). A host of variations to the above formulation exist and those that were used in this study are specified below.

Generally, ANNs are categorized as feedforward neural networks (FNN) and recurrent neural networks (RNN). These have various architectures and those used to study hydrological systems are MLPs, the Elman network (EN), the radial basis function (RBF) networks, and the learning vector quantization (LVQ); for more details see ASCE (2000) and Sharma et al. (2003). MLP and RBF networks are FNNs, and EN and LVQ are RNNs. Each of these four types of ANNs are used in this study. Their inverse problems create yet another set of variations, referred to as back propagation training algorithms, which are implemented as supervised learning approaches; for more details, see Rumelhart et al. (1986) and ASCE (2000).

There are various types of learning algorithms for training ANNs and those used in hydrological studies include the Levenberg-Marquardt (LM) method (ASCE, 2000), the quasi-Newton algorithm (Li and Liu, 2005), the gradient descent with momentum and adaptive learning rate backpropagation (GDX), and the Bayesian regularization (BR) technique; for more details see Anctil et al. (2004). This study employed the following combinations: MLP-LM, EN-LM, MLP-BFGS, EN-BFGS, MLP-GDX, EN-GDX, MLP-BR, EN-BR, and RBF, as well as an additional LVQ model. These are outlined below:

LM: The LM algorithm is an adaptation of the Gauss-Newton algorithm to minimize optimization problems. It approximates the Hessian matrix by linearizing it. It is a fast algorithm when a solution nears a minimum. *BFGS*: The Broyden-Fletcher-Goldfarb-Shanno (BFGS) algorithm is a quasi-Newton (or secant) method, which is another efficient approach that updates an approximate Hessian matrix at each iteration in the algorithm. The update is computed as a function of the gradient. *GDX*: This is also a gradient descent algorithm, which combines adaptive learning rate with momentum training (GDX) and therefore its backpropagation process calculates derivatives of a performance cost function with respect to the weight and bias variables of the network. *BR*: The Bayesian regularization (BR) algorithm searches for optimum parameter values without overfitting the objective function, regardless of the size of the network, in which the weights of the network are assumed to be random variables with specified statistical distributions (Foresee and Hagan, 1997). The regularization parameters are related to the unknown variances associated with these distributions and hence the parameters can be estimated using statistical methods (see Anctil et al., 2004).

RBF networks are nonlinear hybrid networks and typically use one hidden layer based on nonlinear Gaussian transfer functions rather than the standard sigmoidal functions employed by MLPs. Typical RBF networks have similar conventional three-layer FNN topology but operate in a fundamentally different manner. The output layer produces the results from Gaussian functions in a linear fashion. The centers and widths of Gaussian functions are assigned by unsupervised learning methods when a supervised learning method is utilized by the output layer (Haykin, 1999). The faster learning of the RBF network is the main advantage compared with the other networks.

Elman (1990) introduced a partially connected network, now known as the Elman network (EN), which is an RNN as it feeds the outgoing signals of neurons in the hidden layer back to itself by introducing a context layer. ENs consist of an input layer, a hidden layer, a context layer, and an output layer. Each input neuron connects to all the hidden neurons, and each hidden neuron connects to all the context neurons. The number of the hidden neurons is the same as those of the context neurons. The context layer is interconnected with the hidden-layer neurons and plays the role of the network's memory. The outgoing signal from the hidden-layer neurons feeds back to the context-layer neurons to be used as new input data in the input-layer neurons (Medsker and Jain, 1999). For more mathematical details on RNN, see Islam and Kothari (2000).

LVQ is a powerful method for classifying nonlinear patterns within specified boundaries. LVQ networks link to an ANN because specific output classes should be used. The LVQ network models are constructed through the following steps: (i) use supervised training to learn the output neurons and label them for the known classes; (ii) use supervised training to learn refinement of the map according to the known features of the output classes and groups (Kasabov, 1998). The topology of LVQs is composed of two layers: (i) the competitive layer, which uses a competitive transfer function for training, and (ii) the second layer, which is a linear layer and its neurons transform the classification information of defined classes from the first layer. Through these neurons the weighted inputs, net inputs, and biases are calculated. The LVQ algorithm represents clusters of vectors at the hidden neurons and groups the clusters at the output neurons to form the desired classes. In this study, the MATLAB platform (Mathworks, 2009) was used to build all the models and their runs.

19.3 STUDY AREA

The Bazargan-Poldasht Plain stretches over a large area in the north of the west Azarbaijan province, northwest Iran, but the study area was focused on the areas within the vicinity of the city of Maku and is to be referred to as the Maku area. It covers the border with Turkey in the west and north and by the Republic of Azerbaijan in the north and northeast, where the River Araz marks the natural border. The southern boundary of the study area is marked by local mountain ranges. The area is approximately $1600 \, km^2$, of which basaltic lavas cover up to $402 \, km^2$. Three main cities in study area consist of Maku (also known as Dash Maki), Poldasht (also known as Araplar) and Bazargan. The plain is drained by the Zangmar River with its main tributary being Sari Su (Yellow Water), which flows through the study area into the River Araz, as shown in Fig. 19.1.

The Maku area is an arid zone with a yearly mean temperature of 11°C (1996–2013) and annual mean precipitation of 287mm (1992–2013). May and September months have maximum and minimum rainfall, respectively. In this area, groundwater is the main source for the supply of water for drinking, agriculture, and industry. There are 12 springs producing a large amount of water ranging from 20 to 4000 L/s and groundwater withdrawal is the main source of water in the Maku area.

Hydrogeological examination of the study has invoked a perceptual model, according to which the area may be demarcated into two main zones: (i) the aquifer in the nonbasaltic zones is composed of the Qum Formation (Oligo-Miocene), massive limestone and dolomite of the Permian age (the Ruteh Formation), and other lithologies in small amounts, such as marl, shale, and conglomerate; (ii) the aquifer in the basaltic zone consists of mainly basaltic rocks covered by alluvium (basalt-alluvium aquifer). As discussed by Nadiri et al. (2013) and Fijani et al. (2013), in the nonbasaltic aquifer, ion concentrations, electrical conductivity (EC) and alkalinity are lower than those in basaltic aquifer. Also, Asghari Moghaddam and Fijani (2008) reported that the maximum thickness of the basalt-alluvium aquifer in this area is approximately 150m, which is intensively karstified in some places. The basaltic aquifer is the main source of high fluoride groundwater in the Maku area. Fluoride concentration in some places varies due to the mixing of nonbasaltic and basaltic groundwater, which are called the mixing zone. Using groundwater from the basaltic aquifer causes dental fluorosis in the inhabitants. The groundwater residence time in the aquifer, geochemistry of formation, and relative pH amounts are the main factors controling fluoride concentration in the aquifer of the study area (Asghari Moghaddam and Fijani, 2008).

The Maku area contains two main types of aquifers (basaltic and nonbasaltic aquifers), which have a complex relationship (Nadiri et al., 2014). The mixing zone refers to the areas where the water from these two formations are mixed (Fig. 19.2).

19.4 PRELIMINARY INVESTIGATIONS AND SETTING UP DATASETS

19.4.1 Identifying Significant Variables by Factor Analysis

The fluoride concentration in the study area was investigated by collecting 189 water samples from different water sources (springs, wells, and qanats) over a 9-year period between 2004 and 20013. The sampling locations are shown in Fig. 19.1. The water-quality parameters were obtained by analyzing the water samples in the hydrogeological laboratory of Tabriz University to form datasets for the ANNs models at the next stage. Data collection in the study area encountered some

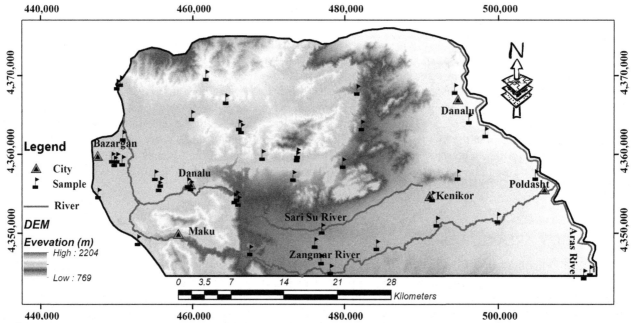

FIG. 19.1 Topographic map of the study area and sample locations.

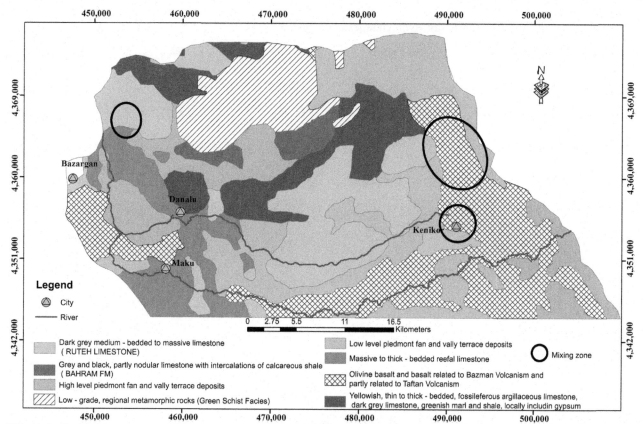

FIG. 19.2 Geological map of the study area—basaltic zones: *hatched*; mixing zones: *circled*.

difficulties due to the geographic conditions, as the Maku area is mountainous and the climate in winter is harsh and cold with heavy snowfall.

The sampling program for fluoride ion concentration (F^-) was part of a larger sampling activity during which measurements were taken for minor ions and major ions (Ca^{2+}, Mg^{2+}, Na^+, K^+, HCO_3^-, CO_3^{2-}, SO_4^{2-} and Cl^-), silicon (SiO_2), pH, and electrical conductivity (EC). These were analyzed in the laboratory as per the standard methods (American Public Health Association, 1998). The acceptable precision of ion analysis is lower than $\pm5\%$. The statistical summary of the 12 water-quality parameters is given in Table 19.1.

Correlation matrices of the sampled data are given in Table 19.2, according to which the fluoride ions are shown to have a positive correlation with Na^+, K^+, HCO_3^-, pH, EC, and SiO_2, and a negative correlation with Ca^{2+}, Mg^{2+}, CO_3^{2-}, and SO_4^{2-}, and the lowest correlation with Cl^-.

The factor analysis (FA) technique (Cattell, 1952) is used to select highly effective water-quality parameters for fluoride contamination as inputs to ANN from the large dataset. The parameters may have different units of measurement and to resolve this problem the raw data were standardized before conducting FA (Nadiri et al., 2013, 2018). The FA technique was carried out using the following steps: (i) generate a linear combination of variables (so-called factors) via the original variables; (ii) select the factors with eigenvalues higher than one; (iii) determine the significance of the correlation between the factors and variables via factor loadings (1 to -1); (iv) adopt the varimax rotation technique (Kaiser, 1958) to rotate the factor loading matrix; and (v) select a set of correlated water-quality parameters in a particular factor as the representative of a special chemical process (contamination, anthropogenic, or genetic).

In this study, the selected three factors are representative of 75.2% of the dataset variance (Fig. 19.3). Factor 1 indicates high positive-factor loadings for F^-, Na^+, K^+, and HCO_3^-, and high negative-factor loading for Ca^{2+}. Factor 2 indicates that SiO_2 has high negative-factor loadings and CO_3^{2-}, SO_4^{2-}, and pH have high positive-factor loadings. Factor 3 indicates that high positive-factor loadings are Na^+, Cl^-, and EC variables.

As per the FA results, factor 1 is a representative of increasing F^- concentrations in water resources, factor 2 is a representative of anthropogenic processes, and factor 3 is a representative of geogenic processes. The F^- concentration is affected by the ion concentrations of HCO_3^-, Na^+, K^+, and Ca^{2+}, which are selected as the input data for the ANN models.

TABLE 19.1 Statistical Characteristics of Hydrochemical Parameters

Parameter	Unit	Minimum	Maximum	Mean	Std. Deviation
F^-	mg/L	0.05	8.1	2.8	1.75
Ca^{2+}	mg/L	4.93	183.3	73.9	29.2
Mg^{2+}	mg/L	<0.1	184.8	47.3	23.6
Cl^-	mg/L	<0.1	442.5	46.4	53.8
CO_3^{2-}	mg/L	<0.1	245	31.6	43.4
HCO_3^-	mg/L	17.08	902.8	568.91	226.6
Na^+	mg/L	0.7	335.67	147.39	101.7
K^+	mg/L	0.4	18.6	8.7	4.9
SiO_2	mg/L	<0.1	129.8	51.42	25.62
SO_4^{2-}	mg/L	<0.1	702.8	118.8	97.6
pH[a]	–	6.5	9.1	7.99	0.59
EC[a]	µS/cm	295	4610	1194	517.46

[a]Measured at 25°C.

TABLE 19.2 Correlation Matrices of Hydrochemical Parameters

Parameter	F^-	Ca^{2+}	Mg^{2+}	Cl^-	CO_3^{2-}	HCO_3^-	Na^+	K^+	SiO_2	SO_4	pH	EC
F^-	1											
Ca^{2+}	−0.68	1										
Mg^{2+}	−0.35	0.24	1									
Cl^-	0	0.14	0.73	1								
CO_3^{2-}	−0.2	0.08	−0.07	−0.39	1							
HCO_3^-	0.29	−0.3	0.2	0.39	−0.8	1						
Na^+	0.64	−0.77	−0.14	0.01	−0.07	0.51	1					
K^+	0.41	−0.61	−0.14	−0.28	0.03	0.3	0.61	1				
SiO_2	0.2	0.05	−0.07	0.23	−0.52	0.39	−0.02	0.06	1			
SO_4^{2-}	−0.21	0.11	0.18	−0.03	0.39	−0.14	0.19	0.13	−0.25	1		
pH	0.5	−0.22	0.36	0.68	−0.29	0.41	0.41	0.12	0.2	−0.02	1	
EC	0.01	−0.29	0.01	−0.28	0.56	−0.09	0.53	0.4	−0.49	0.53	−0.13	1

Normalizing the input and output data is an important procedure to successfully instruct an ANN model, as different hydrochemical parameters have significantly different data ranges. All the input data are scaled in the range between 0 and 1.

19.4.2 Preliminary Setups for Modeling Test Runs

Driven by the FA results above, the ANN input variables considered for each sample point are: Na^+, K^+, HCO_3^-, Ca^{2+}, the universal transverse mercator (UTM) coordinate system for X and Y grid references and the target data is F^-. The first step

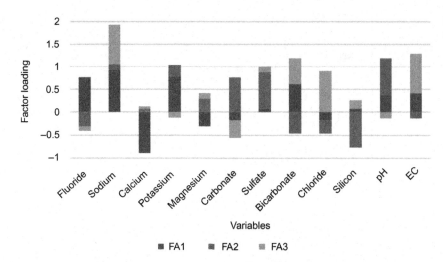

FIG. 19.3 Factor loadings using varimax rotation (loadings larger than 0.6 is selected).

of ANN modeling is training. The aim of ANN training is to optimize the model parameters (e.g., weights, biases, the learning rate, and the number of epochs) that enable an ANN model with a given functional form to best represent the desired input-output relationship. To prevent overtraining, the 189 sample datapoints were divided into proportions of 80% for the training phase and 20% for the testing phase. The stopping criterion suggested for the termination of the model run was set as per the recommendations of ASCE (2000) and Maier et al. (2010), according to which the model run was terminated when further reduction of the training and test errors were not possible.

The topology of the hidden layers was identified in a stepwise approach, according to which one neuron is added at a time until error cannot be further reduced and this is a trial-and-error procedure. The required number of neurons for the input layer is six (Na^+, K^+, HCO_3^-, Ca^{2+}, the UTM X and UTM Y).

A hyperbolic tangent sigmoid function is employed as the transfer function between the input and the hidden layer, and a linear transfer function is employed between the hidden layer and the output layer. The root mean square error (RMSE) (mg/L) and the coefficient of determination (R^2) are used to evaluate the performance of the ANN models. Inter-comparison of the models is based on using two performance measures of RMSE (close to zero) and highest R^2 (close to one).

19.4.3 Spatial Strategy for the Models

The study was focused on the spatial variation of fluoride and other ion concentrations in the study area, which comprise basaltic zones, nonbasaltic zones, and their mixing zone. This spatial variation was accounted for introducing a hybrid ANN (HANN), in which LVQ network is presented as per Kohonen (1990). Thus input data comprises three parallel streams targeting: (i) basaltic, (ii) nonbasaltic, and (iii) mixing zones. The classified inputs are used for the selected ANN models. A HANN model is shown in Fig. 19.4, which combines the LVQ network with a FNN-BFGS model. LVQ has three outputs and each output links to the selected ANN models.

19.5 RESULTS

Spatial HANN models were constructed to investigate the distribution of fluoride concentrations in the study area, similar to Maier and Dandy (2000) and Li and Liu (2005). The following section presents two strategies: strategy 1 uses MLP-LM, EN-LM, MLP-BFGS, EN-BFGS, MLP-GDX, EN-GDX, MLP-BR, EN-BR, and RBF models and strategy 2 uses a hybrid of ANN with LVQ.

19.5.1 Artificial Neural Network Modeling—Strategy 1

Strategy 1 sets out to understand the role of regional variations in aquifers, selects ANN models in steps of accounting for spatial fluoride in the following manner: (i) lumped aquifers studied by constructing nine ANN models, as shown in Section 19.5.1.1; (ii) demarcating basaltic and nonbasaltic zones with the selected ANN model; and (iii) demarcating basaltic, nonbasaltic, and mixing zones with the selected ANN model, which is presented in Section 19.5.1.2.

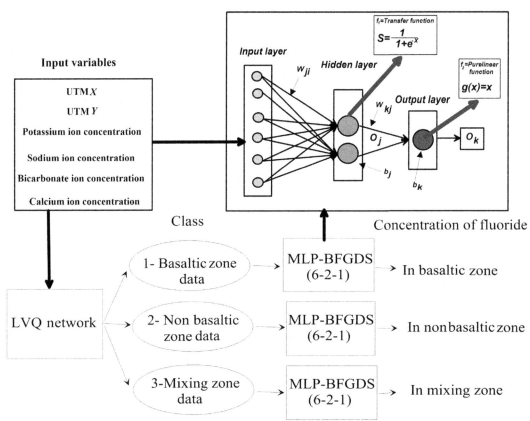

FIG. 19.4 Architecture of the hybrid of LVQ and MLP-BFGS models—the HANN model.

19.5.1.1 Artificial Neural Network Model for the Lumped Spatial Variations of Fluoride

The performance of the nine selected ANN models in handling complex aquifers with respect to their spatial analysis of fluoride concentrations is presented in Table 19.3. According to the table, the overall performance of the MLP-BFGS and MLP-LM lumped models is better than the other alternatives. .EN-BR seems to score the poorest performance measures, which is attributed to lumping the three zones of basaltic, non-basaltic and mixing zones in the study area.

The RMSE for the MLP-BFGS model in the training phase is slightly higher than MLP-LM, as MLP-BFGS performs slightly better in the testing phase. This case study shows that the MLP-BFGS model is better able to adjust its weights to estimate fluoride concentration in the groundwater system. This agrees with some other studies, where MLP-BFGS can be reliable in many groundwater problems, e.g., Li and Liu (2005).

TABLE 19.3 Performance Metrics of the Nine ANN Models for Lumped Complexity

			ANN Models								
			MLP-LM	EN-LM	MLP-BFGS	EN-BFGS	MLP-GDX	EN-GDX	MLP-BR	EN-BR	RBF
Criteria	R^2	Training	**0.81**	0.80	**0.81**	0.80	0.73	0.78	0.62	0.53	0.68
		Testing	**0.78**	0.68	**0.80**	0.71	0.64	0.66	0.58	0.41	0.55
	RMSE	Training	**0.67**	0.83	**0.69**	0.83	0.98	0.94	1.62	1.98	1.21
		Testing	**0.93**	1.13	**0.87**	1.07	1.32	1.32	1.92	2.48	1.68

Topology: MLP and RBF topologies are set to: 6-3-1, identified through preliminary tests; EN topology is set as: input neurons, 6; hidden layer, 3; context layer, 3; output layer, 1. The learning rate = 0.6; and the number of epochs = 500.
"**Bold numbers**: selected two better performing models (the highest rank and the next lower)."

19.5.1.2 Artificial Neural Network Models Allow for Spatial Variations in Fluoride Concentrations

A further study was carried out as part of strategy 1 to investigate the spatial analysis of fluoride concentration using predictive models by accounting through demarcating the basaltic zone, nonbasaltic zone, and mixing zone. To simplify the procedure, MLP-BFGS is investigated alone, as the previously mentioned findings identified it to be a better performing model. Two sets of model runs are presented: *set 1 of runs*—the mixing zone is lumped into the nearest zones, but the model runs distinguish between the sampling locations from basaltic and nonbasaltic zones; *set 2 of runs*—the mixing zone is accounted for in the same ways as those of basaltic and nonbasaltic zones.

The results for MLP-BFGS are presented in Table 19.4. The performance metrics for set 1 of runs show a significant improvement over the corresponding results presented in Table 19.3. It shows that the lumping together of the basaltic and nonbasaltic samples' data was a poor strategy. Moreover, fluoride concentration is better predicted in nonbasaltic zones than in basaltic zones, as the basaltic formation has dual porosity and is more complex than nonbasaltic formation.

The results in Table 19.4 also present the performance metrics for set 2 of model runs, in which the study area takes account of the mixing zone as well as the basaltic and nonbasaltic zones. The table shows that the improvements in the performance metrics are significant. These results provide evidence that accounting for groundwater in mixing zones produces significant improvements and should be treated on its own. Some of the data obtained from the nonbasaltic zones show the hydrochemical characteristics of the basaltic aquifer and vice versa (Asghari Moghaddam and Fijani, 2008).

19.5.2 Hybrid Artificial Neural Network Model Integrating Artificial Neural Networks With Learning Vector Quantization—Strategy 2

A hybrid of the LVQ networks was constructed with the selected MLP-BFGS model, in which the input data was first classified. The authors had already perceptually demarcated the study area into three areas of: basaltic, nonbasaltic, and mixing zones. These were used in the production of the results presented in Table 19.4, but the classification can also be carried out automatically by LVQ. As shown in Fig. 19.4, the input data for the 189 sampled datapoints (Na^+, K^+, HCO_3^-, Ca^{2+}, X, and Y) were fed into LVQ in two sets of 80% for training step and the remining 20% for testing step. The LVQ model confirmed the authors' perceptual model and automatically classified the samples into basaltic zones, nonbasaltic zones, and mixing zones. The data for the mixing zone correspond to five sampling locations (Yarim Qaya, Kolus Bulaghi, Pezik, Qarakhajlu, and Zakirlu) but their hydrochemical characteristics are yet to be studied in any detail.

The running of the HANN model is automatic and once the classification by LVQ was complete, the results were fed into the MLP-BFGS for both training and testing phases. The performance of the HANN model in predicting fluoride concentrations (using MLP-BFGS) in the testing phase is shown in Table 19.5.

A comparison of the results given in Table 19.5 with those given in Tables 19.4 shows that the HANN model performed better than the MLP-BFGS model in predicting fluoride concentrations in the demarcated (classified) study area. The increased accuracy is attributed to the capability of LVQ to classify the input data. However, the prediction accuracy for the fluoride data in the mixing zone is not as high as that for the other two zones but this drop may be attributed to (i) insufficient samples for input data from the mixing zone, and (ii) the high complexity of the mass transport in the mixing zone. Increasing sample sizes and borehole investigations in the mixing zone is likely to improve this prediction model.

TABLE 19.4 MLP-BFGS Model Results for Demarcated Study Area

Area		Basaltic Zones		Nonbasaltic Zones	
ANN Step		Training	Testing	Training	Testing
Set 1 runs: including mixing zone	R^2	0.943	0.884	0.969	0.921
	RMSE (mg/L)	0.417	0.727	0.316	0.513
Set 2 runs: without mixing zone	R^2	**0.987**	**0.941**	**0.991**	**0.966**
	RMSE (mg/L)	**0.213**	**0.402**	**0.181**	**0.322**

Topology: 6-2-1 (through another trial-and-error procedure for the demarcated study area); the learning rate = 0.6; the number of epochs = 500.
"Bold numbers: better performing set of model runs."

TABLE 19.5 Performance Metrics of the HANN Model for the Testing Phase

Area	Basaltic Zones	Nonbasaltic Zones	Mixing Zones
Phase	Testing	Testing	Testing
R^2	0.978	0.989	0.910
RMSE (mg/L)	0.302	0.271	0.510

The results in Table 19.5 support the HANN model and shows it to be capable of predicting groundwater quality concentrations in complex aquifers. The results also confirm those reported by Nourani et al. (2008a), which demonstrated that ANN models are more effective in predicting complex aquifers when their areas are demarcated. However, using the LVQ network to approach the problem makes the procedure more systematic.

19.6 DISCUSSIONS

The results presented in the previous section clearly provide quantitative evidence that lumping a study area is a poor modeling strategy to use to develop models for the spatial prediction of fluoride concentrations in complex aquifers. The results show that even using perceptual models to demarcate the study area into isotropic classes is a positive strategy to improve the prediction capabilities. Automatic classifications by LVQ is more effective.

For comparison purposes, a linear regression analysis was carried out, as a traditional alternative for HANN. This employs four highly correlated parameters (Na^+, K^+, HCO_3^-, and Ca^{2+}); their coordinates were used as the independent variables and the fluoride concentration as the dependent variable. The multivariable linear regression model for fluoride concentration is formulated as:

$$F^- = a + bX + cY - dCa^{+2} + eHCO_3^- + fNa^+ + hK^+ \tag{19.1}$$

where a, b, c, d, e, f, and h are the model coefficients.

Using the least square technique, the estimated coefficients of the multiple linear regression model are shown in the following section, for which the model residuals were checked and found to be homoscedastic and normally distributed at the 5% level. The fitted model is

$$F^- = -3.88 + 0.001X + 0.0005Y - 0.01Ca^{+2} + 0.009HCO_3^- + 0.005Na^+ + 0.158K^+ \tag{19.2}$$

The performance metrics for Eq. (19.2) for the training phase are: $R^2 = 0.68$, RMSE = 1.31 mg/L and for the testing phase: $R^2 = 0.38$, RMSE = 1.99 mg/L. Evidently the fit by the various versions of ANN models are considerable improvements.

Attention is drawn to AI techniques, which are bottom-up data-driven and as such new knowledge is learned from site-specific data, often with little restrictive assumptions. Therefore the findings reported in this chapter still have two restrictions: (i) they are true according to the tools used and some variation is expected if other modeling strategies are formulated, and (ii) the findings are driven by the measured data and therefore do not apply to different sites. However, the methodology is generic and has broader applications.

The authors have a perceptual model on the origins and sources of fluoride in the study area and are currently investigating the identification of the mechanisms responsible for transforming existing geogenic hazards into an aquifer-wide problem. The paper is largely focused on the environmental issue of spatially predictive fluoride concentrations and not on policy issues and their mitigation. The reported adverse situation in the study area is partially the outcome of a lack of planning. However, it remains to be seen if the findings will affect policymakers and health authorities to formulate an appropriate policy to mitigate the problem.

19.7 SUMMARY AND CONCLUSIONS

This chapter presents the development of ANN models for the spatial prediction of fluoride concentrations in complex aquifers. Perceptually, the study area was demarcated into three zones: basaltic, nonbasaltic, and mixing. The data available comprise 189 samples of major, minor, and trace ions collected between the years 2004 and 2013. A basic statistical analysis of these data indicated that the mean fluoride concentration is 2.8 mg/L and the maximum is 8.1 mg/L. This is a clear evidence that the local population is exposed to risk, hence the research undertaken for this chapter.

The study used 10 variations of ANNs through a modeling strategy, in which the complexity of the study area was investigated using two strategies: (i) in *strategy 1*, a series of model runs were formulated to select the topology of the ANN hidden layers, hierarchically increasing the model complexity from a lumped aquifer, considering basalt and non-basalt aquifers but ignoring the mixing zone, and finally considering the three perceptual zones of basaltic, nonbasaltic, and mixing zones—the results show that the perceptual model of treating regional complexity contributed to significant improvement; and (ii) in *strategy 2*, a hybrid of LVQ and the selected ANN model (MLP-BFGS) performs better than the other alternatives. Although the conclusions reported in this chapter are derived from site-specific data, arguably, the truth of demarcating variations in aquifers should generally improve models of spatial analysis.

ACKNOWLEDGMENTS

The study was partially supported by Research Office at University of Tabriz. The authors acknowledge the West Azerbaijan Regional Water Authority and Mr. R. Abdollahzadeh for the help on the fieldwork. Our thanks are also due to Mr. Orooji, who helped with the chemical analysis of the water samples in the Hydrochemistry Laboratory in the Department of Geology, University of Tabriz.

REFERENCES

Almasri, M.N., Kaluarachchi, J.J., 2005. Modular neural networks to predict the nitrate distribution in ground water using the on-ground nitrogen loading and recharge data. Environ. Model Softw. 20 (7), 851–871.

American Public Health Association, 1998. Standard Method for the Examination of Water and Wastewater, 17th ed. American Public Health Association, Washington, DC.

Anctil, F., Perrin, C., Andreassian, V., 2004. Impact of the length of observed records on the performance of ANN and of conceptual parsimonious rainfall-runoff forecasting models. Environ. Model Softw. 19 (4), 357–368.

ASCE Task Committee on Application of Artificial Neural Networks in Hydrology, 2000. Artificial neural networks in hydrology, parts I and II. American Society of Civil Engineering. J. Hydraul. Eng. 5 (2), 115–137.

Asghari Moghaddam, A., Fijani, E., 2008. Distribution of fluoride in groundwater of Maku area, northwest of Iran. Environ. Geol. 56 (2), 281–287.

Asghari Moghaddam, A., Fijani, E., 2009. Hydrogeologic framework of the Maku area basalts, northwestern Iran. Hydrogeol. J. 17 (4), 949–959.

Cattell, R.B., 1952. Factor Analysis. Harper, New York.

Chitsazan, N., Nadiri, A.A., Tsai, F.T-C., 2015. Prediction and structural uncertainty analyses of artificial neural networks using hierarchical Bayesian model averaging. J. Hydrol. 528, 52–62.

Coppola, E., Rana, A.J., Poulton, M., Szidarovszky, F., Uhi, V.W., 2005. A neural networks model for predicting aquifer water level elevation. Ground Water 43, 231–241.

Davraz, A., Sener, E., Sener, S., 2008. Temporal variations of fluoride concentration in Isparta public water system and health impact assessment (SW-Turkey). Environ. Geol. 56 (1), 159–170.

Diodato, N., Tartari, G., Bellocchi, G., 2010. Geospatial rainfall modelling at Eastern Nepalese highland from ground environmental data. Water Resour. Manag. 24, 2703–2720.

Dissanayake, C.B., 1991. The fluoride problem in the groundwater of Sri Lanka—environmental management and health. Int. J. Environ. Stud. 38, 137–156.

Elman, J.L., 1990. Finding structure in time. Cogn. Sci. 14 (2), 179–211.

EPA, 2009. National Primary Drinking Water Regulations. USA Environmental Protection Agency.

Fijani, E., Nadiri, A.A., Asghari Moghaddam, A., Tsai, F., Dixon, B., Iran, 2013. J. In: Optimization of DRASTIC method by supervised committee machine artificial intelligence to assess groundwater vulnerability for Maragheh-Bonab plain aquifer.Iran. J. Hydrol.503, pp. 89–100.

Foresee, F.D., Hagan, M.T., 1997. Gauss–Newton approximation to Bayesian learning. In: Proc. International Joint Conference. Neural Networkspp. 1930–1935.

Garcia, L.A., Shigidi, A., 2006. Using neural networks for parameter estimation in groundwater. J. Hydrol. 318 (1–4), 215–231.

Gizaw, B., 1996. The origin of high bicarbonate and fluoride in the Main Ethiopian Rift Valley, East African Rift system. J. Afr. Earth Sci. 22, 392–402.

Guo, Q., Wang, Y., Ma, T., Ma, R., 2007. Geochemical processes controlling the elevated fluoride concentration in groundwaters of the Taiyuan Basin, Northern China. J. Geochem. Explor. 93, 1–12.

Gupta, S.K., Deshpande, R.D., Agarwal, M., Raval, B.R., 2005. Origin of high fluoride in groundwater in the North Gujarat-Cambay region, India. Hydrogeol. J. 13, 596–605.

Handa, B.K., 1975. Geochemistry and genesis of fluoride contains groundwater in India. Ground Water 13, 275–281.

Haykin, S., 1999. Neural Networks: A Comprehensive Foundation, 2nd ed. Prentice Hall, Upper Saddle River, NJ.

Hudak, P.F., Sanmanee, S., 2003. Spatial patterns of nitrate, chloride, sulfate, and fluoride concentrations in the Woodbine aquifer of north-central Texas. Environ. Monit. Assess. 82 (3), 311–320. (Hydrology 328 (3, 4), 717–725).

Islam, S., Kothari, R., 2000. Artificial neural networks in remote sensing of hydrologic processes. J. Hydrol. Eng. 5 (2), 138–144.

Jacks, G., Bhattacharya, P., Chaudhary, V., Singh, K.P., 2005. Controls on the genesis of high-fluoride groundwaters in India. Appl. Geochem. 20, 221–228.

Kaiser, H.F., 1958. The varimax criterion for analytic rotation in factor analysis. Psychometrika 23, 187–200.

Kasabov, N.K., 1998. Foundations of Neural Networks, Fuzzy Systems, and Knowledge Engineering. The MIT Press, Cambridge, MA; London.

Kharb, P., Susheela, A.K., 1994. Fluoride ingestion in excess and its effects on organic and certain inorganic constituents of soft tissues. Med. Res. Soc. 22, 43–44.

Khatibi, R., Ghorbani, M.A., Akhoni Pourhosseini, F., 2017. Stream flow predictions using nature-inspired firefly algorithms and a multiple model strategy—directions of innovation towards next generation practices. J. Adv.Eng. Inform. http://www.sciencedirect.com/science/article/pii/S1474034617301271.

Kim, K., Jeong, Y.G., 2005. Factors influencing natural occurrence of fluoride-rich ground waters: a case study in the southeastern part of the Korean Peninsula. Chemosphere 58, 1399–1408.

Kohonen, T., 1990. The self-organizing map. Proc. IEEE 789, 1464–1480.

Li, S., Liu, Y., 2005. Parameter identification procedure in groundwater hydrology with artificial neural network. Adv. Intell. Comput. 3645, 276–285.

Maier, H.R., Dandy, G.C., 2000. Neural network for the prediction and forecasting water resources variables: a review of modeling issues and applications. Environ. Model. Softw. 15, 101–124.

Maier, H.R., Jain, A., Dandy, G.C., Sudheer, K.P., 2010. Methods used for the development of neural networks for the prediction of water resource variables in river systems: Current status and future directions. Environ. Model Softw. 25, 891–909.

Mathworks, 2009. Matlab User Manual, Version 7. 8.0.347 (R2009a). www.mathworks.com.

Medsker, L.R., Jain, L.C., 1999. Recurrent Neural Networks: Design and Applications. CRC Press, Boca Raton, Boston, London, New York, Wash.

Merdun, H., Cinar, O., Meral, R., Apan, M., 2006. Comparison of artificial neural network and regression pedotransfer functions for prediction of soil water retention and saturated hydraulic conductivity. Soil Tillage Res. 90 (1–2), 108–116.

Nadiri, A.A., 2007. Water Level Evaluation in Tabriz Underground Area by Artificial Neural Networks. (M.S. theses). University of Tabriz.

Nadiri, A.A., 2015. Application of Artificial Intelligence methods in Geosciences and Hydrology. OMICS Publication.

Nadiri, A.A., Hassan, M.M., 2015. Supervised intelligence committee machine to evaluate field performance of photocatalytic asphalt pavement for ambient air purification. Transport. Res. Rec.: J. Transport. Res. Board 2528, 96–105.

Nadiri, A.A., Asghari Moghaddam, A., Nourani, V., 2007. Forecasting spatiotemporal water levels of Tabriz city underground area by neural kriging. In: *Proceeding of the 10th Symposium of Geological Society of Iran*. Ferdowsi University of Mashhad, Iran.

Nadiri, A.A., Asghari Moghaddam, A., Tsai, F.T-C, Fijani, E., 2013. Hydrogeochemical Analysis for Tasuj Plain Aquifer, Iran. Journals of Earth System 22, 1091–1105. https://doi.org/10.4172/978-1-63278-061-4-062.

Nadiri, A.A., Chitsazan, N., Tsai, F.T.-C., Moghaddam, A.A., 2014. Bayesian artificial intelligence model averaging for hydraulic conductivity estimation. J. Hydrol. Eng. 19, 520–532.

Nadiri, A.A., Gharekhani, M., Khatibi, R., Sadeghfam, S., Asgari Moghaddam, A., 2017a. Groundwater vulnerability indices conditioned by supervised intelligence committee machine (SICM). Sci. Total Environ. 574, 691–706. https://doi.org/10.1016/j.scitotenv.2016.09.093.

Nadiri, A.A., Gharekhani, M., Khatibi, R., Asgari Moghaddam, A., 2017b. Assessment of groundwater vulnerability using supervised committee to combine fuzzy logic models. Environ. Sci. Pollut. Res. 24 (9), 8562–8577. 10.1007/s11356-017-8489-4.

Nadiri, A.A., Sedghi, Z., Khatibi, R., Gharekhani, M., 2017c. Mapping vulnerability of multiple aquifers using multiple models and fuzzy logic to objectively derive model structures. Sci. Total Environ. 593, 75–90. 10.1016/j.scitotenv.2017.03.109.

Nadiri, A.A., Aghdam, F.S., Khatibi, R., Moghaddam, A.A., 2018. The problem of identifying arsenic anomalies in the basin of Sahand dam through risk-based 'soft modelling. Sci. Total Environ. 613, 693–706.

Nourani, V., Asghari Moghaddam, A., Nadiri, A.A., 2008a. An ANN-based model for spatiotemporal groundwater level forecasting. Hydrol. Process. Int. J. 22 (26), 5054–5066.

Nourani, V., Asghari Moghaddam, A., Nadiri, A.A., Singh, V.P., 2008b. Forecasting spatiotemporal water levels of Tabriz aquifer. Trend. Appl. Sci. Res. 3 (4), 319–329.

Oruc, N., 2003. Problems of high fluoride waters in Turkey (hydrogeology and health aspects). The short course on medical geology-health and environment, Canberra, Australia. .

Ozsvath, D.L., 2006. Fluoride concentrations in a crystalline bedrock aquifer Marathon County, Wisconsin. Environ. Geol. 50, 132–138.

Rao Nagendra, C.R., 2003. Fluoride and environment—a review. In: Martin, J., Bunch, V., Suresh, M., Vasantha Kumaran, T. (Eds.), Proceedings of the Third International Conference on Environment and Health, Chennai, IndiaIn: vols. 15–17. pp. 386–399.

Rumelhart, D.E., Hinton, G.E., Williams, R.J., 1986. Learning internal representations by error propagation. In: Rumelhart, D.E., McClelland, J.L. (Eds.), Parallel Distributed Processing: Explorations in the Microstructure of Cognition. In: vol. 1. MIT Press, Cambridge, MA, pp. 318–362.

Saxena, V.K., Ahmed, S., 2001. Dissolution of fluoride in groundwater: a water–rock interaction study. Environ. Geol. 40, 1084.

Saxena, V.K., Ahmed, S., 2003. Inferring the chemical parameters for the dissolution of fluoride in groundwater. Environ. Geol. 4, 731–736.

Sharma, V., Negi, S.C., Rudra, R.P., Yang, S., 2003. Neural networks for predicting nitrate-nitrogen in drainage water. Agric. Water Manag. 63, 169–183.

Shrestha, S., Kazama, F., 2007. Assessment of surface water quality using multivariate statistical techniques: a case study of the Fuji river basin, Japan. Environ. Model. Softw. 22 (4), 464–475.

WHO, 2006. Guidelines for Drinking-Water Quality. World Health Organization. First addendum to third edition, vol. 1, Recommendations.

Xie, Z., Li, J., Wu, W., 2008. Application of GIS and geostatistics to characterize spatial variation of soil fluoride on Hang-Jia-Hu Plain, China. In: Computer and Computing Technologies in Agriculture.vol. 258. pp. 253–266.

Zietsman, S., 1991. Spatial variation of fluorosis and fluoride content of water in an endemic area in Bophuthatswana. J. Dent. Assoc. S. Afr. 46, 11–15.

FURTHER READING

Ablameyko, S., Goras, L., Gori, M., Vincenzo, P., 2003. Neural Networks for Instrument, Measurement and Related Industrial Applications. IOS Press and Kluwer Academic Publishers in conjunction with the NATO Scientific Affairs Division, p. 340.

Anderson, D., McNeill, G., 1992. Artificial Neural Networks Technology. Kaman Sciences Corporation, New York.

Anmala, J., Zhang, B., Govindaraju, R.S., 2000. Comparison of ANNs and empirical approaches for predicting watershed runoff. J. Water Resour. Plan. Manag. 126 (3), 156–166.

Ba, H., Guo, S., Wang, Y., Hong, Z., Zhong, Y., Liu, Z., 2017. Improving ANN model performance in runoff forecasting by adding soil moisture input and using data preprocessing techniques. Hydrol. Res. https://doi.org/10.2166/nh.2017.048.

Bårdsen, A., Bjorvatn, K., Selvig, K.A., 1996. Variability in fluoride content of subsurface water reservoirs. Acta Odontol. Scand. 54 (6), 343–347.

Bowden, G.J., Maier, H.R., Dandy, G.C., 2002. Optimal division of data for neural network models in water resources applications. Water Resour. Res. 38 (2), 1010.

Bowden, G.J., Maier, H.R., Dandy, G.C., 2005. Input determination for neural network models in water resources applications. Part 2. Case study: forecasting salinity in a river. J. Hydrol. 301 (1-4), 93–107.

Brown, C.E., 1998. Applied Multivariate Statistics in Geohydrology and Related Sciences. Speringer-Verlag, Berlin.

Cameron, D., Kneale, P., See, L., 2002. An evaluation of a traditional and a neural net modelling approach to flood forecasting for an upland catchment. Hydrol. Process. 16 (5), 1033–1046.

Dalton, M.G., Upchurch, S.G., 1978. Interpretation of hydrochemical facies by factor analysis. Ground Water 16, 228–233.

Davis, J.C., *1973*. Statistics and Data Analysis in Geology. Wiley, New York.

Dixon, B., 2005. Applicability of neuro-fuzzy techniques in predicting ground-water vulnerability: a GIS-based sensitivity analysis. J. Hydrol. 309, 17–38.

Dragon, K., 2006. Application of factor analysis to study contamination of semi-confined aquifer (Wielkopolska Buried Valley aquifer, Poland). J. Hydrol. 331, 272–279.

Goovaerts, P., 1997. Geostatistics for Natural Resources Evaluation. Oxford University Press, New York.

Govindaraju, R.S., Ramachandra Rao, A., 2000. Artificial Neural Networks in Hydrology. Springer Publication, p. 384.

Grande, A., Gonzalez, R., Beltran, R., Sanchez-Rodas, D., 1996. Application of factor analysis to the study of contamination in the aquifer system of Ayamonte-Huelva (Spain). Ground Water 34, 155–161.

Gutiérrez-Estrada, J.C., de Pedro-Sanz, E., López-Luque, R., Pulido-Calvo, I., 2004. Comparison between traditional methods and artificial neural networks for ammonia concentration forecasting in an eel (Anguilla anguilla L.) intensive rearing system. Aquac. Eng. 31, 183–203.

Hornik, K., Stinchcombe, M., White, H., 1989. Multilayer feedforward networks are universal approximators. Neural Netw. 2, 359–366.

Hu, T.S., Lam, K.C., Ng, S.T., 2001. River flow time series prediction with a range dependent neural network. Hydrol. Sci. J. 46 (5), 729–745.

Hu, T.S., Wu, F.Y., Zhang, X., 2007. Rainfall-runoff modeling using principal component analysis and neural network. Nord. Hydrol. 38 (3), 235–248.

Jeong, D., Kim, Y., 2005. Rainfall-runoff models using artificial neural networks for ensemble streamflow prediction. Hydrol. Process. 19 (19), 3819–3835.

Lin, G., Chen, L., 2004. A non-linear rainfall-runoff model using radial basis function network. J. Hydrol. 289, 1–8.

Lischeid, G., 2001. Investigating short-term dynamics and long-term trends of SO4 in the runoff of a forested catchment using artificial neural networks. J. Hydrol. 243 (1, 2), 31e42.

Maier, H.R., Dandy, G.C., 1999. Empirical comparison of various methods for training feed-forward neural networks for salinity forecasting. Water Resour. Res. 35 (8), 2591–2596.

McGrail, B.P., 2001. Inverse reactive transport simulator (inverts): an inverse model for contaminant transport with nonlinear adsorption and source terms. Environ. Model. Softw. 16 (8), 711–723.

Riad, S., Mania, J., 2004. Rainfall-runoff model using an artificial neural network approach. Math. Comput. Model. 40, 839–846.

Sahoo, G.B., Ray, C., 2005. Flow forecasting for a Hawaii stream using rating curves and neural networks. J. Hydrol. 317, 63–80.

Sahoo, G.B., Ray, C., De Carlo, E.H., 2006a. Use of neural network to predict flash flood and attendant water qualities of a mountainous stream on Oahu, Hawaii. J. Hydrol. 327 (3, 4), 525–538.

Sahoo, G.B., Ray, C., Edward Mehnert, E., Keefer, D.A., 2006b. Application of artificial neural networks to assess pesticide contamination in shallow groundwater. Sci. Total Environ. 367, 234–251.

Solomatine, D.P., Dulal, K.N., 2003. Model trees as an alternative to neural networks in rainfall-runoff modelling. Hydrol. Sci. J. 48 (3), 399–411.

Solomatine, D.P., Xue, Y.P., 2004. M5 model trees and neural networks: application to flood forecasting in the upper reach of the Huai River in China. J. Hydrol. Eng. 9 (6), 491–501.

Stetzenbach, K.J., Hodge, V.F., Guo, C., Farnham, I.M., Johannesson, K.H., 2001. Geochemical and statistical evidence of deep carbonate ground water within overlying volcanic rock aquifers/aquitard of southern Nevada, USA. J. Hydrol. 243, 254–271.

Suen, J.P., Eheart, J.W., 2003. Evaluation of neural networks for modeling nitrate concentrations in rivers. J. Water Resour. Plan. Manag. 129 (6), 505–510.

Tayfur, G., Singh, V.P., 2008. ANN and fuzzy logic models for simulating event-based rainfall-runoff. American Society of Civil Engineering. J. Hydraul. Eng. 134 (9), 1400–1401.

Chapter 20

Spatio-Temporal Variations of Fluoride in the Groundwater of Dindigul District, Tamil Nadu, India: A Comparative Assessment Using Two Interpolation Techniques

N.S. Magesh* and L. Elango[†]

*National Centre for Antarctic and Ocean Research, Vasco da Gama, Goa, India [†]Department of Geology, Anna University, Chennai, India

Chapter Outline

20.1 INTRODUCTION

Fluoride is widely dispersed in nature and it is estimated to be the 13th most abundant element on our planet (Mason and Moore, 1982). Geochemically fluoride is the most electronegative of the elements and occurs primarily as a negatively charged ion in water, with an average crustal concentration of 625 mg/kg (Hem, 1985; Edmunds and Smedley, 2005). Fluoride is classified as an incompatible lithophile element (Faure, 1991), which means that it can preferentially partition into silicate melts as magmatic crystallization proceeds (Xiaolin and Zhenhua, 1998). As a result, late-stage pegmatite granites, hydrothermal vein deposits, and rocks that crystallize from highly evolved pristine magmas often contain fluorite, fluorapatite, and fluoride-enriched micas and/or amphiboles (Nagadu et al., 2003; Taylor and Fallick, 1997; Scaillet and Macdonald, 2004). Some hydrochemical conditions that favor the dissolution of fluoride from silicates include an alkaline pH, an anion exchange (OH^- for F^-) capacity of aquifer materials, a cation exchange capacity (Na^+ for Ca^{2+}), a long residence time of water in a rock-water interaction system, and tropical climatic condition (Saxena and Ahmed, 2001; Ozsvath, 2009). Fluoride is an essential microelement for animals and humans, and its intake under desirable limits protects teeth against caries, especially in childhood, but excessive exposure to fluoride can damage skeletal tissues (bones and teeth) (Brindha and Elango, 2011). The recommended concentration of fluoride in drinking water is between 0.6 and 1.5 mg/L (WHO, 2004). In India, approximately 60–65 million people consume fluoride-contaminated groundwater and the number affected by fluorosis is estimated at 2.5 to 6 million, predominately children (Athavale and Das, 1999; Rao and Devadas, 2003). Nearly 177 districts covering 19 states have a high content of fluoride in their groundwater. It appears that a high fluoride content in the groundwater may exist in most of the districts (Brindha et al., 2010, 2011; Muralidharan et al., 2002; Brindha and Elango, 2013; Jagadeshan et al., 2015a,b). Cost-effective techniques such as managed aquifer recharge (MAR) structures were found to be efficient in managing the fluoride problem in India (Jagadeshan and Elango, 2015; Brindha et al., 2016; Gowrisankar et al., 2017).

Mapping the spatial extent of contaminated groundwater is very important in the management of health risks (Lee et al., 2007; Vetrimurugan et al., 2017). However, only a few representative groundwater samples from the field can be analyzed as the process demands time and money. However, sparse data puts data reliability into question and this uncertainty is associated with sampling and the derived models (Liu et al., 2004). Therefore spatial interpolation methods are used to generate spatial maps from the field data. Generally interpolations works by predicting values from few data points by assuming the fact that things that are close to one another are more alike than those that are farther apart. Such processes can be worked out in a geographical information system (GIS), which is a reliable tool to handle geographically distributed data and predict the surface values through suitable interpolation techniques. The most commonly used interpolation techniques in groundwater-quality mapping are inverse distance weighted (IDW) and kriging (Kumar, 2007; Nas and Berktay, 2010; Arslan, 2012; Gong et al., 2014). IDW estimates the cell values by taking the mean of sample data points in the vicinity of each processing cell. The closer a point is to the center of the cell being estimated, the more control or weight it has in the averaging process. Whereas kriging is an advanced geostatistical method that creates an estimated surface from a set of points having z-values. Moreover, a detailed examination of the data is required before opting for the best estimation method, i.e., universal kriging, ordinary kriging, or cokriging.

Studies suggest that ordinary kriging and cokriging methods were found to be more reliable than IDW (Zare-Mehrjardi et al., 2010). Robinson and Metternicht (2006) used three diverse techniques, i.e., IDW, radial basis function (RBF), and kriging, for predicting the concentration of organic matter, acidity, and soil salinity. Another study by Pang et al. (2011) reported ordinary kriging to be a widely used kriging method that gives a predicted surface of different soil properties. Recent research suggests that empirical Bayesian kriging (EBK) is the most appropriate tool for the prediction of various environmental parameters (Hussain et al., 2014; Mirzaei and Sakizadeh, 2016; Magesh et al., 2016, 2017). Among these discussed techniques, representative techniques from deterministic (IDW) and geostatistical (EBK) interpolations were chosen to estimate the concentration of the fluoride ion. Therefore the aim of this study was to conduct a systematic assessment of the GIS-based interpolation techniques for estimating the spatial distribution of the fluoride ion in the groundwater of the Dindigul district for a period of 7 years and to suggest which technique has better reliability by comparing the derived root mean square error (RMSE) measurements.

20.2 STUDY AREA

The Dindigul district is generally a hard-rock terrain covering an area of about 6267 km^2, which is situated in Tamil Nadu, southern India with the coordinates between latitude 9°91' and 10°84' north and longitude 77°24' and 78°44' east (Fig. 20.1). The administrative setup of the district includes seven taluks, which are further divided into 14 blocks having 539 panchayats. The major river basin/subbasin includes Cauvery, Vaigai, and Pambar Rivers, and other important rivers such as the Shanmuganadhi, Nangangiar, Amaravathi, Kodavanar, Marudhanadhi, and Manjalar drain through the area. These rivers originate from the Palani and Sirumalai hills of the western Ghats. The rivers are ephemeral in nature and flow only during the monsoon period. The drainage patterns in the district are dentritic to subdentritic and they are structurally controlled in most places. The hard-rock terrain is mainly composed of charnockite and hornblende biotite gneiss whereas charnockite formation prevails along the southern zone where structural hills are dominant. Patches of granulite, garnetiferous quartzo-feldspathic gneiss are common and anorthosite is observed in the central and eastern part of the study area. Occurrences of quartzite and migmatite veins are noted in the structurally controlled southern part. Rock types such as pink augen gneiss, granite, garnetiferous sillimanite graphite gneiss, and pyroxene granulites are also noticed in the region. Few structural disturbances are observed in the eastern part of the study area. A tropical climate prevails in the region and the annual rainfall varies between 700 and 1600 mm. The northeast monsoon offers the major share of the rainfall when compared to the southwest monsoon. The temperature in the study area varies between 13°C and 43°C with a relative humidity range between 65% and 85%. The major water-bearing formations in the study area include weathered and fractured charnockite along with granite gneisses. The depth of water level during both the seasons ranges from 0.12 to 15 mbgl (CGWB, 2008). The long-term (10 year) rise in water level is between 0.024 and 0.590 m/year and the fall in water level ranges between 0.041 m/year and 1.523 m/year. The quality of the groundwater in the shallow aquifers is moderate in the study area and parameters such as TH, $CaCO_3$, and NO_3 exceed the permissible limit for drinking water in some places. However, in deep aquifers, it is the fluoride concentration that is high and poses a significant health risk to the population of the district.

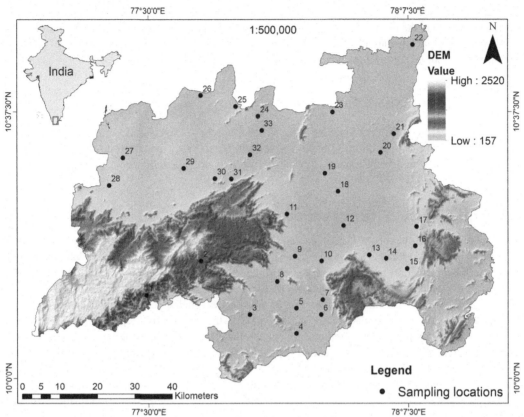

FIG. 20.1 Location map of the study area.

20.3 MATERIALS AND METHODS

20.3.1 Data Collection of Groundwater Quality

The groundwater-quality data was obtained from the Public Works Department (PWD), Chennai for the year 2001 to 2007, representing the pre monsoon season. The monitoring wells were selected based on the data availability for all the years. A total of 32 monitoring wells were selected representing the entire district. The wells were sited on different land use conditions including agricultural fields and settlements. The PWD water quality-testing laboratory had analyzed the water samples in accordance with the guidelines of APHA (1995). The physicochemical parameters, such as total dissolved solids, pH, electrical conductivity, cations (calcium, magnesium, sodium, and potassium), and anions (bicarbonate, chloride, sulphate, carbonate, nitrate, and fluoride) were analyzed. For the present study the focus was given to fluoride ion concentrations for a 7-year period (2001–07).

20.3.2 GIS Analysis

The spatio-temporal distribution of places of high fluoride concentration in the study area was analyzed using geospatial technology. The GPS-located sampling wells were imported into the ArcGIS 10.3 platform for data analysis and visualization. For the present study two types of interpolation techniques were adopted; the first one refers to the deterministic technique in which a interpolated surface was created from the measured points and the second one was a geostatistical technique in which a interpolated surface was produced by utilizing the statistical properties from the measured points. In this study a commonly used deterministic technique (IDW) and geostatistical interpolation technique (EBK) were used to produce data on the spatio-temporal distribution of fluoride in the study area.

20.3.3 Inverse Distance Weighted Method

The IDW method is one of the most commonly used deterministic interpolation techniques for the spatial mapping of groundwater quality. This technique uses the data points, which are near to the point being estimated. Weights are assigned to the interpolating points based on the inverse of the distance from the point being estimated. It assumes that the points that are nearby are more likely than the points that are far apart. Thus the data points that are close to the point being estimated have more weight than the points that are far apart. The fluoride distribution maps for the years 2001–07 were plotted using the IDW algorithm. The general formula for this algorithm is

$$\hat{Z}(S_0) = \sum_{i=1}^{N} \lambda_i Z(S_i)$$

where $\hat{Z}(S_0)$ is the value for predicting the location at S_0, N is the number of measured sampling points close to the prediction location for the purpose of prediction, λ_i is the weights allotted to every measured point, and $Z(S_i)$ is the observed value at the location S_i. The concentration of fluoride in mg/L is assigned in the Z value field for interpolation (Xie et al., 2011).

20.3.4 Empirical Bayesian Kriging Method

EBK automates the complicated side of constructing a valid kriging model by subsetting and simulating the measured data. The major advantage in the EBK model is it uses a large number of semivariogram models rather than just a single one, thus offering a great benefit over other kriging models. EBK estimates the uncertainty level in generating the semivariograms. The steps involved in estimating the values through the EBK method include semivariogram model estimation through field-observed data, followed by simulation of a new value from the estimated semivariogram and estimation of a new semivariogram model from the new simulated data. The weights of the new semivariogram model are calculated based on the Bayes' rule. During this process the prediction of new values and generation of standard errors were calculated at unsampled locations. The process undergoes several repetitions at each sampling point and a semivariogram spectrum is generated. K-Bessel semivariogram type was applied along with an empirical transformation as this method seems to be more flexible and accurate with the only drawback being a longer processing time. Using this method the average standard error was found to be low in comparison with other transformations. A complete explanation of the EBK model and its functionalities was well described by Krivoruchko (2012a,b).

20.4 RESULTS

The minimum, maximum, mean, and standard deviation of fluoride concentration measured in the present study are given in Table 20.1. The pH value of the groundwater ranges from 6.9 to 9.1, and it is found to be slightly basic in nature. The concentration of fluoride will change with respect to changes in alkalinity. The fluoride concentration (from 2001 to 2007) in the study area was found to vary between 0.05 and 2.38 mg/L (Table 20.1). The desirable range of fluoride concentration in drinking water is from 0.6 to 1.2 mg/L (BIS, 1992). However, it is suggested that the maximum can be extended up to 1.5 mg/L. Based on the concentration of fluoride, the study area samples are classified into three groups: low category (<0.5 mg/L), medium category (0.5 to 1.5 mg/L), and high category (1.5 to 3 mg/L; Table 20.2). Out of the 224 samples (from 2001 to 2007), 48.7% of the samples fell into the low category, 47.3% of the samples fell into medium category, and 4% of the samples fell into the high category (Table 20.2).

20.4.1 Spatio-Temporal Variation of Fluoride and Its Geogenic Sources

The spatio-temporal distribution of fluoride ions in the groundwater revealed that a high concentration is observed in the northeastern and southern parts of the study area, whereas a low concentration is observed in the western parts (Table 20.1). Variations in the fluoride concentrations in the study area are generally caused by the geological setup, which is mainly hard rock (Handa, 1975). The hard rock aquifers are composed of granitic and gneissic rocks, which consist of apatite, hornblende, and biotite, and may contribute to the fluoride levels in the groundwater (Chae et al., 2007). Studies suggest that charnockites composed of apatites and fluorapatites contribute fluoride to the groundwater (Chidambaram, 2000; Brindha and Elango, 2011). Other factors, such as pH and the depth of the groundwater table, also affect the dissolution of fluoride in the groundwater (Saxena and Ahmed, 2003). A variation in the concentration pattern is noticed in different years but the overall distribution remains almost the same. This indicates that the geological features are the key factor responsible for the occurrence of fluoride in the study area. The variations in the concentrations may be due to various factors, such as dilution due to rainfall, pH variation, ion exchange, residence time, etc. The detailed hydrochemistry and genesis of the fluoride in the groundwater of the Dindigul district has been studied by Mondal et al. (2002), Viswanathan et al. (2010), Manivannan et al. (2010, 2012), Chidambaram et al. (2013), Amalraj and Pius (2013), and Magesh et al. (2013).

TABLE 20.1 Drinking Water Fluoride Level in Dindigul District, Tamil Nadu, India

Name of the Village	Level of Fluoride (mg/L)							Mean±SD	Ranges
	2001	2002	2003	2004	2005	2006	2007		
Ambathurai	0.75	0.81	0.86	0.13	0.14	0.09	0.26	0.43±0.08	0.09–0.86
Athoor	0.10	0.15	0.83	0.21	0.41	0.19	0.45	0.33±0.08	0.10–0.83
Batlagundu	1.10	0.38	0.33	0.13	0.50	0.54	1.20	0.60±0.20	0.13–1.20
Chatrapatti	0.38	0.09	0.12	0.29	0.24	0.24	0.45	0.26±0.13	0.09–0.45
Chettinaickenpatti	0.73	0.27	0.22	0.56	0.68	1.60	0.87	0.70±0.24	0.22–1.60
Dindigul	0.48	0.52	0.78	0.11	0.23	0.12	0.95	0.46±0.13	0.11–0.95
Eramanaickenpatti	0.23	0.93	0.99	0.20	0.68	0.64	1.00	0.67±0.33	0.20–1.00
Gullalagundu	0.84	1.26	1.38	0.17	1.30	1.50	1.10	1.08±0.63	0.17–1.50
Kalavarpatti	0.70	0.21	0.26	0.33	0.37	0.23	0.67	0.40±0.19	0.21–0.70
Kambiliampatti	0.50	0.52	0.59	0.44	0.80	0.89	0.75	0.64±0.47	0.44–0.89
Kodaikanal	0.96	0.08	0.07	0.05	0.10	0.09	0.10	0.21±0.13	0.05–0.96
Koovanuthu	0.43	0.41	0.36	0.65	0.89	1.50	1.10	0.76±0.34	0.36–1.50
Nilakkottai	1.20	1.30	1.56	0.32	0.60	0.43	0.66	0.87±0.39	0.32–1.56
Odaipatti	0.25	0.37	0.42	0.34	1.00	0.32	0.56	0.47±0.21	0.25–1.00
Palayakannivadi	0.35	0.40	0.93	0.29	0.44	0.29	0.91	0.52±0.23	0.29–0.93
Pallapatti	0.70	0.55	0.53	0.20	0.39	0.61	0.39	0.48±0.31	0.20–0.70
Palvarpatti	0.41	1.36	1.28	0.65	0.80	0.48	1.20	0.88±0.49	0.41–1.36
Pannaikadu	0.23	0.28	1.51	0.24	0.12	0.07	0.19	0.38±0.13	0.07–1.51
Pappampatti	0.35	0.24	0.26	0.35	0.35	0.33	0.42	0.33±0.27	0.24–0.42
Pudupatti	0.25	0.38	0.42	0.20	0.39	0.68	0.58	0.41±0.24	0.20–0.68
R.Kombai	0.62	2.38	0.84	0.32	1.50	1.50	0.99	1.16±0.48	0.32–2.38
Sanarpatti	0.81	1.74	1.82	0.89	1.80	0.74	0.55	1.19±0.63	0.55–1.82
Sitharevu	0.35	0.38	0.31	0.48	0.68	1.60	0.96	0.68±0.21	0.31–1.60
Sithargalnatham	0.81	0.86	0.92	0.16	0.87	0.75	0.43	0.69±0.40	0.16–0.92
Sukkampatti	0.80	0.99	1.02	0.55	1.10	0.72	1.00	0.88±0.68	0.55–1.10
T.M.G.Valasu	0.50	1.20	1.36	0.35	0.52	0.66	0.96	0.79±0.41	0.35–1.36
Thalaiyuthu	0.50	0.72	0.81	0.31	0.28	0.38	0.62	0.52±0.31	0.28–0.81
Thankachiammapatti	0.65	0.49	0.56	0.19	0.22	0.14	0.25	0.36±0.15	0.14–0.65
Thennampatti	0.38	0.42	0.49	0.46	0.80	0.50	1.00	0.58±0.35	0.38–1.00
Vadipatti	0.35	0.35	2.00	0.21	0.09	0.08	0.17	0.46±0.22	0.08–2.00
Vembarpatti	0.52	0.58	0.63	0.36	0.66	0.32	1.00	0.58±0.36	0.32–1.00
Virupatchi	0.58	0.77	0.84	0.24	1.00	0.90	0.20	0.65±0.33	0.20–1.00

SD, standard deviation.

TABLE 20.2 Distribution of Different Fluoride Concentration in Dindigul District, Tamil Nadu, India

		Distribution of Fluoride					
		No. of Samples Below 0.5		No. of Samples Between 0.5 and 1.5		No. of Samples Between 1.5 and 3	
Year	pH Ranges	mg/L	%	mg/L	%	mg/L	%
2001	7.7–9.1	14	12.8	18	17.0	0	0.0
2002	7.2–9.1	16	14.7	14	13.2	2	22.2
2003	6.9–8.8	11	10.1	17	16.0	4	44.4
2004	7.5–8.5	27	24.8	5	4.7	0	0.0
2005	7.2–8.6	14	12.8	17	16.0	1	11.1
2006	70–8.3	16	14.7	14	13.2	2	22.2
2007	7.3–8.4	11	10.1	21	19.8	0	0.0
Total number of samples per year $n=32$ ($n=224$ for 7 years)		109	48.7	106	47.3	9	4.0

20.4.2 Comparison of Deterministic and Geostatistical Methods

Deterministic (IDW) and geostatistical (EBK) methods were assessed to find the most reliable interpolation model for the study area. Table 20.3 illustrates the cross validation of IDW and EBK methods for use in the analysis of fluoride concentrations in the groundwater for the 7-year period of. The RMSE values indicate the precision of the interpolation methods, where the lower RMSE value is more precise than the higher value (Gumiere et al., 2014). Temporal interpolation of fluoride ion concentration reveals an RMSE value of between 0.172 and 0.677 for IDW and between 0.184 and 0.507 for the EBK model. By comparing each year's data, the performance achieved by EBK is shown to be more promising than that for IDW (Table 20.3). However, the performance of the IDW method is very close to that of EBK, but the pattern of distribution varies to a smaller extent (Figs. 20.2 and 20.3). The predicted versus measured graph of fluoride ion concentrations attained using the IDW and EBK methods for a 7-year period is shown in Figs. 20.4 and 20.5, respectively. It is clear from the figures that the data dispersion is higher in IDW compared with the EBK method. A similar observation is also made in relation to the estimated error versus the measured value using the IDW and EBK methods (Figs. 20.6 and 20.7). The abovementioned parameters indicate that the EBK method is a better interpolator when compared with IDW for the estimation of values at unsampled locations.

TABLE 20.3 Cross Validation Results of Inverse Distance Weighting and Empirical Bayesian Kriging Methods [Root Mean Square Error (RMSE)]

Inverse Distance Weighted (IDW)			Empirical Bayesian Kriging (EBK)		
Year	Mean	RMSE	Year	Mean	RMSE
2001	0.008	0.259	2001	0.004	0.242
2002	0.018	0.598	2002	0.027	0.507
2003	0.026	0.677	2003	0.006	0.499
2004	0.0001	0.172	2004	0.004	0.184
2005	0.037	0.461	2005	0.014	0.411
2006	0.041	0.527	2006	0.013	0.468
2007	0.025	0.332	2007	0.004	0.33

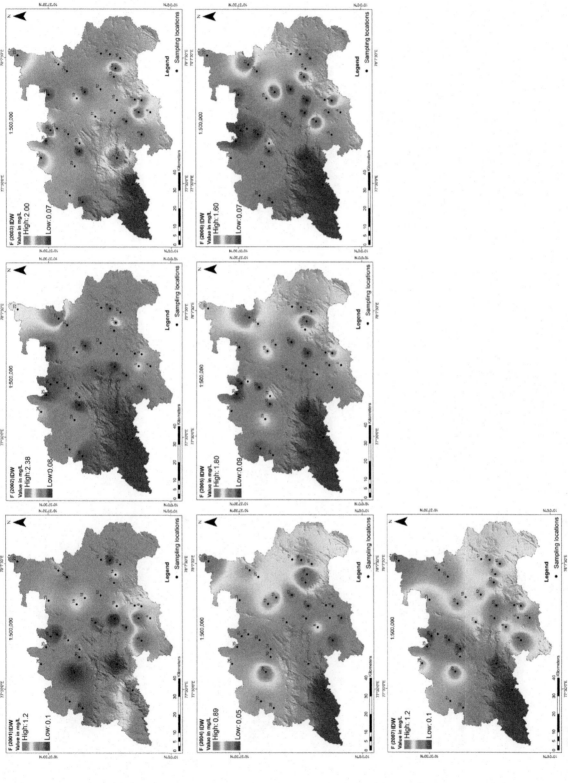

FIG. 20.2 Spatio-temporal distribution of fluoride (2001–07) using inverse distance weighting method.

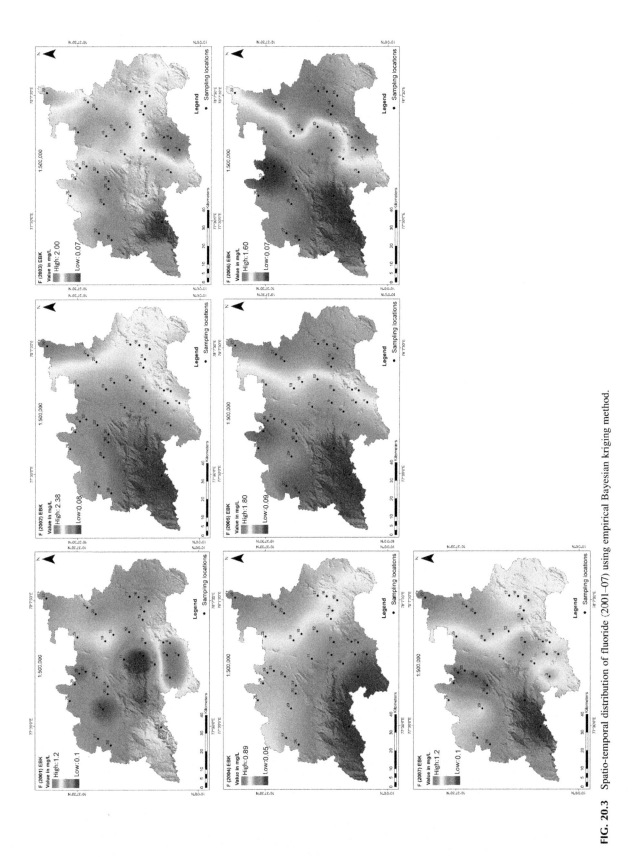

FIG. 20.3 Spatio-temporal distribution of fluoride (2001–07) using empirical Bayesian kriging method.

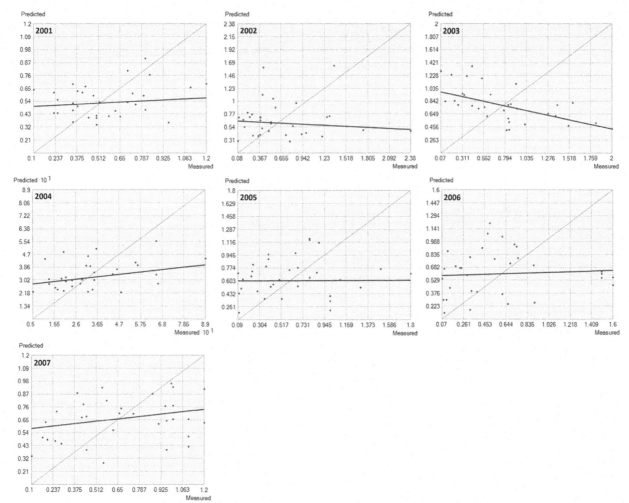

FIG. 20.4 Predicted versus measured value of fluoride concentration (2001–07) using inverse distance weighting method.

20.5 DISCUSSION

Groundwater contamination in a river basin can be effectively visualized using certain interpolation techniques. Conversely, these interpolation techniques are normally associated with certain level of uncertainties. The uncertainties occur due to various factors such as sample density, distance among the sampling locations, and the type of interpolation technique used (Liu et al., 2014). In the present study, two different kinds of interpolation methods (deterministic and geostatistical) were assessed for the study area. The IDW is the commonly used deterministic method for visualizing groundwater-quality, whereas EBK is a geostatistical method that has only recently been used to visualize groundwater quality and this accounts for a level of uncertainty with regards to the model. The reliability of the technique used can be quantified by estimating the cross-validation errors. The lower the cross-validation error, the better the model will be (Falivene et al., 2010). A cross-validation indicator, such as RMSE, was used to evaluate the reliability of the IDW and EBK methods (Wagner et al., 2012). The analysis reveals that EBK was found to be better than IDW, as the RMSE values are low in EBK when compared with IDW. However, RMSE values for IDW do not vary much when compared with the EBK model. This might be due to the distribution of the sampling wells, a gridded sampling with close intervals may have performed well but in the present case the sampling wells are not well distributed. This may lead to uncertainty in the southwestern part of the study area where the well intensity is low due to its hilly terrain. Moreover, the smoothing effect is also an important factor for considering the type of interpolation methods. A low smoothing effect preserves the observed data continuity and it is observed more in the IDW than the EBK method; however, the cross-validation results still point to the EBK method being superior. Furthermore, studies conducted by Liu et al. (2014) and Zhang et al. (2014) suggest that model parameters govern the simulations in a reliable way. For the IDW method model are generally less when compared with EBK but the semivariogram fitting are automated in the later case, which might be the reason for low uncertainty values in adopting the EBK method.

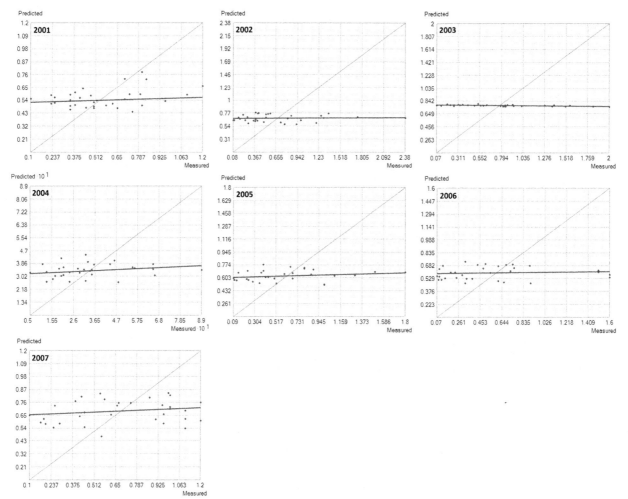

FIG. 20.5 Predicted versus measured value of fluoride concentration (2001–07) using empirical Bayesian kriging method.

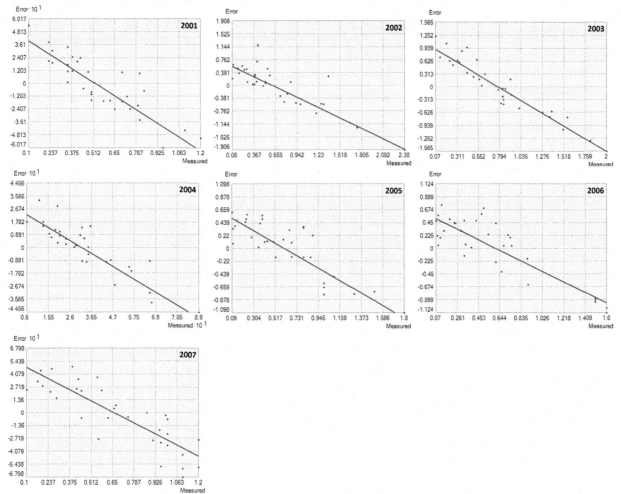

FIG. 20.6   Predicted error versus measured value of fluoride concentration (2001–07) using inverse distance weighting method.

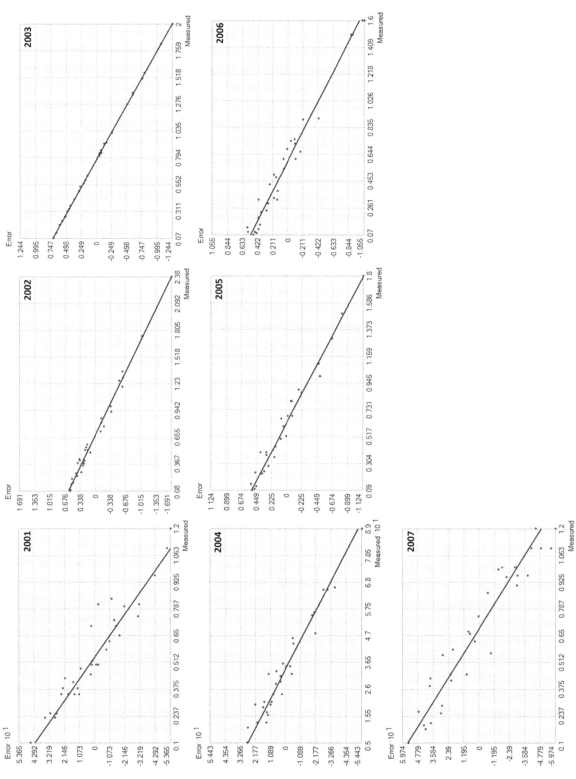

FIG. 20.7 Predicted error versus measured value of fluoride concentration (2001–07) using empirical Bayesian kriging method.

20.6   CONCLUSION

Two commonly used interpolation methods were assessed in the analysis of fluoride contamination in the groundwater of the Dindigul district, southern Tamil Nadu. The interpolation techniques were deterministic (IDW) and geostatistical (EBK) methods and these were compared for their reliable distribution of fluoride ion concentrations in the groundwater. It was found that both methods had similar efficiency but the EBK method's performance was superior and it was estimated using cross-validation. The RMSE value was found to be low in the case of EBK when compared with the IDW. It was also noted that the efficiency of both methods relies upon the density of the distribution of the sampling wells. A network of wells having a gridded pattern with close intervals might have improved the interpolation quality. In addition, variations in the concentrations of fluoride in the sampling wells affect the uncertainty in interpolation. Other factors, such as the smoothing effect and over- and underestimation need to be considered in selecting the appropriate interpolation methods. Since none of the methods will give 100% accuracy, room for error should be given while performing such analyses. Numerous factors, such as topography, sampling pattern, statistical data distribution, the scope of the study, and modeling skills, affect the results and these vary accordingly.

REFERENCES

Amalraj, A., Pius, A., 2013. Health risk from fluoride exposure of a population in selected areas of Tamil Nadu, South India. Food Sci. Human Wellness 2, 75–86.

APHA, 1995. Standard Methods for the Examination of Water and Waste Water, 19th ed. APHA, Washington, DC.

Arslan, H., 2012. Spatial and temporal mapping of groundwater salinity using ordinary kriging and indicator kriging: the case of Bafra Plain, Turkey. Agric. Water Manag. 113, 57–63.

Athavale, R.N., Das, R.K., 1999. Beware! Fluorosis is zeroing in on you. Down Earth 8, 24–25.

BIS, 1992. Indian Standard Specifications for Drinking Water. IS: 10500. http://hppcb.gov.in/EIAsorang/Spec.pdf.

Brindha, K., Elango, L., 2011. Fluoride in groundwater: causes, implications and mitigation measures. In: Monroy, S.D. (Ed.), Fluoride Properties, Applications and Environmental Management. Nova, New York, pp. 111–136.

Brindha, K., Elango, L., 2013. Geochemistry of fluoride rich groundwater in a weathered granitic rock region, southern India. Water Qual Expo Health 5, 127–138.

Brindha, K., Rajesh, R., Murugan, R., Elango, L., 2010. Natural and anthropogenic influence on the fluoride and nitrate concentration of groundwater in parts of Nalgonda district, Andhra Pradesh, India. J. Appl. Geochem. 12, 231–241.

Brindha, K., Rajesh, R., Murugan, R., Elango, L., 2011. Fluoride contamination in groundwater in parts of Nalgonda district, Andhra Pradesh, India. Environ. Monit. Assess. 172, 481–492.

Brindha, K., Jagadeshan, G., Kalpana, L., Elango, L., 2016. Fluoride in weathered rock aquifers of southern India: managed aquifer recharge for mitigation. Environ. Sci. Pollut. Res. 23, 8302–8316.

CGWB, 2008. Annual Report. In: South Eastern Coastal Region C.

Chae, G.T., Yun, S.T., Mayer, B., Kim, K.H., Kim, S.Y., Kwon, J.S., Kim, K., Koh, Y.K., 2007. Fluorine geochemistry in bedrock groundwater of South Korea. Sci. Total Environ. 385, 272–283.

Chidambaram, S., 2000. Hydrochemical Studies of Groundwater in Periyar District, Tamilnadu, India. (PhD thesis). Annamalai Univ.

Chidambaram, S., Manivannan, R., Anandhan, P., Prasad, B.M., Karmegam, U., Prasanna, M.V., Singaraja, C., Manikandan, S., 2013. Environmental hydrogeochemistry and genesis of fluoride in groundwaters of Dindigul district, Tamilnadu (India). Environ. Earth Sci. 68, 333–342.

Edmunds, W.M., Smedley, P.L., 2005. Fluoride in natural waters. In: Selinus, O. (Ed.), Essentials of Medical Geology. Elsevier Academic Press, Burlington, MA, pp. 301–329.

Falivene, O., Cabrera, L., Tolosana-Delgado, R., Sáez, A., 2010. Interpolation algorithm ranking using cross-validation and the role of smoothing effect. A coal zone example. Comput. Geosci. 36, 512–519.

Faure, G., 1991. Principles and Applications of Inorganic Geochemistry. Macmillan Publishing Company, New York, p. 626.

Gong, G., Mattevada, S., O'Bryant, S.E., 2014. Comparison of the accuracy of kriging and IDW interpolations in estimating groundwater arsenic concentrations in Texas. Environ. Res. 130, 59–69.

Gowrisankar, G., Jagadeshan, G., Elango, L., 2017. Managed aquifer recharge by a check dam to improve the quality of fluoride-rich groundwater: a case study from southern India. Environ. Monit. Assess. 189, 200.

Gumiere, S.J., Lafond, J.A., Hallema, D.W., Périard, Y., Caron, J., Gallichand, J., 2014. Mapping soil hydraulic conductivity and matric potential for water management of Cranberry: characterisation and spatial interpolation methods. Biosyst. Eng. 128, 29–40.

Handa, B.K., 1975. Geochemistry and genesis of fluoride containing groundwaters in India. Ground Water 13, 275–281.

Hem, J.D., 1985. Study and interpretation of the chemical characteristics of natural water, third ed. 2254, p. 363. U.S. Geological Survey Water-Supply Paper.

Hussain, I., Shakeel, M., Faisal, M., Soomro, Z.A., Hussain, T., 2014. Distribution of total dissolved solids in drinking water by means of Bayesian kriging and Gaussian spatial predictive process. Water Qual Expo Health 6, 177–185.

Jagadeshan, G., Elango, L., 2015. Suitability of fluoride contaminated groundwater for various purposes in a part of Vaniyar river basin, Dharmapuri district, Tamil Nadu. Water Qual Expo Health 7, 1–10.

Jagadeshan, G., Kalpana, L., Elango, L., 2015a. Hydrogeochemistry of high fluoride groundwater in hard rock aquifer of a part of Dharmapuri district, Tamil Nadu, India. Geochem. Int. 53, 554–564.

Jagadeshan, G., Kalpana, L., Elango, L., 2015b. Major ion signatures for identification of geochemical reactions responsible for release of fluoride from geogenic sources to groundwater and associated risk in Vaniyar River basin, Dharmapuri district, Tamil Nadu, India. Environ. Earth Sci. 74, 2439–2450.

Krivoruchko, K., 2012a. Empirical Bayesian Kriging. ArcUser (Fall), 6–10. http://www.esri.com/news/arcuser/1012/empirical-byesian-kriging.html.

Krivoruchko, K., 2012b. Modeling contamination using empirical Bayesian kriging. ArcUser. (Fall), http://www.esri.com/news/arcuser/1012/modeling-contamination-using-empirical-bayesiankriging.html.

Kumar, V., 2007. Optimal contour mapping of groundwater levels using universal kriging—a case study. Hydrol. Sci. J. 52, 1038–1050.

Lee, J.J., Jang, C.S., Wang, S.W., Liu, C.W., 2007. Evaluation of potential health risk of arsenic-affected groundwater using indicator kriging and dose response model. Sci. Total Environ. 384, 151–162.

Liu, C.W., Jang, C.S., Liao, C.M., 2004. Evaluation of arsenic contamination potential using indicator kriging in the Yun-Lin aquifer (Taiwan). Sci. Total Environ. 321, 173–188.

Liu, R., Chen, Y., Sun, C., Zhang, P., Wang, J., Yu, W., Shen, Z., 2014. Uncertainty analysis of total phosphorus spatial-temporal variations in the Yangtze River Estuary using different interpolation methods. Mar. Pollut. Bull. 86, 68–75.

Magesh, N.S., Krishnakumar, S., Chandrasekar, N., Soundranayagam, J.P., 2013. 2013. Groundwater quality assessment using WQI and GIS techniques, Dindigul district, Tamil Nadu, India. Arab. J. Geosci. 6, 4179–4189.

Magesh, N.S., Chandrasekar, N., Elango, L., 2016. Occurrence and distribution of fluoride in the groundwater of the Tamiraparani River basin, South India: a geostatistical modeling approach. Environ. Earth Sci. 75, 1483.

Magesh, N.S., Chandrasekar, N., Elango, L., 2017. Trace element concentrations in the groundwater of the Tamiraparani river basin, South India: Insights from human health risk and multivariate statistical techniques. Chemosphere 185, 468–479.

Manivannan, R., Chidambaram, S., Karmegam, U., Anandhan, P., Manikandan, S., Shahul, H., 2012. Mapping of fluoride ions in groundwater of Dindigul district, Tamilnadu, India, using GIS technique. Arab. J. Geosci. 5, 433–439.

Manivannan, R., Chidambaram, S., Srinivasamoorthy, K., Anandhan, P., Karmegam, U., Manikandan, S., Dhievanayaki, V., 2010. A statistical approach to study the spatial and temporal variation of Electrical conductivity of groundwater using GIS in Dindigul district. In: Chidambaram, S. (Ed.), Recent Trends in Water Research, Remote Sensing and General Perspectives. IK International, New Delhi, pp. 186–196.

Mason, B., Moore, C.B., 1982. Principles of Geochemistry, fourth ed. Wiley, New York, pp. 386–399.

Mirzaei, R., Sakizadeh, M., 2016. Comparison of interpolation methods for the estimation of groundwater contamination in Andimeshk–Shush Plain Southwest of Iran. Environ. Sci. Pollut. Res. 23, 2758–2769.

Mondal, N.C., Singh, V.S., Sarma, M.R.K., Thangarajan, M., 2002. Impact of tannery effluent in groundwater and its control: a case study in and around Dindigul, Tamil Nadu, India. In: Proceedings of ICHWM. BS Publisher, pp. 596–604.

Muralidharan, D., Nair, A.P., Sathyanarayana, U., 2002. Fluoride in shallow aquifers in Rajgarh Tehsil of Churu district Rajasthan—an arid environment. Curr. Sci. 83, 699–702.

Nagadu, B., Koeberl, C., Kurat, G., 2003. Petrography and geochemistry of the Singo granite, Uganda, and implications for its origin. J. Afr. Earth Sci. 36, 73–87.

Nas, B., Berktay, A., 2010. Groundwater quality mapping in urban groundwater using GIS. Environ. Monit. Assess. 160, 215–227.

Ozsvath, D.L., 2009. Fluoride and environmental health: a review. Rev. Environ. Sci. Biotechnol. 8, 59–79.

Pang, S., Li, T.X., Zhang, X.F., Wang, Y.D., Yu, H.Y., 2011. Spatial variability of cropland lead and its influencing factors: a case study in Shuangliu county, Sichuan province, China. Geoderma 162, 223–230.

Rao, N.S., Devadas, D.J., 2003. Fluoride incidence in groundwater in an area of peninsular India. Environ. Geol. 45, 243–251.

Robinson, T.P., Metternicht, G.M., 2006. Testing the performance of spatial interpolation techniques for mapping soil properties. Comput. Electron. Agric. 50, 97–108.

Saxena, V.K., Ahmed, S., 2001. Dissolution of fluoride in groundwater: a water–rock interaction study. Environmental Geology 40, 1084–1087.

Saxena, K., Ahmed, S., 2003. Inferring the chemical parameters for the dissolution of fluoride in groundwater. Environ. Geol. 43, 731–736.

Scaillet, B., Macdonald, R., 2004. Fluorite stability in silicic magmas. Contrib. Mineral. Petrol. 147, 319–329.

Taylor, R.P., Fallick, A.E., 1997. The evolution of fluorine rich felsic magmas: source dichotomy, magmatic convergence and the origins of Topaz Granite. Terra Nova 9, 105–108.

Vetrimurugan, E., Brindha, K., Sithole, B., Elango, L., 2017. Spatial interpolation methods and geostatistics for mapping groundwater contamination in a coastal area. Environ. Sci. Pollut. Res. 24, 11601–11617.

Viswanathan, G., Gopalakrishnan, S., Ilango, S.S., 2010. Assessment of water contribution on total fluoride intake of various age groups of people in fluoride endemic and non-endemic areas of Dindigul District, Tamil Nadu, South India. Water Res. 44, 6186–6200.

Wagner, P.D., Fiener, P., Wilken, F., Kumar, S., Schneider, K., 2012. Comparison and evaluation of spatial interpolation schemes for daily rainfall in data scarce regions. J. Hydrol. 464–465, 388–400.

WHO, 2004. Guidelines for Drinking Water Quality. World Health Organization, Geneva, p. 515.

Xiaolin, X., Zhenhua, Z., 1998. Partitioning of F between aqueous fluids and albite granite melt and its petrogenetic and metallogenetic significance. Chin. J. Geochem. 17, 303–310.

Xie, Y., Chen, T.B., Lei, M., Yang, J., Guo, Q.J., Song, B., Zhou, X.Y., 2011. Spatial distribution of soil heavy metal pollution estimated by different interpolation methods: accuracy and uncertainty analysis. Chemosphere 82, 468–476.

Zare-Mehrjardi, M., Taghizadeh-Mehrjardi, R., Akbarzadeh, A., 2010. Evaluation of geostatistical techniques for mapping spatial distribution of soil PH, salinity and plant cover affected by environmental factors in Southern Iran. Not. Sci. Biol. 2, 92–103.

Zhang, P., Liu, R., Bao, Y., Wang, J., Yu, W., Shen, Z., 2014. Uncertainty of SWAT model at different DEM resolutions in a large mountainous watershed. Water Res. 53, 132–144.

Chapter 21

Geostatistical Studies for Evaluation of Fluoride Contamination in Part of Dharmapuri District, South India

S. Anbazhagan*, M. Rajendran[†] and A. Jothibasu*

*Centre for Geoinformatics and Planetary Studies, Periyar University, Salem, India [†]Groundwater Division, Tamil Nadu Water Supply and Drainage Board (TWAD), Chennai, India

Chapter Outline

21.1 INTRODUCTION

Fluoride is found in all natural water sources at some concentration. Fluoride-rich groundwater has most often been reported in crystalline basement aquifers and arid sedimentary basins (Edmunds and Smedley, 1996). Evaluating the causes of elevated fluoride concentrations in natural water entails the identification of the fluoride source and requires an understanding of solubility, transport, and sinks for fluoride ion. Thus a variety of geochemical studies have been performed on various aspects of fluoride in groundwater. Fluoride in drinking water has both positive and negative effects on human health. Small concentrations of fluoride have beneficial effects on the teeth by hardening the enamel and reducing the incident of caries (Fung et al., 1999; Shomar et al., 2004; Elango and Kannan, 2007). The subsurface rocks in an area control the zones in which weathering affects the host rocks' minerals (Perez and Sanz, 1999). Saxena and Ahmed (2003) inferred the chemical parameters for the dissolution of fluoride into groundwater. Their study indicated that decomposition, dissociation, and dissolution are the main chemical processes responsible for the occurrence of fluoride in groundwater. Jagadeshan et al. (2015) studied the hydrogeochemistry of high fluoride groundwater in a hard-rock aquifer in a part of the Dharmapuri District, Tamil Nadu. Chae et al. (2007) studied in depth the fluorine geochemistry of bedrock groundwater in south Korea.

Using geostatistical modeling Sajil Kumar (2017) analyzed fluoride enrichment and nitrate contamination in the groundwater of the lower Bhavani Basin in Tamil Nadu. Daniel and Karuppasamy (2008) evaluated fluoride contamination in groundwater using geographical information systems (GIS) techniques in the Virudhunagar District. Anbazhagan and Archana Nair (2003) utilized GIS techniques to pictorially represent groundwater-quality zones in the Panvel Basin, Maharastra. Avishek et al. (2010) assessed water quality with special reference to fluoride in the Majhiaon Block of Garwa District in Jharkhand.

The purpose of this study was to outline the relationships between the fluoride concentrations in groundwater and the geology of the area and to explain the geochemical behavior of fluoride in deep-bedrock groundwater. The study area is mainly composed of hornblende biotite gneiss, charnockite, and granitic gneiss. In the present study, the hydrogeochemistry of the groundwater with special reference to fluoride, as well as the geological formations for enrichment of fluoride, was evaluated. The study examined the concentrations and spatial distribution of fluoride and the possible influence of the

GIS and Geostatistical Techniques for Groundwater Science. https://doi.org/10.1016/B978-0-12-815413-7.00021-3

geological features of the study area. The interrelationship of major ions with fluoride was also studied, since this can help to decipher the actual cause of fluoride enrichment in the aquifer.

21.2 STUDY AREA

The study area, Nallampalli Block, is located in the Dharmapuri District. It is bounded by the Palacode Block in the north, Dharmapuri Block in the east, Salem District in the south and Pennagaram Block in the west (Fig. 21.1). It falls between latitude $11°55'30''$ to $12°11'20''$ north and longitude $77°55'30''$ to $78°11'30''$ east. The total area covered by the block is $469 \, km^2$. There are fourteen minor irrigation tanks in the Block. The southern part of the block is covered by Thoppur reserve forest and the south western part is covered by the Parigam reserve forest. The major river drains in the Block are the Thoppiar and Palar at the border. Most of the Block is undulating in nature with local elevating and low-lying areas. The general course of the river is north to south for the Palar and east to west for the Thoppaiyar. The maximum relief is $450 \, m$ above mean sea level (MSL) with a minimum of $375 \, m$ above MSL.

21.3 METHODOLOGY

21.3.1 Frequency Distribution

The frequency distributions of fluoride concentration in groundwater were plotted for the pre- and postmonsoon periods (Fig. 21.2). The plot shows the variation in fluoride concentrations in the groundwater for both monsoons in the study area. From the frequency distribution it is inferred that the average concentration of fluoride is $1.2 \, mg/L$, which is on the border of the desirable limit. The plot shows that the fluoride concentrations in the groundwater fluctuate during both monsoon periods. The number of wells with fluoride contamination is reduced, which may be due to an increase in rainfall and dilution of the contamination.

21.3.2 Spatial Variation of Fluoride Concentration

The fluoride content in groundwater aquifer varies greatly depending on the geological settings and types of rock. The fluoride content may increase during evaporation if the solution remains in equilibrium with calcite and alkalinity

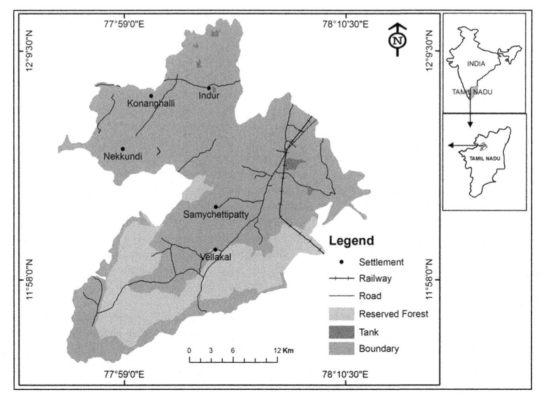

FIG. 21.1 Location map of the study area (Nallampalli Block)

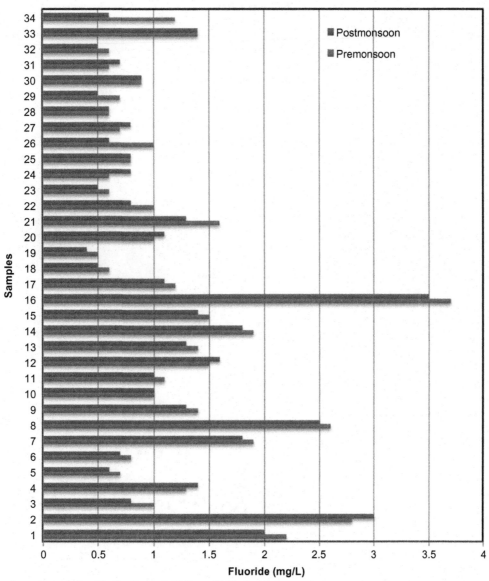

FIG. 21.2   Pre and postmonsoon fluoride distribution in Nallampalli Block.

is greater than hardness. The dissolution of evaporative salts deposited in an arid zone may be an important source for fluoride (Frencken, 1992). With a fluoride content of >1.5 mg/L mottling of teeth may occur to an objectionable degree. Fluoride concentrations in between 3 mg/L and 6 mg/L may cause skeletal fluorosis. Continued consumption of water with fluoride levels in excess of 10 mg/L can result in crippling fluorosis. The reason for such excess fluoride may be attributable to the presence of fluroapatite minerals in granitic gneiss and fluorine contents in biotite gneiss. According to ISI (1983), the Nallampalli Block is classified into three categories, such as a safe zone (0.60 mg/L to 1.2 mg/L), marginal-risk zone (1.2 mg/L to 1.5 mg/L), and risk zone (>1.5 mg/L), to outline the spatial variations in fluoride concentration in the pre- and postmonsoon periods. Fluoride concentrations of <0.60 mg/L are also effective for dental caries (Fig. 21.3).

21.3.3   Variations of Fluoride With Other Ions

The concentration of calcium (Ca), sodium (Na), and hydroxyl ions can alter the concentration of fluoride in the ground-water (Raju et al., 1993). In the present context, the pre and postmonsoon variations of fluoride with respect various ions, such as fluoride vs Ca, fluoride vs magnesium (Mg), fluoride vs Na, fluoride vs potassium (K), fluoride vs chloride (Cl), and fluoride vs bicarbonate, were plotted for a better understating of the fluoride contamination (Fig. 21.4). The linear correlation of fluoride with other ions indicates a mostly positive correlation. However, it is insignificant for ions including Ca, Mg, Na, potassium, Cl, nitrate, and sulphate (coefficient value ranges from 0.025 to 0.259).

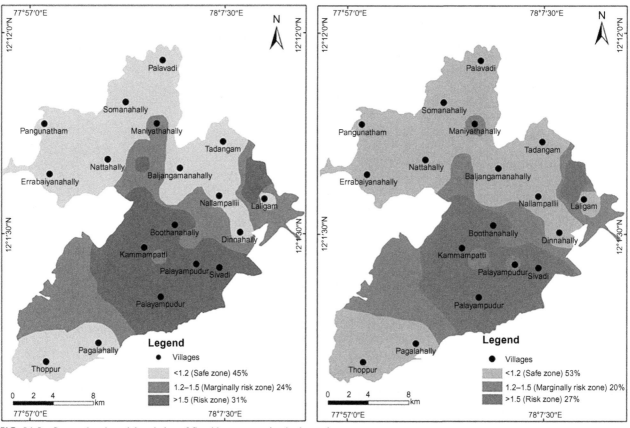

FIG. 21.3 Seasonal and spatial variation of fluoride concentration in the study area.

FIG. 21.4 Fluoride concentration variation with other ions in the study area.

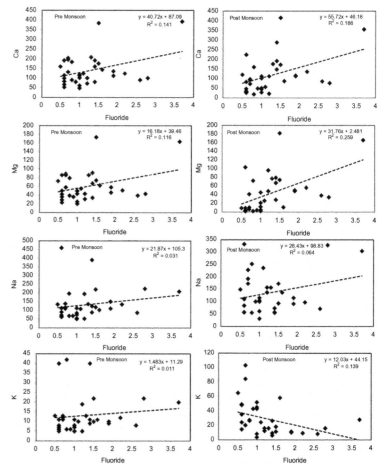

21.3.4 Variation of Chemistry at Selected Traverses

The variations in the chemistry of the groundwater are influenced by rock types, recharge, and flow conditions. It is necessary to assess the elemental variations from the different directions in the block. In order to understand the chemical variations in the groundwater aquifers in the block five traverses were chosen for the present study (Fig. 21.5). The wells fall in five different directions are given in Table 21.1.

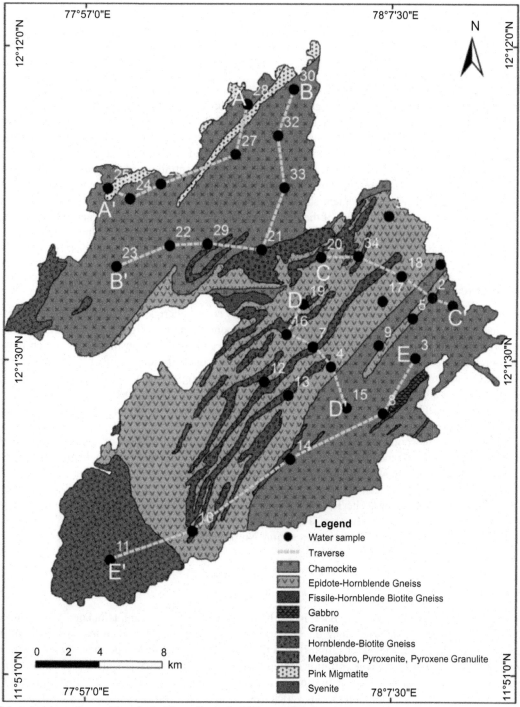

FIG. 21.5 Selected traverses along with geological formation in the Nallampalli Block.

TABLE 21.1 Well number and directions along different traverse in the study area

Traverse	Directions	Well Numbers
Traverse A–A′	Upper NE-SW	28, 27, 26, 24, and 25
Traverse B–B′	Parallel to AA′	30, 32, 33, 21, 29, 22 and 23
Traverse C–C′	E-W	20, 34, 18, 2 and 6
Traverse D–D′	Central NW-NE	19, 16, 7, 4 and 15
Traverse E–E′	Down NE-SW	3, 8, 14, 10 and 11

21.4 RESULTS AND DISCUSSIONS

In the frequency distribution, nearly 50% of the groundwater samples (15 out of 34) have high fluoride contamination, which exceeds the permissible limit (>1.5 mg/L). Moreover, the recharge of groundwater during monsoon indicates dilution of fluoride concentrations in the postmonsoon period. Since the study area has very meager concentrations of fluoride in the groundwater, the safe zone is considered to be represented by a concentration of <1.2 mg/L. In the premonsoon period about 31% of the block is represented by risk-zone concentrations in the groundwater, where as in the postmonsoon period the risk zone is represented by 29%. This indicates that a significant amount of fluoride concentration is reduced by recharge. The bicarbonate content shows a slightly high positive correlation (0.275–0.385). The positive correlation indicating that the higher alkalinity of the water promotes the leaching of fluoride and thus affects the concentrations of fluoride in the groundwater. The slight increase in the positive correlation between fluoride and other ions during the postmonsoon period may be due to an increase of dissolution during the recharge conditions.

21.4.1 Traverse A–A′

There are five wells located along traverse A–A′ in the block, mostly covers charnockite. The fluoride concentration in this groundwater ranges from 0.6 mg/L to 1.0 mg/L. The fluoride concentrations fall slightly in between the pre- and postmonsoon periods. However, in well 26 the concentration is significantly reduced postmonsoon, which may be due to dilution of the groundwater by recharge. In addition to fluoride, the concentrations of other elements, such as Ca, Na, Mg, SO_4, Cl, HCO_3, and total dissolved solids (TDS), were plotted along this traverse (Fig. 21.6). The concentration of Ca along this sector is 20 mg/L to 220 mg/L. Overall a parallel trend is observed between the pre- and postmonsoon periods; however, the concentrations are reduced during postmonsoon. The highest concentration for Ca was observed in well number 27. The concentration of Na in the groundwater ranges from 45 mg/L to 200 mg/L in this traverse. Though the trend is similar in both monsoons, the concentration is increased during postmonsoon period. The concentration of Mg ranges from 190 mg/L to 400 mg/L and has a similar pattern in both monsoon seasons. The SO_4 in groundwater ranges from 15 mg/L to 140 mg/L and the plot shows variations in the concentration for a few wells in different seasons. Cl ranges from 10 mg/L to 500 mg/L and similar concentrations and trends were observed in both seasons. Similarly, bicarbonate also shows the same pattern in both seasons. The TDS in the wells ranges from 400 mg/L to 1700 mg/L. Out of the five wells, three fall within more than permissible limits of TDS concentration. The quality of groundwater was found not to be suitable for domestic purposes in well 27 along this sector. No definite trend was observed between the concentration of fluoride and other major elements in this traverse.

21.4.2 Traverse B–B′

There are seven wells located along traverse B–B′ in the block, which are once again associated with charnockite. The fluoride concentration in the groundwater along this traverse ranges from 0.5 mg/L to 1.6 mg/L, which is slightly higher than the traverse A–A′. The fluoride level falls slightly in both the pre- and postmonsoon periods. Well 21, which is located in pink migmatite gneiss, shows high a concentration of fluoride, which indicates that the gneissic rock has more influence on fluoride contamination. Similarly, the concentrations of the major elements and TDS were plotted for the B–B′ traverse. The Ca content of the groundwater along this sector varies from 30 mg/L to 300 mg/L. It was observed that well 33 has a high concentration during the postmonsoon period. The Na value ranges from 70 mg/L to 400 mg/L in this sector. A high Na value was observed in well 33 during premonsoon. The concentration of Mg ranges from 0 mg/L to 90 mg/L. A high

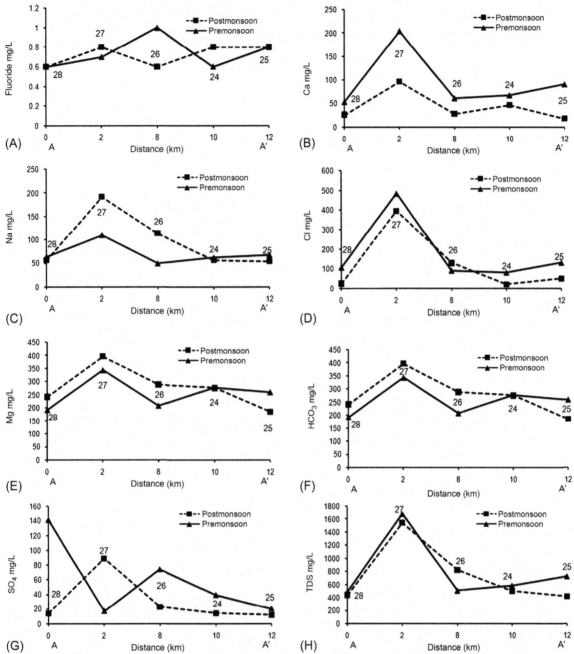

FIG. 21.6   Concentration of fluoride (A) and other major (B–H) elements in groundwater samples along traverse A–A′ in the Nallampalli Block.

concentration of Mg is observed during premonsoon, although this decreased during the postmonsoon season. SO_4 in the groundwater ranges from 30 mg/L to 190 mg/L and the plot shows variations in the concentration for a few wells in different seasons. Cl ranges from 120 mg/L to 550 mg/L and the concentration varies in the pre- and postmonsoon periods. The bicarbonate value ranges from 280 mg/L to 570 mg/L and has different patterns in the pre- and postmonsoon periods. The TDS in the wells ranges from 750 mg/L to 1800 mg/L and the same trend is observed in both seasons. Well 33 has been shown to have poor-quality groundwater in all aspects, as the majority of the elements exceeds the permissible limits (Fig. 21.7).

21.4.3   Traverse C–C′

There are five wells located along traverse C–C′ in the block. This traverse covers heterogeneous rock types such as epidote hornblende biotite gneiss, pink migmatite gneiss, and charnockite. However, most of the wells are located in epidote

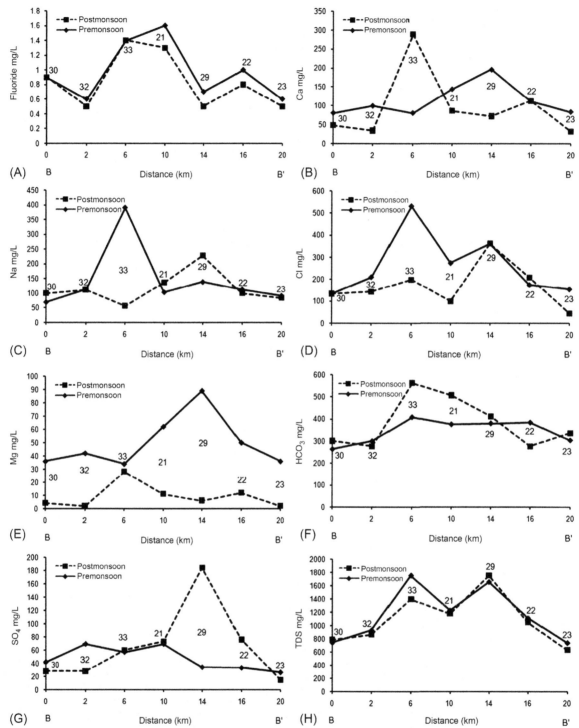

FIG. 21.7 Concentration of fluoride (A) and other major (B–H) elements in groundwater samples along traverse B–B′ in the Nallampalli Block.

hornblende biotite gneiss. The fluoride value ranges from 0.5 mg/L to 3.1 mg/L. The concentration of fluoride in the groundwater is similar in both pre- and postmonsoon periods in the block. Well 2, located at Mathemangalam in gneissic rock, shows a high fluoride concentration (3.1 mg/L). Across traverse C–C′, most of the wells show similar concentrations of the major elements in both seasons (Fig. 21.8). In comparison with fluoride concentration, the major elements show a negative correlation. For example, well number 18 shows.

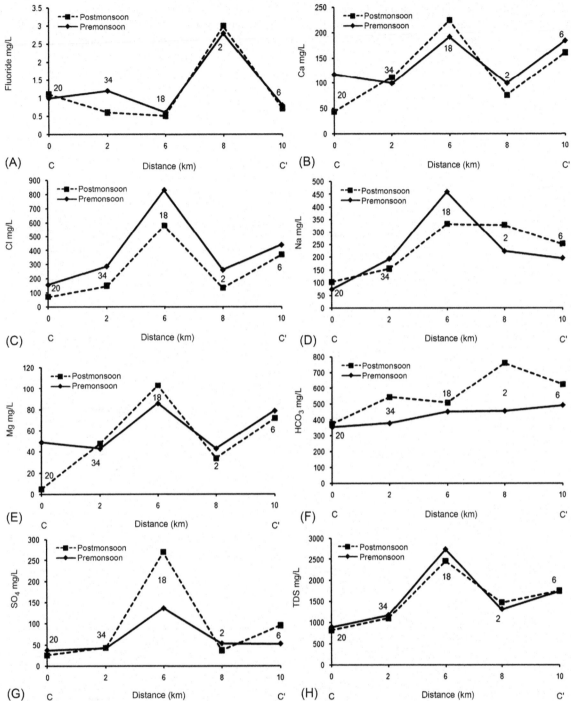

FIG. 21.8 Concentration of fluoride (A) and other major elements (B–H) in groundwater samples along traverse C–C′ in the Nallampalli Block.

High concentrations of the major elements, whereas the fluoride concentration in the groundwater is comparatively low (0.5 mg/L).

21.4.4 Traverse D–D′

There are five wells located along traverse D–D′ in the block. This traverse also covers heterogeneous rock types such as epidote hornblende biotite gneiss, pink migmatite gneiss, and charnockite. Similar to traverse C–C′, most of the wells are located in epidote hornblende biotite gneiss. The concentration of fluoride in the groundwater ranges from 0.4 mg/L to

3.6 mg/L. It is similar in both the pre- and postmonsoon periods in the block. Well 16, located in gneissic rocks at Kombai, shows a high fluoride concentration (3.6 mg/L). Overall, the fluoride contamination along this sector is high. A comparison of the fluoride concentration in the groundwater with the rest of the major elements shows a positive correlation. An increase in the major elements increases the concentration of fluoride in the groundwater (Fig. 21.9). Moreover, along this traverse, the concentration of the major elements in the groundwater shows a similar pattern in both seasons. Anomalous high values were observed in well 16 compared with the rest of the wells in the traverse. Most of the wells show an increase in bicarbonate concentration in the postmonsoon period indicating significant recharge along this sector.

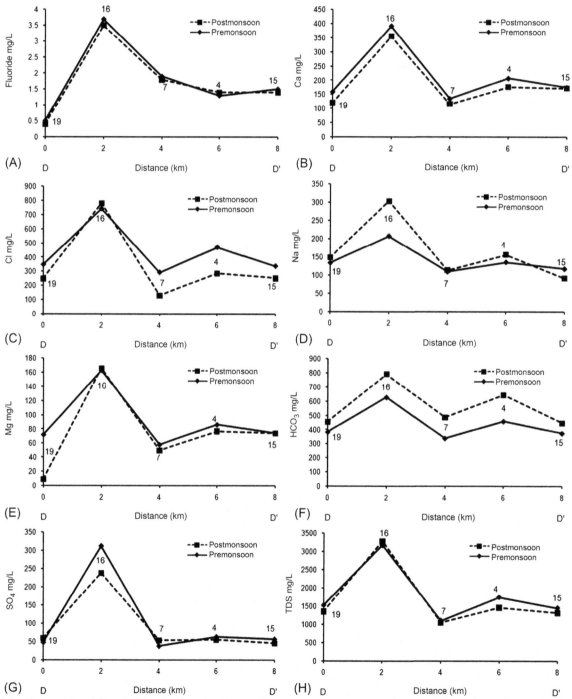

FIG. 21.9 Concentration of fluoride (A) and other major elements (B–H) in groundwater samples along traverse D–D′ in the Nallampalli Block.

21.4.5 Traverse E–E′

Along the traverse E–E′, wells are located in hornblende biotite gneiss, pyroxenite, epidote hornblende gneiss, and charnockite. The fluoride concentration along this stretch ranges from 0.7 mg/L to 2.7 mg/L. The concentration of fluoride in the groundwater is similar in both the pre- and postmonsoon periods. Well 8 located at Sivadi shows a high fluoride concentration (2.7 mg/L). No major variations were noticed in the Ca, Mg, and TDS concentrations in the pre- and postmonsoon periods; however, significant variations were observed in Cl, HCO$_3$, SO$_4$, and Na concentrations in most of the wells (Fig. 21.10).

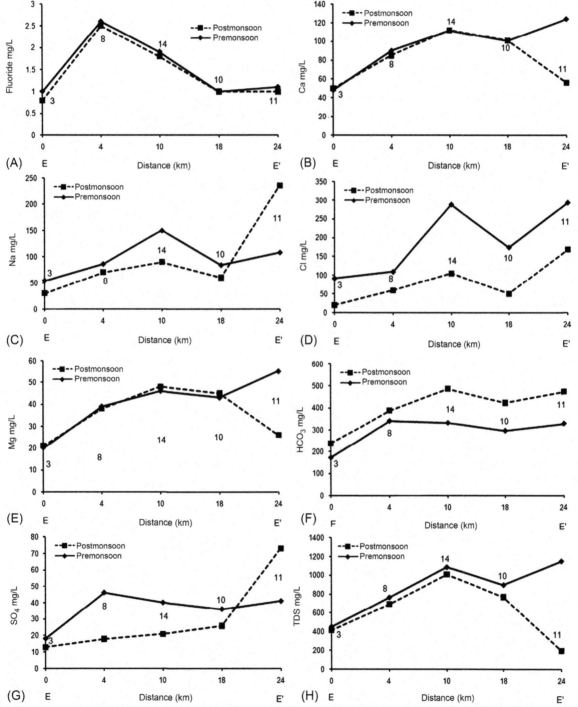

FIG. 21.10 Concentration of fluoride (A) and other major elements (B–H) in groundwater samples along traverse E–E′ in the Nallampalli Block.

21.5 CONCLUSION

Groundwater samples were collected from 34 wells for both the pre- and postmonsoon period and to assess the groundwater quality in the study area. In general, fluoride contamination is the major problem in the Nallampalli Block and the concentration ranges from 0.4 mg/L to 3.7 mg/L. The fluoride contamination in the groundwater is linearly correlated with other major ions, which indicates that most of the ions show a positive correlation. The concentration of fluoride in the block is assessed in different directions through traverse plot. Fluoride concentration shows a positive correlation with most of the elements, particularly with pH and bicarbonate, indicated that alkaline, which provides suitable conditions for the leaching of fluoride from the rocks. High levels of fluoride were observed in the premonsoon period, which indicated that fluoride levels increase when the water level reaches the deeper part of the aquifers and comes into contact with fresh rock. The results of the spatial and temporal variability of the fluoride concentrations in the study area are helpful for sustainable planning and management of the groundwater resource to supply potable water to a dependent population.

REFERENCES

Anbazhagan, S., Archana Nair, M., 2003. Remote sensing and GIS for land and groundwater resource analysis in Panvel basin, India. In: VI Inter-Regional Conference on Environment Water (Envirowater'2003), Albacete, Spain, 3–5 September. Abstractpp. 160–161.

Avishek, K., Pathak, G., Nathawat, M.S., Jha, U., Kumari, N., 2010. Water quality assessment of Majhiaon block of Garwa district in Jharkhand with special focus on fluoride analysis. Environ. Monit. Assess. 167 (1–4), 6. 17–23.

Chae, G.T., Yun, S.-T., Mayer, B., Kim, K.-H., Kim, S.-Y., Kwon, J.-S., Kim, K., Koh, Y.-K., 2007. Fluorine geochemistry in bedrock groundwater of South Korea. Sci. Total Environ. 385 (1–3), 272–283. https://doi.org/10.1016/j.scitotenv.2007.06.038.

Daniel, J., Karuppasamy, S., 2008. Evaluation of fluoride contamination in groundwater using remote sensing and GIS techniques in Virudhunagar District. Int. J. Adv. Remote Sens. GIS 1, 12–19.

Edmunds, W.M., Smedley, P.L., 1996. Groundwater geochemistry and health: an overview. In: Appleton, J.D., Fuge, R., McCall, H.J.G. (Eds.), Environmental Geochemistry and Health. Geological Society Special Publication No. 113, pp. 91–105.

Elango, L., Kannan, R., 2007. Depart chapter 11: Rock–water interaction and its control on chemical composition of groundwater. Dev. Environ. Sci. 5, 229–243. https://doi.org/10.1016/S1474-8177(07)05011-5.

Frencken, J.E., 1992. Endemic fluorosis in developing countries, causes, effects and possible solutions. NIPG-TNO, Leiden, The Netherlands. Publication number 91:082.

Fung, K., Zhang, Z., Wong, J., Wong, M., 1999. Fluoride contents in tea and soil from tea plantations and the release of fluoride into tea liquor during infusion. Environ. Pollut. 104, 197–205.

Indian Standard Institution (ISI), 1983. Indian Standard Specification for Drinking Water IS: 10500. .

Jagadeshan, G., Kalpana, L., Elango, L., 2015. Hydrogeochemistry of high fluoride groundwater in hard rock aquifer in a part of Dharmapuri District, Tamil Nadu, India. Geochem. Int. 53 (6), 554–564.

Perez, E.S., Sanz, J., 1999. Fluoride concentration in drinking water in the province of Soria, Central Spain. Environ. Geochem. Health 21, 133–140.

Raju, D.V.R., Janardhana Raju, N., Kotaiah, B., 1993. Complexation of fluoride ions with alum-flocs at various pH values during coagulation and flocculation. J. Geol. Soc. India 42, 51–54.

Sajil Kumar, P.J., 2017. Fluoride in groundwater- sources, geochemical mobilization and treatment options. Int. J. Environ. Sci. Nat. Res. 1 (4). IJESNR. MS.ID.555567.

Saxena, V., Ahmed, S., 2003. Inferring chemical parameters for dissolution of fluoride in groundwater. Environ. Geol. 43 (6), 731–736.

Shomar, B., Müller, G., Yahya, A., Askar, S., Sansur, R., 2004. Fluorides in groundwater, soil and infused-black tea and the occurrence of dental fluorosis among school children of the Gaza Strip. J. Wat. Health 2, 23–35.

Chapter 22

Fluoride Contamination in Groundwater— A GIS and Geostatistics Reappraisal

Banajarani Panda*,‡, V. Dhanu Radha†, S. Chidambaram†, M. Arindam‡, R. Thilagavathi*, S. Manikandan*, C. Thivya§, A.L. Ramanathan¶ and N. Ganesh*

*Department of Earth Sciences, Annamalai University, Chidambaram, India †Water Research Center, Kuwait Institute for Scientific Research, Safat, Kuwait ‡Water Sciences Lab, University of Nebraska-Lincoln, Lincoln, NE, United States §School of Earth and Atmospheric Sciences, University of Madras, Chennai, India ¶School of Environmental Sciences, JNU, New Delhi, India

Chapter Outline

22.1 INTRODUCTION

Groundwater is one of the most important sustainable resources, but it is currently declining in regions of Asia and North America (Gleeson et al., 2012). The supply of safe drinking water has become an issue with respect to its quality and quantity. According to the World Health Organization (WHO, 2011) a fluoride concentration of above 1.5 mg/L is an indicator of quality deterioration, which is a situation that is affecting over 260 million individuals round the globe (Amini et al., 2008).

Fluoride in groundwater is one of the critical pollutants that is reported in increasing level in Andes, western Brazil, and the Rift Valley in Africa (Chandrajith et al., 2012; Craig et al., 2015; Rafique et al., 2009). Moreover, it has been reported that the concentrations of fluoride are high in countries like China (Wang and Huang, 1995), Mexico (Di'az-Barriga et al., 1997), Argentina (Kruse and Ainchil, 2003), Tanzania (Mjengera and Mkongo, 2003), South Africa (WRC, 2001), India (Ayoob and Gupta, 2006), etc. The fluoride concentration in groundwater is due to geogenic contamination, as evidenced by Edmunds and Smedley (2005) and Ayoob and Gupta (2006).

The main sources of fluoride are a number of minerals, such as fluorite, biotites, topaz, and their corresponding host rocks, like granite, basalt, syenite, shale, etc., which may discharge fluoride into the groundwater and increases its concentration (Apambire et al., 1997; Reddy and Prasad, 2003; Edmunds and Smedley, 2005). Also, some anthropogenic activities, like the use of phosphate fertilizer or aluminum smelting, may introduce significant amounts of fluoride into the environment (Saxena and Ahmed, 2003). Several studies have reported that fluoride contamination in groundwater is from crystalline basement rocks/volcanic bedrocks (Handa, 1975; Apambire et al., 1997; Genxu and Guodong, 2001). However, change in climatic conditions and a long residence time of groundwater play vital role in the elevated concentration of fluoride in groundwater (Saxena and Ahmed, 2001; Jacks et al., 2005; Edmunds and Smedley, 2005; Sreedevi et al., 2006; Shaji et al., 2007; Amini et al., 2008).

Literature from a number of Indian papers show the high levels of fluoride in groundwater are alarming (CGWB, 2010). The various reasons for high fluoride content include the evapotranspiration of groundwater with residual alkalinity

GIS and Geostatistical Techniques for Groundwater Science. https://doi.org/10.1016/B978-0-12-815413-7.00022-5

(Jacks et al., 2005), endemic fluorosis (Shortt et al., 1937), and a longer residence time in deeper aquifers than shallow aquifers (Madhnure and Pandith Malpe, 2007), 325 mg/kg to 3200 mg/kg of fluoride is available through granitic rocks in Nalgonda area (Brindha et al., 2011).

From the literature data on the fluoride contamination of groundwater on a national and international scale, very few reviews are available on the application of GIS (Manikandan et al., 2014; Singaraja et al., 2014) or geostatistical methods (Chidambaram et al., 2013. Thivya et al., 2017) to the assessment of fluoride contamination in groundwater. Hence this chapter addresses a review of GIS and geostatistical aspects to assess fluoride contamination in groundwater and summarizes the health impacts. The chapter also describes the sources, ranges of fluoride level in different aquifers, and its distribution worldwide.

22.2 REVIEW OF STATUS OF FLUORIDE CONTAMINATION IN GROUNDWATER LINKED WITH GIS AND GEOSTATISTICAL ASPECTS

Evaluating the causes of elevated fluoride concentrations in natural water entails not only the identification of the fluoride source, but also the attempt to gain an understanding of solubility, transport, and sinks for fluoride ion. Thus a variety of geostatistical studies along with GIS have been performed on various aspects of fluoride in groundwater. A few researchers have also tried to model the fluoride concentration in groundwater using GIS techniques to determine the extent of fluoride contamination and to guide the localization and remediation of the problem (Podgorski et al., 2017; Raju et al., 2009; Padhi and Muralidharan, 2012).

In particular, some researchers have studied fluoride concentrations and water-rock interactions in various aquifers with different geologic settings with the application of geostatistical techniques (Nordstrom and Jenne, 1977; Edmunds and Smedley, 1996; Nordstrom et al., 1989; Gaciri and Davies, 1993; Saxena and Ahmed, 2003).

In the recent past many investigators have used factor analysis (a tool of geostatistics) to characterize the relationship between fluoride contamination in the groundwater and to identify possible geochemical reactions under the prevailing hydrogeological conditions, as well as their role in mobilizing fluoride concentration (Reghunath et al., 2002; Farnham et al., 2003; Liu et al., 2003; Stüben et al., 2003).

A global probability map using advanced statistical procedures available within ArcGIS (Ver 9.1) was developed to indicate the risk of fluoride contamination in groundwater (Amini et al., 2008). The probability maps were developed for eight "process regions (PR)." The PRs were explained based on the dissolution of fluoride (geological, climatic, and soil properties) as well as statistical testing. The modeled data were compared with measured data and inferred that 60%–70% of fluoride variations could be explained by the models in six PRs. However, the remaining 30% of the measured data variations were explained by two PRs (Amini et al., 2008). The resulting probability map matched well with the fluoride-affected regions, as mentioned in the international literature. Thus, using these probability maps, possible contamination zones can be predicted before proceeding to fluoride testing.

22.3 SOURCES

The main sources of fluoride include the following rock types: igneous, sedimentary metamorphic, such as granite, basalt, shale, and clays; and calcium phosphate, which represents 0.06% to 0.09% of the Earth's crust, but has a low average abundance, i.e., $300 \, mg \, kg^{-1}$ (Wenzel and Blum, 1992; Thy, 1983). Fluorite (CaF_2) is one of the dominant sources of fluoride in groundwater, especially in granitic terrains (Deshmukh et al., 1995), yet its solubility in fresh water and dissolution rate is remarkably slow (Nordstrom and Jenne, 1977). Some researchers have reported that high fluoride concentration in groundwater is due to the dissolution of biotite, which may contain significant fluoride at the OH^- sites of their octahedral sheet (Nordstrom et al., 1989; Li et al., 2003).

Quantitative estimates of fluoride enrichment in natural waters, particularly the study of fluid-mineral equilibria, are sparse (Saxena and Ahmed, 2001). High fluoride concentrations are often found in deep groundwater, especially in granitic bedrocks, as well as in granitic and gneissic aquifers, which leads to a serious issue for exploitation of these sources (White et al., 1963; Yun et al., 1998a). Increases in temperature and residence time with increasing depth may be the cause for high fluoride concentration, which enhances the dissolution of fluorine-bearing minerals from rocks (Nordstrom et al., 1989; Saxena and Ahmed, 2003). Table 22.1 represents the various minerals that contain fluoride within their composition and the rocks in which they present. Minerals that have the greatest effect on the hydrogeochemistry of fluoride are fluorite, apatite, mica, amphiboles, lepidolite, fluorspar, tremolite, actinolite, clays, and villiaumite (Deshmukh et al., 1995).

TABLE 22.1 Minerals Containing Fluoride (Ashok and Parveen, 2010)

S.No.	Mineral	Chemical Composition	Rocks
1	Fluorspar	$[CaF_23Ca_3(PO_4)_2]$	Pegmatite Pneumatolytic deposits
2	Fluorite	CaF_2	CaF_2 Pegmatite Metamorphosed Limestone
3	Lepidolite	$K_2(Li, Al)_5(Si_6Al_2)O_{20}(OHF)_4$	Gabbro, dolerites
4	Tremolite actinolite	$Ca_2(MgFe^{2+})_5(Si_8O_{22}) (OHF)_2$	Clay
5	Rock phosphate	$NaCa_2(MgFe^{2+})_4(AlFe^{3+})(SiAl)_8O_{22}(OHF)_2$	Limestone, fossils

22.4 MECHANISM OF FLUORIDE RELEASE TO GROUNDWATER

22.4.1 From Atmosphere and Surface Water

The main mechanisms of fluorine transfer from the atmosphere are volcanism, evaporation, marine aerosol, and industrial pollution (Edmunds and Smedley, 2013). Water-rock interactions and anthropogenic activities are responsible for the uptake or release of fluorine to the water, in the geosphere (Edmunds and Smedley, 2013). Fluoride in the rainfall releases to the groundwater once it infiltrates the aquifer. The main source of fluoride via rainfall is volcanic emission, marine aerosol, anthropogenic activities of chlorofluorocarbons (CFCs), and industries. Coal, brick melting, and aluminum smelting industries contribute fluoride to the rainfall (Edmunds and Smedley, 2013; Fuge and Andrews, 1988). The burning of domestic coal also leads to fluoride release into the atmosphere. Fluoride intrusion into seawater through marine aerosol precipitates in a coastal region was found to be $0.68 \, \mu g/L^{-1}$ fluoride per every $10 \, mg/L^{-1}$ Cl present. Chloride precipitation of $1 \, mg/L$ or below in continental territories contributes to $0.1 \, \mu g/L^{-1}$ fluoride in streams and groundwater. In fact, the precipitation of fluoride tends to be higher than these assessments, which are due to fractionation take-up of the more unpredictable fluoride on the ocean surface or by air. As the consequences of human activity are likely to predominate in many zones, the concentrations of fluoride in pristine precipitation are difficult to survey. Similar investigations at inland and coastal locales in Virginia, USA, found median values of $4 \, \mu g/L^{-1}$ and $9 \, \mu g/L^{-1}$ individually and marine vaporized sources of information were considered to be sparse (Barnard and Nordstrom, 1982). In surface water fluoride concentration is higher than the rainfall in geothermal areas.

22.4.2 From Minerals

Fluorine is an essential mineral present in biotite and amphiboles. The hydroxyl positions in the mineral are replaced with fluorine. For example:

$$K_2(Mg, Fe)_4(Fe, Al)_2[Si_6Al_2O_{20}](OH)_2(F, Cl)_2$$

Fluoride is leached out from these minerals and driven into the groundwater during the process of weathering.

In marine dregs, fluorine is concentrated by adsorption onto clays and biogeochemical forms by replacing phosphorous. Limestone may contain a limited amount of fluorapatite, particularly "Francolite." As most of the sandstones contain a low amount of fluorine, the fluoride in groundwater may be low.

22.4.3 From Soil

A little amount of fluoride is naturally available in soil, i.e., $<10 \, mg/kg$, which is easily soluble. Anthropogenic activities contribute to soil concentrations; high fluoride concentrations of $8500-38,000 \, mg/kg^{-1}$ are found in phosphate manures (Kabata-Pendias and Pendias, 1984) and $80-1950 \, mg \, kg^{-1}$ in sewage dumps (Rea, 1979). The amount of fluoride adsorbed by soils varies with soil composition, the pH of soil, salinity, and fluoride concentration (Fuhong and Shuquin, 1988; Lavado and Reinaudi, 1979). Wenzel and Blum (1992) noticed that the minimum mobilization of fluoride occurs at pH 6.0–6.5, while it increases at pH <6 because of the arrangement of $[AlF]^{2+}$ and $[AlF_2]^+$ in the solution. Fine-grained soils have more holding capacity for fluoride than sandy soils. Clay minerals with organic matter prove to be very good adsorbent of fluoride compared with other minerals (Wuyi et al., 2002; Fuge and Andrews, 1988).

When water with high fluoride content passes through a soil profile comprised of mostly sand, clay, iron, or aluminum, up to half of the infiltrating fluoride in the water may move through the soil profile (Pickering, 1985). Similarly, the fluoride content from anthropogenic activities may pass through the soil and saturate the water table when it exceeds the soil's retention capacity in that area. Most of the time soil acts as a natural filter of fluoride before it reaches the water table. Sometimes low fluoride concentration is managed in soils through evapotranspiration at the surface. This may elevate fluoride content in the water table by up to five times in mild atmospheres and 10–100 times under semi-arid conditions.

22.4.4 From Solution

The $K_{fluorite}$ mainly controls the activities of fluoride in aqueous solution, as given below (Edmunds and Smedley, 2013):

$$CaF_2 = Ca^{2+} + 2F^- \tag{22.1}$$

$$K_{fluorite} = (Ca^{2+}).(F)^2 = 10^{-10.57} \text{ at } 25^\circ C \tag{22.2}$$

$$\text{Or, } \log K_{fluorite} = Ca^{2+}, 2\log (F^- = 0.57 \tag{22.3}$$

Considering Eq. (22.1), the fluoride concentration in the solution is directly proportional to Ca^{2+} concentration, in the presence of fluorite. Thus the absence of Ca^{2+} in a solution can allow more fluoride to accumulate and the precipitation of carbonates can be directly linked to the loss of fluoride in the solution in the absence of Ca^{2+} ion (Reddy et al., 2010; Turner et al., 2005). Low-Ca groundwater is mostly noted in volcanic terrain comprised predominantly of basic rocks (Ashley and Burley, 1994; Kilham and Hecky, 1973) or where cation exchange occurs naturally (Handa, 1975). The ion-exchange process also leads to low Ca groundwater, as Ca^{2+} is removed by the Na^+ ion in the clay minerals. In both of the above mentioned cases the groundwater will be of a $Na-HCO_3$ type.

As mentioned in Eq. (22.2), fluorite solubility is dependent on temperature. As solubility decreases with a decrease in temperature, this will be reflected in a change in the equilibrium constant. Considering an example, the value of fluorite will change from $10^{-10.57}$ at 25°C to $10^{-10.08}$ at 10°C (Edmunds and Smedley, 2013). The salinity or ionic strength influence the equilibrium constant, providing a varied saturation index of water with respect to different minerals.

The response times of a formation matrix of an area play an essential role in the fluoride concentration of the media (water). A high content of fluoride will occur in groundwater that has had a long residence time within the host aquifer owing to diagenetic responses. Surface waters predominantly have low concentrations, as most shallow groundwater from open wells is recharged by these waters and with direct contact to the atmosphere the water is young, recently infiltrated and fresh with low fluoride. Hence, profound (older) groundwaters from boreholes on these lines are destined to contain a high quantity of fluoride. Exceptions happen in dynamic volcanic territories wherever surface water and shallow groundwater had been influenced by hydrothermal sources.

22.5 A CASE STUDY ON GIS AND GEOSTATISTICAL APPLICATION

Multivariate analyses, such as cluster, factor, and discriminate analysis, aim to interpret the governing processes through data reduction and classification. They are widely applied mainly to spatial data in geochemistry (Papatheodorou et al., 1999; Papatheodorou et al., 2002a), hydrochemistry (Voudouris et al., 1997; Lambrakis et al., 2004), mineralogy (Seymour et al., 2004), and even marine geophysics (Papatheodorou et al., 2002b). The use of these methods in water-quality monitoring and assessment has increased in the last decade, mainly due to the need to obtain appreciable data reduction for analysis and decision (Vega et al., 1998; Helena et al., 2000; Lambrakis et al., 2004). Multivariate treatment of environmental data is widely used to characterize and evaluate groundwater quality (Vengosh and Keren, 1996; Suk and Lee, 1999; Helena et al., 2000; Lambrakis et al., 2004; Panagopoulos et al., 2004) and it is useful for evidencing the temporal and spatial variations caused by the natural and human factors linked to seasonality.

Data reduction was carried out for groundwater samples in the Krishnagiri District of Tamil Nadu, India (Fig. 22.1) by using statistical applications like Correlation and Factor Analysis. The spatial distribution of fluoride level in groundwater of the study area, plotted in the GIS platform, is represented in Fig. 22.2.

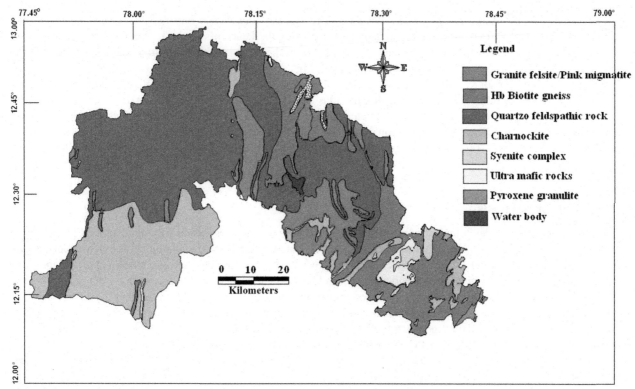

FIG. 22.1 Geology map of the study area.

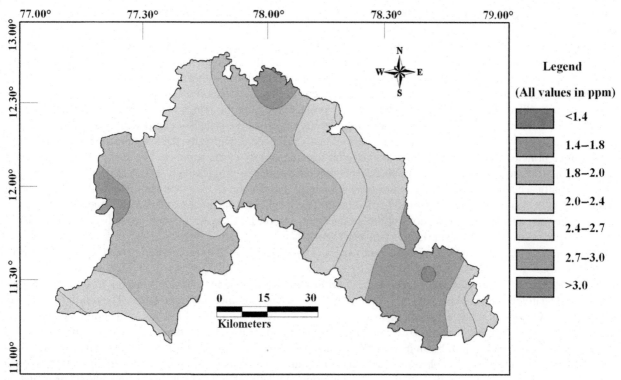

FIG. 22.2 Spatial distribution of fluoride in groundwater of the study area.

22.5.1 Correlation

Correlation analysis was carried out using bivariate correlation using the Pearson's method. This was carried out for different datasets and it indicates the hydrogeochemical complexity of the region with less positive and less negative correlation. Table 22.2 shows a significant correlation of SO_4^- with Mg^{2+}, Na^+, K^+, Cl^-, and NO_3^-; Ca^{2+} with NO_3^-; Mg^{2+} with Cl^-, NO_3^-, and SO_4^-; Na^+ with K^+, HCO_3^-, and SO_4^-, and K with SO_4^- and Ca^{2+}. There is a negative relationship of pH and Ca^{2+}. The major ions exhibiting a correlation with other ions are Cl^-, Na^+, and Ca^{2+}. This is mainly possible due to the impact of leaching of the secondary salts precipitated along the fissures or the permeable zones of the formations (Chidambaram et al., 2009). The negative relation of pH with Mg^{2+} indicates the dominance of the ion-exchange process (Chapelle and Knobel, 1983). Temperature also plays a role in pH, indicating the thermodynamic control over the groundwater chemistry.

22.5.2 Factor Analysis

Factor analysis, which includes principal component analysis (PCA), is a very powerful technique applied to reduce the dimensionality of a dataset consisting of a large number of interrelated variables, while remaining as much as possible the variability present in the dataset. This reduction is achieved by transforming the data set into a new set of variables, the principal components (PCs), which are orthogonal (noncorrelated) and are arranged in a decreasing order of importance. Mathematically, the PCs are computed from covariance or other cross-product matrix, which describes the dispersion of the multiple measured parameters to obtain eigen values and eigenvectors. The principal components are the linear combinations of the original variables and the eigenvectors (Wunderlin et al., 2001). Varifactors (VFs), a new group of variables are obtained by rotating the axis defined by PCA. Varimax rotation distributes the PC loadings such that their dispersion is maximized by minimizing the number of large and small coefficients (Richman, 1986). Besides considerable data reduction, entire dataset variability is described through a few VFs/PCs without losing much information. Further, grouping of the studied variables according to their common features by VFs helps in data interpretation (Vega et al., 1998; Moralesa et al., 1999; Helena et al., 2000; Simeonov et al., 2003; Singh and Saxena, 2004).

PCA results vary considerably depending on whether the covariance or correlation matrix is used when large differences exist in the standard deviation of the variables (Davis, 1986). When the correlation matrix is used, each variable is normalized to unit variance and therefore contributes equally.

The concentrations of the elements studied in the groundwater samples of the research area vary and the PCA is therefore applied to the correlation matrix for the present study. Thus the raw-data matrix was first centered around zero by subtracting the means from each column and then dividing each of the values within each column by the column standard deviation. The eigen vectors of the correlation matrix are the principal components and each original observation is converted to what is called the principal component score by projecting it onto the principal axes. The elements of the eigen vectors that are used to compute the scores of the observations are called the principal component loadings. Because the correlation matrix is symmetrical, the eigen vectors are mutually orthogonal. Typically, the raw-data matrix can be reduced to two or three principal component loadings that account for the majority of the variance. The first principal component loading explains the most variance and each subsequent component explains progressively less. As a result, a small number of factors usually account for approximately the same amount of information as the much larger set of the original observations do. The PC loadings can be examined to provide further insight into the processes that are responsible for the similarities in the trace-element concentrations in the groundwater samples. PC scores of each groundwater sample can be plotted together to investigate the similarities between them.

The groundwater samples show four significant factors (Tables 22.3 and 22.4) representing about 70% of TDV (total dissolved solid).

The representation of fluoride was noted as the fourth factor with moderate positive loadings of Ca^+, HCO_3^-, and fluoride' and negative loadings of pH with Na^+ indicating the dissolution of mica and hornblende. The fourth factor shows that when Na-CO_3 type groundwater is compared with Ca-HCO_3 type, the latter is known to generally contain lower fluoride (Lee et al., 1997), but here HCO_3^- water has high fluoride during monsoon due to dissolution and weathering. Its hydrochemistry is characterized by an increased Ca^{2+} ion concentration with increasing TDS due to the gradual dissolution of carbonate minerals or Ca^{2+}-bearing plagioclase in aquifer materials (Yun et al., 1998a,b; Kim, 2001; Sujatha, 2003). The factor score was used to find out the spatial variation of the factor and to identify the zone of its representation. They are commonly obtained using two approaches: weighted least square method and the regression method. The regression method is used in the study to compute the factor scores. The positive zones indicate the dominance of that, hydrogeochemical regime mostly affected by the respective factor (Chidambaram et al., 2007). The spatial representation

TABLE 22.2 Correlation Analysis of Groundwater Samples Collected

	Ca²⁺	Mg²⁺	Na⁺	K⁺	F⁻	Cl⁻	HCO₃⁻	NO₃⁻	PO₄³⁻	SO₄²⁻	H₄SiO₄	pH	EC	TDS	Temp
Ca²⁺	1.00														
Mg²⁺	0.41	1.00													
Na⁺	0.32	0.23	1.00												
K⁺	0.46	0.13	0.56	1.00											
F⁻	0.27	0.22	-0.03	-0.12	1.00										
Cl⁻	0.49	0.50	0.30	0.31	-0.04	1.00									
HCO₃⁻	0.43	0.39	0.66	0.42	0.20	0.14	1.00								
NO₃⁻	0.55	0.50	0.41	0.36	0.08	0.34	0.29	1.00							
PO₄³⁻	0.22	0.07	0.05	0.04	-0.03	0.37	0.13	0.03	1.00						
SO₄	0.44	0.50	0.49	0.59	-0.03	0.45	0.20	0.55	0.02	1.00					
H₄SiO₄	0.03	0.07	0.08	0.12	-0.09	0.04	-0.25	0.29	-0.10	0.20	1.00				
pH	-0.12	0.19	0.21	0.03	0.07	0.17	0.28	-0.14	0.06	0.03	-0.32	1.00			
EC	0.64	0.50	0.83	0.60	0.10	0.45	0.70	0.70	0.10	0.63	0.05	0.04	1.00		
TDS	0.67	0.63	0.79	0.64	0.10	0.62	0.75	0.65	0.20	0.66	0.06	0.19	0.92	1.00	
Temp	-0.12	0.13	0.28	-0.16	-0.18	0.11	0.11	0.17	0.01	-0.06	-0.16	0.42	0.10	0.13	1.00

TABLE 22.3 Factor Analysis of the Groundwater Samples (Varimax Rotated)

	1	2	3	4
Ca^{2+}	0.561	0.457	−0.209	0.393
Mg^{2+}	0.4	0.62	0.068	0.143
Na^{+}	0.862	−0.044	0.278	−0.116
K^{+}	0.768	−0.036	−0.143	−0.052
F^{-}	0.007	0.093	−0.003	0.806
Cl^{-}	0.318	0.797	0.039	−0.123
HCO_3^{-}	0.697	−0.023	0.434	0.399
NO_3^{-}	0.648	0.355	−0.237	−0.071
PO_4^{3-}	−0.084	0.608	0.131	0.075
SO_4^{2-}	0.682	0.328	−0.229	−0.156
H_4SiO_4	0.159	0.034	−0.632	−0.379
pH	0.056	0.13	0.79	−0.053
EC	0.925	0.213	0.058	0.107
TDS	0.894	0.397	0.132	0.098
Temp	0.07	0.14	0.655	−0.454

TABLE 22.4 Factor Representations and Total Data Variability

Factor	Loadings	TDV (%)
Factor I	Ca^{2+}, Na^{+}, K^{+}, SO_4^{2-}, HCO_3^{-}, and pH	33
Factor II	Ca^{2+}, Mg^{2+}, and PO_4^{3-}	13
Factor III	Temperature, HCO_3^{-}, and pH	12
Factor IV	Ca^{2+}, F^{-}, and HCO_3^{-}	9

of the factor scores of each sample for this season is plotted in the GIS platform. This shows that the positive representation of this factor is represented in the western and southeastern part of the study area (Fig. 22.3). The northwestern part of the study area is chiefly composed of quartzo felspathic gneiss with a few granitic intrusions. The southeastern part of the study area is composed of various rock types, chiefly hornblende biotite gneiss.

It was also inferred that the region with hornblende biotite gneiss (HBG) and biotite undergoes weathering to vermiculite. At a lower pH dissolution takes place, where water with dissolved HCO_3^{-} in the form of H_2CO_3 acts on the mineral and releases H^{+}, fluoride, cation, and HCO_3^{-}.

$$(Na)_2 (Ca, Mg, Fe)_3 Al_2Si_6O_{20} (OH)_4 + H_2O + H_2CO_3 (Ca, Mg_3Fe_3) Al_3Si_5O_{20}(OH) + F^{-} + Na^{+} + H^{+} + HCO_3{}^{-}$$

Manikandan et al. (2014) inferred that this process is more prominent during the monsoon period as the rainwater reacting on the rock matrix is slightly acidic or near neutral. At a higher pH hydrolysis takes place where OH^{-} is released into the system with more cations and fluoride. The release of high concentrations of fluoride in higher pH ranges may also be due to anthropogenic sources.

$$(Na)_2 (Ca, Mg, Fe)_3 Al_2Si_6O_{20} (OH)_4 + H_2O à (Mg_3Fe_3) Al_3Si_5O_{20}(OH) + OH^{-} + F^{-} + Na^{+} + Ca^{2+} + Mg^{2+} + HCO_3{}^{-}$$

It was also inferred by him that the above mentioned process is also a significant contributing factor in the release of F^{-} during the process of hydrolysis and that it is more prominent in the post-monsoon period.

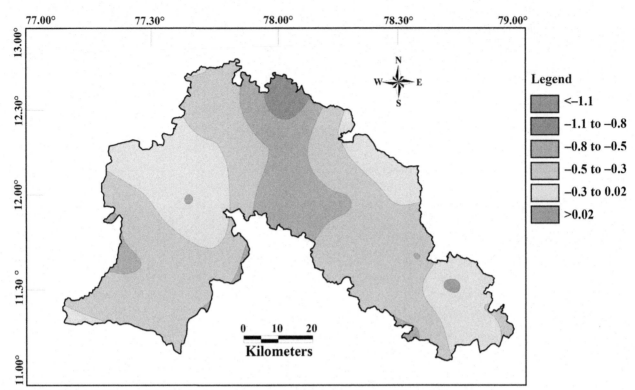

FIG. 22.3   Spatial distribution of the fluoride factor score representations in groundwater of the study area.

22.6   FLUORIDE DISTRIBUTIONS IN GROUNDWATER

The distribution of fluoride in groundwater based on the geology, depth, salinity, and other parameters should be examined before the control or reduction of fluoride concentration. A few investigations and analysis have been carried out based on the data sources of Alberta, Canada (Hitchon, 1995), India (Gupta et al., 1999), Germany (Queste et al., 2001), and the UK (Edmunds et al., 1989). A statistical summary of fluorine distributions in sedimentary basins and other aquifers from different parts of the world are included in Table 22.5.

22.7   HEALTH EFFECTS (REVIEW OF STATUS)

Globally many people suffer from fluorosis due to intake of fluoride-rich groundwater. In India, around 20 million people were severely affected by fluorosis and around 40 million were exposed to its associated risks (Chinoy, 1991). The bioavailability of soluble fluoride ingested along with water was nearly 100%, because soluble fluoride in drinking water was easily absorbed by the gastrointestinal tract without the intervention of interfering elements such as calcium, magnesium, and aluminum (Ekstrand et al., 1978, 1994; Ekstrand and Ehrnebo, 1979; Spak et al., 1982; Rao, 2003; Whitford, 1990). The primary adverse effects associated with chronic excess-fluoride intake are dental and skeletal fluoroses (Susheela, 2003) (Fig. 22.4).

It also adversely affects the fetal cerebral function and neurotransmitters (Zhao et al., 1998; Shi and Dai, 1990; Chen Jr. et al., 1956). Exposure to high fluoride levels through food and drinking water leads to reduced intelligence in children (Li et al., 1995; Zhao et al., 1996; Xiang et al., 2003). Though the fluoride content in groundwater is mainly due to natural contamination, the mechanism of dissolution is unknown (Handa, 1975; Saxena and Ahmed, 2001). In spite of the unverifiable results of fluoride with respect to health impact in drinking water at low concentrations ($0.7\,mg/L^{-1}$ or less), the chronic impacts of exposure to excessive fluoride in drinking water are well established. The most widely recognized health issue is that long-term exposure to fluoride in drinking water at concentrations above $1.5\,mg/L^{-1}$ can result in dental fluorosis ("mottled enamel") as the fluoride reacts with tooth enamel, and eventually, after the stages of darkening and weakening, the loss of teeth occurs. The accumulation of fluoride into the body, i.e., above $4\,mg/L^{-1}$, gradually leads to skeletal fluorosis after a series of changes, such as osteosclerosis, the solidifying and calcifying of bones, and sporadic bone development (Dissanayake, 1991). At the very least daily consumption at these levels leads to bone twisting and incapacity.

TABLE 22.5 Concentration of Fluoride From Various Sources in Different Countries (Edmunds and Smedley, 2013)

Country	Aquifer	Range of F (mg/L)	Average (mg/L)	Reference
Groundwater: crystalline basement rocks				
Norway	Igneous and metamorphic rocks, Hordaland	0.02–9.48	0.30	Bårdsen et al. (1996)
Cameroon	Granite	0.2–15.2		Fantong et al. (2010)
Ghana	Crystalline basement including granite and metasediment	0.09–3.8	1.07	Smedley et al. (1995)
India	Crystalline basement, Nalgonda, Andhra Pradesh	0.5–7.6	3.6	Reddy et al. (2010)
Pakistan	Nagar Parkar, Sindh Province	1.1–7.9	3.33	Rafique et al. (2009)
Sri Lanka	Crystalline basement including granite and charnockite	<0.02–10		Dissanayake (1991)
South Africa	Western Bushveld Complex	0.1–10	143	McCaffrey (1998)
Groundwater: volcanic rocks				
Ethiopia	Volcanic bedrock (Wonji/Shoa area)	6.1–20.0	12.9	Ashley and Burley (1994)
Tanzania	Kimberlites, Shinyanga	110–250		Bugaisa (1971)
Groundwater: sediments and sedimentary basins				
China	Quaternary sands, Hunchun Basin, northeast China	1.0–7.8		Woo et al. (2000)
Argentina	Quaternary loess, Quequen Grande River Basin	0–5.7	1.84	Martinez et al. (2012)
India	Quaternary alluvium Agra, Uttar Pradesh	0.11–12.8	2.1	Gupta et al. (1999)
Canada	Nonmarine Upper Cretaceous sediments, Alberta Basin	0.01–22.0	1.83	Hitchon (1995)
Germany	Cretaceous Chalk Marls	<0.01–8.9	1.28	Queste et al. (2001)
Libya	Miocene, Upper Sirte Basin	0.63–3.6	1.4	Edmunds (1994)
Sudan	Cretaceous, Nubian Sandstone (Butana area)	0.29–6.2	1.8	Edmunds (1994)
Senegal	Paleocene sediments	1.5–12.5		Travi (1993)
Tunisia	Cretaceous to quaternary sediments	0.1–2.3		Travi (1993)
USA	Carboniferous sediments, Ohio	0.05–5.9		Corbett and Manner (1984)

FIG. 22.4 Fluoride concentrations in drinking water and its health effects of (Dissanayake, 1991; Edmunds and Smedley, 2013).

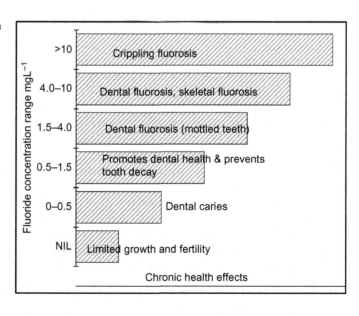

22.8 CONCLUSION

This chapter has summarized the sources, occurrence, and health impacts of fluoride in groundwater. The study also described the application of GIS and geostatistical techniques for the assessment of fluoride contamination in groundwater. Groundwater is more vulnerable to the development of fluoride, despite the low concentrations ($<1\,mg/L^{-1}$). Geology plays a key role in characterizing low or high fluoride concentrations. Territories with conceivably high-fluoride groundwater include crystalline basement rocks, granitic composition in the form of apatite, mica, hornblende, and fluorite and have low calcium content. Such territories are present over extensive parts of India, Sri Lanka, and Africa. High fluoride concentrations are also noted in the sedimentary aquifers with low calcium concentrations and where $Na\text{-}HCO_3$ water dominates.

The concentration of fluoride in groundwater varies due to complex interactions between hydrology, the geoenvironment, and other environmental factors. To guide the localization and remediation of the problem, it is necessary to define regions where similar conditions prevail in order to understand the extent of fluoride contamination. GIS and geostatistical techniques enable the identification of several important chemicals and the climatic, geological, and soil conditions associated with fluoride-rich groundwater. The probability map produced using a geostatistical tool in a GIS platform can be taken as a preliminary attempt in the prediction of a fluoride-contaminated zone before proceeding to field testing.

If the fluoride concentration is high in a territory, remediation techniques may be incorporated based on the residence time and geology. Since the fluoride concentration in groundwater is much variable on a local scale to identify the concentration in each well. Thus for fluoride-prone territories, the testing of each well should be undertaken before utilizing the water for drinking purpose.

ACKNOWLEDGMENTS

The authors wish to express their thanks to Ministry of Environment and Forest (Mo En &F), India for providing the necessary financial support to carry out this study vide letter No.16/2007-CT Dated 07.10.2009.

REFERENCES

Amini, M., Mueller, K., Abbaspour, K.C., Rosenberg, T., Afyuni, M., Møller, K.N., Johnson, C.A., 2008. Statistical modeling of global geogenic fluoride contamination in groundwaters. Environ. Sci. Technol. 42 (10), 3662–3668.

Apambire, W.B., Boyle, D.R., Michel, F.A., 1997. Geochemistry, genesis and health implication of fluoriferous groundwater in the upper regions, Ghana. Environ. Geol. 33, 13–24.

Ashley, P.P., Burley, M.J., 1994. Controls on the occurrence of fluoride in groundwater in the Rift Valley of Ethiopia. In: Nash, H., McCall, G.J.H. (Eds.), Groundwater Quality. Chapman & Hall, London, pp. 45–54.

Ashok, K.Y., Parveen, K., 2010. Fluoride and Flurosis status in groundwater of Todaraisingh area of district Tonk (Rajasthan, India): a case study. ijCEPr 1 (1), 6–11.

Ayoob, S., Gupta, A.K., 2006. Fluoride in drinking water: a review on the status and stress effects. Crit. Rev. Environ. Sci. Technol. 36 (6), 433–487.

Bårdsen, A., Bjorvatn, K., Selvig, K.A., 1996. Variability in fluoride content of subsurface water reservoirs. Acta Odont. Scand. 54, 343–347.

Barnard, W.R., Nordstrom, D.K., 1982. Fluoride in precipitation. II. Implication for the geochemicalcycling of fluorine. Atmos. Environ. 16, 105–111.

Brindha, K., Rajesh, R., Murugan, R., Elango, L., 2011. Fluoride contamination in groundwater in parts of Nalgonda district, Andhra Pradesh, India. Environ. Monit. Assess. 172 (1–4), 481–492.

Bugaisa, S.L., 1971. Significance of fluorine in Tanzania drinking water. In: Tschannerl, G. (Ed.), Proceedings of a Conference on Rural Water Supply in East Africa. Bureau of Resource Assessment and Land Use Planning, Dar Es Salaam, pp. 107–113.

CGWB, 2010. Groundwater Quality in Shallow Aquifers of India. 2010 CGWB, Faridabad.

Chandrajith, R., Padmasiri, J.P., Dissanayake, C.B., Prematilaka, 2012. Spatial distribution of fluoride in groundwater of Sri Lanka. J. Natl. Sci. Found. Sri Lanka 40 (4), 303–309.

Chen Jr., P.S., Smith, F.A., Gardner, D.E., O'brien, J.A., Hodge, H.A., 1956. Renal clearance of fluoride. Proc. Soc. Exp. Biol. Med. 92, 879–883.

Chapelle, F.H., Knobel, 1983. Groundwater geochemistry and calcite cementation of the Aquia aquifer in southern Maryland. Water Resour. 19 (2), 545–548.

Chidambaram, S., Ramanathan, A.L., Prasanna, M.V., Anandhan, P., Srinivasamoorthy, K., Vasudevan, S., 2007. Identification of hydrogeochemically active regimes in groundwaters of Erode District, Tamilnadu – a statistical approach. Asian J. Water Environ. Pollut. 5 (30), 93–102.

Chidambaram, S., Bala Krishna Prasad, M., Manivannan, R., Karmegam, U., Singaraja, C., Anandhan, P., Prasanna, M.V., Manikandan, S., 2013. Environmental hydrogeochemistry and genesis of fluoride in groundwaters of Dindigul district, Tamilnadu (India). Environ. Earth Sci. V68, 333–342.

Chidambaram, S., SenthilKumar, G., Prasanna, M.V., JohnPeter, A., Ramanthan, A.L., Srinivasamoorthy, K., 2009. A study on the hydrogeology and hydrogeochemistry of groundwater from different depths in a coastal aquifer: Annamalai Nagar, Tamilnadu, India. Environ. Geol. 57 (1), 59–73.

Chinoy, N.J., 1991. Effects of fluoride on physiology of animals and humans' beings. Indian J. Environ. Toxicol. 1, 17–32.

Corbett, R.G., Manner, B.M., 1984. Fluoride in ground water of northeastern Ohio. Ground Water 22, 13–17.

Craig, L.L., Stillings, D.L., Decker, J.M., 2015. Thomas Comparing activated alumina with indigenous laterite and bauxite as potential sorbents for removing fluoride from drinking water in Ghana. Appl. Geochem. 56, 50–66.

Davis, J.C., 1986. Statistics and Data Analysis in Geology, second ed. John Willey and Sons, New York.

Deshmukh, A.N., Wadaskar, P.M., Malpe, D.B., 1995. Fluorine in environment: a review. Gondwana Geol. Mag 9, 1–20.

Di'az-Barriga, F., Navarro-Quezada, A., Grijalva, M.I., Grimaldo, M., Loyola-Rodrl'Guez, J.P., Ortiz, M.D., 1997. Endemic fluorosis in Mexico. Fluoride 30 (4), 233–239.

Dissanayake, C.B., 1991. The fluoride problem in the groundwater of Sri Lanka—environmental management and health. Int. J. Environ. Stud. 38, 137–156.

Edmunds, W.M., Smedley, P.L., 1996. Groundwater geochemistry and health: an overview. Geol. Soc. Lond., Spec. Publ. 113 (1), 91–105.

Edmunds, W.M., Smedley, P.L., 2013. Fluoride in natural waters. In: Essentials of Medical Geology. Springer, Dordrecht, pp. 311–336.

Edmunds, W.M., 1994. Characterization of groundwaters in semi-arid and arid zones using minor elements. In: Nash, H., McCall, G.J.H. (Eds.), Groundwater Quality. Chapman & Hall, London, pp. 19–30.

Edmunds, W.M., Cook, J.M., Kinniburgh, D.G., Miles, D.G., Trafford, J.M., 1989. Trace element occurrence in British groundwaters. British Geological Survey Research Report SD/89/3, Keyworth, p. 424.

Edmunds, W.M., Smedley, P.L., 2005. Fluoride in natural waters. In: Selinus, O. (Ed.), Essentials of Medical Geology. Elsevier Academic Press, Burlington, MA, pp. 301–329.

Ekstrand, J., Ehrnebo, M., 1979. Influence of milk products on fluoride bioavailability in man. Eur. J. Clin. Pharmacol. 16, 211–215.

Ekstrand, J., Ehrnebo, M., Boréus, L.O., 1978. Fluoride bioavailability after intravenous and oral administration: importance of renal clearance and urine flow. Clin. Pharmacol. Ther. 23, 329–337.

Ekstrand, J., Fomon, S.J., Ziegler, E.E., Nelson, S.E., 1994. Fluoride pharmacokinetics in infancy. Pediatr. Res. 35, 157–163.

Fantong, W.Y., Satake, H., Ayonghe, S.N., Suh, E.C., Adelana, S.M.A., Fantong, E.B.S., Banseka, H.S., Gwanfogbe, C.D., Woincham, L.N., Uehara, Y., Zhang, J., 2010. Geochemical provenance and spatial distribution of fluoride in groundwater of Mayo Tsanaga River Basin, far north region, Cameroon: implications for incidence of fluorosis and optimal consumption dose. Environ. Geochem. Health 32, 147–163.

Farnham, I.M., Johannesson, K.H., Singh, A.K., Hodge, V.F., Stetzenbach, K.J., 2003. Factor analytical approaches for evaluating groundwater trace element chemistry data. Anal. Chim. Acta 490, 123–138.

Fuhong, R., Shuquin, J., 1988. Distribution and formation of high-fluorine groundwater in China. Environ. Geol. Water Sci. 12, 3–10.

Fuge, R., Andrews, M.J., 1988. Fluorine in the UK environment. Environ. Geochem. Health 10, 96104.

Gaciri, S.J., Davies, T.C., 1993. The occurrence and geochemistry of fluoride in some natural waters of Kenya. J. Hydrol. 143 (3–4), 395–412.

Genxu, W., Guodong, C., 2001. Fluoride distribution in water and the governing factors of environment in arid north-west China. J. Arid Environ. 49, 601–614.

Gleeson, T., Wada, Y., Bierkens, M.F., van Beek, L.P., 2012. Water balance of global aquifers revealed by groundwater footprint. Nature 488 (7410), 197–200.

Gupta, M.K., Singh, V., Rajwanshi, P., Agarwal, M., Rai, K., Srivastava, S., Srivastav, R., Dass, S., 1999. Groundwater quality assessment of Tehsil Kheragarh, Agra (India) with special reference to fluoride. Environ. Monit. Assess. 59, 275–285.

Handa, B.K., 1975. Geochemistry and genesis of fluoride-containing groundwater in India. Groundwater 13, 275–281.

Helena, B., Pardo, R., Vega, M., Barrado, E., Fernandez, J., Fernandez, L., 2000. Temporal evolution of groundwater composition in an alluvial aquifer (Pisuerga River, Spain) by principal component analysis. Water Res. 34 (3), 807–816.

Hitchon, B., 1995. Fluorine in formation waters in Canada. Appl. Geochem. 10, 357–367.

Jacks, G., Bhattacharya, P., Chaudhary, V., Singh, K.P., 2005. Controls on the genesis of some high-fluoride groundwaters in India. Appl. Geochem. 20 (2), 221–228.

Kabata-Pendias, A., Pendias, H., 1984. Trace Elements in Soils and Plants. CRC Press, Boca Raton, FL.

Kilham, P., Hecky, R.E., 1973. Fluoride: geochemical and ecological significance in East Africanwaters and sediments. Limnol. Oceanogr. 18, 932–945.

Kim, S.Y., 2001. Hydrogeochemical, geostatistical and thermodynamically studies on deep thermal groundwater, Korea. Unpub. Ph.D thesis, Korea Univ.

Kruse, E., Ainchil, J., 2003. Fluoride variations in groundwater of an area in Buenos Aires Province, Argentina. Environ. Geol. 44 (1), 86–89.

Lavado, R.S., Reinaudi, N., 1979. Fluoride in salt affected soils of La Pampa (Republica Argentina). Fluoride 12, 28–32.

Lambrakis, N., Antonakos, A., Panagopoulos, G., 2004. The use of multicomponent statistical analysis in hydro geological environmental research. Water Res. 38 (7), 1862–1872.

Lee, J.U., Chon, H.T., John, Y.W., 1997. Geochemical characteristics of deep granitic groundwater in Korea. J. Kor. Soc. Groundwater Environ. 4, 199–211 (in Korean).

Li, X.S., Zhi, J.L., Gao, R.O., 1995. Effect of fluoride exposure on intelligence in children. Fluoride 28 (4), 189–192.

Li, Y.H., Wang, S., Zhang, X., Wei, J., Xu, C., Luan, Z., Wu, D., 2003. Adsorption of fluoride from water by aligned carbon nanotubes. Mater. Res. Bull. 38 (3), 469–476.

Liu, C.W., Lin, K.H., Kuo, Y.M., 2003. Application of factor analysis in the assessment of groundwater quality in a blackfoot disease area in Taiwan. Sci. Total Environ. 313 (1–3), 77–89.

Madhnure, P., Pandith Malpe, D.B., 2007. Fluoride contamination of groundwater in rural parts of Yavatmal District, Maharashtra—Causes and remedies. Gondwana Geol. Mag. 11, 127–135.

Manikandan, S., Chidambaram, S., Ramanathan, A.L., Prasanna, M.V., Karmegam, U., Singaraja, C., Paramaguru, P., Jainab, I., 2014. A study on the high fluoride concentration in the magnesium-rich waters of hard rock aquifer in Krishnagiri district, Tamilnadu, India. Arab. J. Geosci. 7 (1), 273–285.

Martinez, D.E., Quiroz Londono, O.M., Massone, H.E., Palacio Buitrago, P., Lima, L., 2012. Hydrogeochemistry of fluoride in the Quequen river basin: natural pollutants distribution in the argentine pampa. Environ. Earth Sci. 65, 411–420.

McCaffrey, L.P., 1998. Distribution and Causes of High Fluoride Groundwater in the Western Bushveld Area of South Africa. Cape Town.

Mjengera, H., Mkongo, G., 2003. Appropriate deflouridation technology for use in flourotic areas in Tanzania. Phys. Chem. Earth Parts A/B/C 28 (20–27), 1097–1104.

Moralesa, M.M., Marti, P., Llopis, A., Campos, L., Sagrado, S., 1999. An environmental study by factor analysis of surface seawaters in the Gulf of Valencia (Western Mediterranean). Anal. Chim. Acta 394, 109–117.

Nordstrom, D.K., Jenne, E.A., 1977. Fluorite solubility equilibria in selected geothermal waters. Geochim. Cosmochim. Acta 41 (2), 175–188.

Nordstrom, D.K., Ball, J.W., Donahoe, R.J., Whittemore, D., 1989. Groundwater chemistry and water-rock interactions at Stripa. Geochim. Cosmochim. Acta 53 (8), 1727–1740.

Padhi, S., Muralidharan, D., 2012. Fluoride occurrence and mobilization in geo-environment of semi-arid granite watershed in southern peninsular India. Environ. Earth Sci. 66 (2), 471–479. Pakistan J. Hazard. Mater. 171, 424–430.

Panagopoulos, G., Lamprakis, N., Tsolis-Katagas, P., Papoulis, D., 2004. Cation exchange processes and human activities in inconfined aquifers. Environ. Geol. 46, 542–552.

Papatheodorou, G., Hotos, G., Geraga, M., Avramidou, D., Vorinakis, T., 2002a. Heavy metal concentrations in sediments of Klisova lagoon (S.E. Mesolonghi-Aitolikon Lagoon complex) W Greece. Fresen. Environ. Bull. 11 (11), 951–956.

Papatheodorou, G., Lyberis, E., Ferentinos, G., 1999. Use of factor analysis to study the distribution of metalliferous bauxitic tailings in the seabed of the Gulf of Corinth. Greece Nat. Resour. Res. 8 (4), 277–285.

Papatheodorou, G., Mitsis, C., Christodoulou, D., Ferentinos, G., 2002b. A multivariate statistical approach to the investigation of pockmarks growth and activity. An example from a pock-mark field in the Gulf of Patras (W Greece). In: Proceedings of the Eighth Annual Conference of the IAMG, 15–20 September, Abstract Book, Berlin.

Pickering, W.F., 1985. The mobility of soluble fluoride in soils. Environ. Pollut. Ser. B 9, 281–308.

Podgorski, J.E., Eqani, S.A.M.A.S., Khanam, T., Ullah, R., Shen, H., Berg, M., 2017. Extensive arsenic contamination in high-pH unconfined aquifers in the Indus Valley. Sci. Adv. 3 (8), e1700935.

Queste, A., Lacombe, M., Hellmeier, W., Hillerman, F., Bortulussi, B., Kaup, M., Ott, O., Mathys, W., 2001. High concentrations of fluoride and boron in drinking water wells in the muenster region – results of a preliminary investigation. Int. J. Hygiene Environ. Health 203, 221–224.

Rafique, T., Naseem, S., Usmani, T.H., Bashir, E., Khan, F.A., Bhanger, M.I., 2009. Geochemical factors controlling the occurrence of high fluoride groundwater in the Nagar Parkar area, Sindh, Pakistan. J. Hazard. Mater. 171, 424–430.

Raju, N., Dey, S., Das, K., 2009. Fluoride contamination in groundwaters of Sonbhadra District, Uttar Pradesh. India Curr. Sci. 96 (7), 979–985.

Rao, N.C.R., 2003. Fluoride and environment—a review. In: Martin, J., Bunch, V., Suresh, M., Vasantha Kumaran, T. (Eds.), Proceedings of the Third International Conference on Environment and Health, Chennai, India. Department of Geography, University of Madras, and Faculty of Environmental Studies, York University, Madras, India, pp. 386–399.

Reddy, D.V., Nagabhushanam, P., Sukhija, B.S., Reddy, A.G.S., Smedley, P.L., 2010. Fluoride dynamics in the granitic aquifer of the Wailapally watershed, Nalgonda District, India. Chem. Geol. 269, 278–289.

Reddy, N.B., Prasad, K.S., 2003. Pyroclastic fluoride in ground waters in some parts of Tadpatri taluk, Anantapur district, Andhra Pradesh. Indian J. Environ. Health 45 (4), 285–288.

Rea, R.E., 1979. A rapid method for the determination of fluoride in sewage sludges. Water Pollut. Control 78, 139–142.

Reghunath, R., Murthy, T.R.S., Raghavan, B.R., 2002. The utility of multivariate statistical techniques in hydrogeochemical studies: an example from Karnataka, India. Water Res. 36, 2437–2442.

Richman, M.B., 1986. Rotation of principal components. J. Climatol. 6, 293–335.

Saxena, V., Ahmed, S., 2001. Dissolution of fluoride in groundwater: a water-rock interaction study. Environ. Geol. 40 (9), 1084–1087.

Saxena, V., Ahmed, S., 2003. Inferring the chemical parameters for the dissolution of fluoride in groundwater. Environ. Geol. 43 (6), 731–736.

Seymour, S.K., Christanis, K., Bouzinos, A., Papazisimou, S., Papatheodorou, G., Moran, E., Denes, G., 2004. Tephrostratigraphy and tephrochronology in the Philippi peat basin, Macedonia, Northern Hellas (Greece). Quarter. Int. 121, 53–65.

Shaji, E., Bindu, J.V., Thambi, D.S., 2007. High fluoride in groundwater of Palghat District, Kerala. Curr. Sci. 92 (2), 240.

Shi, J., Dai, G., 1990. A study of the effects of fluoride on the human foetus in an endemic fluorosis area. Chung Hua Liu Hsing Ping Hsueh Tsa Chih 9, 10–12.

Shortt, H.E., Pandit, C.G., Raghavachari, T.N.S., 1937. Endemic fluorosis in Nellore district, south India. .

Simeonov, V., Stratis, J.A., Samara, C., Zachariadis, G., Voutsa, D., Anthemidis, A., Sofoniou, M., Kouimtzis, T., 2003. Assessment of the surface water quality in Northern Greece. Water Res. 37, 4119–4124.

Singaraja, C., Chidambaram, S., Anandhan, P., Prasanna, M.V., Thivya, C., Thilagavathi, R., Sarathidasan, J., 2014. Geochemical evaluation of fluoride contamination of groundwater in the Thoothukudi District of Tamilnadu, India. Appl Water Sci. https://doi.org/10.1007/s13201-014-0157-y.

Singh, V.S., Saxena, V.K., 2004. Assessment of utilization ground water resources in a coastal shallow aquifer. In: Proceeding of the 2nd Asia Pacific Association of Hydrology & Water Resources Conference, Singapore. vol. 2, pp. 347–364.

Smedley, P.L., Edmunds, W.M., Pelig-Ba, K.B., 1995. Groundwater vulnerability due to natural geochemical environment: 2: Health problems related to groundwater in the Obuasi and Bolgatanga areas, Ghana. BGS Technical Report.

Spak, C.J., Ekstrand, J., Zylberstein, D., 1982. Bioavailability of fluoride added to baby formula and milk. Caries Res. 16, 249–256.

Sreedevi, P.D., Ahmed, S., Made, B., Ledoux, E., Gandolfi, J.M., 2006. Association of hydrogeological factors in temporal variations of fluoride concentration in a crystalline aquifer in India. Environ. Geol. 50, 1–11.

Stüben, D., Berner, Z., Chandrasekharam, D., Karmakar, J., 2003. Arsenic enrichment in groundwater of West Bengal, India: geochemical evidence for mobilization of As under reducing conditions. Appl. Geochem. 18 (9), 1417–1434.

Sujatha, D., 2003. Fluoride levels in the groundwater of the south-eastern part of Ranga Reddy district, Andhra Pradesh, India. Environ. Geol. 44, 587–591.

Suk, H., Lee, K., 1999. Characterization of a ground water hydro-chemical system through multivariate analysis: clustering into ground water zones. Ground Water 37 (3), 358–366.

Susheela, A.K., 2003. Treatise on Fluorosis. Fluorosis Research and Rural Development Foundation, New Delhi, India.

Turner, B.D., Binning, P., Stipp, S.L.S., 2005. Fluoride removal by calcite: evidence for fluorite precipitation and surface adsorption. Environ. Sci. Technol. 39, 9561–9568.

Thy, T., 1983. Relationship between Natural Water Quality and Health. .

Thivya, C., Chidambaram, S., Rao, M.S., Thilagavathi, R., Prasanna, M.V., Manikandan, S., 2017. Appl. Water Sci. 7, 1011. https://doi.org/10.1007/s13201-015-0312-0.

Travi, Y., 1993. Hydrogéologie et hydrochimie des aquifères du Sénégal. Sciences Géologiques, Memoire 95, Université de Paris-Sud, Paris.

Vega, M., Pardo, R., Deban, L., 1998. Assessment of seasonal and polluting effects on the quality of river water by exploratory data analysis. Water Res. 32 (12), 3581–3592.

Vengosh, A., Keren, R., 1996. Chemical modifications of groundwater contaminated by recharge of treated sewage effluent. Contam. Hydrol. 23, 347–360.

Voudouris, K., Lambrakis, N., Papatheodorou, G., Daskalaki, P., 1997. An application of factor analysis for the study of the hydro-geological conditions in Plio-Pleistocene aquifers of NW Achaia (NW Peloponnesus, Greece). Math. Geol. 29 (1), 43–59.

Wang, L.F., Huang, J.Z., 1995. Outline of control practice of endemic fluorosis in China. Soc. Sci. Med. 41, 1191–1195.

Wenzel, W.W., Blum, W.E., 1992. Fluorine speciation and mobility in F-contaminated soils. Soil Sci. 153 (5), 357–364.

White, D.E., Hem, J.D., Warming, G.A., 1963. Chemical composition of subsurface water. USGS professional papers 440-F, U.S. Geological Survey.

Whitford, G.M., 1990. The physiological and toxicological characteristics of fluoride. J. Dent. Res. 69 (Suppl), 539–549.

WHO, 2011. Guidelines for Drinking-Water Quality, fourth ed. World Health Organization.

Woo, N.C., Moon, J.W., Won, J.S., Hahn, J.S., Lin, X.Y., Zhao, Y.S., 2000. Water quality and pollution in the Hunchun Basin. China. Environ. Geochem. Health 22, 1–18.

WRC, 2001. Distribution of fluoride-rich groundwater in Eastern and Mogwase region of Northern and North-west province. WRC Report, No. 526/1/011.1-9.85 Pretoria.

Wunderlin, D.A., Díaz, M.P., Amé, M.V., Pesce, S.F., Hued, A.C., Bistoni, M.A., 2001. Pattern recognition techniques for the evaluation of spatial and temporal variations in water quality. A case study: Suquía river basin (Córdoba-Argentina). Water Res. 35 (12), 2881–2894.

Wuyi, W., Ribang, L., Jian'an, T., Kunli, L., Lisheng, Y., Hairong, L., Yonghua, L., 2002. Adsorptionand leaching of fluoride in soils of China. Fluoride 35, 122–129.

Yun, S.T., Chae, G.T., Koh, Y.K., Kim, S.R., Choi, B.Y., Lee, B.H., Kim, S.Y., 1998a. Hydrogeochemical and environmental isotope study of ground-waters in the Pungki area. J. Kor. Soc. Groundwater Environ. 5, 177–191.

Yun, S.T., Koh, Y.K., Choi, H.S., Youm, S.J., So, C.S., 1998b. Geochemistry of geothermal waters in Korea: environmental isotope and hydrochemical characteristics. II. Jungwon and Munkyeong areas. Econ. Environ. Geol. 31, 201–213.

Xiang, Q., Liang, Y., Chen, L., Wang, C., Chen, B., Chen, X., et al., 2003. Effect of fluoride in drinking water on children's intelligence. Fluoride 36 (2), 84–94.

Zhao, L.B., Liang, G.H., Zhang, D.N., Wu, X.R., 1996. Effect of a high fluoride water supply on children's intelligence. Fluoride 29 (4), 190–192.

Zhao, W., Zhu, H., Yu, Z., Aoki, K., Misumi, J., Zhang, X., 1998. Long-term effects of various iodine and fluorine doses on the thyroid and fluorosis in mice. Endocr. Regul. 32 (2), 63–70.

FURTHER READING

Binbin, W., Baoshan, Z., Hongying, W., Yakun, P., Yuehua, T., 2005. Dental caries in fluorine exposure areas in China. Environ. Geochem. Health 27, 285–288.

Grimaldo, M., Borja-Aburto, V.H., Ramírez, A.L., Ponce, M., Rosas, M., Diaz-Barriga, F., 1995. Endemic fluorosis in San Luis Potosí, Mexico. Environ. Res. 68, 25–30.

Meenakshi, Maheshwari, R.C., 2006. Fluoride in drinking water and its removal. J. Hazard. Mater. B 137, 456–463.

Ramamohana Rao, N.V., Rao, N., Rao, K.S.P., Schuiling, R.D., 1993. Fluorine distribution in waters of Nalgonda District, Andhra Pradesh, India. Environ. Geol. 21, 84–89.

Shomar, B., Müller, G., Yahya, A., Askar, S., Sansur, R., 2004. Fluorides in groundwater, soil and infused-black tea and the occurrence of dental fluorosis among school children of the Gaza Strip. J. Wat. Health 2, 23–35.

Teotia, S.D.S., Teotia, M., Singh, R.K., 1981. Hydrogeochemical aspects of endemic skeletal fluorosis in India—an epidemiological study. Fluoride 4, 69–74.

Wold, S., Esbensen, K., Geladi, P., 1987. Principal component analysis. Chemometr. Intell. Lab. Syst. 2, 37–52.

Zhang, B., Hong, M., Zhao, Y., Lin, X., Zhang, X., Dong, J., 2003. Distribution and risk assessment of fluoride in drinking water in the western plain region of Jilin province, China. Environ. Geochem. Health 25, 421–431.

Chapter 23

Arsenic Contamination

S. Selvam*, K. Jesuraja* and G. Gnanachandrasamy[†]

*Department of Geology, V.O. Chidambaram College, Thoothukudi, India [†]School of Earth Sciences and Engineering, Sun Yat-sen Univeristy, Guangzhou, China

Chapter Outline

23.1 INTRODUCTION

Heavy metals present in trace elements play a chief role in the metabolism and healthy escalation of animals and plants. Auxiliary amounts may have a number of toxicological effects on people. The two main sources of heavy metals concentrations in natural water are natural and anthropogenic. The natural sources consist of the release of metals from the weathering of rocks and leaching into groundwater through rock-water interaction. The anthropogenic sources (human activity) include the release of heavy metals into the atmosphere through the burning of fossil fuels or other industrial activities and thereafter into the streams through rainfall and of the release of industrial effluents and sewage water into streams and surface-water bodies (Handa, 1981; Leung and Jiao, 2006; Selvam et al., 2015).

The accretion of metals in groundwater is corollated to man and to the ecological unit. There is a delicate balance in metals like copper (Cu) and zinc (Zn), which are necessary for the metabolism of organisms, between their essentiality and toxicity. Other elements like cadmium (Cd), aluminium (Al), and lead (Pb) demonstrate intense toxicity even at trace levels (Vanloon and Duffy, 2005). Water is one of the largest prerequisites that supports all forms of plant and animal life (Vanloon and Duffy, 2005) and is obtained usually from two primary natural sources: surface-water bodies, such as freshwater, lakes, rivers, streams, etc., and groundwater, such as lined and unlined well water and borehole water (Bachmat, 1994; Carter and Fernando, 1979). Only a small portion (about 2.5%) of Earth's water is fresh and appropriate for human utilization. Of this small portion 13% is groundwater, which is a significant resource for drinking water for many citizens worldwide (Mendie, 2005). Among rural and small communities, groundwater serves as the only source of drinking water. >50% of the world inhabitants depend on groundwater for household use (Marcovecchio et al., 2007).

Arsenic is an omnipresent element in the environment that originates from the atmosphere, rocks, soil, groundwater, and marine organisms. Among the pollutants, arsenic is an important toxic heavy metal because it acts in humans as a carcinogen. The populace are exposed to eminent levels of arsenic through spoiled drinking water, the use of polluted water for the irrigation of crops, and industrial practices. Arsenic levels are liable to be higher in drinking water that comes from land sources, such as wells and ponds. The MDCH (Michigan Department of Community Health), MDEQ (Michigan Department of Environmental Quality), and the USGS (United States Geological Survey) have indicated that numerous counties in southeast Michigan exceed the EPA (Environmental Protection Agency's) maximum containment level (MCL) of 10 μg/L for drinking water from groundwater and well water, making this an issue that must be addressed. The IARC (International Agency for Research on Cancer) classifies inorganic arsenic as carcinogenic to people, predominantly putting individuals in danger of lung, bladder, and skin cancer. Subsequently, inhabitants of southeast Michigan are at higher risk for bladder, lung, and skin cancers due to levels of arsenic in the groundwater and well water greater than the EPA MCL of 10 μg/L (Singaraja et al., 2015).

GIS and Geostatistical Techniques for Groundwater Science. https://doi.org/10.1016/B978-0-12-815413-7.00023-7

Trace metals are major toxic containments that drastically limit the favorable use of water for household or industrial purposes (Nouri et al., 2006). Groundwater contamination over the past decades due to pollutant leaking from the disposal sites is a major problem in many places. Industries such as painting; glass and ceramic manufacture; mining; and battery manufacture and recycling are the chief sources of trace metals in local water systems, which ultimately pollute groundwater with trace metals. Landfill leachate is an additional source of trace-metal contamination of groundwater (Sang et al., 2008). Add to this human activity, such as the practices of industry, accompanied with overpopulation and increasing ambient temperatures, and we are faced with the major ecological issues that have become apparent in recent years. Heavy metals like cadmium, lead and mercury have been exposed to have engorgement effects on humans, since there is no homeostatic apparatus that can control to normalize the levels of these poisonous substances (Carter and Fernando, 1979). The quality of water has now developed into a significant topic worldwide, particularly with reference to drinking water. Although water plays a crucial role in human life, it has an immense potential for transmitting an extensive assortment of diseases and illnesses. Spoiled water can be responsible for the transmission of ring worm, cholera, typhoid fever, dysentery, and any other illnesses. The present study reveals the levels of dissolved trace elements and heavy metals in a groundwater system. The coastal area supports a swiftly growing population and there are concerns regarding the water quality of the groundwater system. The main uses of water in the catchment area are household and agricultural (livestock watering) (Carter and Fernando, 1979). Consequently, the presence of high levels of heavy metals in the environment presents a possible hazard to human health due to their tremendous toxicity (Fatoki et al., 2012). One of the objectives was to measure the level of the heavy metal arsenic in the groundwater samples of Tuticorin Corporation and its relation to the highly urbanized industrial processes. The results obtained will establish baseline data for potential reference.

23.2 STUDY AREA

Tuticorin coastal area extends over roughly 154 km^2 and lies between latitude 8°43′ to 8°51′ north and longitude 78°5′ to 78°10′ east in the south most part of Tamil Nadu, India (Fig. 23.1). Tuticorin City and the neighboring areas are flourishing with several large- and small-scale industries. The big industries in the study area are SPIC (fertilizers and chemicals), Sterlite (copper), and Kilburn Chemicals (titanium dioxide) (in addition to infrastructure providers, i.e., the port and the power plant). Furthermore, cottage industries like the textile mills are also present in the district. The occurrence of large industries can lead to the establishment of small-scale industries that are supportive to these units (Selvam et al., 2013). A number of medium and small industries, together with the time-honored ones, are located in Tuticorin and its neighborhood. The main segments include salt industries and domestic, marine merchandise, heavy minerals industries, dry flowers exports, edible oil pulling out, and garment exports. The present study area supports a swiftly increasing population and there are concerns regarding the water quality of the groundwater system (Selvam et al., 2014).

23.3 MATERIALS AND METHODS

A total number of 36 groundwater samples were collected from 16 bore wells and 20 open wells with the depths that vary between 8 mbgl and 80 mbgl in premonsoon season 2015. The coordinates of the sampling stations are identifiable with the help of a handheld GPS (Garmin, etrex). The groundwater samples were collected in pre-cleaned, acid washed (1 N HCl) HDPE bottles of 750 mL capability. The groundwater samples were filtered, acidified with nitric acid (1 N), and analyzed for trace elements at the geochemistry laboratory of National Geophysical Research Institute (NGRI), Hyderabad using Inductively Coupled Plasma Mass Spectrometer (ICP-MS) Model ELAN DRC II, Perkin-Elmer Sciex Instrument, USA. The acidified groundwater samples were fed into the ICP-MS instrument by conservative pneumatic nebulization, using a peristaltic pump with a solution uptake rate of about 1 mL/min. The nebulizer gas flow, sample uptake rate, detector voltage, and lens voltage was optimized for a sensitivity of about 50,000 counts for a 1 ng/mL solution. Calibration was performed using the certified reference material NIST 1640a (National Institute of Standards and Technology, USA) to minimize matrix and other associated interference effects. Relative standard deviation (RSD) was found to be better than 6% in the majority of cases, which indicates that the precision of the analysis is reasonably good (Selvam, 2013). The results were evaluated in accordance with the drinking water quality standards given by the World Health Organization (WHO, 2004). Statistical analyses were carried out using SPSS version 19.0. Factor analysis was used for data interpretation and source identification. Inverse distance weighting (IDW) interpolation technique was used to visualize the contamination extent in the study area. The analytical results of 20 selected trace-element concentrations are reported in mg/L.

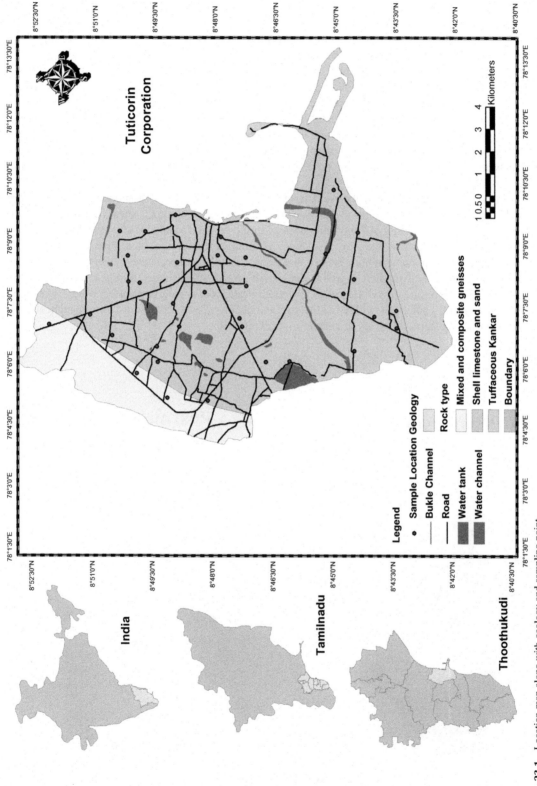

FIG. 23.1   Location map along with geology and sampling point.

23.4 RESULTS AND DISCUSSION

23.4.1 Arsenic (As)

In water, arsenic has no taste or smell and can only be detected during a chemical test. Arsenic absorption in the groundwater varies from 0.00087 mg/L to 0.0837 mg/L with an average absorption of 0.015796 mg/L during premonsoon and 0.00309 mg/L to 0.04743 mg/L with an average absorption of 0.017139 mg/L during postmonsoon. The upper range of allowable limit of arsenic ion absorption in groundwater is 0.01 mg/L as per WHO 2004 categorization. The spatial circulation map of arsenic ion absorption in the groundwater of the study area is shown in Fig. 23.2. In premonsoon, the not permissible limit of arsenic was observed within the northwest, northeast, and central segments, but in the premonsoon it was observed that the intact study area was affected due to industrial-waste discharge or percolation through the subsurface. Sterlite is one of the copper industries located within the study area that extract/produce copper from copper concentrates and it is situated near Tuticorin Town. Arsenic trioxide is obtained as a byproduct from the dust and residues formed during the treatment of other metal ores such as copper and gold. The high arsenic absorption is due to anthropogenic processes like brick making, poultry waste, and agricultural activity (Selvam, 2013).

Elevation to high levels of arsenic can cause short-term or acute symptoms in the population, as well as long-term or chronic health effects. Symptoms of exposures to rising levels of arsenic may include diarrhea, vomiting, stomach pain, and impaired nerve function and the feeling of "pins and needles" in the feet and hands. Arsenic can also create a pattern of changes in the skin, which includes wart-like growths—most commonly found on the soles or palms. Since children tend to drink larger quantities of water per unit of body weight than adults, they may experience greater exposure to arsenic from drinking water and as a result be at augmented risk to unpleasant effects when concentrations of arsenic are high. Long-standing (years to decades) exposure to even relatively low concentrations of arsenic in drinking water can increase the risk of certain embryonic cancers in addition to kidney, skin, lung, and bladder. Cancer is the most prominent health effect noted in the WHO guidelines for acceptable levels of arsenic in drinking water.

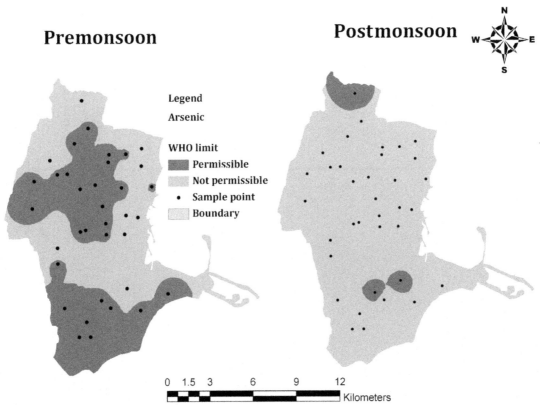

FIG. 23.2 Spatial variation of arsenic in the study area.

23.5 STATISTICAL ANALYSIS

Factor analysis describes the processes relevant to the investigated region and recognition of these factors allows an accurate explanation of the environment. Consequently, it is a helpful tool to describe the factors that affect the groundwater quality and its hydrogeochemical processes. The results of this procedure are low-factor loadings (close to 0) obtained for the remaining variables and high-factor loadings (close to 1 or −1) obtained for the variables concurrent in the factor. The factors that best portray the discrepancy of the analyzed data have an eigen value of >1 and consequently can be rationally interpreted. The sum of the squared factor loadings shows how factors were obtained and portrays the discrepancy of the variables and the factor scores were thereafter calculated for each groundwater sample. Five factors illuminate 75.40% of the total variance of the original data set, which is adequate to give a high-quality idea of data organization (Table 23.1).

The first factor explains the major part of variance and accounts for 40.62% of the total variance with positive loadings on As, Al, Ag, V, Fe, Zn, Ni, Sr, Cd, and Mo, and stumpy factor loadings for the enduring parameters. High factor loadings point to a strong rapport between the variable and the factor recitation for this variable. This organization powerfully suggests factor 1 represents an industrial activity variable, representative of the anthropogenic activity.

TABLE 23.1 Factor Analysis Results of Trace Elements in Groundwater

Parameters	Factors				
	1	2	3	4	5
Al	0.245	0.629	−0.105	−0.089	0.216
Si	−0.693	−0.149	−0.320	−0.049	0.137
V	0.136	−0.184	0.479	0.663	0.147
Cr	−0.044	0.473	0.764	−0.069	−0.258
Mn	−0.582	0.596	0.429	−0.140	−0.066
Fe	0.831	0.392	−0.055	−0.134	0.112
Ni	0.859	0.326	−0.115	0.189	0.008
Co	−0.698	0.173	−0.066	0.122	0.245
Cu	−0.894	0.226	0.068	0.028	0.155
Zn	0.608	0.069	0.158	−0.098	0.442
As	0.908	0.109	−0.046	0.005	0.140
Se	−0.951	0.121	0.090	−0.025	0.140
Rb	−0.850	−0.016	−0.241	0.012	0.252
Sr	0.581	−0.222	−0.003	−0.281	−0.297
Mo	0.556	−0.119	0.161	0.637	0.128
Ag	0.326	0.092	0.007	−0.015	0.670
Cd	0.428	0.501	−0.472	0.339	−0.290
Sb	−0.680	0.242	0.395	0.045	−0.038
Ba	0.495	0.121	0.085	−0.575	0.182
Pb	−0.209	0.731	−0.448	0.224	−0.122
Total	**8.124**	**2.335**	**1.844**	**1.537**	**1.241**
% of Variance	40.622	11.676	9.218	7.684	6.204
Cumulative %	40.622	52.298	61.516	69.200	75.404

Arsenic have high factor loading 0.908% due to an industrial wastes acquit represent that the geogenic movement that is the natural phenomenon representative. This demonstrates that manmade activity is one of the most significant factors affecting groundwater chemical composition in the present study area. The arsenic factor corresponds to the role of the imprudent use of sulfate, potassium, nitrate, and phosphate fertilizers. Conversely, the nature and spatial allocation of contamination observed in the coastal aquifer indicates that the main contamination sources are the agricultural activities and unsewered rural and urban areas.

The second factor accounts for 11.67% of the total discrepancy with soaring loadings for Cd, Mn, and Pb. This demonstrates that the manipulate of natural movement is one of the most important factors affecting the groundwater's chemical balance in the study area. This factor corresponds to the role of the imprudent use of phosphate and manganese fertilizers. The nature and spatial allocation of contamination observed in the coastal aquifer indicates that the main contamination sources are the agricultural activities. The third factor accounts for 9.21% of the total variance with high loadings for V, Cr, and Sb. This may result from industrial movement. The fourth factor represents 7.68% of the total variance with high loadings for Mo, V, and Cd. This factor represents endorsed to man-made activities like agricultural runoff, domestic and industrial wastes exonerate indicates that the geogenic activity that is the natural occurrence representative. According to the above argument, the first and second factors describe hydrochemical progression as a result of human activity and the third and fourth factors demonstrate geogenic and anthropogenic phenomenon in the study area. The extracted factor loading indicates that the source of the trace element absorption is probably anthropogenic activity.

23.5.1 Impact of Arsenic on Human Health

Humans are exposed to arsenic in three main ways: respiratory, gastrointestinal, and dermal absorption. The level of arsenic absorption relies on not only on the types of absorption, but also on the chemical/physical properties of the arsenic compounds and the period of exposure. However, in general it is noticed that gastrointestinal absorption is regarded as the most dangerous, as it is likely to lead to the sorption of higher doses of arsenic compared to the others (Hughes et al., 2009).

Arsenic affects human health in two principal ways: acute and chronic poisoning. In the case of acute arsenic poisoning, an adult human can die within a few days of ingesting inorganic arsenic in doses of approximately 1–3 mg As/kg. Some common symptoms of the poison include nausea, vomiting, and diarrhea (Hughes et al., 2009). Chronic arsenic poisoning takes a long time to show effects, depending on the general health of the subject and the dose and frequency of intake. Chronic arsenic poisoning is usually ascribed to contamination of drinking water at concentrations above 0.05 μg/L—the previous WHO standard for arsenic concentration in drinking water. Some diseases attributed to chronic arsenic poison include cardiovascular disease, peripheral vascular disease, ischemic heart disease, cerebrovascular disease, hypertension, diabetes, hepatic disease, skin cancer, bladder cancer, and lung cancer (Hughes et al., 2009; Ravenscroft et al., 2009).

23.6 CONCLUSION

The statistical results of the factor analysis recommend that there are five factors that can be used to describe the causes of groundwater pollution in the study area. The anthropogenic contamination of the Tuticorin coastal area has augmented very swiftly, as suggested by groundwater eminence changes. The results of values obtained for industrial-development areas were above the nonindustrial areas, indicating the contaminating effects of industrial activities on the ecosystem. Even though well synchronized in some areas, industry has been the source of many pollutants and contaminants in the groundwater. Major industrial performance has probably been the source of waste-water effluents, air emanation, and solid wastes, which have entered water bodies. In view of these conclusions, there is a need to scrutinize more intimately the environment and put in position proper checks and balances to conserve the health of the communities in the neighborhood of the industrial areas, predominantly as the effects of trace elements are bio-accumulative and pose immense danger to human, animal, and plant health. From the results of the present study it is suggested that the government should introduce the use of treatment technologies to the study areas to curtail this trace element pollution of groundwater and surface water to ensure that the water is safe for drinking and other public usages.

ACKNOWLEDGMENTS

Dr. S. Selvam is thankful to Department of Science and Technology (DST), Government of India, New Delhi for awarded INSPIRE Fellowship to carry out this study (Ref. No. DST/INSPIRE FELLOWSHIP/2010/308).

REFERENCES

Groundwater Contamination and control. Bachmat, Y., Zoker, U. (Eds.), 1994. Ground Water Contamination and Control. Marcel Dekker, Inc, New York.

Carter, D.E., Fernando, Q., 1979. Chemical toxicology. J. Chem. Educ. 56 (8), 491.

Fatoki, O.S., Okoro, H.K., Adekola, F.A., Ximba, B.J., Synman, R.G., 2012. Bioaccumulation of metals in black mussels (*Mytillus galloprovincialis*) in Cape Town Harbour, South Africa. Environmentalist 32, 48–57.

Handa, B., 1981. Hydro Geochemistry Water Quality and Water Pollution in UP. Tech. Report. CGWB, Min. of Irrigation, p. 317.

Hughes, M.F., Thomas, D.J., Kenyon, E.M., 2009. Toxicology and Epidemiology of Arsenic and Its Compounds. In: Henke, K. (Ed.), Arsenic: Environmental Chemistry, Health Threats and Waste Treatment. John Wiley & Sons, Ltd, West Sussex, p. 588.

Leung, C.M., Jiao, J.J., 2006. Heavy metal and trace element distributions in groundwater in natural slopes and highly urbanized spaces in min-levels area, Hong Kong. Water Res. 40, 753–767.

Marcovecchio, J.E., Botte, S.E., Freije, R.H., 2007. Heavy metals, major metals, trace elements. In: Nollet, L.M. (Ed.), Handbook of Water Analysis. 2nd ed. CRC Press, London, pp. 275–311.

Mendie, U., 2005. The nature of water. In: The Theory and Practice of Clean Water Production for Domestic and Industrial Use. vol. 14. Neurotoxicology, Lagos, pp. 16–166.

Nouri, J., Mahvi, A.H., Babaei, A.A., Jahed, G.R., Ahmadpour, E., 2006. Investigation of heavy metals in groundwater. Pak. J. Biol. Sci. 9 (3), 377–384.

Ravenscroft, P., Brammer, H., Richards, K., 2009. Arsenic Pollution: A Global Synthesis. Wiley-Blackwell.

Sang, Y., Fasheng, L., Qingbao, G., Cunzhen, L., Jiaqing, C., 2008. Heavy metal-contaminated groundwater treatment by a novel nanofiber membrane. Desalination 223, 349–360.

Selvam, S., 2013. Environmental Impact assessment, Human Health and Evaluation of Groundwater Pollution in and Around Tuticorin City, Tamil Nadu. Unpublished Ph.D thesis. Manonmaniam Sundranar University, pp. 1–258.

Selvam, S., Manimaran, G., Sivasubramanian, P., 2013. Hydrochemical characteristics and GIS-based assessment of groundwater quality in the coastal aquifers of Tuticorin corporation, Tamilnadu, India. Appl. Water Sci. 3, 145–159.

Selvam, S., Manimaran, G., Sivasubramanian, P., Balasubramanian, N., Seshunarayana, T., 2014. GIS-based evaluation of water quality index of groundwater resources around Tuticorin coastal city, South India. Environ. Earth Sci.. https://doi.org/10.1007/s12665-013-2662-y.

Selvam, S., Venkatramanan, S., Singaraja, C., 2015. A GIS-based assessment of water quality pollution indices for heavy metal contamination in Tuticorin corporation, Tamilnadu, India. Arab. J. Geosci.. https://doi.org/10.1007/s12517-015-1968-3.

Singaraja, C., Chidambaram, S., Srinivasamoorthy, K., Anandhan, P., Selvam, S., 2015. A study on assessment of credible sources of heavy metal pollution vulnerability in groundwater of Thoothukudi districts, Tamilnadu, India. Water Qual Expo Health. https://doi.org/10.1007/s12403-015-0162-x.

Vanloon, G.W., Duffy, S.J., 2005. The Hydrosphere. In: Environmental Chemistry: A Global Perspective. 2nd ed. Oxford University Press, New York, pp. 197–211.

WHO, 2004. Guidelines for Drinking Water Quality, vol 1. Recommendations, third ed WHO, Geneva, p. 515.

FURTHER READING

Selvam, S., 2014. Irrigational feasibility of groundwater and evaluation of hydrochemistry facies in the SIPCOT industrial area, south Tamilnadu, India: a GIS approach. Water Qual Expo Health 7 (3), 265–284. https://doi.org/10.1007/s12403-014-0146-2.

Chapter 24

Evaluation of Heavy-Metal Contamination in Groundwater using Hydrogeochemical and Multivariate statistical Analyses

Sang Yong Chung*, R. Rajesh† and S. Venkatramanan*,‡,§,¶

*Department of Earth and Environmental Sciences, Pukyong National University, Busan, South Korea †Department of Civil Engineering, Indian Institute of Science, Bangalore, India ‡Department for Management of Science and Technology Development, Ton Duc Thang University, Ho Chi Minh City, Vietnam §Faculty of Applied Sciences, Ton Duc Thang University, Ho Chi Minh City, Vietnam ¶Department of Geology, Alagappa University, Karaikudi, India

Chapter Outline

24.1 INTRODUCTION

Heavy metals in trace concentrations have the most important function in the metabolism and healthy growth of plants and animals. However, if the concentration is increased it may have several toxicological effects on human beings. Two processes comprise the main sources of heavy metals: natural and anthropogenic. The natural source involves the release of metals as a result of rock weathering and their ultimate leaching into groundwater through rock-water interactions. The anthropogenic sources are the discharge of heavy metals through mining, construction and industrial activities, which thereafter enter streams around mines, the pulverization of construction materials discharge of industrial effluents and landfill leachate into groundwater and surface water bodies; and other activities (Handa et al., 1981; Leung and Jiao, 2006).

Water-quality indices are helpful in estimating the combined effect of different parameters. Water-contamination indices are practical and comparatively simple approaches for evaluating the influence and source of overall contamination. The contamination indices are calculated to give useful and approachable guiding tools for water-quality administrators, environmental managers, decision-makers, and prospective users of a prospective water system. Specific contamination indices have also been used to evaluate the extent of contamination with respect to certain metals (Prasad and Jaiprakas, 1999; Prasad and Bose, 2001). Statistical methods, particularly multivariate techniques, are adequate to resolve this lack of contamination indices, and are helpful for environmental data reduction and the interpretation of multiple parameters. Principal component analysis (PCA) and correlation matrix (CM) are widely used in water-quality assessment and monitoring (Venkatramanan et al., 2015; Bhuiyanet al., 2010). PCA and CM help to arrange metals or analyzed parameters into different factors/groups on the basis of contribution from the expected source. Depending on the co-variance of the analyzed parameters, the multivariate technique is able to provide useful information on the origin of groundwater contamination. The hydrogeochemical species of an element is important regarding its environmental chemistry. The species also provides information on the mobility and, consequently, the availability of the metal to living things and its potential toxicity (Fergusson, 1990).

The Nakdong and its tributaries serve as a major source of drinking water for the inhabitants of the river basin. Rapid industrialization, urbanization, and population increases over the last few decades have caused a dramatic increase in the demand for river water, as well as a significant deterioration in water quality, especially near the large industrial complexes and in the lower basin, and water contamination remains a serious concern of this area. Few studies have been published in recent years on groundwater quality and its sources in Nakdong River Basin (Chun et al., 2001). Thus the current case study will deliver an updated scientific basis of heavy-metal contamination and its source identification and also may propose important directions for future research. The main objective of this case study is to assess the groundwater quality in a confined aquifer using the combined methods of contamination indices, multivariate statistical approaches, and geochemical modeling to identify the contamination status and probable sources of pollutants in the study area.

24.2 STUDY AREA

The Nakdong River (Fig. 24.1) is the largest river in South Korea. The length and total watershed of the Nakdong River are 525 km and 24,000 km², respectively. It is divided into eight main tributaries. The study area is located in a part of the Nakdong River Basin in Busan Metropolitan City, Korea. It covers a 4 km² area. This area lies in the transitional zone between a continental and subtropical climate. It has four distinct seasons. The annual average temperature is 15°C and average annual precipitation is 1397 mm. Agricultural activity depends on the climatic conditions and the availability of water sources. Greenhouses are popular in the study area, and vegetables are mainly cultivated here. However, the study area comprises a large parkland and natural environmental conditions are conserved, with the exception of an exercise area.

24.3 GEOLOGY AND HYDROGEOLOGY

The deltaic plain consists of a lozenge-shaped delta and flat areas with elevations of approximately 2 m to 3 m above the mean sea level (MSL). The basement rocks of the Nakdong River Plain are composed of Cretaceous sedimentary and volcanic rocks, as well as late Cretaceous and Paleocene intrusive granites in ascending order. The basement rocks are unconformably overlain by Quaternary formations. Quaternary deposits form the Samrak Park region (Fig. 24.1). The geological relation between the Quaternary deposits and the Cretaceous rocks is defined by unconformities. The topography of the

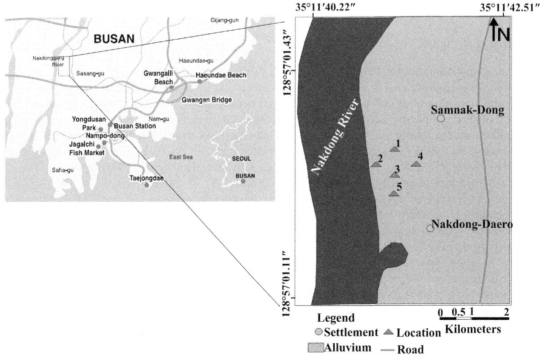

FIG. 24.1 Sampling location and geology of the study area.

study area is 0.1 m to 3 m above the MSL. The aquifer system of the Samrak park region is confined. Porosity and hydraulic conductivity of the area is 37%–59% and 1.7–300 m/day, respectively. The depth of the wells ranges from 69 m to 77 m. The groundwater level of the region is 0.2 m to 3 m MSL.

24.4 METHODOLOGY

24.4.1 Sampling

A detailed preliminary field investigation was done in the Nakdong River Basin and five boreholes were drilled (Fig. 24.1) in the area for the purpose of aquifer storage and recovery project. There are no other boreholes in the surrounding area. The samples were collected after pumping for 10 min. This was done to remove the groundwater stored in the well. Groundwater samples were collected in clean polyethylene bottles of 600 mL capacity. The sampling bottles were soaked in 1:1 diluted HCl solution for 24 h previous to sampling and then washed with distilled water. They were washed again before each sampling with the filtrates of the sample. Each sample was instantly filtered on site through 0.45 μm filters in cellulose nitrate membrane. The pH and electrical conductivity (EC) were measured in the field using a portable meter (Horiba U-51, Japan). Subsequently, the samples were analyzed in the laboratory for their major ions, such as calcium, magnesium, sodium, potassium, chloride, bicarbonate, sulfate, silica, and aluminum. This was attained using standard methods, as recommended by the American Public Health Association (APHA, 1995). The accuracy of the chemical analysis was confirmed by computing ion–balance errors where the errors were usually around ±5%. The filtered and acidified samples were analyzed for nine elements (Al, As, Ba, Cu, Fe, Mn, Sr, V, and Zn) using ICP-MS. Instrument blank was used to test for preference from possible contamination of blank water, which consisted of distilled water. This is to validate that decontamination measures and laboratory protocols are suitable (Koterba et al., 1995).

24.4.2 Evaluation Indices

In order to estimate the amount of heavy-metal contamination of water for specific use, different contamination indices are studied. Evaluation indices are classified as the heavy metal contamination index (HPI), heavy metal evaluation index (HEI) and the degree of contamination (Cd), which are used to estimate the quality of water in the study area. The HEI and HPI methods used the World Health Organization's (WHO, 2011) maximum admissible concentrations for drinking water to monitor the values of the metals (Prasanna et al., 2012). In the Cd method, the quality of water is calculated as the sum of the contamination factors of each component exceeding the upper permissible limit. Therefore the Cd summarizes the combined effects of a number of quality parameters regarded as adverse for household water (Prasanna et al., 2012).

24.4.3 Heavy-Metal Contamination Index

The HPI method was developed by assigning a rating or weightage (Wi) for each chosen parameter. The rating is an arbitrary value between zero and one and its selection reflects the relative importance of individual quality considerations. It can be defined as inversely proportional to the standard permissible value (Si) for each parameter (Horton, 1965; Reddy, 1995; Mohan et al., 1996; Prasanna et al., 2012). In the present study, the concentration limits (i.e., Si and the highest desirable (ideal) value (Ii) for each parameter) were taken from the WHO (2011) standard (Table 24.6). The uppermost permissive value for drinking water (Si) refers to the maximum allowable concentration in drinking water in the absence of any alternate water source. The desirable maximum value (Ii) indicates the standard limits for the same parameters in drinking water (Prasanna et al., 2012; Bhuiyan et al., 2010). The HPI is determined using the following equation proposed by Mohan et al. (1996) and used by Prasanna et al. (2012), Venkatramanan et al. (2014):

$$HPI = \frac{\sum_{i=1}^{n} W_i Q_i}{\sum_{i=1}^{n} W_i Q_i} \tag{24.1}$$

where Q_i and W_i are the subindex and unit weight of the ith parameter, correspondingly, and n is the number of parameters considered. The subindex (Q_i) is calculated by:

$$Q_i = \sum_{i=1}^{n} \frac{|M_i - I_i|}{|S_i - I_i|} \times 100 \tag{24.2}$$

where M_i, I_i, and S_i are the monitored heavy metal, ideal, and standard values of the ith parameter, respectively. The sign $(-)$ indicates the numerical difference of the two values, neglecting the algebraic sign.

24.4.4 Heavy Metal Evaluation Index

HEI provides an overall quality of the water with respect to heavy metals (Eder and Offiong, 2002) and is expressed as:

$$HEI = \sum_{i=1}^{n} \frac{H_c}{H_{mac}} \tag{24.3}$$

where H_C and H_{mac} are the monitored value and maximum admissible concentration (MAC) of the ith parameter, respectively.

24.4.5 Degree of Contamination (C_d)

The contamination index (C_d) defines the combined effects of several quality parameters analyzed for their harmful effects on domestic water (Backman et al., 1997) and it is calculated as follows:

$$C_d = \sum_{i=1}^{n} Cr_i \tag{24.4}$$

where

$$Cr_i = \frac{C_{Ai}}{C_{Ni}} - 1$$

where C_{ri}, C_{Ai}, and C_{Ni} represent the contamination factor, analytical value, and upper permissible concentration of the ith component, respectively. N denotes the "normative value" and C_{Ni} is taken as MAC.

24.4.6 Multivariate Statistical Analysis

Multivariate statistical analysis is a quantitative and independent method of groundwater classification allowing the grouping of groundwater samples and correlations to be made between metals and groundwater samples (Cloutier et al., 2008). In this study, two multivariate methods were applied using STATISTICA, factor analysis (FA), hierarchical cluster analysis (HCA), and correlation analysis.

The percentages of eigenvalues are computed since the eigenvalues quantify the contribution of a factor to the total variance (i.e., the sum of the eigenvalues). The factor extraction is done using a minimum acceptable eigenvalue that is greater than 1.0 (Kaiser, 1960). The factor-loading matrix is rotated to an orthogonal simple structure, according to varimax rotation, which results in the maximization of the variance of the factor loading of the variables. This procedure renders a new rotated factor matrix in which each factor is described in terms of only those variables and affords a greater ease for interpretation. Factor loading is the measure of the degree of closeness between the variables and the factor (Dalton and Upchurch, 1978).

HCA was applied by Ward's method for sample classification, because it dealt with a small space changing result and used more information on cluster substances than other approaches. This approach was applied to standardized data using squared Euclidean distances as a measure of similarity. It is a particularly powerful grouping mechanism and uses the analysis of variance approach to estimate distances between clusters, attempting to minimize the sum of squares of any clusters that can be formed at each step (Willet, 1987). Cluster analysis requires the standardization of sample data, as each component of water quality has a different unit. The standardization of data ensures that each variable has the same effect in the investigation. Data were identical to the Z score ($m=0$ and $S=1$) by applying the following equation, $Z=(X-m)/S$, where Z is the standardized value, X is the value of sample data, m is the mean, and S is the standard deviation.

Correlation analysis was used to find the relationships between the physicochemical parameters of the groundwater samples, which can expose the source of solutes, and the process that generated the determined water compositions (Azaza et al., 2011; Parizi and Samani, 2013). The result of the correlation analysis was measured in the consequent interpretation. A high correlation coefficient (near 1 or 1) means a good positive relationship between two variables and a value of around zero means no relationship at a significant level of $P < .05$. More exactly, it can be said that parameters showing $r > .7$ are considered to be strongly correlated whereas r of 0.5–0.7 shows a moderate correlation (Manish et al., 2006).

24.5 RESULTS AND DISCUSSION

24.5.1 Hydrochemistry and Metals

The statistical parameters of the physicochemical characteristics and metal concentrations of the groundwater of this area are given in Table 24.1. The pH of the groundwater samples of this area ranges from 7.16 to 7.68 with an average of 7.40, which is slightly neutral to alkaline. EC ranges from 40.66 to 41.55 average of 41.13 ms/cm due to the effect of saline-water intrusion. The DO concentrations of >5 mg/L represent an oxygenated groundwater environment, but the dissolved oxygen of the area ranges from 0.86 to 4.24 with an average of 2.22 mg/L, which indicates that the groundwater is aerobic. The mean concentrations of metals in groundwater are, in descending order, $Sr > Fe > Ba > Mn > As > Al > Cu > Zn > V$. The descriptive statistics including maximum admissible concentration (MAC) and world standards are given in Table 24.1. The method of Ficklin et al. (1992), as modified by Caboi et al. (1999), based on the relationship between metal load and

TABLE 24.1 Descriptive Statistical Data of Physicochemical Parameters and Metals

Parameter	Units	Minimum	Maximum	Mean	SD	MAC	USEPA (2012)	Adverse Effect Beyond the Desirable Limit
pH		7.16	7.68	7.40	0.23	6.5–8.5[a]	6.5–8.5	Taste disorder, effects mucus membrane
EC	mS/Cm	40.66	41.55	41.14	0.32	1400[a]	–	
DO	mg/L	0.86	4.24	2.22	1.57	–	–	
Al	μg/L	80.00	220.00	122.00	60.17	200[a]	200	Loss of memory, severe trembling, and dementia
As	μg/L	97.67	236.63	147.25	61.62	50[a]	10	Kidney, skin, blood, and liver disorders
Ba	μg/l	0.00	5651.00	2478.60	2491.80	700[a]	2000	Increase in blood pressure
Cu	μg/L	0.00	444.00	97.60	193.96	1000[a]	1300	Liver damage and anemia
Fe	μg/L	3642.86	10,928.57	6492.86	2735.66	200[a]	300	Staining of cloths and plumbing material
Mn	μg/L	1280.97	1661.06	1505.79	150.91	50[a]	50	Staining and discoloration
Sr	μg/L	0.00	15,120.00	7684.00	7232.24	–	–	Lung cancer and problem in bone growth
V	μg/L	0.00	3.20	1.76	1.61	–	–	Irritation of lungs, throat, eyes, and damage of nervous system
Zn	μg/L	0.00	11.00	2.20	4.92	5000[a]	5000	Corrosion of plumbing material and industrial contamination

MAC, maximum admissible concentration (adapted from Seigel, 2002).
[a]WHO (2011).

FIG. 24.2 Classification of groundwater samples based on the plot of metal load and pH.

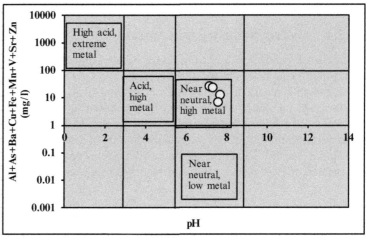

pH, was used to categorize the water in the area. Metal load was defined as the sum of metals in milligram per liter. However, in this study the metal load was calculated as $Al + As + Ba + Cu + Fe + Mn + Sr + V + Zn$ (Fig. 24.2). Based on this method, groundwater samples are present in the near neutral to the high metal category. The heavy metals were classified as follows: toxic, transition, alkaline, alkali, and metallic. In this study, the transition and alkali metals are not present in the study area.

24.5.2 Toxic Heavy Metals (As and V)

Toxic heavy metals such as arsenic (As) and vanadium are present in the study area. As concentration in the groundwater ranges from 97.67 µg/L to 236.63 µg/L with an average of 147.25 µg/L (Table 24.1). All the samples exceed the acceptable limit of 10 µg/L set by WHO (2011) and USEPA (2012). The high concentration of As is mainly due to the marine sediments, fossil shell fragments, agriculture, and irrigation practices (Hunt and Howard 1994). Vanadium concentration ranges from 0 µg/L to 3.20 µg/L with an average of 1.76 µg/L. The USEPA (2012) and WHO (2011) has not prescribed any guideline value for vanadium concentration in groundwater.

24.5.3 Alkaline Heavy Metals (Sr and Ba)

A high concentration of Sr is the result of seawater influence on the aquifer system and the Sr content may be solely marine in origin. The concentration of strontium in the groundwater ranges from 0 µg/L to 15,120 µg/L with an average of 7684 µg/L (Table 24.1). The USEPA (2012) and WHO (2011) has no recommended guidelines for strontium concentration in groundwater. Barium ranges from 0 µg/L to 5651 µg/L with an average of 2478.60 µg/L. Groundwater samples of barium exceed the permissible limits of 2000 µg/L set by USEPA (2012) and 700 µg/L set by WHO (2011). The primary source of the barium is saline water in the study area, because the water was trapped in the formation at the time the material was deposited.

24.5.4 Metallic Heavy Metals (Cu, Fe, Mn, Zn, and Al)

In the study area copper concentration ranges from 0 µg/L to 444 µg/L with an average of 97.60 µg/L (Table 24.1). All the groundwater samples of the study area are within the permissible limit of 1000 µg/L set by WHO (2011) and USEPA's (2012) limit of 1300 µg/L. The concentration of copper in the study area is chiefly a byproduct of industry, i.e., wastes from electroplating and copper salts, and of anthropogenic activities. The iron concentration of groundwater samples ranges from 3642.86 µg/L to 10,928.57 µg/L, with an average of 6493 µg/L (Table 24.1). The concentration of Fe in all groundwater samples exceeds the permissible limits of 200 µg/L set by WHO (2011) and the USEPA's (2012) limit of 300 µg/L. A high concentration of Fe is essentially due to pyroxenes, amphiboles, biotites, and magnetites (Hem, 1985) and also possibly as a result of the removal of dissolved oxygen by organic matter, which leads to the reduced conditions. Under reducing conditions, the solubility of Fe-bearing minerals (siderite, marcacite, etc.) increases, leading to enrichment of dissolved iron in the groundwater (Applin and Zhao, 1989; White et al., 1991). Manganese ranges from 1281 µg/L to 1661 µg/L, with an average of 1506 µg/L (Table 24.1). Groundwater samples from all locations exceed the permissible limit of 50 μg/L set by WHO (2011) and USEPA (2012). A high concentration of Mn is found, the likely origin of which is the marine environment. Zn ranges from 0 μg/L to 11 μg/L, with an average of 2 μg/L (Table 24.1), and all of the samples fall within the permissible limit of 5000 μg/L set by WHO (2011) and USEPA (2012). Al ranges from 80 μg/L to 200 μg/L, with an average of 122 μg/L. The permissible limits of WHO (2011) and USEPA (2012) are 200 μg/L to 500 μg/L.

24.5.5 Comparison of Heavy Metals Worldwide and in Korea

Jahanshahi and Zare (2015) explained that the mean metal concentration in the groundwater of the Golgohar iron ore mine area, Iran was low and within the range of natural background values. Bhuyian et al. (2010) reported that the metal concentrations for all their samples were within the acceptable limits, except for Fe. The results from this study represent the mean metal concentration, except for Zn, of Samrak Park groundwater compared with other studies (Table 24.2), such as Stamatis et al. (2000), Krishna et al. (2009), Abiye and Leshomo (2014), Arkoc (2014), Venkatramanan et al. (2014), and Ahn (2011).

24.5.6 Contamination Evaluation Indices

The contamination evaluation index is a method of calculating and placing into arithmetical criterion the chemical quality of water in a specific area of interest (Edet and Offiong, 2002). These indices are calculated based on WHO outlines (2004, 2011). The range and mean values of HPI are 90 to 92 and 90.6, respectively. The results of the indices indicated that the heavy-metal contamination index for all the samples are below the critical limit of 100 for drinking water (Mohan et al., 1996). The HPI is estimated with a mean value of 90.6 for all sampling locations, which is also below the critical limit of 100.

The degree of contamination values (C_d) range from 19 to 57 with a mean value of 34. Based on this, the C_d values for all samples exceed the value of 3, which indicates that the area has a high degree of contamination. The results of C_d is different from the HPI value, all the locations in the study area are below the critical limit of 100 (Prasanna et al., 2012).

The HEI values range from 50 to 84 with a mean value of 65.62. The mean deviations and percentage deviation for all the indices were computed for each sampling point (Table 24.3). Remarkably, the Cd and HPI values of these same samples were below the respective mean value of the indices. The values of Cd and HPI are fall below the mean value and relatively compared to percent deviation values indicates the better water quality of groundwater, as observed by Prasad and Bose (2001), Edet and Offiong (2002), and Venkatramanan et al. (2014).

In order to evaluate the heavy-metal contamination level of the groundwater in the study area, the present water-quality strategies for indices were revised using the approach presented by Edet and Offiong (2002). The contamination indices values are thus classified and represented in Table 24.4 as low for samples with indices below the mean value, medium for those having values between mean and two times the mean, and high when the value exceeds two times the mean values (Jahanshahi and Zare, 2015). Heavy-metal contamination index samples (Fig. 24.3A) having values below 50 are of low level (0%), between 50 and 100 are of medium level (80%), and above 100 are of high level (20%). The HEI values (Fig. 24.3B) with indices of <40 are classified as low level (0%), medium level is between 40 and 80 (80%), and high level is >80 (20%). The Cd index values (Fig. 24.3C) show that 100% of the samples present as the high level with a value of >12; the medium level is between 6 and 12 and the low level is <6.

The values of the HPI, HEI, and C_d indices are plotted against each other and represented in Fig. 24.4. Based on this figure, there is a strong correlation between HPI and C_d with $R^2 = .82$, HEI and C_d with $R^2 = .96$, and HPI and HEI with $R^2 = .67$. HEI and modified criteria of HPI and C_d indicate comparable results and the HEI approach may be used as a simple criteria to estimate the quality of water in the groundwater. Thus all samples are found to be highly contaminated by metal pollutants (Prasanna et al., 2012).

24.5.7 Contamination Source Identification

A multivariate statistical method yields the general relationship between measured metal contamination by showing multivariate patterns that may help to classify the original data. FA can be used to identify several contamination factors practically but the interpretation of these factors in terms of actual controlling sources and processes is highly subjective (Matalas and Reiher, 1967; Bahar and Reza, 2010; Edet et al., 2013). Two factors were extracted from groundwater samples with the eigenvalues >1. The estimated factor loadings are combined with cumulative percentage and the percentages of variance for each factor are represented in Table 24.5. The factors in samples led to decreases the preliminary data set, which explained of about 86% of the total variance. Evaluate the factor loadings of the samples are shown in Table 24.5 Positive

TABLE 24.2 Comparison of Physicochemical Parameters and Heavy Metals in Groundwater Samples of Worldwide and the Korea Region

Study Area	pH	EC	DO	Al	As	Ba	Cu	Fe	Mn	Sr	V	Zn
Groundwater, Greece (March 1996: n=31)[a,b]	7.52	3.732	–	–	–	–	0.03c	0.012c	0.23c	–	–	0.57c
Groundwater, India (Jun 2008: n=31)[a,d]	8.2	1888.33	–	–	146.5c	246.5c	–	117.0c	745.4c	2916.0c	–	89.7c
Groundwater, Bangladesh (Oct 2009: n=10)[a,e]	8.85	177.54	9.82	–	–	–	–	5400	26	–	–	290
Groundwater, Iran (March 2011: n=70)[a,f]	7.40	18,681.31	–	148.6	70.3	338.3	14	4106.9	896.6	80,238.7	–	191.9
Groundwater, South Africa (Dec 2013: n=30)[a,g]	6.87	5438	–	–	–	37.40	31.17	2346.50	65.96	2764.07	15.28	688.91
Groundwater, Turkey (Dec 2013: n=17)[a,h]	7.5	737	–	–	–	–	0.005c	0.012c	–	–	–	0.083a
Korea region												
Groundwater, Korea (May 2010: n=40)[a,i]	6.95	1638	8.5	–	–	–	23	9	25	–	–	57
Groundwater, Korea (Aug 2010: n=150)[a,j]	6.54	117	4.29	–	7.37	3.52	–	–	–	63.9	–	5.75
Groundwater, (Present Study) (July 2014: n=5)[a]	7.40	41,140	2.22	122.0	147.25	2478.6	97.60	6492.86	1505.79	7684.00	1.76	97.60

EC values in µS/cm; all heavy metals values in µg/L.

[a]Mean calculated values.
[b]Stamatis et al. (2000).
[c]mg/L; (n=total number of samples).
[d]Krishna et al. (2009).
[e]Bhuyian et al. (2010).
[f]Jahanshahi and Zare (2015).
[g]Abiye and Leshomo (2014).
[h]Arkoc (2014).
[i]Venkatramanan et al. (2014).
[j]Ahn (2011).

TABLE 24.3 Results of Contamination Evaluation Indices

Sample No	C_d	Mean Deviation	% Deviation	HPI	Mean Deviation	% Deviation	HEI	Mean Deviation	% Deviation
1	57	4.53	2.49	91.10	0.20	4.31	84.00	3.68	2.23
2	33	−0.20	1.15	90.66	0.13	4.79	68.24	0.52	1.42
3	33	−0.21	1.18	90.66	0.24	0.53	66.53	0.18	1.33
4	19	−3.02	0.96	90.39	0.24	0.48	50.29	−3.07	0.88
5	29	−1.10	0.00	90.39	0.25	0.00	59.06	−1.31	0.00

TABLE 24.4 Classification of Water Quality Using the Modified Contamination Indices (the Method Suggested by Edet and Offiong, 2002)

Indices	Class of Contamination	No. of Locations	Percentage (%)
HPI	Low (<50)	0	
	Medium (50–100)	5	100
	High (>100)	0	
HEI	Low (<40)	0	
	Medium (40–80)	4	80
	High (>80)	1	20
C_d	Low (<6)	0	
	Medium (6–12)	0	
	High (>12)	5	100

loading in factor show that water samples are changed by the presence of the parameters that are significantly loaded on a distinct factor, whereas negative loading propose that water quality is significantly unchanged by those parameters.

Factors 1 and 2 describe about 61% and 21% of the total variance in the groundwater data. Factor 1 is highly loaded and describes the geogenic source of components such as pH, EC, Fe, Mn, Al, As, V, and Sr. Fe and Mn is released by the leaching of parent rock from the soil zone into the water. The solubility of Fe and Mn minerals is strongly controlled by redox particularly at near-neutral pH (Lorite-Herrera et al., 2008). The aluminum in the groundwater may be due to dissolution of the element from clays and other alumino-silicate minerals found in the aquifer materials. Sulfide minerals are a common source of natural geogenic As. As mobilization is associated with oxidation processes that lead to the leaching of As, causing its release into the groundwater. V concentrations are commonly related to oxic and alkaline groundwater while V concentrations in anoxic groundwater were generally low regardless of pH. Sr content is due to the geochemical alteration/weathering of sulfate minerals present in the aquifer materials.

Factor 2 is highly loaded on pH, Do, Mn, As, V, Zn, Ba, and Cu. DO results in a reducing environment for groundwater with a long residence time and in parts of the study area where there is rock weathering and the consumption of DO by organic matter. Corrosion of pump parts may also lead to Zn being released into the groundwater. The groundwater is naturally enriched with Cu, probably signifying enrichment from aquifer materials such as feldspar, biotite, and muscovite minerals. The concentration of Ba^{2+} is mainly controlled by sulfate concentration, which promotes barite precipitation. The precipitation of barite may be an important mechanism of the removal of Ba^{2+}, and barite solubility is usually reported as the factor that controls dissolved barium in the groundwater.

R-mode cluster analysis is used to understand the physicochemical and create elemental groupings in the dataset and the results are shown in Fig. 24.5. R-mode cluster analysis was used in the groundwater samples and two clusters were produced. Cluster 1 consist of As, Ba, Cu, Fe, and Sr; cluster 2 consists of Al, Ec, Zn, V, DO, and pH. This method reveals the

FIG. 24.3 Contamination evaluation indices (A–C) of groundwater samples representing the low, medium and high contamination level.

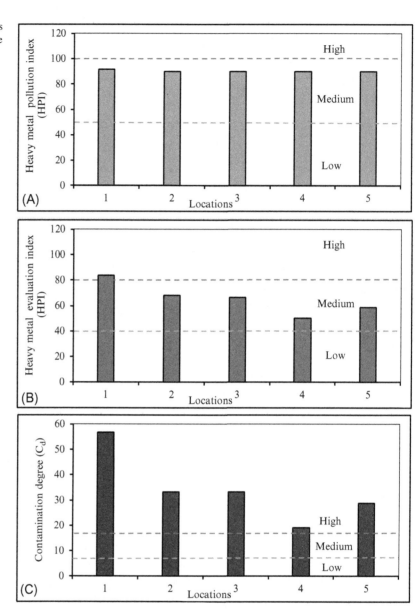

influence of natural hydrogeochemical processes, i.e., the leaching of constituents from the soils and saline water intrusion. Based on FA and cluster analysis there is some difference in the results, but effective concord between the two statistical methods is noticeable in all samples of the study area.

The Pearson correlation matrices interpret the interelement relationships between the metals in the groundwater. The interelement relationships confirm the results found from FA, and the CM has also been useful in indicating a few new associations in the metals that have not been properly expressed in the FA analysis. Strong correlations (Table 24.6) are obtained from the samples. pH strongly correlates with EC ($r = .69$), DO ($r = .75$), Mn ($r = .91$), As ($r = .88$), Sr (0.92), and Ba (0.92). DO correlates with Mn ($r = .81$) and Ba (.76). EC also strong correlation with Fe ($r = .94$), Mn ($r = .63$), As (.58), Al (.82), V (.77), and Sr (.81). These results are similar to those for factor 1 in the previous section. Metal pairs, such as Fe-Mn, Fe-Al, Fe-V, and Fe-Sr, correlate strongly with respective correlation coefficient (r) values of 0.57, 0.64, 0.54, and 0.57; Mn-As, Mn-V, Mn-Sr, Mn-Zn, Mn-Ba, and Mn-Cu with correlation coefficient (r) values of .65, .78, .72, .58, .90, and .61; As-Al, As-V, and As-Ba with values of .57, .96, .94, and .84; and Al correlates with V ($r = .98$) and Sr (.74). These values indicate a similar source to factor 1. V correlates with Sr ($r = .98$) and Ba (.87); Sr correlates with the Ba (.78); and Zn-Ba and Ba-Cu with correlation coefficient (r) values of .71 and .75, respectively. The source of these correlation values matches with the factor 2 results as explained in the previous section.

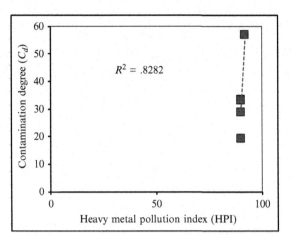

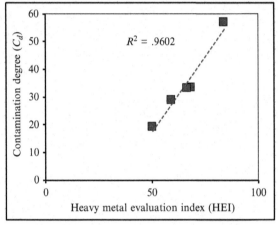

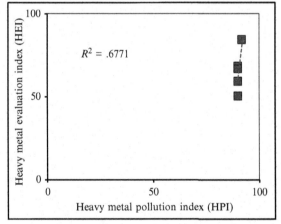

24.5.8 Geochemical Modeling

PHREEQC is a software program for simulating chemical reactions and transport processes in natural or polluted water. The program works on the equilibrium chemistry of aqueous solutions interacting with minerals, gases, solid solutions, exchangers, and sorption surfaces. It is based on an ion-association aqueous model and has capabilities for speciation and saturation-index calculations.

The saturation index (SI) is defined as the logarithm of the ratio of the ion activity product (IAP) of the component ions of the solid in solution to the solubility product (K_{sp}) for the solid:

$$SI = \log \frac{IAP}{K_{sp}}$$

TABLE 24.5 Varimax Rotated Factor Analysis for the Groundwater

Parameters	Factor 1	Factor 2
pH	**0.73**	**0.64**
EC	**0.95**	0.05
DO	0.16	**0.79**
Fe	**0.78**	0.02
Mn	**0.55**	**0.76**
As	**0.72**	**0.53**
Al	**0.92**	0.33
V	**0.83**	**0.52**
Sr	**0.90**	0.38
Zn	−0.14	**0.90**
Ba	0.48	**0.87**
Cu	−0.08	**0.91**
Eigenvalue	7.30	3.02
% Total	60.82	25.14
Cumulative	60.82	85.95

The bold values indicate the strong and moderate significant loadings of factor analysis in the groundwater of the study area.

FIG. 24.5 Hierarchical cluster analysis for metals.

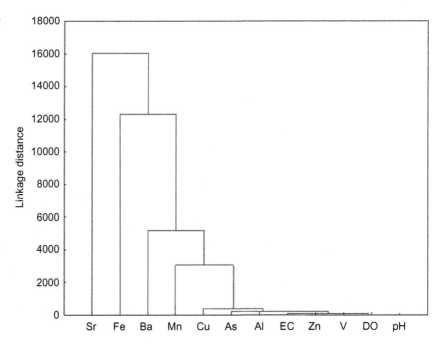

If the SI is zero the water composition reveals the solubility equilibrium with respect to the mineral phase. A negative value shows undersaturation and a positive value shows oversaturation. The saturation index (Fig. 24.6) of the study area is oversaturated with respect to hematite (Fe_2O_3), goethite (FeO(OH)), gibbsite (Al(OH)$_3$), rhodochrosite ($MnCO_3$), and siderite ($FeCO_3$). The SIs of hematite and goethite are about 20 and 10, respectively, and those of gibbsite, rhodochrosite, and siderite are <3. The undersaturated samples are alunite (KAl$_3$(SO$_4$)$_2$(OH)$_6$), hausmannite (Mn_2O_4), manganite (MnO(OH)), melanterite ($FeSO_4·H_2O$), pyrochroite (Mn(OH)$_2$), and pyrolusite (MnO_2). The SIs of hausmannite and pyrolusite

TABLE 24.6 Correlation Analysis of the Heavy Metals

Parameters	pH	EC	DO	Fe	Mn	As	Al	V	Sr	Zn	Ba	Cu
pH	1.00											
EC	**0.69**	1.00										
DO	**0.75**	0.15	1.00									
Fe	0.49	**0.94**	−0.07	1.00								
Mn	**0.91**	**0.63**	**0.81**	0.57	1.00							
As	**0.88**	**0.58**	−0.46	0.29	**0.65**	1.00						
Al	−0.45	**0.82**	−0.20	**0.64**	0.18	**0.57**	1.00					
V	**0.94**	**0.77**	0.49	**0.54**	**0.78**	**0.96**	**0.63**	1.00				
Sr	**0.92**	**0.81**	0.43	**0.57**	**0.72**	**0.94**	**0.74**	**0.98**	1.00			
Zn	−0.39	−0.01	0.47	0.00	**0.58**	−0.36	−0.39	0.36	0.19	1.00		
Ba	**0.92**	0.47	**0.76**	−0.30	**0.90**	**0.84**	0.17	**0.87**	0.78	0.71	1.00	
Cu	−0.43	0.04	0.49	−0.04	**0.61**	−0.41	−0.34	0.41	0.25	1.00	**0.75**	1.00

The bold values indicate the significant correlation at the level of 0.05.

are less than −10, and for melanterite, pyrochroite, and manganite are less than −5. Barite (BaSO4) falls within both the oversaturated and undersaturated categories. The high content of aluminum, iron, and manganese is ascribed to the weathering of silicate minerals, which is accelerated by the dissolution and precipitation of the aquifer materials in the study area.

The ionic strength has also been estimated with the software, PHREEQC. The ionic strength of a solution is a measure of the ion protection that occurs around charged dissolved species (Deutsch, 1997). The ionic strength is estimated as follows:

$$I = 0.5 \sum \left(C_i Z_i^2 \right)$$

where C_i is the concentration in mol/L (M) of ionic strength (I) and Z_i is the charge on ion.

The ionic strength (Fig. 24.7) of the groundwater ranges between 5.56×10^{-01} and 5.58×10^{-01} with mean of 5.65×10^{-01}. Appelo and Postma (1999) suggested that the ionic strength for fresh water is normally <0.02 while seawater has an ionic strength of about 0.7. Deutsch (1997) reported that the ionic strength of dilute groundwater is in the range of 10^{-2} to 10^{-3}. The values of ionic strength indicated that the groundwater samples from the area are saline. It has been noted that ion protecting lowers the activity of dissolved species; therefore the higher the ionic strength, the greater the protection and the greater the solubility of the mineral in contact (Deutsch, 1997). Hence, the result of the ionic strength indicates greater solubility and the mobility of the dissolved species in the study area.

24.6 CONCLUSION

Contamination evaluation indices, FA, cluster analysis and CM have been used to assess the intensity and sources of contamination in the Samrak Park of Nakdong River Basin. The mean concentrations of metals in groundwater are, in the descending order, Sr > Fe > Ba > Mn > As > Al > Cu > Zn > V. Only the mean concentrations of Al, Fe, As, and Mn are above the MAC values for drinking water. The groundwater of this area is differentiated as near-neutral high metal. Cd explains that all samples of groundwater are highly polluted. HPI shows that all samples fall within the critical limit of 100. An improved water quality grouping of the samples is accomplished by using HEI. The HEI of the samples is 0% low (<40), 80% medium (40–80), and 20% high (>80) contamination. There is a strong correlation between HPI and HEI ($R^2 = .67$), HPI and C_d ($R^2 = .82$), and HEI and C_d ($R^2 = .96$). A reclassification of the types using a multiple of the mean indicate comparable results.

Multivariate statistical analysis of FA supports the cluster analysis for the identification of weathering and leaching of parent rocks and interaction with groundwater, and for the nonpoint source, such as fertilizers, herbicides and insecticides from agricultural lands and residential areas; and bacteria and nutrients from livestock, pet wastes, and faulty septic systems, which are responsible for the changeability of the physicochemical parameters and metal concentrations in

FIG. 24.6 Saturation index of metal species in groundwater of the area.

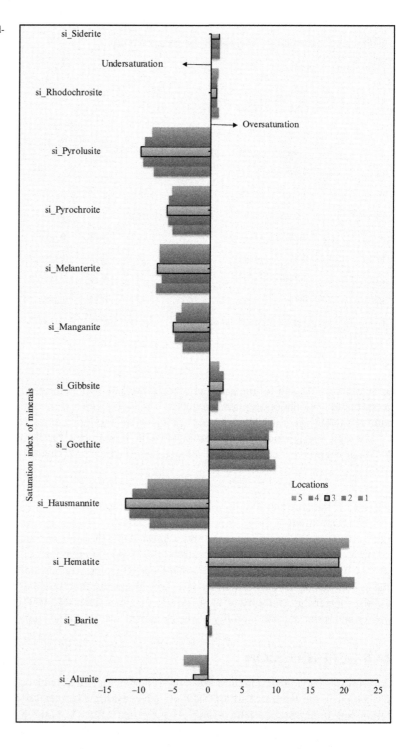

the groundwater. The statistical tools of correlation analysis and FA support the effective interpretation of all the datasets for the area. The saturation index (Fig. 24.5) of the study area shows oversaturation with respect to hematite (Fe_2O_3), goethite ($FeOOH$), gibbsite ($Al(OH)_3$), rhodochrosite ($MnCO_3$), and siderite ($FeCO_3$). The undersaturated samples are alunite [$KAl_3(SO_4)_2(OH)_6$], hausmannite (Mn_2O_4), manganite ($MnO(OH)$), melanterite ($FeSO_4H_2O$), pyrochroite ($Mn(OH)_2$), and pyrolusite (MnO_2). Barite ($BaSO4$) falls within both the oversaturated and undersaturated categories. Groundwater contamination by heavy metals suggests a critical limit of groundwater quality in the area. Hence, this work gives a

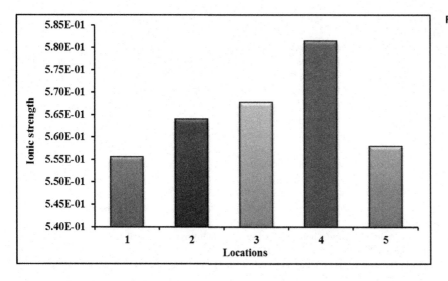

FIG. 24.7 Ionic strength of the metals.

background information on toxic metals and the possible sources in the Nakdong River Basin. This work also emphasizes the importance of the combined studies for contamination evaluation indices and multivariate statistical methods in studying the sources and processes of groundwater contamination. Further studies are in progress to determine the level of contaminants in groundwater sources using other techniques such as modeling or isotopic studies.

ACKNOWLEDGMENT

This research was supported by the Basic Science Research Program through the National Research Foundation of Korea (NRF) funded by the Ministry of Education (No. 2016R1D1A3B03934558).

REFERENCES

Abiye, T., Leshomo, J., 2014. Metal enrichment in the groundwater of the arid environment in South Africa. Environ. Earth Sci. 72, 4587–4598.

Ahn, J.S., 2012. Geochemical occurrences of arsenic and fluoride in bedrock groundwater: a case study in Geumsan County, Korea. Environ. Geochem. Health 34, 43–54.

APHA, 1995. Standard methods for the examination of water and wastewater, 19th edn. American Public Health Association, New York.

Appelo, C.A.J., Postma, D., 1999. Geochemistry, groundwater and pollution. AA Balkema, Rotterdam. 536.

Applin, K.R., Zha, 1989. The kinetics of Fe (II) oxidation and well screen encrustation. Ground Water 27, 168–174.

Arkoc, O., 2014. Heavy Metal Concentrations of Groundwater in the East of Ergene Basin, Turkey. Environ. Cont. and Toxic 93, 429–433.

Azaza, F.H., Ketata, M., Bouhlila, R., Gueddari, M., Riberio, L., 2011. Hydrochemical characteristics and assessment of drinking water quality in Zeuss-Koutine aquifer, southeastern Tunisia. Environ. Monit. Assess. 174, 283–298.

Backman, B., Bodis, D., Lahermo, P., Rapant, S., Tarvainen, T., 1997. Application of a groundwater contamination index in Finland and Slovakia. Environ. Geol. 36, 55–64.

Bahar, M.M., Yamamuro, M., 2008. Assessing the influence of watershed land use patterns on major ion chemistry of river waters in the Shimousa Upland Japan. Chem. Ecol. 24 (5), 341–355.

Bhuiyan, M.A.H., Parvez, L., Islam, M.A., Dampare, S.B., Suzuki, S., 2010. Heavy metal pollution of coal mine affected agricultural soils in the northern part of Bangladesh. Hazard. Mater. 173, 384–392.

Caboi, R., Cidu, R., Fanfani, L., Lattanzi, P., Zuddas, P., 1999. Environmental mineralogy and geochemistry of the abandoned Pb-Zn Montevecchio-Ingurtosu mining district, Sardinia, Italy. Chronicle of mineral research & exploration 534, 21–28.

Chun, K.C., Chang, R.W., Williams, G.P., Chang, Y.S., Tomasko, D., LaGory, K., Ditmars, J., Chu, H.D., Lee, B.K., 2001. Water quality issues in the Nakdong River Basin in the Republic of Korea. Environ. Eng. Policy 2, 131–143.

Cloutier, V., Lefebvre, R., Therrien, R., Savard, M.M., 2008. Multivariate statistical analysis of geochemical data as indicative of the hydrogeochemical evolution of groundwater in a sedimentary rock aquifer system. J. Hydrol. 353 (3), 294–313.

Dalton, M.G., Upchurch, S.B., 1978. Interpretation of hydrochemical facies by factor analysis. Ground Water 16 (4), 228–233.

Deutsch, W.J., 1997. Groundwater geochemistry. Fundamentals and applications to contamination. Lewis, New York, p. 221.

Edet, A.E., Offiong, O.E., 2002. Evaluation of water quality pollution indices for heavy metal contamination monitoring. A study case from Akpabuyo-Odukpani area, Lower Cross River Basin (southeastern Nigeria). Geo. Journal. 57, 295–304.

Edet, A., Ukpong, A., Nganje, T., 2013. Hydrochemical studies of cross River Basin (southeastern Nigeria) river systems using cross plots, statistics and water quality index. Environ. Earth Sci. 70, 3043–3056.

Fergusson, J.E., 1990. The heavy metals: chemistry, environmental Impacts and health effects. Pergamon Press, Oxford, p. 613.

Ficklin, W.H., Plumlee, G.S., Smith, K.S., McHugh, J.B., 1992. Geo- chemical classification of mining drainages and natural drain- ages in mineralised areas. In: Kharaka, Y.K., Maest, A.S. (Eds.), Water-rock interaction, Balkema, Rotterdam, 7, 381–384.

Handa, B., 1981. Hydro geochemistry water quality and water pollution in U.P. Tech. Report, CGWB, Min. of Irrigation, p. 317.

Hem, J.D., 1991. Study and interpretation of the chemical characteristics of natural water, 3rd ed. US Geological Survey Water-Supply, p. 2254.

Horton, R.K., 1965. An index system for rating water quality. J Water Poll. Cont. Feder. 37, 300–306.

Hunt, L.E., Howard, A.G., 1994. Arsenic speciation and distribution in the CarnonEstuary following the acute discharge of contaminated water from a disused mine. Mar. Pollut. Bull. 28 (1), 33–38.

Jahanshahi, R., Zare, M., 2015. Assessment of heavy metals pollution in groundwater of Golgohar iron ore mine area, Iran. Environ. Earth Sci. https://doi.org/10.1007/s12665-015-4057-8.

Kaiser, H.E., 1960. The application of electronic computers to factor analysis. Edu. Psychol. Measure. 20, 141–151.

Koterba, M.T., Wilde, F.D., Lapham, W.W., 1995. Groundwater data collection protocols and procedures for the national water quality assessment program—collection and documentation of water quality samples and related data. US Geological Survey Open- File Report 95–399, p. 113.

Krishna, A., Satyanarayanan, P., Pradip, K.G., 2009. Assessment of heavy metal pollution inwater using multivariate statistical techniques in an industrial area: A case study from Patancheru, Medak District, Andhra Pradesh, India. Hazard. Mater. 167, 366–373.

Leung, C.M., Jiao, J.J., 2006. Heavy metal and trace element distributions in groundwater in natural slopes and highly urbanized spaces in min-levels area, Hong Kong. Water Res. 40, 753–767.

Lorite-Herrer, M., Jimenez-Espinosa, R., Jimeneze-Millan, J., Hiscock, K.M., 2008. Integrated hydrochemical assessment of the Quaternary alluvial aquifer of the Guadalquivir River, Southern Spain. Appl. Geochem. 23, 2040–2054.

Manish, K., Ramanathan, A., Rao, M.S., Kumar, B., 2006. Identification and evaluation of hydrogeochemical processes in the ground- water environment Of Delhi. India. Environ. Geol. 50, 1025–1039.

Matalas, C.N., Reiher, J.B., 1967. Some comments on the use of factor analysis. Water Resour. Res. 3, 213–223.

Mohan, S.V., Nithila, P., Reddy, S.J., 1996. Estimation of heavy metal in drinking water and development of heavy metal pollution index. Environ. Sci. Health. 31, 283–289.

Parizi, H.S., Samani, N., 2013. Geochemical evolution and quality assessment of water resources in the Sarcheshmeh copper mine area (Iran) using multivariate statistical techniques. Environ. Earth Sci. 69, 1699–1718.

Prasad, B., Bose, J.M., 2001. Evaluation of the heavy metal pollution index for surface and spring water near a limestone mining area of the lower Himalayas. Environ. Geol. 41, 183–188.

Prasad, B., Jaiprakas, K.C., 1999. Evaluation of heavy metals in ground water near mining area and development of heavy metal pollution index. Environ. Sci. Health. A34 (1), 91–102.

Prasanna, M.V., Praveena, S.M., Chidambaram, S., Nagarajan, R., Elayaraja, A., 2012. Evaluation of water quality pollution indices for heavy metal contamination monitoring: a case study from Curtin Lake, Miri City, East Malaysia. Environ. Earth Sci. 67, 1987–2001.

Venkatramanan, S., Chung, S.Y., Ramkumar, T., Kim, B.W., 2015. An assessment of selected hydrochemical parameter trend of the Nakdong River water in South Korea, using time series analyses and PCA. Environ. Monit. Assess. 187, 4192.

Venkatramanan, S., Chung, S.Y., Kim, T.H., Prasanna, M.V., Hamm, S.Y., 2014. Assessment and Distribution of Metals Contamination in Groundwater: a Case Study of Busan City, Korea. Water Qual. Expo. Health. https://doi.org/10.1007/s12403-014-0142-6.

Stamatis, G., Voudouris, K., Karefilakis, F., 2001. Groundwater pollution by heavy metals in historical mining area of Lavrio, Attica, Greece. *Water, Air, Soil Pollu.* 128, pp. 61–83.

USEPA (US Environmental Protection Agency), 2012. 2012 Edition of the drinking water standards and health advisories. EPA, Washington.

White, A.F., Benson, S.M., Yee, A.W., Woolenberg, H.A., Flexser, S., 1991. Groundwater contamination at the Kesterson reservoir, California—Geochemical parameters influencing selenium mobility. Water Resour. Res. 27, 1085–1098.

WHO, 2011. Guidelines for drinking water quality, 4th edn. World Health Organisation, Geneva.

Willet, P., 1987. Similarity and clustering in chemical information systems. Wiley, New York.

Section D

Application of GIS and Geostatistical Techniques in Groundwater Resources Management

Chapter 25

Delineation of Groundwater Potential Zones in Hard Rock Terrain Using Integrated Remote Sensing, GIS and MCDM Techniques: A Case Study From Vamanapuram River Basin, Kerala, India

R.G. Rejith[*,†], S. Anirudhan[‡] and M. Sundararajan[*,†]

*Minerals Section, Materials Science and Technology Division, National Institute for Interdisciplinary Science and Technology (CSIR-NIIST), Council of Scientific & Industrial Research, Thiruvananthapuram, Kerala, India, †Academy of Scientific and Innovative Research (AcSIR), CSIR-NIIST, Thiruvananthapuram, Kerala, India, ‡Department of Geology, University of Kerala, Kariavattom Campus, Thiruvananthapuram, Kerala, India

Chapter Outline

25.1 INTRODUCTION

Groundwater is the largest available fresh-water resource in the world. In India, 35% of the total usable water comes from a replenishable groundwater source. About 30% of the urban and 90% of the rural populations depend on groundwater for meeting their basic needs. Data on water consumption show that during the year 2006 in India it was 829 billion m^3, which is likely to increase to 1093 billion m^3 by 2025, and 1047 billion m^3 by 2050, as estimated by the Government of India (2009). As the potential for increasing the volume of water to meet the demand only amounts to 5%–10%, India is bound to face a severe shortage of water in the near future. Despite Kerala receiving the highest rainfall among the Indian States, the unfavorable topography (high slopes) and geology of the state (hard rocks) act as negative factors resulting in low groundwater potential if monsoon rains fail or no rain falls during summer time. In the coastal area there are problems like salt water intrusion and submergence during monsoon. In general, valleys are the most dependable sources of drinking water for many people in the country.

Numerous studies exist on the delineation of groundwater potential zones in the literature with reference to India (Sharma and Jugran, 1992; Prakash and Mishra, 1993; Obi Reddy et al., 1994; Krishnamurthy and Srinivas, 1995; Tiwari and Rai, 1996; Pal et al., 1997; Ravindran and Jeyaram, 1997; Jain, 1998; Kumar et al., 1999; Rao and Reddy, 1999; Pratap et al., 2000; Rao et al., 2001; Sankar, 2002), including some for Kerala (Dinesh Kumar et al., 2007; Nair

GIS and Geostatistical Techniques for Groundwater Science. https://doi.org/10.1016/B978-0-12-815413-7.00025-0

et al., 2017; Preeja et al., 2011; Vijith, 2007). Currently, Remote Sensing (RS) and Geographic Information System (GIS) techniques are used to extract terrain details such as geomorphology, slope, land use/land cover, drainage patterns, lineaments, etc., which directly or indirectly control the occurrence and movement of groundwater (Chatterji et al., 1978; Jackson and Mason, 1986; Das et al., 1997; Nag, 1998; Rao et al., 2001). For locating suitable sites for groundwater exploration, multicontrolling factors need to be considered, especially in a hard-rock terrain. Remote sensing and GIS have advanced the prospecting strategies of groundwater far ahead in the modern scientific era. Usual procedures now in vogue include integrating thematic layers on GIS (Bahuguna et al., 2003). The relative importance of each thematic layer used for delineating groundwater potential zones varies from region to region. Generally, thematic layers, i.e., rainfall, lithology, drainage density, lineament density, and slope, are chosen as the active factors (Kaliraj et al., 2014; Nair et al., 2017). These thematic layers serve as the foundation for building a reliable picture of the groundwater distribution pattern in an area. Therefore assigning weight to each thematic layer is undoubtedly the most important act, which nowadays is solely based on expert decisions and is certainly more subjective. Recently, Chowdhury et al. (2013) and Jha et al. (2010) have successfully used the multi-criteria decision making (MCDM) technique for assigning weights to each parameter for mapping groundwater potential zones. The MCDM technique, like analytic hierarchy process (AHP), uses a pair-wise comparison matrix for analyzing different thematic layers (Chowdhury et al., 2013).

In this study RS, GIS, and MCDM techniques were used for the first time in a hard-rock terrain for delineating potential zones and groundwater budget estimation during the summer months because it is during this time that the country witnesses severe drinking-water shortage. Further, nine thematic layers were used for GIS integration by assigning the appropriate weightage according to AHP analysis.

25.2 STUDY AREA

The State of Kerala is traversed from east to west by number of rivers (Rahmati et al., 2015) originating from the Western Ghats. The Vamanapurm River is one such river located in the southern part. The long axis of the basin (Fig. 25.1) trends in an east-to-west direction covering all of the physiographic divisions of the state. The main channel length is 54 km, starting

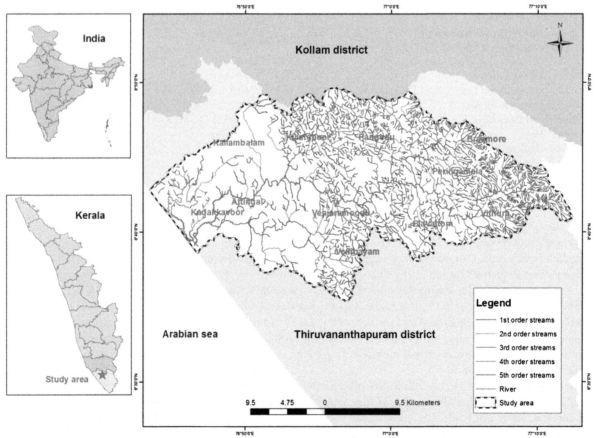

FIG. 25.1 Study area—Vamanapuram River Basin, Kerala, India.

its course from the Chemmunji Mottai (1717 m, msl) of the Western Ghats and ending up in the Anjengo Lake near Chirayinkizhu in the Thiruvananthapuram District. The basin has a catchment area of 767.32 km², located mainly in the Thiruvananthapuram District and a small part falls in the Kollam District of Kerala State. It includes Nedumangad Taluk of the Thiruvananthapuram District in the south, Kottarakkara Taluk of the Kollam District in the north, the State of Tamil Nadu in the east, and the Arabian Sea in the west. A large area of Vamanpuram River basin falls in the midland terrain of the state, which is characterized by lateritic uplands with undulating topography and intervening valleys. The altitudes vary from 40 m in the northwestern parts to about 300 m in the eastern and southeastern parts. The stream originates from the foothills of the Ponmudi hills and ends up in the Arabian Sea at Mudalapallipozhi near Perumathura in the Thiruvananthapuram District.

25.3 MATERIALS AND METHODS

Thematic layers were prepared from conventional maps and remote sensing data. Nine spatial parameters, i.e., geology, geomorphology, slope, land use/land cover, lineament density, drainage density, soil, curvature, and rainfall, were analyzed using the AHP approach to explore the potential zone for groundwater recharge in the Vamanapuram River basin. The complete process of the groundwater potential prediction is shown in Fig. 25.2.

25.3.1 Groundwater Potential Index

The groundwater potential index (GWPI) is a dimensionless quantity estimated using a weighted linear combination method (Malczewski, 1999; Shekhar and Pandey, 2015). Mathematically, the GWPI is expressed as:

$$\text{GWPI} = \sum_{j=1}^{m} \sum_{i=1}^{n} \left(w_j \times x_i \right) \tag{25.1}$$

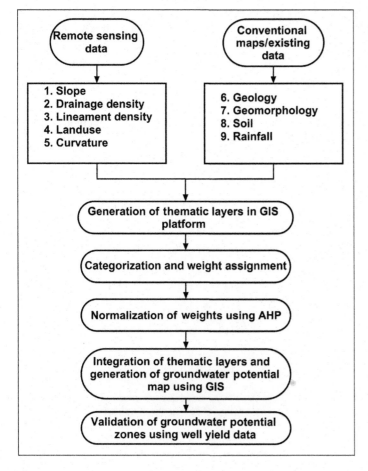

FIG. 25.2 Flowchart showing the methodology adopted in this study.

where, w_j is the normalized weight of the jth thematic layer, x_i is the normalized weight of the ith class with respect to the j layer, m is the total number of themes, and n is the total number of classes in a theme. The summer groundwater potential map was prepared based on the GIS-based AHP technique. GWPI values were grouped into five classes of very poor, poor, moderate, good, and very good.

25.3.2 Preparation of Thematic Layers

The thematic layers were prepared from various sources and projected with UTM-WGS 84 projection and coordinate system. Published maps of Geological Survey of India were digitized in ArcGIS software to make the thematic layers of geology and geomorphology. The soil map was created from the published map of the Kerala Soil Survey Organization, 2001. The slope, profile curvature, and the drainage layers were prepared from ASTER DEM. The lineament map of the area was produced from a Landsat ETM+ image and ASTER DEM. The lineament-density and drainage-density layers were calculated using the density tool in the ArcGIS software (Suresh et al., 2015; Nair et al., 2017). The land-use map was prepared from Landsat ETM+ satellite using the maximum likelihood classification algorithm with selected training samples from known locations (Kaliraj et al., 2014). Monthly rainfall data (2011–15) from two India Meteorological Department stations at Nedumangad and Varkala within the Vamanapuram River Basin were used to prepare the rainfall distribution map using inverse distance weighting (IDW) interpolation technique available in the ArcGIS software.

25.3.3 Assignment of Normalized Weights to Thematic Layers

The GIS-based AHP method was used to determine the weights of the thematic layers for demarcating the potential groundwater recharge zones (Saaty, 1980). Saaty's AHP is a widely accepted MCDM technique for analyzing complex spatial decision problems in the field of natural resources, groundwater, and environmental management. All the nine thematic layers are compared against each other using Saaty's 1–9 scale in a pair-wise comparison matrix (Table 25.1) and the AHP method was used to estimate the relative weights according to the importance of the thematic layers and to test for consistency ratio (CR). These weightings are actually normalized values of the eigenvectors associated with maximum eigenvalues of the ratio matrix (Jha et al., 2010; Adiat et al., 2012) (Table 25.2).

The CR is calculated using the following equation:

$$CR = \frac{CI}{RI}$$

(25.2)

where CI is the consistency index and RI is the random index. The value of random index depends on the order of the matrix. The value of CR must be < 0.1 (Saaty, 1980; Malczewski, 1999).

TABLE 25.1 Pair-Wise Comparison Matrix

Theme	GE	GM	LU	SO	DD	SL	LD	CU	RF
GE	1	1/4	1/4	1/3	1/4	1/4	1/4	1/4	2
GM	4	1	3	3	1/2	1/2	3	1/2	4
LU	4	1/3	1	4	1/2	1/3	3	1/3	4
SO	3	1/3	1/4	1	1/3	1/4	1/3	1/3	4
DD	4	2	2	3	1	1/2	2	1/2	5
SL	4	2	3	4	2	1	3	2	6
LD	4	1/3	1/3	3	1/2	1/3	1	1/2	4
CU	4	2	3	3	2	1/2	2	1	4
RF	1/2	1/4	1/4	1/4	1/5	1/6	1/4	1/4	1

GE, geology; GM, geomorphology; LU, land use/land cover; SO, soil; DD, drainage density; SL, slope; LD, lineament density; CU, curvature; RF, rainfall.

TABLE 25.2 Normalized Weights for Thematic Layers

Theme	GE	GM	LU	SO	DD	SL	LD	CU	RF	Normalized Weights
GE	0.04	0.03	0.02	0.02	0.03	0.07	0.02	0.04	0.06	0.04
GM	0.14	0.12	0.23	0.14	0.07	0.13	0.20	0.09	0.12	0.14
LU	0.14	0.04	0.08	0.19	0.07	0.09	0.20	0.06	0.12	0.11
SO	0.11	0.04	0.02	0.05	0.05	0.07	0.02	0.06	0.12	0.06
DD	0.14	0.24	0.15	0.14	0.14	0.13	0.13	0.09	0.15	0.15
SL	0.14	0.24	0.23	0.19	0.27	0.26	0.20	0.35	0.18	0.23
LD	0.14	0.04	0.03	0.14	0.07	0.09	0.07	0.09	0.12	0.09
CU	0.14	0.24	0.23	0.14	0.27	0.13	0.13	0.18	0.12	0.18
RF	0.02	0.03	0.02	0.01	0.03	0.04	0.02	0.04	0.03	0.03

Consistency ratio (CR) = 0.067 < 0.1.
GE, geology; GM, geomorphology; LU, land use/land cover; SO, soil; DD, drainage density; SL, slope; LD, lineament density; CU, curvature; RF, rainfall.

The consistency index (CI) is calculated using the following equation:

$$CI = \frac{\lambda_{max} - n}{n - 1} \tag{25.3}$$

where λ is the largest eigenvalue of the matrix calculated from the matrix and n is the number of thematic layers. Here the value of n is 9.

The feature classes of each thematic layers were quantitatively weighted as very poor (weight = 1), poor (weight = 2), moderate (weight = 3), good (weight = 4), and very good (weight = 5). All the feature classes were assigned with a suitable weight and integrated with the geometric mean of the corresponding layer to derive a normalized weight for each of the parameters (Machiwal et al., 2011; Chowdary et al., 2013). The ranks assigned to different features of the individual themes and their normalized weights are presented in Table 25.3.

TABLE 25.3 Assigned and Normalized Weights of Different Features of Thematic Layers

Sl. No.	Criteria	Classes	Assigned Weight	Normalized Weight
1	Slope (degrees)	0–6.06	5	0.33
		6.07–11.1	4	0.27
		11.11–17.66	3	0.20
		17.67–27.25	2	0.13
		27.26–64.34	1	0.07
2	Lineament density (km/km²)	0–0.15	2	0.17
		0.16–0.43	2	0.17
		0.44–0.73	2	0.17
		0.74–1.11	3	0.25
		1.12–2.1	3	0.25
3	Drainage density (km/km²)	0–0.94	5	0.33
		0.95–1.88	4	0.27
		1.89–2.73	3	0.20
		2.74–3.72	2	0.13
		3.73–6.16	1	0.07

Continued

TABLE 25.3 Assigned and Normalized Weights of Different Features of Thematic Layers—cont'd

Sl. No.	Criteria	Classes	Assigned Weight	Normalized Weight
4	Geology	Khondalite group of rocks	3	0.13
		Basic rocks	2	0.09
		Charnockite group of rocks	2	0.09
		Laterite	4	0.17
		Migmatite complex	2	0.09
		Sand and silt	5	0.22
		Sandstone and clay with lignite interc	5	0.22
5	Mean summer rainfall (mm)	366.08–366.72	1	0.07
		366.73–367.6	2	0.13
		367.61–368.51	3	0.20
		368.52–369.48	4	0.27
		369.49–370.48	5	0.33
6	Land use/land cover	Paddy	4	0.14
		Mixed crops	4	0.14
		Plantations	2	0.07
		Fallow land	2	0.07
		Built-up land	4	0.14
		Barren rock/stone quarry	1	0.03
		Forest	2	0.07
		Water body	5	0.17
		Beach/sandy area	5	0.17
7	Geomorphology	Pediplain Weathered/buried	3	0.11
		Coastal plain	5	0.18
		Denudational hills	2	0.07
		Denudational structural hills	2	0.07
		Lower plateau (lateritic)	4	0.14
		Piedmont zone	3	0.11
		Residual hill	4	0.14
		Water body	5	0.18
8	Soil	Alluvium	4	0.22
		Coastal alluvium	5	0.28
		Forest soil	3	0.17
		Hill soil	2	0.11
		Laterite soil	4	0.22
9	Curvature	Convex	1	0.17
		Flat	2	0.33
		Concave	3	0.50

25.3.4 Assessment of Groundwater Potential Map

The accuracy of the groundwater potential map derived using a GIS-based AHP method was determined using the receiver operating characteristic (ROC) method (Pradhan, 2009; Mohammady et al., 2012; Pradhan, 2013; Pourtaghi and Pourghasemi, 2014; Rahmati et al., 2015; Pourghasemi et al., 2013). The well-yield data for 60 open wells were used for verifying the accuracy of groundwater potential map. During the summer, these 60 wells experience a severe decline in the water depth (Fig.25.3). The area under curve (AUC) determines the prediction accuracy. AUC values of 0.5–0.6, 0.6–0.7, 0.7–0.8, 0.8–0.9, and 0.9–1 represent a prediction accuracy of poor, average, good, very good, and excellent (Yesilnacar, 2005).

25.3.5 Groundwater Budget

The summer groundwater storage and water demand were calculated to analyze the groundwater budget for the potential zones during the season of summer. The groundwater storage (GWS) available during summer for different zones was estimated to find out how far the demand for drinking water is met by following the procedure adopted by Raghunath (Raghunath, 2007):

$$GWS = WC_T \times SC \times A \qquad (25.4)$$

where WC_T is the average thickness of the water column during summer, SC is the storage coefficient, and A is the area of summer groundwater potential zones derived using GIS-based AHP and remote sensing methods.

The summer water demand (WD) for each of the potential zones were estimated as follows:

$$WD = PCC \times PD \times A \qquad (25.5)$$

where PCC is the per capita consumption of water for domestic use, PD is the population density, and A is the area of the groundwater potential zones. The per capita consumption of water for domestic use for a single day is taken as 150 Lpd.

25.4 RESULTS AND DISCUSSION

Nine thematic layers are integrated with one another in accordance with their relative importance using a GIS-based AHP technique. The resultant groundwater potential zones are classified into (i) very good, (ii) good, (iii) moderate, (iv) poor, and (v) very poor classes.

25.4.1 Slope

The slope of a physical feature or topography refers to the amount of inclination of that surface to the horizontal. Slope analysis is an important parameter for the identification of groundwater conditions (Al Saud, 2010; Ettazarini, 2007). The degree of slope exhibited by the watershed varies from 0 to 64.34 degrees and is grouped into five classes: 0–6.06, 6.07–11.1, 11.11–17.66, 17.67–27.25, and 27.26–64.34 degrees (Fig. 25.4A). A higher degree of slope results in rapid runoff and an increased erosion rate with feeble recharge potential, i.e., infiltration is inversely related to the slope.

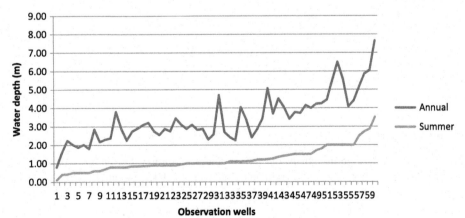

FIG. 25.3 Trend of water depth during annual and summer rainfall conditions.

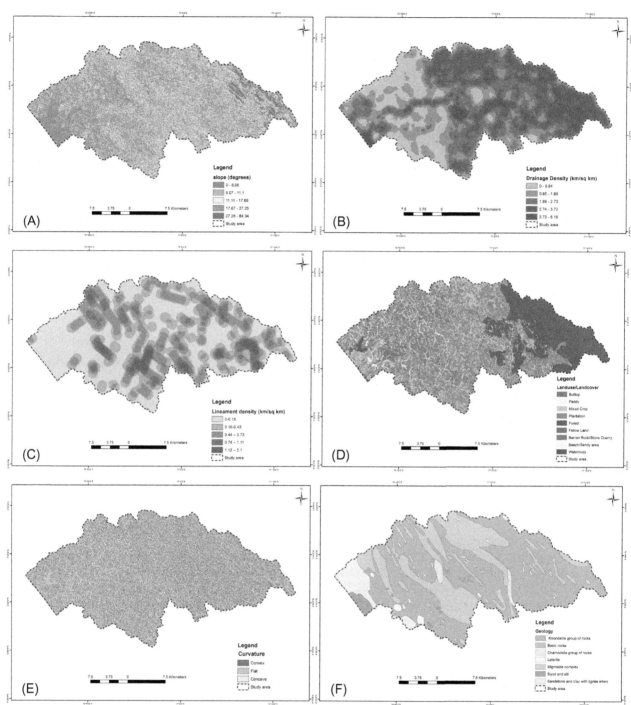

FIG. 25.4 Thematic layers. (A) Slope, (B) drainage density, (C) lineament density, (D) land use/land cover, (E) curvature, (F) geology,

(Continued)

25.4.2 Drainage Density

The drainage system of an area is determined by the nature and structure of the bedrock, kind of vegetation, rainfall absorption capacity of soils, infiltration, and slope gradient (Manap et al., 2013). An area with low-drainage-density causes more infiltration and decreased surface runoff and is suitable for groundwater development (Dinesh Kumar et al., 2007; Magesh et al., 2012). The drainage density is the ratio of the sum of the lengths of the streams to the size of the area of the grid under consideration (Adiat et al., 2012; Mogaji et al., 2015). Based on the surface-drainage density, the study area is

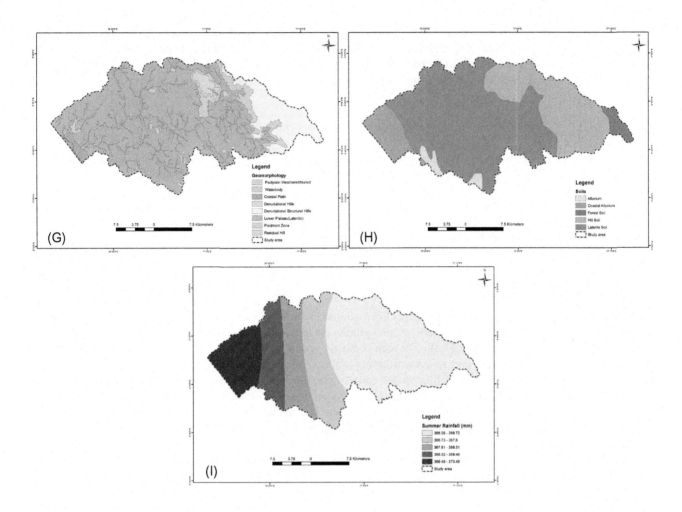

FIG. 25.4, CONT'D (G) geomorphology, (H) soil, (I) summer rainfall.

grouped into five classes: 0–0.94, 0.95–1.88, 1.89–2.73, 2.74–3.72, and 3.73–6.16 km/km^2 (Fig. 25.4B). Very high drainage density is scattered in the central and eastern part and low drainage densities cover the western part, which favors groundwater occurrence.

25.4.3   Lineament Density

The lineaments are linear features on the Earth's surface that reflect a general surface expression of underground fractures (Pradhan et al., 2006; Pradhan and Youssef, 2010). These fractures facilitate infiltration of surface runoff into the subsurface and are of great relevance to the storage and movement of groundwater (Devi et al., 2001). The lineament density was defined as the total length of all the recorded lineaments divided by the area under consideration (Edet et al., 1998). The lineament density of the study area can be grouped into five classes: 0–0.15, 0.16–0.43, 0.44–0.73, 0.74–1.11, and 1.12–2.1 km/km^2 (Fig. 25.4C). Higher values of lineament density favor groundwater potential.

25.4.4   Land Use/Land Cover

The following classes of land-use/land-cover types were distinguished: (i) built-up, (ii) paddy crops, (iii) mixed crops, (iv) plantation, (v) forest, (vi) fallow land, (vii) barren rock/stone quarry, (viii) beach/sandy area, and (ix) water body (Fig. 25.4D). Most of the area is represented by plantation (33.16%) and mixed crops (24.01%) followed by forest

(22.31%), built-up land (14.76%), paddy (3.38%), water body (1.02%), barren rock/stone quarry (0.69%) and fallow land (0.66%). Water bodies and beach/sandy area were given the highest ranking over other land-use features because of their continuous recharge to ground, followed by paddy crops, mixed crops, and built-up land. The lowest rank was given to barren rock/stone quarry, as it lacks vegetation cover.

25.4.5 Curvature

The curvature displays the shape or curvature of the slope of the drainage basin where it helps to demonstrate the erosion and runoff processes. The profile curvature values represent the morphology of the topography with a positive curvature attesting to an upwardly concave and a negative curvature, which indicates upwardly convex, and a value zero, which indicates the flat surfaces (Fig. 25.4E). A concave slope retains more water for a longer period of rainfall (Lee and Pradhan, 2007; Pothiraj and Rajagopalan, 2013; Manap et al., 2014). Therefore concave surfaces are more favorable for groundwater occurrence compared to the convex slope.

25.4.6 Geology

Major geological formations of the study area were classified into seven categories: (1) khondalite group of rocks, (2) basic rocks, (3) charnockite group of rocks, (4) laterite, (5) migmatite complex, (6) sand and silt, and (7) sandstone and clay with lignite interference (Fig. 25.4F). Geologically the area is mostly occupied by pre-Cambrian crystalline rocks, i.e., khondalite and charnockites, which have undergone weathering to form laterite at varying depths, and also a part of the area is occupied by migmatite complex. Khondalites are essentially graphite-bearing garnetiferous biotite gneisses. There are patches of charnokites in the khondalite exposures. Hard crystalline rocks occupy a major portion of the basin except for the coastal area, which is made up of sedimentary rocks of the Tertiary age known as Warkallai formations. Hard rocks are aquicludes that do not hold or transmit water due to an absence of porosity but these become productive by virtue of secondary porosities like fractures, joints, or fault planes. The entire hard-rock region is covered by weathered materials of varying thickness, primarily the residual weathered products laterite and alluvium of fluvial origin from the Quaternary to Recent age. Groundwater in the crystalline terrain occurs under unconfined conditions in the weathered portion, i.e., laterite zone. The depth of water table depends upon the thickness of the laterite profile. Sedimentary formation of Warkallai groups of rocks, i.e., sand, silt, sandstone, and clay with lignite and overlying laterite were assigned the higher ranks while the basic, khondalite, and charnockite group of rocks were assigned the lowest ranks.

25.4.7 Geomorphology

A geomorphologic map depicting important geomorphic units and land-form features indicates the nature of groundwater occurrence in a terrain as these features are formed over a long period of surface processes affecting the diverse lithology and underlying structures. Geomorphologically, the study area consists of pediment, pediplain weathered/buried, coastal plain, denudational hills, denudational structural hills, lower plateau (lateritic), and residual hill (Fig. 25.4G). Denudational hills in the region are developed in mountain peaks or ridges, which are either continuous or breached in some parts by an antecedent stream, formed mainly by a slope retreat process. These slopes are essentially weathering limited slopes where in situ weathering product seldom develops leaving the peaks of the hills with barren rocks and slopes with debris of varied sizes. Isolated hills and elevated landforms that are undergoing dilapidation due to weathering processes are categorized as residual hills (RH). These landforms exhibit soil cover and regolith of varying thickness that depends upon its evolutionary stage. The next important geomorphic unit of the slope down from the hilly terrain and consisting of wash slope and toe or base (Wallace, 1977) is collectively designated here as a pediment slope. A pediment is mainly a broad gently-sloping, rock-floored, erosional surface of low relief between the hills and plains. It is comprised of varied lithology and criss-crossed by fractures and faults. A buried pediment is a pediment covered essentially with relatively thicker alluvial, colluvial, or weathered materials. A pediment is a gently-inclined slope of transportation and/or erosion that truncates rock and connects eroding slopes or scarps to the areas of sediment deposition at lower levels (Oberlander, 1989). Structural hills are linear to arcuate hills showing definite trend-lines with varying lithology associated with folding, faulting, etc. Coastal plain, lower plateau (lateritic) and residual hill are good for groundwater occurrence followed by pediplain and peidmont zones.

25.4.8   Soil

Five soils, i.e., alluvium soil, coastal alluvium soil, forest soil, hill soil, and laterite soil (Fig. 25.4H) are found in the study area. Laterite soils are the weathering products of rock in which several forms of weathering and mineral transformations take place. A large part of the study area is covered by laterite soil. Coastal alluvium soils are of marine origin and are seen along the coastal plains and basin lands as a narrow strip. Alluvium soils are formed from fluvial sediments of lacustrine or riverine sediments. Forest soils are seen under forest cover and are formed from crystalline rocks of Archaean age. The hill soils are mostly seen at an elevation of above 80 m MSL. Coastal alluvium favors the occurrence of groundwater followed by alluvium and laterite soils. Hill soil is least favorable to the occurrence of groundwater.

25.4.9   Rainfall

Rainfall is the ultimate source for all potable waters in land. A region with a high rainfall levels is considered as an area with higher water potential. But this statement is not valid in hard-rock terrain where only the weathered portion acts as an aquifer. Here the groundwater resource, especially one that exists in an unconfined condition, depends on the nature and frequency of the rainfall (Magesh et al., 2012; Shekhar and Pandey, 2015). Although the Kerala State enjoys a tropical climate with an average annual rainfall of 2500 mm, a large volume of water flows as surface run off into the sea due to the terrain's steep slopes. Further, unfavorable groundwater storage conditions exist in this area due to the widespread presence of hard crystalline rocks. Therefore the frequency and duration of the rainfall significantly affect the groundwater potential (Adiat et al., 2012). Kerala has nonrainy and rainy seasons. The nonrainy season starts in December and lasts up to May, and there is a severe shortage of water in many places from March to May (Krishnakumar et al., 2009). During the summer time, Kerala receives the least rainfall and this causes a sudden depletion in the amount of available groundwater. The rainy season lasts for the other half of the year (June to November) during which time heavy rainfall occurs due to the southwest monsoon (locally known as Edavapathi) and the northeast monsoon (locally known as Thulavarsham (Dili et al., 2010). The resulting maps showing the distribution of summer rainfall were classified into five major classes (Fig. 25.4I). Based on the maps, the average summer rainfall in the elevated areas is relatively low compared with the areas with low elevation.

25.4.10   Groundwater Potential Map

The summer groundwater potential map was prepared based on the GIS-based AHP technique (Fig. 25.5). The GWPI values were classified into five zones of very poor (VPZ=0.145–0.182), poor (PZ=0.183–0.205), moderate (MZ=0.206–0.222), good (GZ=0.223–0.239), and very good (VGZ=0.24–0.269) classes. The results also showed that 69.55 km^2 (9.08%), 108.60 km^2 (14.18%), 230.02 km^2 (30.03%), 163.02 km^2 (21.28%), and 194.68 km^2 (25.42%) of the area represent VPZ, PZ, MZ, GZ, and VGZ respectively.

The result reveals that around 357.70 km^2 (46.70%) of the total area has been assessed as VGZ and GZ for groundwater occurrence when average summer rainfall is used for analysis. However, during the summer season around 53.29% of the total areal extent is identified as VPZ, PZ, and MZ. The zones GZ and MZ are found as small isolated pockets distributed in the middle part of the basin (Fig. 25.5).The zones VGZ and major portions of GZ are seen on the western parts of the study area where suitable topography and geology exist to hold groundwater for a longer time.

Unlike other zones, the areas showing VPZ remain same for the whole year at the extreme eastern side of the study area. The denudation hills located in this part of the study area are less suitable for groundwater occurrence because of hard rock at shallow depths in these parts. Moreover, these forest landforms have maximum lineament density ranges of between 1.12 km/km^2 and 2.1 km/km^2, maximum drainage density from 3.73 km/km^2 to 6.16 km/km^2, and steep slopes range from 27.26 to 64.34 degrees. The eastern part of the basin with steep slopes and a high drainage density is categorized as a poor to very poor prospect zone. Even though this area receives the maximum annual rainfall of about 1774.24 mm, its low groundwater potential is due to the influence of other factors like geomorphology, slope, drainage density, etc. On comparing the geomorphology with that of the groundwater prospect map it is noticed that most of the valley fills are of VGZ to GPZ whereas the structural hills are MZ to VPZs.

The groundwater potential map was validated using receiver operating characteristic (ROC) analysis. The prediction curves obtained for groundwater assessment using AHP are shown in Fig. 25.6. The value of AUC obtained from ROC analysis is 0.824, which corresponds to the prediction accuracy of 82.4%. It implies that the GIS-based AHP model shows very good accuracy in predicting the groundwater potential. Moreover, it is concluded that the AHP model can be used as

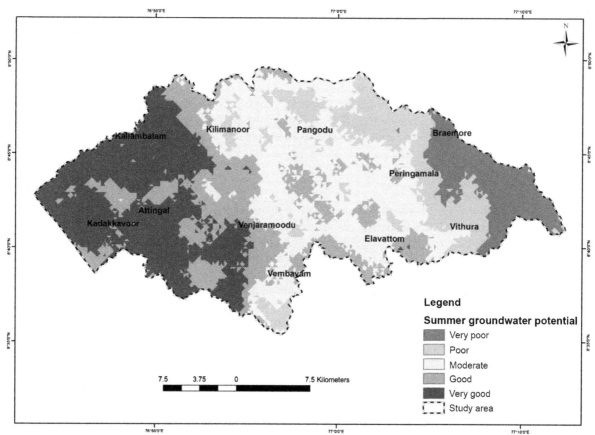

FIG. 25.5 Summer groundwater potential zone map of the study area.

a simple tool for the assessment of seasonal groundwater potential in a region with a tropical climate and hard rock geology.

25.4.11 Zone-Wise Groundwater Budgeting

The Table 25.4 shows the zone-wise groundwater storage and water demand for the summer season. The "very good" potential zone, i.e., the area with <5 degrees, mainly consists of valley fill aquifer (VFA) and foot slope aquifers (FSA). The total area for this category is 194.68 km^2, mostly towards the western part but with northwest-southeast trending residual hills in the middle of the basin. The average depth to water table in this category comes to around 5 m and the fluctuation of the water table is 3 m. The water column during peak summer was measure as 0.75 m. The groundwater storage is 14.60 × 10^6 m^3 where the water demand is estimated as 5.78 × 10^6 m^3. According to this estimate during the summer months in the valleys and foothills (area <5 degrees slopes) there is sufficient water available. But this quantity is not assured for long dry conditions because a large quantity drains out into nearby streams and channels daily.

The "good" and "moderate" zones have higher slopes (5–10 degrees) so that the storage depletes quickly due to seepage and daily use. Therefore many wells dry up early in this segment. The "poor" and "very poor" zones falls in land with high slope angle (>10 degrees). This part is the most critical area regarding water availability during the summer months. Saturated thickness of the aquifer depends upon the extent and depth of the weathered material. Most of the wells go dry during peak summer except those in a thick laterite profile where the water column is maintained at 1.0 m. The water column in this zone ranges from 0 m to 1.0 m only in summer. A schematic of the terrain characteristics of the midland and high land topography is shown in Fig. 25.7.

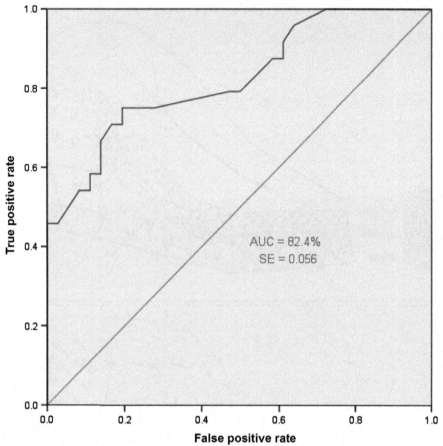

FIG. 25.6 Receiver operating characteristic curve for summer groundwater potential zones.

TABLE 25.4 Groundwater Budget in Summer Months for Vamanapuram River Basin

Potential Zone	Area (km²)	Av. Water Column Depth (m)	Storage Coefficient	GW Storage (m³)	Population Density (km⁻²)	Water Demand (m³)	Deficient/ Surplus (m³)
Very Good	194.68	0.75	0.10	14.60×10^6	2200.81	5.78×10^6	$+8.82 \times 10^6$
Good	163.02	0.50	0.05	4.08×10^6	2079.14	4.58×10^6	-0.5×10^6
Moderate	230.02	0.25	0.05	2.88×10^6	1108.99	3.44×10^6	-0.57×10^6
Poor	108.60	0.10	0.03	0.33×10^6	864.38	1.27×10^6	-0.94×10^6
Very Poor	69.55	0.05	0.03	0.10×10^6	661.98	0.62×10^6	-0.52×10^6

In an uneven hard rock terrain like that of the Vamanapuram River Basin, groundwater occurrence depends mainly on the frequency of rainfall, thickness of weathered stuff, and slope of the terrain. The lower slopes are composed of alluvial and colluvium matter, which is a mixture of fine-grained particles in coarse matrix. The specific yield of such material will be around 10%–15%. Therefore places with low slopes and plains are considered to be most suitable for groundwater occurrence in an uneven topography.

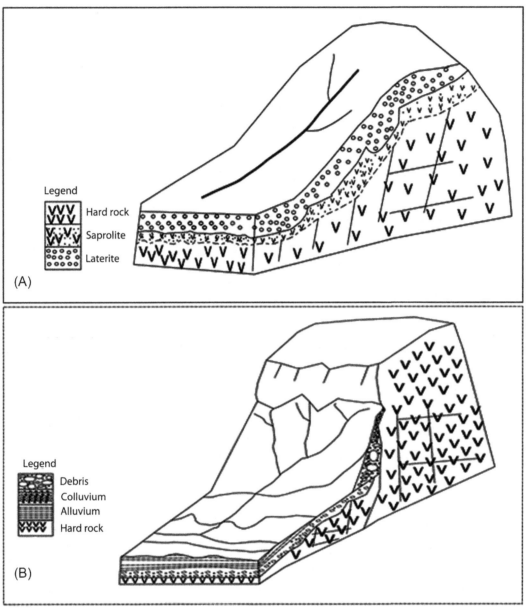

FIG. 25.7 Schematic of the terrain characteristics. (A) Midland topography. (B) High land topography.

25.5 CONCLUSION

GIS and AHP approaches were adopted to obtain spatial and seasonal variation of groundwater potential zones in the Vamanapuram River Basin in India. Nine thematic layers, i.e. geology, geomorphology, drainage density, lineament density, slope, curvature, landuse, soil, and rainfall, were integrated onto the GIS platform. The rainfall data for the summer season was considered for assessing the seasonal variation of groundwater occurrence in the basin against the usual procedure of considering the average annual rainfall for groundwater potential zonation. It identified around $357.70 \, \text{km}^2$ (46.70%) of the total area as VGZ and GPZ after considering the summer rainfall, which clearly indicates that the groundwater potential mapping is determined mostly by rainfall followed by geology, lineament density, drainage density, slope, etc. Almost half of the drainage basin area (53.29%) seems to be water stressed during summer months. Finally, the validation of results using ROC curve demonstrated that the AHP has fairly good predication accuracy. In summary, the results of this study proved that the GIS-based AHP approach could be successfully applied for the groundwater potential mapping. The potential zones can be prioritized for the construction of artificial recharge structures, such as check dams, farm ponds, etc. Thus the study provides valuable tools for the planners in water resource management to take mitigating measures to overcome the water stress especially during the summer time.

ACKNOWLEDGMENTS

This work would not have been possible without the fund granted to the second author. The second author is grateful to Kerala State Council for Science Technology and Environment (KSCSTE) Govt. of Kerala for funding for this research work (Council order No. 1178/2015/KSCSTE). He is also thankful to Dr. Kamalashan Kokal of KSCSTE for valuable and critical comments and suggestions during the course of this work.

REFERENCES

Adiat, K.A.N., Nawawi, M.N.M., Abdullah, K., 2012. Assessing the accuracy of GIS-based elementary multi criteria decision analysis as a spatial prediction tool – a case of predicting potential zones of sustainable groundwater resources. J. Hydrol. 440, 75–89.

Al Saud, M., 2010. Mapping potential areas for groundwater storage in Wadi Aurnah Basin, western Arabian Peninsula, using remote sensing and geographic information system techniques. Hydrogeol. J. 18 (6), 1481–1495.

Bahuguna, I.M., Nayak, S., Tamilarsan, V., Moses, J., 2003. Groundwater prospective zones in basaltic terrain using remote sensing. J. Indian Soc. Remote Sens. 31 (2), 101–105.

Chatterji, P.C., Singh, S., Qureshi, Z.H., 1978. Hydrogeomorphology of the Central Luni Basin, Western Rajasthan, India. Geoforum 9 (3), 211–224.

Chowdary, V.M., Chakraborthy, D., Jeyaram, A., Murthy, Y.K., Sharma, J.R., Dadhwal, V.K., 2013. Multi-criteria decision making approach for watershed prioritization using analytic hierarchy process technique and GIS. Water Resour. Manag. 27 (10), 3555–3571.

Das, S., Behera, S.C., Kar, A., Narendra, P., Guha, S., 1997. Hydrogeomorphological mapping in ground water exploration using remotely sensed data—a case study in keonjhar district, Orissa. J. Indian Soc. Remote Sens. 25 (4), 247–259.

Devi, P.S., Srinivasulu, S., Raju, K.K., 2001. Hydrogeomorphological and groundwater prospects of the Pageru river basin by using remote sensing data. Environ. Geol. 40 (9), 1088–1094.

Dili, A.S., Naseer, M.A., Varghese, T.Z., 2010. Passive control methods of Kerala traditional architecture for a comfortable indoor environment: comparative investigation during various periods of rainy season. Build. Environ. 45 (10), 2218–2230.

Dinesh Kumar, P.K., Gopinath, G., Seralathan, P., 2007. Application of remote sensing and GIS for the demarcation of groundwater potential zones of a river basin in Kerala, southwest coast of India. Int. J. Remote Sens. 28 (24), 5583–5601.

Edet, A.E., Okereke, C.S., Teme, S.C., Esu, E.O., 1998. Application of remote-sensing data to groundwater exploration: a case study of the Cross River state, southeastern Nigeria. Hydrogeol. J. 6 (3), 394–404.

Ettazarini, S., 2007. Groundwater potentiality index: a strategically conceived tool for water research in fractured aquifers. Environ. Geol. 52 (3), 477–487.

Jackson, M.J., Mason, D.C., 1986. The development of integrated geo-information systems. Int. J. Remote Sens. 7 (6), 723–740.

Jain, P.K., 1998. Remote sensing techniques to locate ground water potential zones in upper Urmil River basin, district Chhatarpur—Central India. J. Indian Soc. Remote Sens. 26 (3), 135–147.

Jha, M.K., Chowdary, V.M., Chowdhury, A., 2010. Groundwater assessment in Salboni block, West Bengal (India) using remote sensing, geographical information system and multi-criteria decision analysis techniques. Hydrogeol. J. 18 (7), 1713–1728.

Kaliraj, S., Chandrasekar, N., Magesh, N.S., 2014. Identification of potential groundwater recharge zones in Vaigai upper basin, Tamil Nadu, using GIS-based analytical hierarchical process (AHP) technique. Arab. J. Geosci. 7 (4), 1385–1401.

Krishnakumar, K.N., Rao, G.P., Gopakumar, C.S., 2009. Rainfall trends in twentieth century over Kerala, India. Atmos. Environ. 43 (11), 1940–1944.

Krishnamurthy, J., Srinivas, G., 1995. Role of geological and geomorphological factors in ground water exploration: a study using IRS LISS data. Int. J. Remote Sens. 16 (14), 2595–2618.

Kumar, A., Tomar, S., Prasad, L.B., 1999. Analysis of fractures inferred from DBTM and remotely sensed data for groundwater development in Godavari sub-watershed, Giridih, Bihar. J. Indian Soc. Remote Sens. 27 (2), 105–114.

Lee, S., Pradhan, B., 2007. Landslide hazard mapping at Selangor, Malaysia using frequency ratio and logistic regression models. Landslides 4 (1), 33–41.

Machiwal, D., Jha, M.K., Mal, B.C., 2011. Assessment of groundwater potential in a semi-arid region of India using remote sensing, GIS and MCDM techniques. Water Resour. Manag. 25 (5), 1359–1386.

Magesh, N.S., Chandrasekar, N., Soundranayagam, J.P., 2012. Delineation of groundwater potential zones in Theni district, Tamil Nadu, using remote sensing, GIS and MIF techniques. Geosci. Front. 3 (2), 189–196.

Malczewski, J., 1999. GIS and Multicriteria Decision Analysis. John Wiley & Sons.

Manap, M.A., Sulaiman, W.N.A., Ramli, M.F., Pradhan, B., Surip, N., 2013. A knowledge-driven GIS modeling technique for groundwater potential mapping at the upper Langat Basin, Malaysia. Arab. J. Geosci. 6 (5), 1621–1637.

Manap, M.A., Nampak, H., Pradhan, B., Lee, S., Sulaiman, W.N.A., Ramli, M.F., 2014. Application of probabilistic-based frequency ratio model in groundwater potential mapping using remote sensing data and GIS. Arab. J. Geosci. 7 (2), 711–724.

Mogaji, K.A., Lim, H.S., Abdullah, K., 2015. Regional prediction of groundwater potential mapping in a multifaceted geology terrain using GIS-based Dempster–Shafer model. Arab. J. Geosci. 8 (5), 3235–3258.

Mohammady, M., Pourghasemi, H.R., Pradhan, B., 2012. Landslide susceptibility mapping at Golestan Province, Iran: a comparison between frequency ratio, Dempster–Shafer, and weights-of-evidence models. J. Asian Earth Sci. 61, 221–236.

Nag, S.K., 1998. Morphometric analysis using remote sensing techniques in the Chaka sub-basin, Purulia district, West Bengal. J. Indian Soc. Remote Sens. 26 (1), 69–76.

Nair, H.C., Padmalal, D., Joseph, A., Vinod, P.G., 2017. Delineation of groundwater potential zones in river basins using geospatial tools—an example from southern Western Ghats, Kerala, India. J. Geovisual. Spat. Anal. 1 (1–2), 5.

Oberlander, T.M., 1989. Slope and pediment systems. In: Thomas, D.S.G. (Ed.), Arid Zone Geomorphology. Belhaven, London, pp. 58–59.

Obi Reddy, G.P., Suresh Babu, R., Sambasiva Rao, M., 1994. Hydrogeology and hydrogeo-morphological conditions of Anantapur district using remote sensing data. Indian Geog. J. 69 (2), 128–135.

Pal, D.K., Khare, M.K., Rao, G.S., Jugran, D.K., Roy, A.K., 1997. Demarcation of Groundwater Potential Zones Using Remote Sensing and GIS Techniques: A Case Study of Bala Valley in Parts of Yamunanagar and Sirmaur Districts. Indian Society of Remote Sensing-National Natural Resources Management System, Dehradun, India, pp. 395–402.

Pothiraj, P., Rajagopalan, B., 2013. A GIS and remote sensing based evaluation of groundwater potential zones in a hard rock terrain of Vaigai sub-basin, India. Arab. J. Geosci. 6 (7), 2391–2407.

Pourghasemi, H.R., Moradi, H.R., Aghda, S.F., 2013. Landslide susceptibility mapping by binary logistic regression, analytical hierarchy process, and statistical index models and assessment of their performances. Nat. Hazards 69 (1), 749–779.

Pourtaghi, Z.S., Pourghasemi, H.R., 2014. GIS-based groundwater spring potential assessment and mapping in the Birjand township, southern Khorasan Province, Iran. Hydrogeol. J. 22 (3), 643.

Pradhan, B., 2009. Groundwater potential zonation for basaltic watersheds using satellite remote sensing data and GIS techniques. Open Geosci. 1 (1), 120–129.

Pradhan, B., 2013. A comparative study on the predictive ability of the decision tree, support vector machine and neuro-fuzzy models in landslide susceptibility mapping using GIS. Comput. Geosci. 51, 350–365.

Pradhan, B., Youssef, A.M., 2010. Manifestation of remote sensing data and GIS on landslide hazard analysis using spatial-based statistical models. Arab. J. Geosci. 3 (3), 319–326.

Pradhan, B., Singh, R.P., Buchroithner, M.F., 2006. Estimation of stress and its use in evaluation of landslide prone regions using remote sensing data. Adv. Space Res. 37 (4), 698–709.

Prakash, S.R., Mishra, D., 1993. Identification of groundwater prospective zones by using remote sensing and geoelectrical methods in and around said-nagar area, dakor block, jalaun district, Uttar Pradesh. J. Indian Soc. Remote Sens. 21 (4), 217–227.

Pratap, K., Ravindran, K.V., Prabakaran, B., 2000. Groundwater prospect zoning using remote sensing and geographical information system: A case study in Dala-Renukoot area, Sonbhadra District, Uttar Pradesh. J. Indian Soc. Remote Sens. 28 (4), 249–263.

Preeja, K.R., Joseph, S., Thomas, J., Vijith, H., 2011. Identification of groundwater potential zones of a tropical river basin (Kerala, India) using remote sensing and GIS techniques. J. Indian Soc. Remote Sens. 39 (1), 83–94.

Raghunath, H.M., 2007. Groundwater, third ed. New Age International Publishers. 504 p.

Rahmati, O., Samani, A.N., Mahdavi, M., Pourghasemi, H.R., Zeinivand, H., 2015. Groundwater potential mapping at Kurdistan region of Iran using analytic hierarchy process and GIS. Arab. J. Geosci. 8 (9), 7059–7071.

Rao, N.S., Reddy, R.P., 1999. Groundwater prospects in a developing satellite township of Andhra Pradesh, India, using remote sensing techniques. J. Indian Soc. Remote Sens. 27 (4), 193–203.

Rao, N.S., Chakradhar, G.K.J., Srinivas, V., 2001. Identification of groundwater potential zones using remote sensing techniques in and around Guntur town, Andhra Pradesh, India. J. Indian Soc. Remote Sens. 29 (1–2), 69.

Ravindran, K.V., Jeyaram, A., 1997. Groundwater prospects of Shahbad tehsil, Baran district, eastern Rajasthan: a remote sensing approach. J. Indian Soc. Remote Sens. 25 (4), 239–246.

Saaty, T.L., 1980. The Analytic Hierarchy Process: Planning, Priority Setting, Resource Allocation. McGraw-Hill, New York.

Sankar, K., 2002. Evaluation of groundwater potential zones using remote sensing data in upper Vaigai river basin, Tamil Nadu, India. J. Indian Soc. Remote Sens. 30 (3), 119–129.

Sharma, D., Jugran, D.K., 1992. Hydromorphogeological studies around Pinjaur-Morni-Kala Amb area, Ambala district (Haryana), and Sirmur district (Himachal Pradesh). J. Indian Soc. Remote Sens. 20 (4), 187–197.

Shekhar, S., Pandey, A.C., 2015. Delineation of groundwater potential zone in hard rock terrain of India using remote sensing, geographical information system (GIS) and analytic hierarchy process (AHP) techniques. Geocarto Int. 30 (4), 402–421.

Suresh, D., Colins Johnny, J., Jayaprasad, B.K., 2015. Identification of artificial recharge sites for Neyyar River basin. Int. J. Remote Sens. Geosci. 4 (3), 20–27.

Tiwari, A., Rai, B., 1996. Hydromorphogeological mapping for groundwater prospecting using landsat-MSS images—a case study of part of Dhanbad District, Bihar. J. Indian Soc. Remote Sens. 24 (4), 281–285.

Vijith, H., 2007. Groundwater potential in the hard rock terrain of Western Ghats: a case study from Kottayam district, Kerala using Resourcesat (IRS-P6) data and GIS techniques. J. Indian Soc. Remote Sens. 35 (2), 163–171.

Wallace, R.E., 1977. Profiles and ages of young fault scarps, north-central Nevada. Geol. Soc. Am. Bull. 88 (9), 1267–1281.

Yesilnacar, E.K., 2005. The application of computational intelligence to landslide susceptibility mapping in Turkey. University of Melbourne, Department. 200.

Index

Printed in the United States
By Bookmasters